Titles in This Series

History of
Mathematics
Volume 8

The Emergence of the American Mathematical Research Community, 1876–1900: J. J. Sylvester, Felix Klein, and E. H. Moore

Karen Hunger Parshall
David E. Rowe

American Mathematical Society
London Mathematical Society

1991 *Mathematics Subject Classification.*
Primary 01A55, 01A72, 01A73; Secondary 01A60, 01A74, 01A80.

Photographs on the cover are (clockwise from right) the Göttingen Mathematische Gesselschafft, Felix Klein, J. J. Sylvester, and E. H. Moore. The background photograph is the Court of Honor at the Chicago World's Fair in 1893.

A list of photograph and figure credits is included at the beginning of this volume.

Library of Congress Cataloging-in-Publication Data

Parshall, Karen Hunger, 1955–
The emergence of the American mathematical research community, 1876–1900: J. J. Sylvester, Felix Klein, and E. H. Moore/Karen Hunger Parshall, David E. Rowe.
p. cm.
Includes bibliographical references and index.
ISBN 0-8218-9004-2 (acid-free)
1. Mathematics—Research—United States—History—19th century. 2. Mathematics—Study and teaching (Higher)—United States—History—19th century. I. Rowe, David E., 1950– . II. Title.
QA13.P37 1994
510′.720973′09034—dc20
94-2218
CIP

♾ The paper used in this book is acid-free and falls within the guidelines
established to ensure permanence and durability.
♻ Printed on recycled paper.

The London Mathematical Society is incorporated under Royal Charter
and is registered with the Charity Commissioners.

This publication was typeset by the authors with technical
and editorial assistance from the American Mathematical Society.

10 9 8 7 6 5 4 3 2 1 99 98 97 96 95 94

Contents

Preface

James Joseph Sylvester, a British algebraist, worked as an actuary and then taught at the Royal Military Academy in Woolwich until his forced retirement in 1870. In neither of these posts did he have an opportunity to train his countrymen, much less aspiring Americans, in research-level mathematics. Preparation at that level simply did not form part of the university mission in nineteenth-century Britain. The geometer Christian Felix Klein taught and conducted his mathematical research at various educational institutions in his native Germany. These schools and their mathematical traditions were obviously remote geographically from late nineteenth-century America. They were also intellectually far-removed from a country where institutions of higher education functioned primarily at a collegiate—as opposed to a university—level and where basic mathematical research received little encouragement. The young and unproven American Eliakim Hastings Moore served as the first Professor of Mathematics at the newly established University of Chicago beginning in 1892. At this time, his own development into a major researcher was by no means assured, and he had never taught students at the graduate level. Furthermore, his youth, inexperience, and Midwestern vantage point all seemed to militate against his becoming a major voice in a community of mathematicians which, insofar as it was discernible at all, had begun to coalesce on the East Coast around the fledgling New York Mathematical Society.

On the surface, these three men—separated by generation, mathematical training, and cultural background—would appear to be unlikely protagonists in the story of the emergence of a mathematical research community in the United States between 1876 and 1900. This period would also seem too early for the detection of significant contributions to higher mathematics from a country known more for its "Yankee ingenuity" than for the cultivation of abstract ideas. Nevertheless, within this relatively brief span of time, the United States succeeded in putting itself on the mathematical map, and it did so spurred primarily by the efforts of Sylvester, Klein, and Moore.

This book looks at the confluence of historical trends and events which enabled this disparate trio to emerge as *the* dominant figures in the creation of a community of mathematical researchers on American shores in the years from 1876 to 1900. It also delves into the mathematical achievements of these three principals in order to analyze how their research inspired the subsequent work of many of their students. By casting this research activity in a broader scientific and educational context, it both addresses issues of potential interest to a varied audience—and mathematicians, historians of science, historians, among others—and redresses a serious omission in the literature on the history of American science.

In 1986 the book *Historical Writing on American Science: Perspectives and Prospects* appeared as a result of an initiative taken by the History of Science Society. Representing the collective effort of over a dozen specialists, this work aimed to survey the various areas of the history of American science as well as to suggest fruitful avenues for further research.[1] Indeed, in the volume's preface, editors Sally Gregory Kohlstedt and Margaret Rossiter explained that by the mid-1980s, "[s]everal Americanists were ready to assess the current state of various specialties and to indicate what 'needs and opportunities' remained after more than a decade of significant activity."[2] In their collective assessment, institutional history along with science in medicine, religion, and the federal government constituted the four so-called "classical themes"; the "newer areas" of native American scientific knowledge in addition to science and technology, war, and public policy were highlighted; and the history of the scientific specialties of geology, astronomy, chemistry, biology, physics, and the social sciences received individual attention.[3] Notably absent from the specialties treated? Mathematics.

To be sure, the history of mathematics in general and the history of American mathematics in particular have been relatively neglected in the last several decades by American historians of science.[4] The "glory days" of the 1930s and 1940s have long since passed when George Sarton, the father of the history of science in America, declared the primacy of mathematics within the

[1] Sally Gregory Kohlstedt and Margaret W. Rossiter, *Historical Writing on American Science: Perspectives and Prospects* (Baltimore: The Johns Hopkins University Press, 1986). This work was originally published as *Osiris*, 2d ser., 1 (1985).

[2] *Ibid.*, p. 7.

[3] *Ibid.*, pp. 9–15.

[4] Uta Merzbach surveyed work done on the history of mathematics in America, including research on the history of American mathematics *per se*, in her article "The Study of the History of Mathematics in America: A Centennial Sketch," in *A Century of Mathematics in America—Part* III, ed. Peter Duren *et al.* (Providence: American Mathematical Society, 1989), pp. 639–666. Although she dealt with the historical contributions from 1969 to the present cursorily, Merzbach discussed in some detail the work done from the publication of Florian Cajori's *The Teaching and History of Mathematics in the United States* (Washington, D. C.: Government Printing Office, 1890), through the 1930s, when David Eugene Smith and Jekuthiel Ginsburg produced their *A History of Mathematics in America before* 1900 (Chicago: The Mathematical Association of America, 1934; reprint ed., New York: Arno Press Inc., 1980), and up to the founding by Clifford Truesdell (in 1960) of the *Archive for History of Exact Sciences* and Kenneth O. May (in 1974) of *Historia Mathematica*, two journals which publish high-level research articles in the history of mathematics.

history of science from his lofty positivist heights. "[T]he history of mathematics should really be the kernel of the history of culture," he asserted in 1937. "Take the mathematical developments out of the history of science, and you suppress the skeleton which supported and kept together all the rest. Mathematics gives to science its innermost unity and cohesion, which can never be entirely replaced with props and buttresses or with roundabout connections, no matter how many of these may be introduced."[5] Yet despite such pronouncements, even during Sarton's era, the history of *American* mathematics and, in fact, the history of American science failed to satisfy the prescripts of a generation of internalist historians of science which primarily adhered to a "great name" approach to the discipline.

As historians of science know well, the discipline has increasingly broadened its purview since the 1950s to embrace issues like the impact of philosophical and religious ideas on science, the role of external, social factors in the formation of scientific thought, and the interrelations between science and society at large.[6] This changed historiographical climate has encouraged many new interests, but of greatest importance to the present study has been the increasing concentration on American science. In particular, those sciences perceived as impinging most upon society—medicine, biology, and physics (particularly in the twentieth century with relativity, quantum theory, and nuclear technologies)—have lent themselves naturally to this sort of analysis. Mathematics, however, with its abstruse language and arcane symbolism, with its seemingly insulated practitioners, and with its unapparent impact on public policy or daily life, seems uncongenial—and so uninteresting—within such an historiographical framework.[7] While this explanation may shed some light on the relative indifference accorded the history of American mathematics over the last twenty years, a period characterized by Kohlstedt and Rossiter as one of "significant activity" in the

---

[5] George Sarton, *The Study of the History of Mathematics* (Cambridge, MA: Harvard University Press, 1937; reprint ed., New York: Dover Publications, Inc., 1957), p. 4. The reprinted edition also contains Sarton's *The Study of the History of Science* (Cambridge, MA: Harvard University Press, 1937).

[6] These shifts in emphasis came in response initially to the diverse points of view reflected in works such as Alexandre Koyré, *Études galiléennes*, 3 parts, 1935–1939; reprinted in one volume (Paris: Hermann, 1939); and Koyré *From the Closed World to the Infinite Universe* (Baltimore: The Johns Hopkins University Press, 1957); and Thomas Kuhn, *The Structure of Scientific Revolutions* (Chicago: The University of Chicago Press, 1962). For a concise and cogent discussion of these changes within the history of science, see Allen G. Debus, *Science and History: A Chemist's Appraisal* (Coimbra: Serviço de Documentação e Publicações da Universidade de Coimbra, 1984), pp. 17–33.

[7] Here, we refer to the naive perception of pure mathematics. It goes without saying that applied probability and statistics have affected public policy in key and obvious ways. Perhaps due to their relatively short histories in the United States, not even these areas of the history of American mathematics have received much attention from the general population of historians of science. See, however, Stephen Stigler, "Mathematical Statistics in the Early States," *Annals of Statistics* 6 (1978):239–265; and Stephen Stigler, ed., *American Contributions to Mathematical Statistics in the Nineteenth Century*, 2 vols. (New York: Arno Press, 1980).

history of American science, it fails to illuminate some of the deeper aspects of the relation of the history of mathematics to the history of science as a whole.

In his 1990 article, entitled "Does History of Science Treat of the History of Science? The Case of Mathematics," Ivor Grattan-Guinness stated the case bluntly: the history of mathematics has been largely ignored, making it one of the least developed subdisciplines within the history of science.[8] In his view, historians of science have neglected to examine the historical development both of mathematics *per se* and of mathematics as related to other sciences due to a fundamental fear of the subject. Moreover, the other constituency which might have advanced the subdiscipline, the mathematicians, have generally failed to approach the subject with historiographical sophistication or sensitivity. As Grattan-Guinness claimed, the neglect of the historians of science coupled with the mostly ahistorical approach of the mathematicians have rendered the history of mathematics "a classical example of a ghetto subject: too mathematical for historians and too historical for mathematicians."[9]

Grattan-Guinness's critique, however, broaches only some of the complex issues which have led to the marginalization of the history of mathematics within the history of science and certainly within the wider historical discipline. Part of the blame for this lamentable situation also rests squarely on the historians of mathematics themselves, many of whom have tended to pursue their work independently of the trends that have shaped and altered the historiography of the history of science over the last several decades.[10] On the one hand, this has made their mostly internalistic research seem increasingly outmoded and uninteresting to the wider community of historians of science; on the other, it has deprived it of some of the insights issuing

[8] Ivor Grattan-Guinness, "Does History of Science Treat of the History of Science? The Case of Mathematics," *History of Science* 28 (1990):149–173.

[9] *Ibid.*, p. 158.

[10] The recent interest in empiricist approaches among philosophers of mathematics has opened up new possibilities for placing historical studies in a broader context. See, for example, the essays in *The Space of Mathematics: Philosophical, Epistemological, and Historical Explorations*, ed. Javier Echeverria *et al.* (Berlin: Walter de Gruyter, 1992). For a related study, see *Revolutions in Mathematics*, ed. Donald Gillies (Oxford: Clarendon Press, 1992). This collection of new and reprinted essays deals with the issue of the legitimacy of the Kuhnian model of revolutions in the mathematical setting. This theme received considerable attention from historians of science at least two decades ago, yet historians of mathematics have largely ignored its implications for their subdiscipline. As Herbert Mehrtens described the situation, "[t]he complaint ... [of] historians of mathematics, that their field is too isolated from and too little recognized by colleagues in the history of natural sciences, marks a problem, not of the inaccessibility of mathematics, but rather of the inability of its historians to relate to issues of interest in general history of science." *Ibid.*, p. 42. An important counterexample to this trend is Mehrtens's own recent study, *Moderne–Sprache–Mathematik* (Frankfurt am Main: Suhrkamp Verlag, 1990). This work represents a bold attempt to understand the debates over the foundations of mathematics in the early twentieth century in terms of a power struggle over the rules of mathematical discourse within the (mostly German) mathematical community.

from the newer historiography.[11] This is certainly not to say that historians of mathematics should abandon the technical analyses so crucial for a deeper understanding of the history of mathematical ideas. It does suggest that more than the ideas involved in this pivotal scientific discipline merit historical illumination and that approaches other than the standard internalistic techniques will be required to confront these diverse historical issues. The present book aims to move beyond—but not to ignore—technical aspects of mathematics through its exploration of a dynamic process that stimulated the growth of the discipline in America and, along with it, the emergence of a community of mathematical researchers during the last quarter of the nineteenth century.[12]

The notion of periodization inherent in the preceding statement is central both to the argument and to the overall structure of this study. By targeting the *period* from 1876 to 1900, it explicitly delimits the boundaries of two other periods in the history of American mathematics.[13] Thus, the book's discussion of this key quarter-century is motivated by an examination of the prior period, and its argument is solidified by mapping the contours of the one which followed.

In the first period in the history of American mathematics, the century from 1776 to 1876, the field evolved not as a separate discipline but rather within the context of the general structure-building of American—as opposed to colonial—science. The colleges formed a primary locus of scientific activity, but, by and large, they did little to encourage the pursuit of research for the advancement of science. At the same time, the concept of research in American science—as in other academic disciplines—emerged as scientists looked toward Europe as their model and measured themselves against the yardstick of European scientific achievement. By 1876, these and other

[11]The documentary foundation for much of the history of American mathematics has not yet even been laid. As a result, the history of American mathematics, unlike other areas of American history or even of the history of American science, has not yet reached a stage of historiographical development which would make appropriate or prudent the examination of metathemes like "moral judgment." See Paul Forman's call for the study of this issue in "Independence, Not Transcendence, for the Historian of Science," *Isis* 82 (1991):71–86.

[12]Because of the deficiency of the extant literature, we have also attempted to give the reader as complete a guide as possible to the pertinent sources both in mathematics and in the history of mathematics and American science. (Due to the scope of our study, however, we do not feel it is appropriate to go beyond this into, for example, the literature in American history in general.) We accomplish this in the extensive footnotes to the various chapters of our book, although some of these are also explanatory in nature.

[13]Thomas Fiske actually suggested a periodization of the history of American mathematics as early as 1905 in his article, "Mathematical Progress in America," *Bulletin of the American Mathematical Society* 11 (1905):238–246; reprinted in *A Century of Mathematics in America—Part* I, ed., Peter Duren *et al.* (Providence: American Mathematical Society, 1988), pp. 3–11. While he, too, defined a first period through 1876, Fiske saw the second period as "extend[ing] from the establishment of the Johns Hopkins University up to 1891, when the New York Mathematical Society took on a national character and began the publication of its *Bulletin*" and the third as stretching "from 1891 to the present [that is, 1905]." (See p. 238 of his article.)

factors, such as the support of certain types of scientific research within the federal government, had resulted in the formation of an American *scientific* community which, loosely characterized, earned its living primarily through undergraduate teaching but which defined itself by the extracurricular research it presented before general scientific societies and published in books or general scientific journals.

As the essays in *Historical Writing on American Science* show, a substantial amount of work has been done over the last two decades on precisely these aspects of nineteenth-century American science. Unfortunately, little of this research touches even lightly on mathematics, in spite of the crucial and recognized role that American mathematics has played in, for instance, the emergence of the American research communities in physics and chemistry.[14] This neglect is mirrored in the work done during that same time period on the history of American mathematics *per se*, virtually none of which was informed by the contemporaneous studies issuing from the history of science community. Chapter 1 addresses this false dichotomy by anchoring the history of American mathematics firmly in the history of American scientific and educational institutions.[15] By viewing mathematics as a special case within this broader context, it provides an historical baseline for studying the changes which define the next period in the history of American mathematics, the two-and-a-half decades from 1876 to 1900.

As noted above, three men and the institutions within which they worked largely shaped this second period: the Englishman, James Joseph Sylvester, at the Johns Hopkins University;[16] the German, Felix Klein, first at Leipzig but more crucially at Göttingen; and the American, Eliakim Hastings Moore, at the University of Chicago. Their respective periods of involvement in

---

[14]In "Mathematics and the Physical Sciences in America, 1880–1930," *Isis* 77 (1986):611–629, John Servos argued that the level and character of American mathematical training had a large—and primarily negative—impact on the character of American research in the other sciences.

[15]One of the issues we explore is professionalization. It goes without saying that there is an extensive literature on professionalization in the other scientific, academic, and even non-academic areas within American culture. For an idea of its extent, see, for example, Daniel J. Kevles, *The Physicists: The History of a Scientific Community in Modern America* (Cambridge, MA: Harvard University Press, 1971); Peter Novick, *That Noble Dream: The "Objectivity Question" and the American Historical Profession* (New York: Cambridge University Press, 1988); Dorothy Ross, *The Origins of American Social Science* (New York: Cambridge University Press, 1991); and Olivier Zunz, *Making America Corporate* 1870–1920 (Chicago: University of Chicago Press, 1990). In keeping with one of the goals of our book, namely, the establishment of a firm documentary foundation for one process and one era in the history of American mathematics, we confine ourselves primarily to the literature of mathematics and the history of American science. We would hope that, informed by our work, subsequent studies might draw well-founded generalizations concerning the developments in American mathematics and those in other areas of American culture.

[16]Throughout this book, we adopt the convention used in the nineteenth century, as well as today, of distinguishing the Johns Hopkins University (or simply the Johns Hopkins) from its benefactor, Johns Hopkins. We will, however, use the short form "Hopkins" from time to time in reference to the university, when no confusion will result.

the American scene clearly define three distinct phases centered around three programs of mathematical research. An analysis of these programs and of the nature of the influence of their progenitors reflects the progressive deepening of research standards and output in the United States throughout the twenty-five-year period as well as the generational differences separating Sylvester, Klein, and Moore. It further accentuates, by focusing on the students and research issuing from these schools, the process of maturation of an American mathematical research community which had fully emerged by 1900.

Imported from Britain in 1876, Sylvester assumed the first professorship in mathematics in 1876 at Hopkins, an institution pivotal in the history of higher education in America and, by extension, in the history of American mathematics. With Sylvester's return to Britain in 1883, however, Hopkins lost its momentum as a center of advanced mathematical training, and Americans desirous of such training turned elsewhere for their inspiration. They chose as their mentor another foreigner, but this time a foreigner abroad, Felix Klein. For roughly a decade following Sylvester's departure, Klein actively served as the standard-bearer for a dynamic group of American mathematicians, many of whom studied under him in Germany. His influence reached its climax in 1893 during his seven-week visit to the United States, which began with a lavish reception at the Mathematical Congress held in conjunction with Chicago's World's Columbian Exposition . Somewhat ironically perhaps, Klein's appearance at the latter event foreshadowed his desire to step back from his involvement in American developments and heralded the arrival of a new figure, E. H. Moore. Together with his Chicago colleagues Oskar Bolza and Heinrich Maschke, Moore established an environment at Chicago conducive to the training of research-level mathematicians back on American soil. Moore, in particular, also consciously directed his efforts toward strengthening several institutional and organizational structures that ultimately played crucial roles in cementing a national community.

Chapters 2, 4, and 9 center on Sylvester, Klein, and Moore—on the formative influences in their lives, on their approaches to mathematics, and on their actual research—in order to motivate and trace their influences on the first generation of American mathematical researchers and to chart the growing maturity of the community they helped define. These chapters also place these three men within the institutional settings that actively encouraged them to train students at an advanced level and to pursue their own personal research objectives. In a key shift which serves to delineate partially the first and second periods in the history of American mathematics, a number of influential academic institutions in the United States, led by the Johns Hopkins, adopted and adapted the research ethic which had become firmly entrenched in German higher education earlier in the nineteenth century. This translated into an increasing emphasis on research as an officially sanctioned and supported endeavor in the emergent university setting and, by intimate association, in the emergent mathematical community.

In Chapters 3, 5, and 9, the emphasis shifts to the students of Sylvester, Klein, and Moore and to the actual training they received. An examination first of their teachers' commitment to research and of their varied, even idiosyncratic, styles of instilling this commitment sets the stage for a closer look at the research produced by these young men and women.[17] As often as feasible, this work is placed within the larger context of the history of nineteenth- and early twentieth-century mathematics in order to assess the rapid changes made in the United States between 1876 and 1900.

Although the training of mathematicians at the research-level represents a critical ingredient in the historical process documented here, the actual formation of a *community*—an interacting group of people linked by common interests—required more than just advanced training in mathematics. Chapters 6 through 9 describe how members of this first generation of research-oriented mathematicians reformed the academic environments at a number of American universities. These chapters also investigate the organizational activities which engendered publication outlets for research-level mathematics, a specialized professional society, the first major mathematical meeting which also drew influential foreign participation, and the first research-level colloquium lectures open to and designed for a wide variety of mathematical listeners. The historical conjunction during the last quarter of the nineteenth century of these various innovations and their innovators, of changed attitudes as to the value and desirability of research at both an individual and institutional level, and of the existence of a critical mass numbering some 1000 practitioners had produced, by 1900, a firmly grounded mathematical research community in the United States.[18]

As this last statement suggests, this community clearly extended beyond the spheres defined by Sylvester, Klein, and Moore. Thus, while the present study centers on the influence of these three "great men," it is by no means limited to their contributions. On the contrary, it uses their careers as the points of departure for an examination of a broad spectrum—although by no means the complete roster—of participants in the field at the turn of the twentieth century. Chapters 3, 5, 9, and 10 identify their students and, to some extent, their students' students and colleagues. Who studied and taught mathematics? Where did they come from and where did they go? What did they do mathematically? These are among the questions examined. By penetrating beneath the top layer, the investigation underscores the existence of an

[17]One subtheme which recurs throughout our book concerns women and the obstacles they faced in breaking into mathematics at the research level.

[18]Della Dumbaugh Fenster and Karen Hunger Parshall have recently completed a quantitative study of the American mathematical research community during the period from 1891 to 1906, or the first fifteen years of publication of the *Bulletin of the American Mathematical Society*. They uncovered some 1000 distinct members of this community at all levels of participation. See Della Dumbaugh Fenster and Karen Hunger Parshall, "A Profile of the American Mathematical Research Community: 1891–1906," in *The History of Modern Mathematics*, vol. 3, ed. Eberhard Knobloch and David E. Rowe (Boston: Academic Press, 1994), to appear.

extended population of mathematicians—some talented and some not, some well known and some obscure—based at colleges and universities throughout the United States who actively shared an interest in mathematics at the research level. Clearly much more could—and should—be done to flesh out this community fully, to examine it completely from the top down and from the bottom up. This book, however, aims to explore and document the *emergence* of the community. It leaves to future researchers the study of it as a community *perse.*[19]

Likewise, the tenth and final chapter mainly suggests lines of inquiry for further historical research. It outlines the fuller conception of the periodization of the history of American mathematics and provides some indication of the parameters which delineate a third period of consolidation and growth from the second emergent period. With this sketch of our book's underlying structure and overall argument, we now turn to examine the American mathematical landscape during the first period in its historical development, the century from 1776 to 1876.

[19]For an indication of what such research might entail, see the comparative study of the structure and function of physics communities during this period by Paul Forman, John L. Heilbron, and Spencer Weart, "Physics circa 1900: Personnel and Productivity of the Academic Establishments," *Historical Studies of the Physical Sciences* 5 (1975):1–185.

Acknowledgments

The centennial of the founding of the American Mathematical Society in 1988 gave us the initial idea of writing a history of the dynamic events that forged a vibrant community of mathematical researchers in the United States during the last quarter of the nineteenth century. As we have since come to appreciate, a joint effort of this kind and scope, which strives to give a coherent presentation of a complex phenomenon that took place on two continents, poses considerable intellectual difficulties as well as organizational problems for the collaborators. Had we anticipated all of these complexities and difficulties at the outset, this book might have appeared sooner. Instead, it evolved through various forms over a five-year period during which we had the opportunity to reorient our own perspectives as we familiarized ourselves with new sources pertinent to important ideas and individuals connected with the American mathematical scene. In offering the present study as the culmination of our collaborative efforts, we take satisfaction in knowing that, after much brainstorming and countless revisions, this book could never have been written by either of us alone.

It also could never have been completed without the assistance and support of many others. Indeed, so many people have taken an interest in this project that it would be nearly impossible to mention them all here. First and foremost, however, we would like to thank the entire Publication Committee of the joint AMS/LMS series on the history of mathematics, but particularly Peter Duren, Dick Askey, Ed Edwards, Jeremy Gray, and Charlie Curtis, for not only their encouragement but also their carefully considered comments and critiques of our manuscript. If they sometimes differed with us regarding how the history of mathematics should be written and what issues it should involve, their flexibility has allowed us to produce a book which, we hope, will be of interest to mathematicians and historians alike.

In addition to the members of the Publication Committee, several of our friends and colleagues read and commented on all or part of our manuscript. Bob Cross, Joe Dauben, Ted Feldman, Della Dumbaugh Fenster, Susann Hensel, "Mike" Hunger, Joe Kett, Arnold Koslow, John McCleary, Elliott Mendelsohn, Brian Parshall, and Paul Wolfson helped us to appreciate better the points of view of the mathematicians, historians of science, historians, and more general readers who might chance to open our book. Given the demands of their own work, we sincerely appreciate the time that they devoted to their careful readings of our text.

We would also like to acknowledge the hospitality of the many libraries we worked in during the course of our research on this project. In particular, we extend our thanks to the Johns Hopkins University; the University of Chicago; the Library of Congress; the Smithsonian Institution; Smith College; the University of Texas at Austin; the Master and Fellows of St. John's College, Cambridge; the Niedersächsische Staats- und Universitätsbibliothek, Göttingen; the Mathematisches Institut der Universität Göttingen; and the Universitetsbiblioteket, Oslo for permission to quote from documents held in their archival collections.

Much of the research for this book was conducted in conjunction with grants from various sources. The first author is indebted to the National Science Foundation for support in the form of grants SES 8509795 and DIR 9011625, which allowed her to gather much of the material on the Hopkins and Chicago schools, and to a Fourth-Year Research Fellowship from the University of Virginia. The second author received support from National Science Foundation grant DIR 8821421 as well as a two-year stipend from the Alexander von Humboldt Foundation, which gave him the opportunity to do archival research in Germany.

Finally, the first author would like to express her debt of gratitude to her parents, "Mike" and Maurice Hunger, for all of their encouragement, support, and advice; to her advisers, I. N. Herstein and Allen G. Debus, for all that they taught her about scholarship and friendship; and to her husband, Brian Parshall, for his tireless and selfless devotion. The second author would like to thank his wife, Hilde, his son, Andy, and his other good friends who have helped to remind him that there is more to life than scholarship.

CHARLOTTESVILLE, VIRGINIA AND MAINZ, GERMANY
July 1993

Photograph and Figure Credits

The American Mathematical Society gratefully acknowledges the kindness of these institutions in granting the following permissions:

Columbia University Archives

Photo of Frank Nelson Cole; p. 193; Courtesy of the David Eugene Smith Collection, Rare Book Library, Columbia University

Photo of George Bruce Halsted; p. 114; Courtesy of the David Eugene Smith Collection, Rare Book Library, Columbia University

Göttingen University

Figure 5.1 from Henry White's seminar lectures; p. 226; Courtesy of the Mathematics Institute of Göttingen University

Figure 5.2 from Henry White's seminar lectures; p. 227; Courtesy of the Mathematics Institute of Göttingen University

Johns Hopkins University Archives

Photo of James Joseph Sylvester; p. 60 and cover; Courtesy of the Ferdinand Hamburger, Jr. Archives of the Johns Hopkins University

Photo of the Mathematical Seminary, Johns Hopkins University; p. 87; Courtesy of the Ferdinand Hamburger, Jr. Archives of the Johns Hopkins University

Photo of the old campus, Johns Hopkins University; p. 83; Courtesy of the Ferdinand Hamburger, Jr. Archives of the Johns Hopkins University

Photo of Thomas Craig; p. 115; Courtesy of the Ferdinand Hamburger, Jr. Archives of the Johns Hopkins University

Peabody & Essex Museum

Oil portrait of Nathaniel Bowditch by Charles Osgood; p. 10; Courtesy of the Peabody & Essex Museum, Salem, Massachusetts

Smith College

Photo of Johnson, Winston, and MacKinnon; p. 245; Courtesy of the Sophia Smith Collection, Smith College

Photo of Winston, Chisholm et al.; p. 248; Courtesy of the Sophia Smith Collection, Smith College

Springer-Verlag, Berlin

Figure 4.1; p. 180; from Felix Klein, *Gesammelte Mathematische Abhandlungen*, 3 vols., Springer-Verlag, Berlin, 1921–1923, 3:508; Courtesy of Springer-Verlag

Figure 4.2; p. 181; from Felix Klein, *Gesammelte Mathematische Abhandlungen*, 3 vols., Springer-Verlag, Berlin, 1921–1923, 3:508; Courtesy of Springer-Verlag

Figure 8.2; p. 341; from Felix Klein, *Gesammelte Mathematische Abhandlungen*, 3 vols., Springer-Verlag, Berlin, 1921–1923, 3:508; Courtesy of Springer-Verlag

Figure 8.3; p. 343; from Felix Klein, *Gesammelte Mathematische Abhandlungen*, 3 vols., Springer-Verlag, Berlin, 1921–1923, 3:508; Courtesy of Springer-Verlag

Springer-Verlag, New York

Photo of Felix Klein; p. 149 and cover; from Constance Reid, *Hilbert*, Springer-Verlag, 1970, p. 291; Courtesy of Springer-Verlag

Photo of David Hilbert; p. 441; from Constance Reid, *Hilbert*, Springer-Verlag, 1970, p. 231; Courtesy of Springer-Verlag

Photo of Benjamin Pierce; p. 11; from Rowe and McCleary, eds., *The History of Modern Mathematics*, Academic Press; Courtesy of Springer-Verlag

Photo of Julius Plücker; p. 160; from Rowe and McCleary, eds., *The History of Modern Mathematics*, Academic Press; Courtesy of Springer-Verlag

Photo of Alfred Clebsch; p. 158; from Rowe and McCleary, eds., *The History of Modern Mathematics*, Academic Press; Courtesy of Springer-Verlag

University of Chicago

Photo of Eliakim Hastings Moore; p. 280 and cover; Courtesy of the Department of Mathematics, University of Chicago

Photo of Oskar Bolza; p. 288; Courtesy of the Department of Mathematics, University of Chicago

Photo of Heinrich Maschke; p. 291; Courtesy of the Department of Mathematics, University of Chicago

Photo of Ryerson Physical Laboratory; p. 366; Courtesy of the University of Chicago Archives

Photo of L. E. Dickson; p. 380; Courtesy of the University of Chicago Archives

The American Mathematical Society gratefully acknowledges the kindness of these individuals in granting the following permissions:

Elin Strom

Photo of Sophus Lie; p. 162; Courtesy of Mrs. Elin Strom

Günther Meinhardt

Photo of Auditorium Building; p. 218; from *Günther Meinhardt, Göttingen in alten Ansichten*, Europäisch Bibliothek, Zaltbommel/Niederlande, 1979; Courtesy of Günther Meinhardt

Gerd Fischer

Figure 8.1; p. 339; from Gerd Fischer, ed., *Mathematical Models*, Friedr. Vieweg & Sohn, 1986, p. 14; Courtesy of Gerd Fischer

The following photos are in the public domain:

Models at Technische Hochschule (1917); p. 171

Göttingen Mathematische Gesellschaft (1902); p. 219 and cover

Court of Honor (1893); p. 299 and cover

Giant Ferris Wheel (1893); p. 301

Mathematics Section of German University Exhibit (1893); p. 308

Participants, open plenary session (1893); p. 314

Congress of Mathematicians, World's Col. Exposition (1893); p. 316

The American Mathematical Society holds the copyright to all other photographs in this volume.

Photo of George Birkoff; p. 391

Photo of Gilbert Bliss; p. 397

Photo of Maxime Bocher; p. 206

Photo of Henry Fine; p. 195

Photo of Josiah W. Gibbs; p. 25

Photo of George Hill; p. 37

Photo of Robert Moore; p. 388

Photo of Simon Newcomb; p. 36

Photo of William F. Osgood; p. 205

Photo of Virgil Snyder; p. 216

Photo of William Story; p. 109

Photo of Oswald Veblen; p. 385

Photo of Edward Van Vleck; p. 215

Photo of Henry White; p. 332

Chapter 1
An Overview of American Mathematics: 1776–1876

In 1776, America did not yet support a *scientific* research community, much less a *mathematical* research community. Educationally and professionally, American scientists of the colonial era came from disparate backgrounds and moved in a variety of social circles.[1] Some were self-taught, in keeping with America's long-standing amateur tradition; some had studied abroad; and some had attended the colonial colleges. They earned their livings as farmers, teachers, ministers, statesmen, physicians, surveyors, printers, and clockmakers, among other professions. Since they enjoyed virtually no patronage either from governmental or private sources, they were forced to undertake their various scientific inquiries during their leisure time and to finance them with their own personal funds. As they themselves frequently lamented in private correspondence, these conditions worked at cross purposes to their desires to make important research contributions.[2]

At the colleges, in Philadelphia, in Boston, and in other scattered locales, those interested in science met or corresponded to communicate their findings and ideas, but their numbers were small and their interests diverse. The observational sides of botany, zoology, and geology attracted most of these scientific devotees since local flora, fauna, and rock formations came most easily under scrutiny by the naked eye. The theoretical and experimental aspects of science, however, proved less accessible for both educational and financial reasons. An understanding of the intricacies of the mathematical physics of Euler and Lagrange, for example, required a mastery of both mathematics and physics far beyond that of the college education of the day, and all but the simplest scientific instruments cost more than most private incomes would allow.

[1]On colonial science, see, for example, Dirk Struik, *Yankee Science in the Making* (Boston: Little, Brown and Company, 1948); and Brooke Hindle, *The Pursuit of Science in Revolutionary America,* 1735–1789 (Chapel Hill: University of North Carolina Press, 1956; reprint ed., New York: Norton, 1974).

[2]For numerous examples of such complaints in the correspondence of American scientists in this era, consult John C. Greene, *American Science in the Age of Jefferson* (Ames: Iowa State University Press, 1984), pp. 3–36. Greene quotes various mathematical scientists in Chapter 6, pp. 128–157.

In what follows, we regard the development of mathematics in America during the century from 1776 to 1876 in light of both the evolving professionalization of American science and the concurrent trends in higher education.[3] This wider perspective brings into view explanations for the scarcity of mathematical research in the United States prior to 1876 as well as for the rapid emergence of an American mathematical research community from 1876 to 1900. Yet, while overarching educational and social concerns illuminate some key facets of these developments, they shed less light on the mathematics actually produced by the members of that evolving community.

Throughout the nineteenth century, mathematical research in Europe expanded at an unprecedented rate and in surprising directions. Prior to 1850, the Americans were all but closed out of these developments due to a complex combination of factors including, among others, the absence of adequate educational opportunities at home and the lack of a broader institutional structure supportive of research. As changes on these fronts took place in the century's second half, Americans increasingly found themselves in a position to draw inspiration directly from the active areas of European mathematics and to make original contributions to them. In sketching these background developments, we highlight the impact of three reasonably distinct national traditions in science and mathematics—first those of Britain and France, later that of Germany—and the roles played by their leading institutions and practitioners in directing the course of American mathematics in the period from 1776 to 1876.

British and French Influences

Not surprisingly, as British subjects before the War of Independence, colonists had looked toward Great Britain for inspiration in setting up their colleges and in modeling their curricula. Thus, arithmetic, some trigonometry, as well as the rudiments of algebra and geometry constituted the mathematical curriculum during the latter part of the eighteenth century.[4] Given their common

[3]Obviously, mathematics was not the only science to professionalize around the turn of the twentieth century, nor were the sciences the only disciplines undergoing these sociological changes. As we mentioned in the preface (see note 15), physicists, historians, social scientists, and even the American business community all professionalized at roughly the same time as the American mathematical community. Indeed, the full literature on professionalization is vast, but a discussion of it falls beyond the scope of our book.

[4]According to Florian Cajori in his book, *The Teaching and History of Mathematics in the United States* (Washington: Government Printing Office, 1890), some American colleges did teach fluxions, that is, calculus in the Newtonian style, at least at some time during the colonial period. Thomas Clap, Professor of Mathematics and Natural Philosophy and then President of Yale until 1766, recorded having taught fluxions to his classes. However, his successor, Nehemiah Strong, did not. Also, at the University of Pennsylvania in 1758, fluxions were included in the curriculum for second-year students with Colin Maclaurin's *Treatise on Fluxions* as a recommended text. See Cajori, pp. 31–32 and 36–37.

language, Americans also imported British textbooks such as Robert Simson's translation of Euclid's *Elements*, John Ward's *Young Mathematician's Guide*, and Sir Isaac Barrow's *Geometrical Lectures*.[5] Since these texts represented the height of American training before 1776 and since Great Britain lagged far behind the Continent scientifically, it is hardly surprising that nothing comparable to the output of eighteenth-century Europe issued from the New World.

Immediately after the war, the situation barely improved. The United States achieved political independence as a result of the conflict, but the cultural dependence engendered by common ancestry and language proved much more difficult to shake.[6] Relative to mathematics, Americans continued to import texts, but they increasingly brought out American editions or compendia of British works. Of the new American editions, the English translations by Robert Simson and John Playfair of the *Elements* were particularly popular.[7] Of the compendia, *Mathematics, Compiled from the Best Authors and Intended to Be a Text-book of the Course of Private Lectures on These Sciences in the University at Cambridge* [i.e., Harvard] was typical. Readied for publication in 1801 by Samuel Webber, the fourth Hollis Professor of Mathematics and Natural Philosophy at Harvard, this work in "two volumes, each of 460 pages and in large print, ... embraced the following subjects: arithmetic, logarithms, algebra, geometry, plane trigonometry, mensuration of surfaces, mensuration of solids, gauging, heights and distances, surveying, navigation, conic sections, dialing, spherical geometry, and spherical trigonometry."[8] It targeted a Harvard audience, which, only after standards were raised in 1802, was required to have some knowledge of arithmetic prior to admission.[9] Little wonder, then, that while native productions such as Webber's made elementary and certain applied mathematics somewhat more accessible, they hardly introduced any of the newer eighteenth-century Continental ideas into

[5] See *ibid.*, pp. 25–26, 37, and 55–57. The texts are: Robert Simson, *The Elements of Euclid, viz. the First Six Books together with the Eleventh and Twelfth* (Glasgow: R. & A. Foulis, 1756); John Ward, *The Young Mathematician's Guide, Being a Plain & Easie Introduction to the Mathematicks... with an Appendix of Practical Geometry* (London: J. T. & J. Woodward, 1707); and Isaac Barrow, *The Geometrical Lectures Explaining the Generation, Nature & Properties of Curve Lines* (London: S. Austen, 1735). The latter is one of several translations from the original Latin texts of the lectures, which were delivered in the 1660s.

[6] For a broader discussion of science in the American context in this period, see Greene, *American Science in the Age of Jefferson*.

[7] Cajori, p. 55. Cajori also says that an edition appeared in Worcester as early as 1784, but we have been unable to verify this. Among the early American editions are: Robert Simson, *The Elements of Euclid...* (Philadelphia: Conrad & Co., 1803); and John Playfair, *Elements of Geometry Containing the First Six Books of Euclid with a Supplement on the Quadrature of the Circle and the Geometry of Solids* (Philadelphia: F. Nichols, 1806).

[8] Cajori, p. 60. See Samuel Webber, *Mathematics, Compiled from the Best Authors and Intended to Be a Text-book of the Course of Private Lectures on These Sciences in the University at Cambridge* (Boston: Harvard College, 1801).

[9] Cajori, p. 60. This knowledge extended only up to the so-called "Rule of Three," namely, if $\frac{a}{b} = \frac{c}{x}$, then $x = \frac{bc}{a}$.

the United States. Thus, even after the Revolutionary War, America continued to take its lead from a country which had fallen out of step in science and mathematics.

Although America's cultural dependence on England initially retarded mathematical development in the new nation, Great Britain itself began to shift, mathematically speaking, toward the more progressive Continent during the first two decades of the nineteenth century. The turn of the century, which had occasioned assessments in France of the accomplishments of French mathematicians during the eighteenth century, had witnessed similar appraisals by the British.[10] Unlike their French counterparts, however, British mathematicians did not have a string of stunning successes to enumerate. In an 1808 review of Laplace's *Traité de Mécanique céleste* and in an 1810 review of the English translation of his *Exposition du Système du Monde*, the Scottish mathematician John Playfair was struck by the stark contrast between the magisterial work he was reviewing and the contemporaneous scientific productions of his own countrymen. With reference to the *Mécanique céleste*, he ventured that

> Another reflection of a very different kind from the preceding, must present itself, when we consider the historical details concerning the progress of physical astronomy that have occurred in the foregoing pages. In the list of mathematicians and philosophers, to whom that science, for the last sixty or seventy years has been indebted for its improvements, hardly a name from Great Britain falls to be mentioned. What is the reason of this? and how comes it, when such objects were in view, and when so much reputation was to be gained, that the country of Bacon and Newton looked silently on, without taking any share in so noble a contest?[11]

Playfair obviously had answers in mind to these rhetorical questions. To his way of thinking, English mathematics had suffered owing to three separate and distinctly negative factors.

---

[10] For one such French assessment, see Jean-Baptiste Delambre, *Rapport historique sur les Progrès des Sciences mathématiques depuis 1789, et sur leur État actuel* (Paris: Imprimerie impériale, 1810; reprint ed., Amsterdam: N. V. Boekhandel & Antiquariaat B. M. Israël, 1966), pp. 131–132. Even earlier, in 1781, Joseph-Louis Lagrange felt that mathematics was finished. He expressed his *fin-de-siècle* pessimism when he wrote to his friend, Jean d'Alembert, that "the mine [of mathematics] is already very deep and that unless one discovers new veins it will be necessary sooner or later to abandon it. ... [I]t is not impossible that the chairs of geometry in the Academy will one day become what the chairs of Arabic presently are in the universities." See Joseph-Louis Lagrange to Jean d'Alembert, 21 September 1781, as quoted in Morris Kline, *Mathematical Thought from Ancient to Modern Times* (New York: Oxford University Press, 1972), p. 623.

[11] John Playfair, Review of "*Traité de Méchanique Céleste.* Par P. S. LaPlace, Membre de l'Institut National de France, et du Bureau des Longitudes. Paris. Vol 1. An 7. Vol 3 & 4. 1805," *The Edinburgh Review* 11 (1808):249–284 on pp. 279–280.

First, the British had steadfastly followed the example set by Newton at the end of the seventeenth century of approaching mathematics from a synthetic, geometrical point of view rather than adopting the new, eighteenth-century analytical and algebraic methods which had developed on the other side of the Channel. This placed the British at a decided disadvantage relative to advances on the Continent. As Playfair wryly noted, one of his countrymen

> may be perfectly acquainted with every thing on mathematical learning that has been written in this country [i.e., Great Britain], and may yet find himself stopped at the first page of the works of Euler or D'Alembert. He will be stopped, not from the difference of fluxionary notation, (a difficulty easily overcome) nor from the obscurity of these authors, who are both very clear writers... but from the want of knowing the basic principles and the methods which they take for granted as known to every mathematical reader.[12]

Playfair recognized that mathematics had changed dramatically over the course of the eighteenth century and that while Newton's work still commanded respect, it was far from the vanguard of contemporary research.

The second cause underlying Britain's mathematical inferiority, according to Playfair, was the British system of higher education itself. In his view, even at Cambridge, where mathematical studies formed the backbone of the curriculum, instruction consisted of presentation followed by rote memorization and regurgitation year after year of a limited number of texts. He felt that such a system failed miserably to instill the sense of intellectual adventurousness so critical to original research. "The laws of periodical revolution, and of returning continually to the same tract, may, as we have seen, be excellently adapted to a planetary system," Playfair wrote with tongue in cheek, "but [they] are ill calculated to promote the ends of an educational institution."[13] These ends, from his point of view, should be the fostering of intellectual independence and the preparation of young minds for the furtherance of knowledge, but, unfortunately, neither Cambridge nor Oxford embraced them.

Finally, Playfair blamed the Royal Society of London for failing to promote and encourage mathematics. Looking across the Channel, he viewed with obvious and justified envy the situation in France. There, the *Institut de France*—with its *première classe* devoted to the mathematical and physical sciences—sponsored mathematical competitions, bestowed honors and awards, offered modest subsidies, and generally glorified the pursuit of mathematics. No wonder the mathematical sciences in France attained "a state of

[12] *Ibid.*, p. 281.
[13] *Ibid.*, p. 284.

unexampled prosperity"[14] while in England they languished.

Indeed, France established a network of scientific institutions during the Napoleonic era that subsequently fostered some of the most important mathematical achievements of the early nineteenth century. Supported by educational institutions like the *École polytechnique* and the *École des Mines*, the prestigious *Institut de France*, and periodicals such as the *Journal de l'École polytechnique* (founded in 1794) and Joseph-Diez Gergonne's *Annales de Mathématiques pures et appliquées* (begun in 1810 and the first periodical devoted exclusively to mathematics), this network stimulated advances in both pure and applied mathematics.[15] At the *École polytechnique*, Gaspard Monge institutionalized the new field of descriptive geometry, which he himself had developed to aid students of fortification design and construction.[16] This subject ultimately inspired his disciple, Victor Poncelet, to write the *Traité des Propriétés projectives des Figures*, which effectively launched what would become one of the nineteenth century's most active areas of research, projective geometry.[17] Also associated with the *École polytechnique*, Joseph Fourier, and later Siméon-Denis Poisson and Augustin Cauchy, focused on applied mathematics but made seminal contributions to pure mathematics in such areas as real and complex analysis.[18] Playfair was certainly right; the state of science and mathematics in Great Britain paled in comparison with a dynamic, early nineteenth-century French tradition which was already firmly in place by 1810.

And Playfair's views were by no means isolated. During the opening two

[14]John Playfair, Review of "*The System of the World.* By P. S. LaPlace, Member of the National Institute of France. Translated from the French by J. Pond, F. R. S. 2 vol. 8vo. London, Phillips, 1809," *The Edinburgh Review* 15 (1810):396–417 on p. 398.

[15]For a detailed study of the French mathematical community at this time, consult Ivor Grattan-Guinness, *Convolutions in French Mathematics,* 1800–1840*: From the Calculus and Mechanics to Mathematical Analysis and Mathematical Physics*, 3 vols. (Basel: Birkhäuser, 1990). Note, here, that prior to the French Revolution, and then again after 1816, the *première classe* of the *Institut de France* was known as the *Académie des Sciences.*

[16]The prehistory of descriptive geometry traces back to ancient and medieval technological traditions. Monge, however, succeeded in elevating a core of centuries-old practices of engineers and architects into an independent, theoretical discipline. See René Taton, *L'Oeuvre scientifique de Gaspard Monge* (Paris: Presses universitaires de France, 1951). Taton provides an extensive bibliography of works dealing with Monge and the historical development of his ideas on pp. 396–425.

[17]Victor Poncelet, *Traité des Propriétés projectives des Figures* (Paris: Bachelier, 1822).

[18]On these developments, see, for example, Bruno Belhoste, *Cauchy: Un Mathématicien légitimiste au XIXe Siècle* (Paris: Librairie Classique Eugène Belin, 1985); and *Augustin-Louis Cauchy: A Biography*, trans. Frank Ragland (New York: Springer-Verlag, 1991); Umberto Bottazzini, *The Higher Calculus: A History of Real and Complex Analysis from Euler to Weierstrass*, trans. Warren Van Egmond (New York: Springer-Verlag, 1986); Judith V. Grabiner, *The Origins of Cauchy's Rigorous Calculus* (Cambridge, MA: MIT Press, 1981); Ivor Grattan-Guinness, *The Development of the Foundations of Mathematical Analysis from Euler to Riemann* (Cambridge, MA: MIT Press, 1970); Grattan-Guinness, *Convolutions*; and Amy Dahan Dalmedico, *Mathématisations: Augustin-Louis Cauchy et L'École française* (Argenteuil: Editions du Choix and Paris: Albert Blanchard, 1993).

decades of the nineteenth century, British mathematicians not only recognized the problem but they also took steps to correct it. Perhaps the earliest manifestation of this process appeared in 1812 at Cambridge.

In discussing with a friend the formation of a section of the British and Foreign Bible Society, a society which had the distribution of Bibles as its mission, Charles Babbage facetiously suggested that they form a society to distribute Silvestre Lacroix's *Traité élémentaire de Calcul différentiel et de Calcul intégral*, a work in the French analytic style.[19] While the resulting Cambridge Analytical Society—a group of a dozen or so students, among them George Peacock and John Herschel—had no intention of handing out Lacroix's text on street corners, it met regularly to study, to discuss, and, it was hoped, to make original contributions to the new analytics.[20] Like Playfair, the Society's membership faulted what it viewed as a fundamentally outmoded educational system at Cambridge.

The Analytical Society may not have enjoyed enormous success in its mission to advance the Continental brand of analytical, algebraic mathematics, but its activities were symptomatic of the growing dissatisfaction of British mathematicians with their educational system and with its head-in-the-sand attitude toward the rest of Europe. By the 1820s, a general spirit of reform had overcome Cambridge, which, coupled with the ascension of Analytical Society members and other like-minded young men to influential positions within the Cambridge superstructure, brought about some of the desired changes. Most notably, Continental mathematics became part of the curriculum.[21]

Apparently independent of, but nearly concomitant with, the onset of an active mathematical reform movement at Cambridge, John Farrar, the fifth Hollis Professor of Mathematics and Natural Philosophy at Harvard College, had also set about translating Continental texts for use in his courses. In

---

[19] Silvestre Lacroix, *Traité élémentaire de Calcul différentiel et de Calcul intégral, précédé de Réflexions sur la Manière d'Enseigner les Mathématiques, et d'Apprécier dans les Examens de Savoir de ceux qui les ont étudiées* (Paris: Duprat, 1802).

[20] The usual discussion of the Analytical Society (as in Kline, *Mathematical Thought*, pp. 622–623), which portrays it as a society actively involved in effecting mathematical reforms at Cambridge as early as 1812, is misleading. In his doctoral thesis as well as in subsequent published work, Philip Enros has set the record straight. See Philip C. Enros, "The Analytical Society: Mathematics at Cambridge in the Early Nineteenth Century," (Ph.D. Dissertation, University of Toronto, 1979); "Cambridge University and the Adoption of Analytics in Early Nineteenth-Century England," in *Social History of Nineteenth Century Mathematics*, ed. Herbert Mehrtens, Henk Bos, and Ivo Schneider (Boston: Birkhäuser Verlag, 1981):135–148; and Philip C. Enros, "The Analytical Society (1812–1813): Precursor of the Revival of Cambridge Mathematics," *Historia Mathematica* 10 (1983):24–47.

[21] While Continental mathematics was adopted at Cambridge by the 1820s, it was abandoned by the 1840s due to its perceived inducement of a spirit of professionalism. Such a spirit was viewed as antithetical to the Cambridge liberal arts tradition. For more on this, see Enros, "Cambridge University and the Adoption of Analytics," pp.143–146. See, also, Joan Richards, *Mathematical Visions*: *The Pursuit of Geometry in Victorian England* (Boston: Academic Press, Inc., 1988), Chapter 1, pp. 13–59.

1818, he produced an Americanized version of Lacroix's *Traité élémentaire d'Arithmétique* (1797), followed in 1819 by Legendre's *Éléments de Géométrie—Avec des Notes* (1794), and in 1820 by Lacroix's *Traité élémentaire de Trigonométrie rectiligne et sphérique* (1798) with selections from Étienne Bézout's work in algebra.[22] By 1824, Farrar, together with George Emerson, had brought out perhaps their most important text, the *First Principles of Differential and Integral Calculus.*[23] Based on the writings of Bézout, who was active as a mathematician before the French Revolution, this work introduced Leibnizian as opposed to a Newtonian notation into the United States. Farrar fully realized this fact, of course, but justified his choice of text less on philosophical and more on utilitarian grounds in the preface to his translation. Since he needed a book at a beginning level for his Harvard students, he opted for Bézout's work "on account of the plain and perspicuous manner for which the author is so well known, as also on account of its brevity and adaptation in other respects to the wants of those who have but little time to devote to such studies."[24] This may have been mid-eighteenth-century mathematics, but it was in the Continental as opposed to the British style. So, although it certainly would not have brought a student to the forefront of mathematical research, it would have served to lay a firmer foundation. At least one of Farrar's students, Benjamin Peirce, built upon his early studies to attain a level of mathematical sophistication equal to that of many productive European researchers. However, Peirce's achievements, which we shall discuss in a somewhat different context shortly, owed more perhaps to another American who aimed to make French ideas accessible to his countrymen, Nathaniel Bowditch.

Bowditch was born into a less than well-to-do family in Salem, Massachusetts in 1773.[25] Due to his family's long sea-going tradition and strained

---

[22]Silvestre Lacroix, *Traité élémentaire d'Arithmétique, à l'Usage de l'École des Quatres-nations* (Paris: Duprat, 1797); and *Traité élémentaire de Trigonométrie rectiligne et sphérique, et d'Application de l'Algèbre à la Géométrie* (Paris: Duprat, 1798); and Adrien-Marie Legendre, *Éléments de Géométrie—Avec des Notes* (Paris: F. Didot, 1794). On the many American translations and adaptations of French mathematical texts, see Lao G. Simons, "The Influence of French Mathematicians at the End of the Eighteenth Century upon the Teaching of Mathematics in American Colleges," *Isis* 15 (1931):104–123. Helena M. Pycior discusses the reflection of the British and French styles in American algebra textbooks in "British Synthetic vs. French Analytic Styles of Algebra in the Early American Republic," in *The History of Modern Mathematics*, ed. David E. Rowe and John Mcleary, 2 vols. (Boston: Academic Press, Inc., 1989), 1:125–154. On calculus texts, see George M. Rosenstein, Jr., "The Best Method. American Calculus Textbooks of the Nineteenth Century," in *A Century of American Mathematics—Part* III, ed. Peter Duren *et al.* (Providence: American Mathematical Society, 1989):77–109.

[23]John Farrar and George Emerson, *First Principles of Differential and Integral Calculus, or The Doctrine of Fluxions . . . Taken Chiefly from the Mathematics of Bézout* (Cambridge: Hilliard & Metcalf, 1824).

[24]Cajori, *History of Mathematics*, p. 130. See, also Rosenstein, pp. 78–80.

[25]The most comprehensive contemporary biographical study of Bowditch and his work was written by his son. See Henry Ingersoll Bowditch, "Memoir," in Pierre Simon de la Place, *Celestial Mechanics*, trans. Nathaniel Bowditch, 4 vols. (New York: Chelsea Publishing Co.,

financial situation, Bowditch was apprenticed out at the age of twelve to a ship's chandler and so received only minimal schooling. He did, however, study voraciously during his free hours before and after work, drawing notably from the library of the Irish chemist, Robert Kirwan, which several Salem citizens purchased at auction and allowed Bowditch to use. There, the young Bowditch had access to the *Philosophical Transactions of the Royal Society* as well as to works in mathematics, physics, astronomy, and, of course, chemistry. After teaching himself Latin, he also secured and mastered Newton's *Principia.* Bowditch continued his scientific studies from 1795 to 1804 during the lengthy sea voyages of the Salem ships he had finally come to serve. In particular, he studied the French approach to the calculus as explained by Lacroix in his *Traité élémentaire*, the same text that would attract the attention of Cambridge undergraduate Charles Babbage. Bowditch's familiarity with this text especially prepared him for his most ambitious undertaking, a translation with detailed explanations of Laplace's *Mécanique céleste.*

Although most widely known for his *American Practical Navigator* first published in 1802 and still in print today, Bowditch made his most significant contribution to the early development of mathematics in America through his work on Laplace.[26] Between 1814 and 1817, he completed a translation of the first four volumes of Laplace's classic work. (In 1825, a fifth volume of this treatise appeared, which Bowditch did not ultimately translate.) The fruits of Bowditch's labor, however, were not published until 1829.

After abandoning his seagoing life in 1804, Bowditch assumed the presidency of a marine insurance company in Salem. He changed jobs again in 1823, this time moving to Boston where he served as an actuary at the Massachusetts Hospital Life Insurance Company. While neither of these posts made him a millionaire, Bowditch had managed to save up some money, and he resolved to bring out his work on Laplace himself as soon as his savings would allow. He took advantage of the intervening years by honing, polishing, and extending his explanatory notes as new research came out. Since his intention had always been for the notes to serve as a completion and fuller explanation of Laplace's often sketchy proofs, it also seemed reasonable to

---

1966), 1:9–168. Shorter accounts appear in, for example, Cajori, pp. 86–91 and 103–105; Charles C. Gillispie, ed., *Dictionary of Scientific Biography*, 16 vols., 2 supp. (New York: Charles S. Scribner's Sons, 1970–1990), s.v. "Bowditch, Nathaniel," by Nathan Reingold (hereinafter abbreviated *DSB*); Greene, pp. 144–157; Simons, pp. 106–108; and David Eugene Smith and Jekuthiel Ginsburg, *A History of Mathematics in America before* 1900, The Carus Mathematical Monographs, vol. 5 (n.p.: Mathematical Association of America, 1934; reprint ed., New York: Arno Press, 1980), pp. 92–95.

[26] Nathaniel Bowditch, *American Practical Navigator* (Washington: Government Printing Office, 1802; reprint ed., n. p.: Defense Mapping Agency Hydrographic/Topographic Center, 1984). Benjamin Banneker produced an earlier and also notable almanac. On Banneker's life and place in American science, consult Silvio A. Bedini, *The Life of Benjamin Banneker* (New York: Charles S. Scribner's Sons, 1971); and James A. Donaldson, "Black Americans in Mathematics," in *A Century of Mathematics in America—Part* III, ed. Peter Duren *et al.* (Providence: American Mathematical Society, 1989):449–469.

Nathaniel Bowditch (1773–1838)

Benjamin Peirce (1809–1880)

use them as a vehicle for elucidating more recent discoveries. In particular, he expanded on work in the first four volumes by drawing from the material Laplace included in the fifth and final tome. Shortly after incorporating these latest developments into his own work, Bowditch was financially ready to begin publication, and he enlisted the services of one of his son's friends, the Harvard undergraduate Benjamin Peirce, to help with the proofreading. Peirce continued to assist Bowditch even after finishing his studies and going on to teach first at his old preparatory school and then at Harvard. He also helped see the second volume through the press in 1832 and the third in 1834 by making corrections and revisions. Bowditch died in 1838, just short of finishing his fourth and final volume, and Peirce loyally brought the project to completion in 1839.

Thus, Laplace's *Mécanique céleste*, the work which had occasioned John Playfair's musings on the backward state of mathematics and science in England in 1808, had been made available for the further advancement of mathematics and science in America. This noble idea had motivated the undertaking, at least to some extent, from the very start. In the introduction to his translation, Bowditch explained that while Laplace had admirably succeeded in realizing his stated goal, namely, "to reduce all the known phenomena of the system of the world to the law of gravity, by strict mathematical principles," he had so abridged the calculations that the work "has been found difficult to be understood by many persons, who have a strong and decided taste for mathematical studies, on account of the time and labour required, to insert the intermediate steps of the demonstrations, necessary to enable them easily to follow the author in his reasoning."[27] For this reason, Bowditch took as his own goal as translator to fill in the details of this tersely and often obscurely written text. In so doing, he hoped "that the facility, arising from having the work in our own language, with the aid of these explanatory notes, will render it more accessible to persons who have been able to prepare themselves for this study by a previous course of reading, in those modern publications, which contain the many important discoveries in analysis, made since the time of Newton."[28] Bowditch was convinced that through careful explanatory comments and elaborations, Continental texts like Laplace's and the mathematics they covered would help his countrymen catch up to the Europeans in science. Like Farrar, he recognized that the Continent, and especially France, had to replace Great Britain as the source of inspiration for American science.

Another early nineteenth-century American who independently came to this same conclusion was Major Sylvanus Thayer. Thayer, who assumed the superintendency of the United States Military Academy in 1817, consciously transformed the West Point curriculum according to the model of the *École*

---

[27] Nathaniel Bowditch, "Introduction by the Translator," to Laplace, *Celestial Mechanics*, 1:v.
[28] *Ibid.*

polytechnique. While in France on a fact-finding mission for the United States Government after the War of 1812, Thayer had witnessed firsthand the superior French system of education and had appreciated the strong mathematical component of the course of study for future engineers and army officers. He resolved to bring as much of the French program to West Point as possible. To this end, he secured the services of the former *polytechnicien* Claude Crozet as Professor of Engineering.[29] Crozet quickly realized, however, that before he could instruct his students in engineering he would have to teach them the requisite mathematics. One of the students in Crozet's first American class, recognizing his professor's dismay, wrote: "[Imagine] [t]he surprise of the French engineer, instructed at the Polytechnique, ... when he commenced giving his class certain problems and instructions which not one of them could comprehend and perform."[30] This lack of appropriate mathematical preparation resulted in Crozet's introduction into the Academy's curriculum of the Mongian descriptive geometry then standard in the curriculum of the *École polytechnique.* Furthermore, in the absence of a textbook in English, Crozet wrote *A Treatise on Descriptive Geometry for the Use of the Cadets of the United States Military Academy*, which appeared in 1821.[31] From 1817 until the time his text was ready for his students' use, he demonstrated the supreme utility of another French innovation, blackboard and chalk, which quickly became an indispensable aid in mathematics teaching at West Point and elsewhere.

After Crozet left the Academy in 1823 to assume the post of Chief Engineer for the Commonwealth of Virginia, the emphasis on French mathematics continued unabated. Charles Davies, who assumed the Chair of Mathematics in the shuffle following Crozet's departure, produced a series of textbooks based on French works which quickly supplanted Farrar's outdated books nationwide.[32] Davies's texts, unlike Crozet's book, enjoyed tremendous popularity and served to introduce several generations of American undergraduates to the fundamentals of Continental mathematics. His version of Legendre's *Geometry* (1794) went through at least twenty-eight editions between 1834 and 1890, while his rendition of Louis P. M. Bourdon's *Éléments d'Algèbre*

---

[29] On West Point and Claude Crozet, see Cajori, pp. 114–127; Greene, p. 131; and Simons, p. 115.

[30] Cajori, p. 116.

[31] Claude Crozet, *A Treatise on Descriptive Geometry for the Use of the Cadets of the United States Military Academy* (New York: A. T. Goodrich & Co., 1821). Crozet was not the first to bring French-style mathematical training to the Academy. As early as 1807–1810, Ferdinand Hassler, the Swiss-born scientist whom Thomas Jefferson later chose to plan and set up the United States Coast Survey, had tried unsuccessfully to institutionalize Continental mathematics at West Point. Later in the 1820s under Thayer's regime, Crozet's text was complemented by John Farrar's 1818 translation of Lacroix's *Algebra* (1797) as well as his 1819 rendering of Legendre's *Geometry* (1794). See Simons, pp. 111 and 121. Compare note 22 above.

[32] On Charles Davies, see Cajori, pp. 118–121; and Smith and Ginsburg, pp. 77–79. After leaving West Point in 1837, Davies held academic posts at Trinity College in Hartford, Connecticut, at the University of New York, and finally at Columbia College from 1857 to 1865.

(1817) saw at least twenty-two editions between 1836 and 1873.[33] Davies described the vision which had made it possible for him to bring French mathematics in translation to the United States when he dedicated one of his texts to Superintendent Thayer. He wrote:

> In the organization of the Military Academy under your immediate superintendence, the French methods of instruction, in the exact sciences, were adopted; and near twenty years experience has suggested few alterations in the original plan.
>
> The introduction of these methods is considered an improvement worthy to form an era in the history of education in this country; and public opinion has justly appreciated the benefits which you have conferred at once on the Military Academy, and on the cause of science.[34]

Yet the Continental mathematics which Davies, Crozet, Farrar, and others like William Chauvenet[35] gradually introduced to the students in America's colleges from the 1830s to the 1870s, was largely the mathematics of the turn of the previous century. Furthermore, it tended to be mathematics at a beginner's—and not at a research—level. Consider, for example, Davies's *Elementary Algebra: Embracing the First Principles of the Science* (the translation of Bourdon's *Éléments d'Algèbre*).[36] There, the student learned the basics of algebra—how to manipulate fractions involving unknowns, how to solve equations of the first and second degree, how to raise polynomials to powers—in addition to receiving a first introduction to arithmetic and geometric progressions and logarithms. Clearly, it would take much more than the introduction of French texts at this level to bring the United States into serious competition with Europe in mathematical research. Among other things, Americans would have to come to the realization that mathematics constituted a worthwhile professional pursuit. This process of professionalization, which had begun in American science as early as the first decade of the nineteenth century, was inextricably bound to changes taking place within the system of higher education, to emergent centers of scientific research within the federal government, and to attempts to sustain lines of

---

[33]Louis Pierre Marie Bourdon, *Éléments d'Algèbre* (Paris: Mme Ve Courcier, 1817). See Simons, pp. 114, 117, and 121. Pycior discusses Davies's version of Bourdon's book in "British Synthetic vs. French Analytic Styles of Algebra in the Early American Republic," pp. 137–142.

[34]Simons, pp. 105–106.

[35]William Chauvenet (1820–1870) published several textbooks, notably *Treatise on Plane and Spherical Trigonometry* (Philadelphia: Hogan, Perkins & Co., 1850) and *A Treatise on Elementary Geometry* (Philadelphia: J. B. Lippincott, 1870). He was instrumental in founding the United States Naval Academy in 1845 and remained there until 1860, when he assumed a position at Washington University in St. Louis, Missouri. Today, he is best remembered through the Mathematical Association of America's Chauvenet Prize for mathematical exposition.

[36]Charles Davies, *Elementary Algebra: Embracing the First Principles of the Science* (New York: A. S. Barnes & Burr, 1862).

communication such as journals and societies. For there to be an American "mathematician" as opposed to a "teacher" or a "surveyor," mathematical sympathizers in the United States had to conceive of a discipline extending beyond the confines of the undergraduate curriculum. On one level at least, this enlarged perception developed within the colleges themselves.

Mathematics and the American College: The Cases of Harvard and Yale

Prior to 1800, if mathematics education at America's colleges was not relegated to mere Tutors, it fell under the aegis of the Professor of Mathematics and Natural Philosophy. Since faculties were generally quite small, sometimes numbering only five or six, these chairholders were entrusted not only with all of the college's mathematical instruction but also with all of its instruction in the sciences. While the science component of the eighteenth-century curriculum was limited to a bit of physics and astronomy *à la* Newton, the nineteenth-century offerings gradually increased to include other sciences such as chemistry, natural history, botany, and geology.[37] Like the concurrent redirection of mathematical sights from Britain to Continental Europe, this expansion of the curriculum resulted from a conscious attempt to break out of the Newtonian straitjacket and to escape into the larger and ever-growing world of Continental science. As Playfair suggested in his review of the *Mécanique céleste*, the British and, by association, the Americans had been left behind in *all* of the exact sciences, not just in mathematics.

From the turn of the century up to the mid-1840s, the sciences steadily worked their way into the classical liberal arts curriculum. This infiltration required not only a restructuring of faculty positions but also an increase in faculty size. As early as 1802, Yale created a chair of chemistry, natural history, and mineralogy for Benjamin Silliman, later the founder of the *American Journal of Science and Arts* and an early promoter of American science. In so doing, the college freed Jeremiah Day, the holder of the chair of mathematics and natural philosophy, from responsibility for those other sciences and left him more time to devote to his forte, mathematics.[38] This

[37]On curricular expansion in general, consult Frederick Rudolph, *Curriculum: A History of the American Undergraduate Course of Study since* 1636 (San Francisco: Jossey-Bass Publishers, 1978). On the expansion of the curriculum with respect to the sciences in particular, see Stanley M. Guralnick, "The American Scientist in Higher Education: 1820–1910," in *The Sciences in the American Context: New Perspectives*, ed. Nathan Reingold (Washington: Smithsonian Institution Press, 1979), pp. 99–141 on pp. 103–113.

[38]Day, who stepped down from his chair in 1817 to assume the presidency of Yale, wrote numerous texts for his students, for example, *An Introduction to Algebra, Being the First Part of a Course of Mathematics, Adapted to the Method of Instruction in the American Colleges* (New Haven: Howe & Deforest, 1814); *A Treatise of Plane Trigonometry to Which is Prefixed a Summary View of the Nature and Use of Logarithms, Being the Second Part of a Course of Mathematics, Adapted to the Method of Instruction in the American Colleges* (New Haven: Howe & Deforest, 1815); *A Practical Application of the Principles of Geometry to the Mensuration of*

sort of differentiation and specialization continued well into the century until each separate subject had its own distinct chair. In 1836, for example, the chair Day had held was further subdivided into separate professorships of mathematics and natural philosophy. Then, when Silliman retired in 1853, his position split into chairs of chemistry and natural history. Finally, in 1864 natural history was supplanted by distinct professorships of geology and botany.[39]

While this chain of events at Yale reflected a definite nineteenth-century trend in American education, change occurred at varying rates at different institutions. As late as the 1870s, the biologist David Starr Jordan taught "classes in natural science, political economy, evidences of Christianity, German, Spanish, and literature, and pitched for the baseball team" at Illinois's Lombard College.[40] As a well-established college with a sizable enrollment, Yale had the demand and the resources necessary to effect such changes in its faculty. Such innovations necessarily came more slowly at younger, smaller, and less wealthy schools like Lombard. Regardless of individual variations between colleges, the differentiation and specialization of college faculties which took place during the nineteenth century contributed to an increasing sense of self-identity among the different brands of scientists. Professors of chemistry or geology or mathematics gradually began to see themselves not merely as teachers but as chemists or geologists or mathematicians. The process of delineating academic chairs thus brought with it a concomitant sense of disciplinary cohesion, a necessary ingredient of professionalization.

This academic specialization also allowed the individual science professors to extend somewhat the offerings in their own particular branches of science. Consider the extreme case of Harvard during the second quarter of the nineteenth century. John Farrar, that inveterate translator and adaptor of French texts for American college students, held, as noted above, the Hollis Professorship of Mathematics and Natural Philosophy from 1807 to 1836. As a result of his efforts, by 1830 much of the old British style of mathematics had been replaced by his newer version of French mathematics, and the cal-

---

Superficies and Solids, Being the Third Part of a Course of Mathematics, Adapted to the Method of Instruction in the American College (New Haven: Oliver Steele, 1814); and *The Mathematical Principles of Navigation and Surveying, with the Mensuration of Heights and Distances, Being the Fourth Part of a Course of Mathematics, Adapted to the Method of Instruction in the American Colleges* (New Haven: Steele & Gray, 1817). Written at a very elementary level, but in imitation of the style, if not the content, of Euler and Lacroix, the texts proved very popular and went through numerous editions. On Day, see Cajori, pp. 63–64. For a discussion of Day's algebra textbook, see Pycior, "British Synthetic vs. French Analytic Styles of Algebra in the Early American Republic," pp. 126–129.

[39]George H. Daniels, *American Science in the Age of Jackson* (New York: Columbia University Press, 1968), p. 35.

[40]Richard Hofstadter, "The Revolution in Higher Education," in *Paths of American Thought*, ed. Arthur M. Schlesinger, Jr. and Morton White (Boston: Houghton Mifflin Company, 1970):269–290 on p. 284. Lombard opened in Galesburg, Illinois in the mid-1850s and survived into the twentieth century.

culus had entered the curriculum. Thus, in the highly prescribed curriculum of 1830, all freshmen read Legendre's *Geometry* (1794) in addition to algebra and solid geometry; the sophomores had trigonometry with applications, topography, and calculus; every junior took natural philosophy, mechanics, electricity, and magnetism; and the senior class continued in natural philosophy while picking up optics.[41] (The curricula adopted at American colleges at this time are difficult to characterize in general terms. Many had a prescribed or "fixed" classical component but offered so-called "parallel courses of study" which featured modern languages and the "newer" sciences. Experiments with electives and other optional courses began early in the century with varying degrees of success.[42] Electives became institutionalized at the various colleges at different points throughout the latter part of the nineteenth century.) Although the introduction of the calculus surely represented a significant improvement in the level of mathematical training, this curriculum still targeted the undergraduate. True graduate education, that is, an education designed to raise students to the level of contemporaneous research, remained a thing of the future. In Harvard's case, however, the future was not that far off.

In 1831, the same Benjamin Peirce who had proofread Bowditch's edition of the *Mécanique céleste* became a Tutor in mathematics under Farrar. On receiving his M.A. in 1833, Peirce assumed the University Professorship of Mathematics and Natural Philosophy and began reformulating Harvard mathematics according to his own vision. Following Farrar, he immediately set about writing textbooks for his students, but, unlike his teacher, Peirce produced original texts not merely annotated translations.[43] Universally judged as elegant but virtually impenetrable to all but the most gifted readers, Peirce's books became the bane of the Harvard student. In 1838, though, a struggling elective system liberated all but the freshmen from the rigors of Peirce and his texts.[44] At the same time, it liberated Peirce from

[41] Cajori, p. 133.

[42] On an elective scheme at Harvard as early as 1825, see Rudolph, pp. 76–78. This was drastically rolled back beginning in 1843.

[43] Peirce wrote textbooks such as *An Elementary Treatise on Plane and Spherical Trigonometry with its Applications to Heights and Distances, Navigation and Surveying* (Boston: J. Munroe & Co., 1835); *First Part of an Elementary Treatise on Spherical Trigonometry* (Boston: J. Munroe & Co., 1837); *An Elementary Treatise on Algebra: To Which Are Added Exponential Equations and Logarithms* (Boston: J. Munroe & Co., 1837); *An Elementary Treatise on Plane and Solid Geometry* (Boston: J. Munroe & Co., 1837), and *An Elementary Treatise on Curves, Functions, and Forces*, 2 vols. (Boston: J. Munroe & Co., 1841–1862), among others. For Peirce's complete bibliography, see Raymond C. Archibald *et al.*, "Benjamin Peirce," *American Mathematical Monthly* 32 (1925):1–30 on pp. 20–30. Pycior deals with Peirce's algebra text in "British Synthetic vs. French Analytic Styles of Algebra in the Early American Republic," pp. 143–145; and Rosenstein discusses his calculus book on pp. 84–85.

[44] Peirce basically proposed three different elective tracks in mathematics: (1) a year-long practical course, (2) a more theoretical year-long course for prospective teachers, and (3) a three-year-long course for future mathematicians. See Rudolph, p. 78.

all but the freshmen! Furthermore, the establishment at Harvard in 1847 of the Lawrence Scientific School—for scientific and engineering education at an essentially graduate level—gave Peirce additional incentive to present more serious mathematics. Founded to fill the growing sense of need for advanced training in the sciences, the Lawrence School proposed to give instruction "to graduates and others, in the various branches of exact and physical science... [in t]he pure and mixed Mathematics; Astronomy, theoretical and practical; Chemistry in its various branches, theoretical and operative; Civil Engineering, and generally the application of science to the arts of life and the great industrial interests of the community; with the several branches of Natural Science."[45] Teaching only the most talented and motivated, Peirce guided his Lawrence School students through the most rigorous course of mathematical studies yet offered in the United States. (See Table 1.1.)

Even though many of his contemporaries had made great strides by introducing the translations of eighteenth-century French texts into their colleges' curricula, Peirce himself had brought Harvard's curriculum close to the research frontier in 1848. In particular, he had introduced Cauchy's work, including the *Cours d'Analyse* of 1821; Sir William Rowan Hamilton's researches on quaternions (dating from the mid-1840s); and the 1846 work of John Couch Adams and Urbain Leverrier on the mathematical deduction of the existence of the planet Neptune.[46] Although a stunningly ambitious and sophisticated course of studies, it unfortunately enlightened precious few.

As a teacher, Peirce had little understanding of the average student's difficulties and even less ability to explain obscure points. As one member of the class of 1851 wrote: "I am no mathematician, but that I am so little of one is due to the wretched instruction at Harvard. Professor Peirce was admirable for students with mathematical minds, but had no capacity with others."[47] The mathematically inclined student, however, tended to have quite a different appreciation of Peirce, the instructor:

> As soon as he had finished the problem or filled the blackboard he would rub everything out and begin again. He was impatient of detail, and sometimes the result would not come out right; but instead of going over his work to find the error, he would rub it out, saying that he had made a mistake in a sign somewhere, and that we should find it when we went over our notes.

---

[45]As quoted from the 1847–1848 catalogue of Harvard College, in Russell H. Chittenden, *History of the Sheffield Scientific School of Yale University:* 1846–1922, 2 vols. (New Haven: Yale University Press, 1928), 1:41.

[46]The latter work, in fact, inspired Peirce's own controversial research on the orbit of Neptune. (On this story, see, for example, Nathan Reingold, *Science in Nineteenth-Century America: A Documentary History* (Chicago: University of Chicago Press, 1964), pp. 139–143.) This, together with his advances in the mathematical theory of comets, secured his reputation in mathematical astronomy.

[47]Letter from William F. Allen to Florian Cajori, quoted in Cajori, p. 140.

> Described in this way it may seem strange that such a method of teaching should be inspiring; yet to us it was so in the highest degree. We were carried along by the rush of his thought, by the ease and grasp of his intellectual movement. The inspiration came, I think, partly from his treating us as highly competent pupils, capable of following his line of thought even through errors, which reached a result with the least number of steps in the process, attaining thereby an artistic or literary character; and partly from the quality of his mind which tended to regard any mathematical theorem as a particular case of some more comprehensive one, so that we were led onward to constantly enlarging truths.[48]

In spite of testimonials such as this, the number of students to whom Peirce revealed the beauty of mathematics was always small. In 1849, he taught the Lawrence School curriculum to only two, and this number represented the rule rather than the exception.[49] The Lawrence School itself started feebly, with barely enough money to meet expenses, until a new administration began to turn things around in the late-1860s. Enrollments in Peirce's college courses were hardly better, even after 1850 when the freedoms of the elective system were curtailed to make mathematics again mandatory for sophomores. Among the class of 1854, for instance, only a dozen juniors opted to continue in mathematics under Peirce, and the number had dwindled to four by their senior year.[50]

While Peirce himself valued research and made important mathematical (see below) as well as astronomical contributions, the institutional framework within which he worked did not begin to encourage like interests in its faculty or in its students until the 1870s and 1880s.[51] Relative to mathematics, then, Peirce's ideal of preparing and encouraging future researchers went largely unrealized, with the exception of his talented but eccentric son, the logician and philosopher Charles S. Peirce, and his distant relative Benjamin Osgood Peirce. He thus did not found a school of mathematical *research*, but he did establish a *tradition* in applied mathematics at Harvard which would be

---

[48] Reminiscences of Peirce by A. Lawrence Lowell, in Archibald *et. al.*, "Benjamin Peirce," pp. 4–5.

[49] Cajori, p. 141.

[50] Letter from Truman Henry Safford to Florian Cajori, quoted in Cajori, pp. 142–143.

[51] Peirce set down his ideals for university education in a four-page pamphlet printed in 1856 and entitled "Working Plan for the Foundation of a University." In particular, he believed that "[t]he most feasible plan is that which is most elastic, and which may be the smallest in its germ, while it is most comprehensive in its full development. Its professors must be the ablest men in their respective departments; it must be connected with a fine library, a well equipped observatory, and complete collections and laboratories for the elucidation, illustration, and investigation of every species of knowledge [p. 1]." A copy of this pamphlet may be found in the Daniel C. Gilman Papers Ms. 1, Special Collections Division, Milton S. Eisenhower Library, The Johns Hopkins University.

maintained there into the twentieth century by those he had trained: his son James Mills Peirce, William Byerly, and B. O. Peirce.[52]

The offerings at mid-century Harvard in mathematics and in the mathematical sciences such as astronomy, which prepared these three for their subsequent careers on the Harvard faculty, by no means reflected the state of affairs nationwide. Most colleges required only an understanding of arithmetic and a bit of algebra and geometry for admission by the 1850s, and they taught from increasingly outmoded translations of eighteenth-century French works by writers like Davies. Thus, in 1848 when Harvard's Lawrence School exposed students to the mathematical course of study indicated in Table 1.1, Yale offered only the following undergraduate program: "Freshmen, Day's Algebra, Playfair's Euclid; Sophomores, Day's Mathematics, Bridge's Conic Sections, and Stanley's Spherical Geometry and Trigonometry; Juniors, Olmsted's Natural Philosophy, Mechanics, Hydraulics, Hydrostatics, Olmsted's Astronomy, Analytical Geometry or Fluxions (optional);"[53] while seniors had no mathematical requirements at all. By and large, students proved less than receptive even to the most elementary mathematical instruction. At Yale into the 1860s, for example, students carried on the ritual of "burying" Euclid at the end of their required year of geometry.[54] Due to its lack of popularity and to the woeful shortage of mathematical visionaries like Peirce, mathematics remained suspended in time in the eighteenth century at many American colleges until well into the second half of the 1800s.[55]

Harvard also proved exceptional in that its scientific school had provided at mid-century the vehicle for Peirce's introduction of research-level mathematics into the curriculum. The 1860s and 1870s saw the growing importance of scientific schools as well as the increasing development of engineering curricula at other institutions across the country.[56] While these sorts of additions to the classical American college may have served to promote the applied and some of the pure sciences at both an undergraduate and a graduate level, they generally did little to improve the state of mathematics, as exemplified by the case of Yale College.

[52]Although B. O. Peirce (Leipzig doctorate, 1879) was an active researcher in mathematical physics, the strengths of J. M. Peirce and Byerly lay more in teaching than in research.

[53]Cajori, p. 158.

[54]A vivid description of this ritual may be found in R. E. Langer, "Josiah Willard Gibbs," *American Mathematical Monthly* 46 (1939):75–84 on pp. 78–79.

[55]On the problem of mathematical sophistication and its relationship to performance in the other sciences, see John W. Servos, "Mathematics and the Physical Sciences in America, 1880–1930," *Isis* 77 (1986):611–629.

[56]As we have seen, a serious engineering curriculum developed at West Point as early as the 1820s. Similarly, Renssalaer Polytechnic Institute, founded in 1824, stressed engineering. During the thirty years from 1845 to 1875, and particularly after the Morrill Act of 1862 provided for the establishment of land-grant colleges, a growing emphasis was placed on "practical" studies such as engineering, mining, and agriculture. Scientific schools thus came into vogue. On this period in the development of American science, see Guralnick, "The American Scientist in Higher Education: 1820–1910," pp. 114–125, and Robert V. Bruce, *The Launching of Modern American Science*: 1846–1876 (New York: Alfred A. Knopf, 1987).

In 1847, the same year in which the Lawrence Scientific School was founded at Harvard for graduate instruction in the sciences, Yale instituted a Department of Philosophy and the Arts to provide graduate studies in "philosophy, literature, history, the moral sciences other than law and theology, the natural sciences excepting medicine, and their application to the arts."[57] While in principle this new academic department existed to house graduate studies in all areas (except law, medicine, and theology which already had their own schools), in practice it initially provided an institutional umbrella for two new professors whose specialties lay outside the usual liberal arts curriculum, namely, applied and agricultural chemistry. Drawing eight of eleven graduate students in the department's first year, the School of Applied Chemistry quickly became the dominant force in graduate instruction at Yale. By 1852, it had been joined by a School of Engineering, and this expansion continued, gradually encompassing such areas as physics, astronomy, metallurgy, and mining. Named the Sheffield Scientific School in 1861, the collection of scientists and their students comprised the scientific component of the graduate school.

The Sheffield School, unlike its Harvard counterpart, offered no specialized instruction in mathematics until 1873 when John Emory Clark became the first chairholder in the subject.[58] Prior to Clark's appointment, the Professors of Engineering had dictated the school's mathematical offerings while leaving the actual teaching to Instructors and Assistants. Mathematics thus played merely a service role in the Sheffield School, satisfying the special needs of students of engineering and the applied sciences. In 1852, for example, the Professor of Engineering included only analytical geometry together with differential and integral calculus in the prescribed course of study for students in his new school.[59] Furthermore, mathematics fell under the category of "arts" within the graduate school itself and so lay outside the Sheffield School anyway. Until the mid-1870s, when the gradual expansion of Yale College began to allow for advanced instruction in its formerly undergraduate departments, graduate study in mathematics was essentially nonexistent in New Haven.

To say that Yale did not awaken mathematically until the last quarter of the nineteenth century, however, is to place it in a league with all of the best institutions of higher education in America, with the exception of Peirce's Harvard. It is not to say that mathematics went unappreciated there. On the contrary, Hubert Anson Newton, Yale's Professor of Mathematics from 1855 until his death in 1896, astutely guided the mathematically talented.

---

[57]Chittenden, 1:41.

[58]Clark graduated from the University of Michigan in 1856 and subsequently served as Professor of Mathematics there. In 1859, he left the United States for two years abroad, studying in Berlin, Heidelberg, and Munich. Although not a researcher, Clark was a gifted and patient teacher. For more on his academic career, see *ibid.*, pp. 194–195.

[59]*Ibid.*, p. 58.

Developing into more of an astronomer in the latter part of his career, Newton nevertheless began as a mathematician.[60] He graduated from Yale College in 1850 and then studied mathematics independently for over two years before returning to his *alma mater* in 1853 as a Tutor. Later that same year, he was given complete command of Yale mathematics owing to the death of Anthony D. Stanley, the mathematician since 1836. After being offered and accepting the professorship in 1855 at the age of twenty-five, Newton immediately took a year's leave of absence to deepen his own understanding of the area, spending the 1855–1856 academic year in Paris studying under the influential geometer Michel Chasles. Newton learned well from his experiences abroad and published three papers based on his geometrical research during the first few years after his return to New Haven. By 1861, however, his research interests had been permanently diverted from mathematics to astronomy. In an obituary notice in the *American Journal of Science* in 1897, one of Newton's students, none other than Josiah Willard Gibbs, provided a very penetrating explanation for this shift. Gibbs wrote:

> In the attention which has been paid to astronomy in this country we may recognize the history of the world repeating itself in a new country with respect to the order of the development of the sciences, or it may be enough to say that the questions which nature forces on us are likely to get more attraction in a new country and a bustling age, than those which a reflective mind puts to itself, and that the love of abstract truth which prompts to the construction of a system of doctrine, and the refined taste which is a critic of methods of demonstration are matters of slow growth. At all events, when Professor Newton was entering upon his professorship, the study of the higher geometry was less consonant with the spirit of the age in this country than the pursuit of astronomical knowledge, and the latter sphere of activity soon engrossed his best efforts.[61]

Regardless of what actually prompted Newton to abandon his geometrical studies, he spent the remainder of his lengthy career publishing the original work on meteors and comets that won him a reputation internationally as a notable American astronomer.

---

[60]On Newton's life and work, see the anonymous article written at the time of his election as the President of the American Association for the Advancement of Science, "Biographical Sketch of the President of the Association," *Science* 6 (1885):161–162; J. Willard Gibbs, "Hubert Anson Newton," *American Journal of Science*, 4th ser., 3 (1897):358–378, which also contains Newton's bibliography; and Gillispie, ed., *DSB*, s.v. "Newton, Hubert Anson," by Richard Berendzen.

[61]Gibbs, "Hubert Anson Newton," pp. 360–361. Gibbs's reflections here about the reception of abstract theories in America may have been occasioned as much by his own experiences as by Newton's. See below.

Newton thus did not set an example as a strong *mathematical* researcher to his mathematically talented students; rather he imparted important values to them. His own year of European study as well as his professional ties abroad had instilled in Newton a desire to follow the research being done on the other side of the Atlantic. As a man well connected within the developing American scientific community, he also contributed to and kept up with the research being done on his own shores. He must have recognized the limitations of the system of higher education in the United States for training future scientists, for he evidently influenced the decisions of two of his best students to further their mathematical studies abroad. Of one of those students, Eliakim Hastings Moore, we will have much more to say later. (See Chapters 6, 7, and 9.) A mathematician who earned international recognition for his original contributions to algebra and the foundations of geometry and analysis, Moore was also the motivating force behind the establishment of the Mathematics Department at the University of Chicago and an active participant in the development of mathematics at a national level. An energetic and dynamic personality, E. H. Moore was one of a handful of early leaders of the emergent American mathematical research community. Newton's other distinguished student, however, was quite the opposite of Moore. A shy and retiring man, J. Willard Gibbs preferred quiet reflection to national activism. He thus left his mark on American science in general and on mathematics in particular in a different way.

GERMAN INFLUENCES: J. WILLARD GIBBS AT HOME AND ABROAD

Justly claimed by physicists, chemists, and mathematicians alike as one of their own, Willard Gibbs ranks as perhaps the most important American scientist of the last quarter of the nineteenth century.[62] As an undergraduate at Yale from 1854 to 1858, Gibbs excelled in Latin and mathematics, winning

[62]There are several good sources on Gibbs's life and work. See, for example, Lynde Phelps Wheeler, *Josiah Willard Gibbs: The History of a Great Mind* (New Haven: Yale University Press, 1952); Muriel Rukeyser, *Willard Gibbs* (Garden City: Doubleday, Doran & Company, Inc., 1942); Gillispie, ed., *DSB*, s.v. "Gibbs, Josiah Willard," by Martin Klein; Josiah Willard Gibbs, *The Scientific Papers of J. Willard Gibbs, Ph.D., LL.D. formerly Professor of Mathematical Physics in Yale University*, 2 vols. (New York: Longmans, Green, & Co., 1906; reprint ed., New York: Dover Publications, Inc., 1961); Arthur Haas, ed., *A Commentary on the Scientific Writings of J. Willard Gibbs Ph.D., LL.D. formerly Professor of Mathematical Physics in Yale University*, 2 vols. (New Haven: Yale University Press, 1936); Michael J. Crowe, *A History of Vector Analysis: The Evolution of the Idea of a Vectorial System* (Notre Dame: University of Notre Dame Press, 1967); Martin J. Klein, "The Scientific Work of Josiah Willard Gibbs," in *Springs of Scientific Creativity: Essays on Founders of Modern Science*, ed. R. Aris, H. T. Davis, and R. H. Stuewer (Minneapolis: University of Minnesota Press, 1983):142–162; reprinted in *A Century of Mathematics in America—Part* II, ed. Peter Duren *et al.* (Providence: American Mathematical Society, 1989):99–119; and D. G. Caldi and G. D. Mostow, ed., *Proceedings of the Gibbs Symposium: Yale University, May* 15–17, 1989 (Providence: American Mathematical Society, 1990).

prizes in both. He proceeded into the Yale Scientific School's advanced program in engineering upon graduation and earned the first American Ph.D. in that subject in 1863 for a thesis on gear design. This doctorate aside, Gibbs's future development as a creative scientist depended crucially on the three years from 1866 to 1869, during which he undertook post-graduate studies in Europe. This educational journey coincided with the advent of a new era in which the German universities exerted a profound influence both on the Americans who studied there and on those who sought to reform and elevate higher education in the United States.

Unlike the French universities which had difficulty redefining their mission in the wake of the French Revolution, those in the German states successfully adapted to the very different political and cultural conditions that prevailed east of the Rhine. Led by Wilhelm von Humboldt, the University of Berlin, which opened under French occupation in 1810, spearheaded a neohumanist movement in German higher education that served as the dominant ideology in Prussia for the next several decades. The Humboldtian emphasis on the pursuit of pure research for its own sake was largely intended as an antidote to the utilitarianism Humboldt and his contemporaries identified with the tradition of the French Enlightenment and the excesses of the Revolution. The influence of this ideology reverberated throughout German scholarship and extended beyond the humanities into mathematics and the other sciences. In fact, the purism that characterized German mathematics throughout much of the nineteenth century can, to a considerable degree, be traced to the neohumanist ideals and philosophical idealism that animated both German scholarship and German higher education during the early decades of the nineteenth century.[63]

Neohumanism and idealism were only two of the many aspects which shaped the nineteenth-century German universities, however. These institutions—although administered by the individual states and framed by their own respective histories and traditions—also shared numerous internal policies and structures that facilitated the movement from place to place of students and faculty within the "system" and, hence, the transmission of ideas. First, as semi-autonomous, self-governing corporations, they enjoyed limited protection from the whims of autocratic rulers. This overarching institutional freedom found further expression in the twin ideals of *Lehr- und Lernfreiheit*—the freedom to teach and to learn—which gave professors, in particular,

[63]For an analysis of the background to this general reform movement in German science education, see Frederick Gregory, "Kant, Schelling, and the Administration of Science in the Romantic Era," in *Science in Germany: The Intersection of Institutional and Intellectual Issues*, ed. Kathryn Olesko, *Osiris*, 2nd ser., 5 (1989):17–35. Lewis Pyenson discusses the case of mathematics *per se* in his *Neohumanism and the Persistence of Pure Mathematics in Wilhelmian Germany* (Philadelphia: American Philosophical Society, 1983), pp. 4–15. Consult, also, Charles McClelland, *State, Society, and University in Germany,* 1700–1914 (Cambridge: University Press, 1980); and R. Steven Turner, "The Growth of Professional Research in Prussia, 1818-1848—Causes and Context," *Historical Studies in the Physical Sciences* 3 (1971):137–182.

Josiah Willard Gibbs (1839–1903)

the opportunity to teach more or less whatever and whenever they chose. Second, and perhaps most significant for the institutionalization of research, the notion of the seminar evolved within the German universities during the late eighteenth and early nineteenth centuries. Seminars functioned originally as mechanisms for promoting specialized studies in classical languages and literature and played a central part in the modernized curricula at the Hanoverian university in Göttingen, but especially at the University of Berlin.

The first important seminar for mathematics and physics was established at Königsberg in 1834 by Carl Gustav Jacob Jacobi and the physicist Franz Neumann.[64] Jacobi, who had studied both mathematics and philology at Berlin, joined the Königsberg faculty as a *Privatdozent* in 1826 and taught there for the next seventeen years.[65] Through the sheer force of his personality, quick wit, and legendary virtuosity as an algorist, he engendered an entire school of mathematicians within the seminar context. Among others, it included the master's closest disciple and successor at Königsberg, Friedrich Richelot; the long-time editor of *Crelle's Journal*, Carl Wilhelm Borchardt;[66] and the algebraic geometer, Otto Hesse. Beyond its broader impact on mathematics and mathematics education in Prussia, the Königsberg seminar served as an influential model for seminars at several other German universities, like those at Giessen (founded in 1863), Heidelberg, and Tübingen (the latter both inaugurated in 1869). By 1880, such seminars had been established at all the major universities, with the first purely mathematical seminar being founded at Berlin in 1864.[67]

In light of the programs and activity at Königsberg, Berlin, and elsewhere, Americans entering the field beginning in the 1860s saw Germany as the paragon of mathematical excellence and the progenitor of some of the deepest and most difficult ideas ever to penetrate the discipline. They also saw a support system for original research which encompassed not only the universities and the local academies but also the academies' transactions and various specialized journals. Among the latter, *Crelle's Journal* and Alfred Clebsch's *Mathematische Annalen* (founded in 1868) provided important lines of communication for mathematicians internationally. Through these and other

---

[64] See Wilhelm Lorey, *Das Studium der Mathematik an den deutschen Universitäten seit Anfang des* 19. *Jahrhunderts*, Abhandlungen über den mathematischen Unterricht in Deutschland, Band III, Heft 9 (Leipzig: B. G. Teubner, 1916), pp. 59–63, 71–79, 101–105, and 112–115. For a detailed study of the physics tradition associated with the Königsberg seminar, see Kathryn M. Olesko, *Physics as a Calling: Discipline and Practice in the Königsberg Seminar for Physics* (Ithaca: Cornell University Press, 1991).

[65] For the details of Jacobi's life, consult Leo Königsberger, *Carl Gustav Jacob Jacobi* (Leipzig: B. G. Teubner, 1905).

[66] August Leopold Crelle founded the *Journal für die reine und angewandte Mathematik* in 1826 and served as its first editor. The journal is commonly referred to simply as *Crelle's Journal.* Borchardt succeeded Crelle as editor in 1856.

[67] Kurt-R. Biermann, *Die Mathematik und ihre Dozenten an der Berliner Universität,* 1810–1933 (Berlin: Akademie-Verlag, 1988), pp. 97–100.

sources, the Americans came to recognize Germany as the world's center of mathematical research. By the 1880s, they would flock there—to Berlin, but more importantly to Göttingen, and also to Leipzig—in their attempts to penetrate the mysteries of mathematics. (We will have much more to say about Göttingen and its role in both the history of nineteenth-century mathematics and the emergence of the American mathematical research community in Chapters 4 and 5.)

Willard Gibbs's study tour took him first to Paris, where, on Newton's advice, he heard Chasles's lectures on conic sections. As his Paris notebook reveals, Gibbs also attended lectures given by Jean Marie Constant Duhamel on infinite series, Joseph Liouville on number theory, Joseph Serret on elliptic function theory and celestial mechanics, and Gaston Darboux on mathematical physics.[68] This rigorous course of study caused Gibbs to suffer a breakdown before the end of the first semester and forced him to spend the winter and spring recuperating in Italy. With his Parisian study plans thwarted to some extent by his overzealousness, he carried a lighter course load while in Germany.

Gibbs arrived in Berlin for the Winter Semester of 1867–1868. As it happened, he was joined there by another American student, William James, who later distinguished himself in psychology and philosophy. James had only recently arrived in Berlin from an educational experience rather different from Gibbs's Parisian sojourn—an expedition to the Amazon with the comparative anatomist and natural historian Louis Agassiz. Although James no doubt found Berlin less exotic than the wilds of South America, he nevertheless saw it as full of *wissenschaftliche* mysteries and recognized, with genuine frustration, his seeming inability to penetrate them. In particular, he lamented the fact that his background in mathematics, physics, and chemistry prevented him from participating profitably in any of the laboratory or seminar work that formed the focus of the German scientists' training.[69]

Gibbs presumably experienced some of these same difficulties, but he clearly rose to meet the challenge. In Leopold Kronecker's course on number theory and quadratic forms, he was exposed to the law of quadratic reciprocity and the connections between number theory and Jacobi's theory of elliptic functions. He attended Karl Weierstrass' lectures on determinants and on complex analysis as well as Ernst Kummer's class on probability theory. In August Kundt's physics course, he learned about Hermann von Helmholtz's monumental new treatise, *Die Lehre von den Tonempfindungen*, which established the physiological foundation for the theory of music.[70]

---

[68]Wheeler, p. 40.

[69]Rukeyser, p. 153. William James was also the brother of novelist Henry James. For a discussion of the formation of the theoretical physics community in Germany, consult Christa Jungnickel and Russell McCormach, *Intellectual Mastery of Nature: Theoretical Physics from Ohm to Einstein*, 2 vols. (Chicago: University of Chicago Press, 1986).

[70]Hermann von Helmholtz, *Die Lehre von den Tonempfindungen als physiologische Grundlage*

The following academic year found Gibbs in Heidelberg, where Helmholtz and his distinguished colleagues Gustav Kirchhoff and Robert Bunsen taught physics and where the mathematics faculty included Leo Königsberger, Otto Hesse, and Paul Du Bois-Reymond.[71] When Gibbs returned to New Haven in 1869, he brought back with him a far more comprehensive overview and a much deeper understanding of current research in both mathematics and physics than he could have readily obtained anywhere in the United States.

Back at Yale, Gibbs was offered an unsalaried professorship in mathematical physics in the graduate school in 1871, owing largely to both his post-graduate training and his personal connections.[72] The creation of this new, if unfunded, chair came in the wake of the July, 1871 report of a faculty-appointed, self-study committee. At this time, Yale was about to get its first new President in a quarter-century, and the committee members, among them Daniel Coit Gilman (later the first President of the Johns Hopkins University), Hubert Newton, and Gibbs's brother-in-law, Addison Van Name, wanted to redirect the University onto a course more in keeping with the changing times. In addition to calling for an increase in the endowment and an augmentation of the library's holdings, the committee specifically targeted physics as well as several other areas for the establishment of new chairs. As the report explained, "a field so vast as that of Physics, and one in which the onward march of science is so astonishingly rapid, demands the labors of a professor who shall be permanently and exclusively devoted to it."[73] Willard Gibbs accepted the responsibility of filling that need two days after the committee presented its recommendations to the Yale Corporation.

Gibbs had published nothing in mathematical physics at the time he was hired, and, in fact, his only real research had been in engineering and not in physics at all. Nevertheless, some in New Haven, and especially Newton, realized where Gibbs's true interests lay and so secured his appointment. By 1873, he had proven their judgment correct in his first two research forays into the subject. Published in the *Transactions of the Connecticut Academy of Arts and Sciences*, Gibbs's papers took the mathematical analysis of thermodynamics to new heights.[74]

für die Theorie der Musik (Braunschweig: F. Vieweg und Sohn, 1863).

[71] If Gibbs attended the lectures of either Königsberger or Kirchhoff, then he probably met the young Russian woman, Sofia Kovalevskaya, who attended their courses that year. It has evidently never been established exactly which courses Gibbs attended during his year in Heidelberg, since he did not keep a scientific diary during his third year abroad.

[72] In 1880, the Johns Hopkins University tried to steal Gibbs away from Yale, offering him a salary of $3,000. As a result of this attempted raid, Yale finally offered Gibbs a salary, but only of $2,000, and he chose to stay in New Haven. Thus, for eight years he held an unsalaried chair and lived off of his modest inheritance. See Wheeler, pp. 90–93 for the correspondence relating to this offer.

[73] *Ibid.*, p. 57.

[74] J. Willard Gibbs, "Graphical Methods in the Thermodynamics of Fluids," *Transactions of the Connecticut Academy of Arts and Sciences* 2 (1873):309–342; and "A Method of Geometrical Representation of the Thermodynamic Properties of Substances by Means of Surfaces,"

As developed in the work of such physicists as Rudolf Clausius and William Thomson (later Lord Kelvin), thermodynamics hinged on two basic laws, which, stated in global terms, are: (1) the total energy of the universe is constant, and (2) the entropy of the universe tends to a maximum. In their mathematical formulation, these laws involve the fundamental quantities of volume, pressure, absolute temperature, energy, and entropy as well as work done and heat received by a body in passing from one state to another. Despite the form of the statements of the two laws (in terms of energy and entropy), their traditional mathematization centered upon the quantities of work and heat. Gibbs elegantly shifted the focus away from the latter two concepts to the notions of energy and entropy in his papers of 1873. For him, the crux of thermodynamics lay in the equation which expressed energy as a function of entropy and volume. Thus, whereas Clausius, Thomson, and others understood thermodynamics as the mechanical theory of heat, Gibbs presented it rather as the study of the properties of matter in equilibrium.[75]

When he published his work, the idea of analyzing the thermodynamic properties of a material system by means of an associated volume-pressure diagram had become standard among physicists. This had also been extended to three dimensions, generating a surface based on a volume-pressure-temperature coordinate system. In his papers, Gibbs looked not only at the mathematical effects of choosing different parameters for the coordinate axes but also at the physical interpretations of this new mathematization. In particular, he showed that the mathematics of a volume-entropy plane coordinate system or of a volume-entropy-energy three-dimensional system readily generated important and useful physical information which the traditional coordinate systems obscured. In a volume-entropy-energy coordinate system, for example, the tangent plane at any point on the surface representing the fundamental thermodynamic equation determined the temperature and pressure of the body in any state at the corresponding point. Furthermore, the relationships between the physical states of a substance can be interpreted in terms of the geometry of that surface. Based on this graphical, mathematical foundation, Gibbs went on in 1876 and 1878 to revolutionize the study of physical chemistry in a two-part paper "On the Equilibrium of Heterogeneous Substances."[76] By 1902, Gibbs had further applied his considerable talents

---

Transactions of the Connecticut Academy of Arts and Sciences 2 (1873):382–404.

[75]For an elegant presentation of the ideas Gibbs developed in these papers, see Martin J. Klein, "The Scientific Style of Josiah Willard Gibbs."

[76]J. Willard Gibbs, "On the Equilibrium of Heterogeneous Surfaces," *Transactions of the Connecticut Academy of Arts and Sciences* 3 (1875–1878):108–248 and 343–524. In this work, Gibbs looked at questions involving chemical substances in contact and calculated, among other things, conditions for their coexistence, equilibrium, and stability in the various phases. He also gave his celebrated phase rule, a mathematically derived formula relating the number n of components of a system, the number r of coexistent phases in that system, and the number f of thermodynamic degrees of freedom. In this notation, Gibbs's phase rule says that $r \leq n+2$ and $f = n + 2 - r$.

as a theoretical physicist to the study of statistical mechanics and had produced his ground-breaking *Elementary Principles of Statistical Mechanics.*[77] One outgrowth of his earlier physical and chemical researches had also established Gibbs's reputation as a mathematician (as opposed to a physicist), namely, the development of a new vector analysis.[78]

In 1843, after searching for many years for a three-dimensional analogue to the multiplication of the complex numbers, Sir William Rowan Hamilton discovered the four-dimensional algebra of quaternions. For Hamilton, a quaternion was an expression of the form $Q = a + bi + cj + dk$, where $a, b, c, d \in \mathbb{R}$ and where i, j, k satisfy the familiar relations

$$ij = -ji = k, \quad jk = -kj = i, \quad ki = -ik = j, \quad \text{and} \tag{1.1}$$

$$i^2 = j^2 = k^2 = -1. \tag{1.2}$$

If $R = a' + b'i + c'j + d'k$ is another quaternion, then these relations easily yield the product

$$\begin{aligned} QR = (aa' - bb' - cc' - dd') &+ (ab' + ba' + cd' - dc')i \\ + (ac' + ca' + db' - bd')j &+ (ad' + da' + bc' - cb')k. \end{aligned} \tag{1.3}$$

In his subsequent work on this new species of mathematical creature, he recognized the quaternion Q as the sum $SQ + VQ$ of a real or scalar part $SQ := a$ and an imaginary or vector part $VQ := bi + cj + dk$.[79] The analogy between the vector part of a quaternion and the usual three spatial coordinates proved strongly suggestive to Hamilton as well as to his contemporaries. Thus, thanks to the efforts most notably of Peter Guthrie Tait, vectorial methods in the quaternionic style infiltrated the physical sciences.

In 1844, Hermann Grassmann, a geographically and intellectually isolated *Gymnasium* teacher, independently published his own vectorial system in a book generally referred to succinctly as the *Ausdehnungslehre.*[80] There, he underscored the potential applications of this new mathematical system to physical problems arising in such areas as statics, mechanics, magnetism, and

[77] J. Willard Gibbs, *Elementary Principles of Statistical Mechanics* (New York: C. Scribner's Sons, 1902).

[78] On Gibbs's vector analysis, see Crowe, pp. 150–162.

[79] Hamilton wrote numerous papers on his new discovery, entitled "On Quaternions, or on a New System of Imaginaries in Algebra" and appearing in the *Philosophical Magazine.* This distinction between the scalar and the vector part of a quaternion appeared in 3rd ser., 29 (1846):26–31. For a modern treatment of the geometry of quaternions, see H. S. M. Coxeter, *Regular Complex Polytopes* (Cambridge: University Press, 1974), pp. 64–73. On Hamilton's life and work, see Thomas Hankins, *Sir William Rowan Hamilton* (Baltimore: Johns Hopkins University Press, 1980). On Hamilton's place in the history of vector analysis, see Crowe, pp. 17–46.

[80] Hermann Grassmann, *Die lineale Ausdehnungslehre, ein neuer Zweig der Mathematik dargestellt und durch Anwendungen auf die übrigen Zweigen der Mathematik, wie auch auf die Statik, Mechanik, die Lehre vom Magnetismus und die Krystallonomie erläutert* (Leipzig: O. Wigand, 1844).

even crystallography. Although this first edition of Grassmann's work sank into virtual oblivion, the second, revised edition of 1862 did draw some attention.[81] In particular, it attracted Willard Gibbs.

Drawing especially from the work of Hamilton and Tait, but influenced also by Grassmann and James Clerk Maxwell, Gibbs developed a system of vector analysis prior to 1880 which Tait described as "a sort of hermaphrodite monster, compounded of the notations of Hamilton and Grassmann."[82] Gibbs had read Maxwell's *Treatise on Electricity and Magnetism* sometime after its publication in 1873 and had gotten his first introduction to the language and methods of quaternions there.[83] Although Maxwell preferred, for pedagogical reasons, to write his book using mainly Cartesian and not Hamiltonian notation, he nevertheless advocated the quaternion viewpoint in the text and incorporated some quaternion notation.[84] Gibbs studied this new mathematical system in detail on his own and judged that it fell short, geometrically speaking. He thus introduced his students to his own version of vector analysis in his 1879 course on electricity and magnetism and subsequently wrote up a treatment of his ideas—in two installments privately printed in 1881 and 1884—to aid them in their studies.[85] After abandoning the somewhat artificial starting point of the four-dimensional quaternions, Gibbs isolated and emphasized two distinct vector products. As explained in the introduction to his pamphlets: "The manner in which the subject is developed is somewhat different from that followed in treatises on quaternions, since the object of the writer does not require any use of the conception of the quaternion, being simply to give a suitable notation for those relations between vectors, or between vectors and scalars, which seem most important, and lend themselves most readily to analytical transformations, and to explain some of these transformations."[86]

Gibbs understood a vector as simply "a quantity which is considered as possessing direction as well as magnitude,"[87] and he set up a presentation now

---

[81]On Grassmann's *Ausdehnungslehre*, see Crowe, pp. 54–96. See, also, Jean Dieudonné, "The Tragedy of Grassmann," *Linear and Multilinear Algebra* 8 (1) (1979–1980):1–14.

[82]Peter Guthrie Tait, *Elementary Treatise on Quaternions*, 3rd ed. (Cambridge: University Press, 1890), p. vi.

[83]James Clerk Maxwell, *Treatise on Electricity and Magnetism*, 2 vols. (Oxford: Clarendon Press, 1873).

[84]Crowe, pp. 132–138.

[85]In spite of the fact that this work was privately printed, the ideas that Gibbs presented in it did not go unnoticed internationally. Gibbs personally sent copies of it to many of the leading mathematical scientists of the day, such as Arthur Cayley, James Joseph Sylvester, Felix Klein, Lord Kelvin, and Helmholtz. See Wheeler, pp. 236–248 for a tabulation of Gibbs's mailing lists for reprints. When Peter Guthrie Tait attacked his use of vectors, Gibbs entered into a debate with Tait and his followers on the pages of *Nature*. See, for example, J. Willard Gibbs, "On the Role of Quaternions in the Algebra of Vectors," *Nature* 43 (1891):511–513. Several other papers also appeared in *Nature* as part of this debate between 1891 and 1893.

[86]Gibbs, "Elements of Vector Analysis," *Scientific Papers*, 2:17, quoted in Crowe, p. 155.

[87]Edwin Bidwell Wilson, *Vector Analysis: A Text-book for the Use of Students of Mathematics*

standard in many modern mathematics and physics textbooks. Relative to the three-dimensional Cartesian system, if i, j, and k represent the standard unit vectors on the x-, y-, and z-axes, respectively, a vector α can be written in the form $\alpha = xi + yj + zk$. Following Grassmann, Gibbs viewed i, j, and k more abstractly as basis vectors of $\mathbb{R}^3$, to use the modern terminology. He then defined two products. If $\beta = x'i + y'j + z'k$, the *direct* (now called the *dot*) *product* is

$$\alpha \cdot \beta := xx' + yy' + zz', \text{ a scalar,} \tag{1.4}$$

while the *skew* (now *cross*) *product* is

$$\alpha \times \beta := (yz' - zy')i + (zx' - xz')j + (xy' - yx')k, \text{ a vector.} \tag{1.5}$$

Gibbs based his version of vector analysis on these two separate and distinct notions of vector products.[88] Comparing equations (1.4) and (1.5) with (1.3), however, the kinship with Hamilton's quaternions is virtually inescapable. In his book *A History of Vector Analysis*, Michael Crowe assessed the significance of Gibbs's mathematical research in this way: "If Gibbs cannot be given credit for originality of methods yet he deserves praise for the sensitivity of his judgment as to what deletions and alterations should be made in the quaternionic system in order to make a viable system. His symbolism, now in the main accepted, also constitutes a significant contribution."[89]

From the point of view of the emergence of a mathematical research community in America, however, Gibbs played a less important role. It is true that he, like Benjamin Peirce, taught *bona fide* graduate courses. In addition to his courses in physics, he also lectured regularly on his vector analysis and on the more general theory of algebras. Yet, like Peirce, he had few students, partly because of his subject matter but partly because of his personality. Whereas Peirce lacked tolerance for those unable to follow the train of his often scattered mathematical thoughts, Gibbs was kind but introverted. "This was his chief limitation as a teacher of advanced students," according to one of those students. "[H]e did not take them into his confidence with regard to his current work, and even when he lectured upon a subject in advance of its publication... the work was really complete except for a few

---

and Physics Founded upon the Lectures of J. Willard Gibbs... (New York: Charles Scribner's Sons, 1902), p. 1. One of Gibbs's students and later Professor of Statistics at Harvard, Wilson had Gibbs's permission to write up the vector analysis in book form. He transformed Gibbs's eighty-three, tersely written pages into a text of over four hundred pages.

[88] His work on vector analysis also overlapped with the work on linear associative algebras done by Benjamin Peirce in 1870 and by J. J. Sylvester in 1884. For $n = 3$, Gibbs's *dyadics* were simply the nine-dimensional algebra Sylvester called the nonions. See Chapter 3 below. For a discussion of Gibbs's theory of dyadics, see Percey F. Smith, "Josiah Willard Gibbs, Ph.D., LL.D.: A Short Sketch and Appreciation of His Mathematics," *Bulletin of the American Mathematical Society*, 2nd. ser., 10 (1903–1904):34–39, and Edwin B. Wilson, "The Contributions of Gibbs to Vector Analysis and Multiple Algebra," in *Commentary on the Scientific Writings of J. Willard Gibbs*, pp. 127–160 on pp. 137–144.

[89] Crowe, pp. 157–158.

finishing touches. Thus his students were deprived of seeing his great structures in process of building, of helping him in the details, and of being in such ways encouraged to make for themselves attempts similar in character, however small their scale."[90] Gibbs presented terse, well thought out lectures on research-level material in the classroom, but he found it difficult to guide his students to the point of doing research themselves. Furthermore, Gibbs was in no sense an activist for mathematics (or even physics for that matter) in late nineteenth-century America.[91] His reclusive nature prevented him from participating even minimally in the broader organizational activities underway by the latter part of the nineteenth century in either American mathematics or in American physics. After years of urging, for example, Gibbs only joined the American Mathematical Society on 28 February 1903, one month prior to his death.

As we shall see in the next chapter, the qualities Gibbs lacked were precisely those which James Joseph Sylvester possessed so abundantly and which he used to build America's first school of mathematical research at the Johns Hopkins University. Before Sylvester's arrival in Baltimore in 1876, though, changes other than the redefinition and refinement of academic chairs, other than the deepening of the curriculum with French and later American texts, other than the development of graduate courses in the mathematical sciences by Peirce and Gibbs, in short, changes other than those in higher education also paved the way for the emergence of a mathematical research community in America. Factors such as the federal government's support of basic research, the establishment of societies and journals for the advancement of the sciences, and the concurrent delineation of the various scientific disciplines also contributed fundamentally to the emergence of mathematics as a discipline and of "mathematician" as a viable profession.

Mathematics in the Federal Government

In spite of a strict-constructionist stance toward the Constitution in the early years of the republic, the federal government set a precedent of allocating public funds for scientific research most notably during the presidency of Thomas Jefferson. In 1807, for example, the Congress approved a princely $50,000 appropriation for the establishment of a coastal survey on the basis of the arguments put forth by the expatriot Swiss geodesist Ferdinand Rudolph Hassler.[92] An *ad hoc* advisory committee next screened various plans and

[90] H. A. Bumstead, "Josiah Willard Gibbs," in Gibbs, *Scientific Papers*, 1:xi–xxvi on p. xxiv. This was a reprint with some additions of Bumstead's obituary of Gibbs in the *American Journal of Science*, ser. 4, 16 (1903):187–202.

[91] On Gibbs's minimal role in helping to establish a research physics community, see Daniel J. Kevles, *The Physicists: The History of a Scientific Community in Modern America* (Cambridge, MA.: Harvard University Press, 1987), pp. 31–34 and particularly p. 34.

[92] On the history of the Coast Survey, see A. Hunter Dupree, *Science in the Federal Government: A History of Policies and Activities* (Baltimore: Johns Hopkins University Press, 1986), pp. 29–33.

proposals for the survey and, not surprisingly, chose Hassler's blueprint as well as the man himself to head the project. After several years of dormancy, the survey finally moved forward in 1811 when Hassler went to England to order and to supervise the making of the necessary instruments. He ultimately returned to begin the actual collection of data, but only after five years and a ten percent budget overrun.

A scientist trained in the best European tradition, Hassler adhered to the most exacting standards of research, desiring to produce the best geodetic survey yet done of any coastline. Unfortunately, the standards of the scientist for perfection and the standards of the legislator for results clashed, and Hassler was forced out of the project in 1818 by an act prohibiting all but military personnel from participating in the survey. After languishing in incapable hands for twelve years, the Coast Survey revived in 1830 when Hassler was permitted to return to its helm.[93] But he had not changed. He aimed to insure the high quality of the project's results, and to this end, he personally instructed the civilians and Army and Navy officers under him in the physics and mathematics of surveying as well as in the proper and accurate use of the instruments at their disposal. Under Hassler, the Survey functioned more as a small scientific school than as a producer of coastline maps. Thus, once again, with the high-minded advancement of science and not the utilitarian charting of coastal shipping lanes as his foremost goal, Hassler perpetually irritated the Congress. His seeming lack of political savvy notwithstanding, he managed to work the Survey's budget up from $20,000 in 1832 to $100,000 in 1840 and to set precedents for scientific training as well as for scientific output within a governmentally funded agency.

When Hassler died in 1843, his successor as head of the Survey, Alexander Dallas Bache, more than upheld these precedents.[94] Unlike Hassler, Bache masterfully sold pure science to the Congress by stressing and providing utilitarian results, and his efforts were generously rewarded. By the mid-1850s, for example, his budget had exceeded $500,000.[95] Like Hassler, however, Bache set high scientific standards for the Coast Survey and had the advancement of pure science as his principal objective. Bache's rare combination of political, organizational, and scientific acumen established him as "the most important single person in the evolution of the government's policy toward

---

[93] *Ibid.*, pp. 52–56. On Ferdinand Hassler, see Gillispie, ed. *DSB*, s.v. "Hassler, Ferdinand Rudolph," by Nathan Reingold, as well as Clark C. Elliott, ed., *Biographical Dictionary of American Science: The Seventeenth Through the Nineteenth Centuries* (Westport, Conn.: Greenwood Press, 1979).

[94] For the details of Bache's life, see Gillispie, ed. *DSB*, s.v. "Bache, Alexander Bache," by Nathan Reingold, and Elliott, ed., *Biographical Dictionary*.

[95] Robert Post, "Science, Public Policy, and Popular Precepts: Alexander Dallas Bache and Alfred Beach as Symbolic Adversaries," in *The Sciences in the American Context: New Perspectives*, ed. Nathan Reingold (Washington: Smithsonian Institution Press, 1979), pp. 77–98 on p. 89.

science and technology,"[96] and made the Survey under him what one contemporary called "the best school in the world"[97] for research-level training in various scientific fields. Bache's Survey not only continued its coastal work, expanding its operations to include the Pacific and Gulf coasts, but also enlarged its scientific domain to encompass marine biology, seismology, oceanography, terrestrial magnetism (Bache's own area of research expertise), mathematics, and astronomy. For help in the latter fields, Bache turned not to the federal government's own Naval Observatory, started in 1842 under Matthew Fontaine Maury, but rather to Benjamin Peirce at Harvard.

As noted above, Peirce had established a solid reputation as a mathematical astronomer like his mentor Nathaniel Bowditch, but unlike Bowditch, Peirce had also earned notoriety through his graduate-level program in mathematics and astronomy at Harvard. While he had few students there, those he did have were available for making observations and calculations. Cambridge became *the* recognized center of mathematical and astronomical studies in the United States after 1847, when the Harvard Observatory secured its new and more powerful telescope through the urgings of Peirce and others. During the 1850s, this center also lured the Coast Survey's longitude department away from Washington D.C. and attracted the government's newly formed Nautical Almanac. This move further assured its undisputed claim as the principal American training ground for not only observational but also mathematical astronomy.[98] Thus, the first head of the Nautical Almanac, Lieutenant Charles Henry Davis, qualified for that appointment in 1849 as a result of his formal training under Peirce at Harvard and his practical training in the Coast Survey.[99] The Nautical Almanac later helped prepare mathematical astronomers Simon Newcomb and George William Hill for their distinguished scientific careers.

Born in Nova Scotia in 1835, Newcomb went to the Nautical Almanac as a calculator in 1857.[100] He supplemented his informal mathematical education while in Cambridge with courses under Benjamin Peirce and earned his B.S. degree from the Lawrence Scientific School in 1858. Moving on to the Naval Observatory in Washington, D. C. in 1861, Newcomb instituted a series of observational innovations which led to his reformation of the theory and computations behind the *American Ephemeris*. Sixteen years later, in 1877, he returned to the by then Washington-based Nautical Almanac Office in the

[96]Nathan Reingold, "Alexander Dallas Bache: Science and Technology in the American Idiom," *Technology and Culture* 11 (1970):163–177 on p. 165. See, also, Bruce, p. 172.

[97]J. Dobbin to Alexander Dallas Bache, 2 June 1857, as quoted in Bruce, p. 172.

[98]Dupree, pp. 105–109; and Bruce, pp. 178–180.

[99]The fact that Davis was Peirce's brother-in-law may also have influenced Davis's choice of Cambridge as headquarters for the Nautical Almanac.

[100]On Newcomb's life, see Raymond Clare Archibald, *A Semicentennial History of the American Mathematical Society* 1888–1938 (New York: American Mathematical Society, 1938; reprint ed., New York: Arno Press, 1980), pp. 124–139; and Gillispie ed., *DSB*, s.v. "Newcomb, Simon," by Brian G. Marsden.

SIMON NEWCOMB (1835–1909)

George William Hill (1838–1914)

post of Superintendent. The energetic and opinionated Newcomb capitalized on his various connections in both the government and academe to become a central figure in the developing American scientific community.[101] At the time of his death in 1909, in fact, Newcomb had won distinction as the most honored North American scientist of his era. He had made key advances in mathematical astronomy—most notably his development of what would come to be called *Newcomb* operators and which were crucial in the calculation of elliptical orbits—but he had won his international reputation more for his work in the observational and computational than in the theoretical mathematical aspects of astronomy.[102] His contemporary Hill, on the other hand, made his principal contributions in mathematical astronomy.

George William Hill was born in New York City in 1838 but was raised on his family's farm in nearby West Nyack.[103] He attended Rutgers College, where Theodore Strong, the Professor of Mathematics and a friend of Bowditch, introduced him to such classics of the French analytical tradition as Lagrange's *Mécanique analytique* and Laplace's *Mécanique céleste*.[104] In light of this education, Hill possessed an unusually strong mathematical background for an American when he earned his B. A. in 1859. He took this with him to Cambridge in 1861, where he followed Newcomb on the staff at the Nautical Almanac Office.

Always a solitary figure, Hill spent only a few years in Massachusetts before receiving permission to continue his work at home in West Nyack. At the time of Newcomb's assumption of the superintendency of the Almanac Office, Hill had just completed the work on lunar orbits and the three-body problem which later attracted the attention and the praise of Henri Poincaré.[105] In

[101]On Newcomb's broader role in American science, consult Albert E. Moyer, *A Scientist's Voice in American Culture: Simon Newcomb and the Rhetoric of Scientific Method* (Berkeley: University of California Press, 1992). As we shall see, Newcomb served as a chief, early scientific adviser to the first President of the Johns Hopkins University, Daniel Coit Gilman. He also succeeded J. J. Sylvester as editor of the *American Journal of Mathematics* (1885–1893, 1899–1900) and held the presidencies of the American Association for the Advancement of Science (1876) and the American Mathematical Society (1897–1898).

[102]Newcomb, in fact, was unable to deal with some critical problems introduced into the determinations of orbital constants by the calculational method he had chosen. For the lunar theory, these difficulties were resolved in the mid-1890s by Ernest William Brown, then on the faculty of Haverford College. Brown continued to produce seminal work on the lunar theory after his move to Yale in 1907. For more on his life and work, see Archibald, *Semicentennial History*, pp. 173–183.

[103]On Hill, consult Archibald, *Semicentennial History*, pp. 117–124, and Gillispie, ed., *DSB*, s.v. "Hill, George William," by Carolyn Eisele. On his work, see Kline, pp. 730–732, and George William Hill, *Collected Mathematical Works of George William Hill*, 4 vols. (Washington: Carnegie Institution, 1905–1907).

[104]On Theodore Strong, see Edward R. Hogan, "Theodore Strong and Ante-Bellum American Mathematics," *Historia Mathematica* 8 (1981):439–455. Of the fifty-five American scientists whom George Daniels labeled "leading" scientists between 1815 and 1845, Strong was the only mathematician. See Daniels, *American Science in the Age of Jackson*, pp. 201–228 (Appendix I).

[105]George W. Hill, *On the Part of the Motion of the Lunar Perigee Which Is a Function of*

this 1877 work, Hill gave his novel techniques for solving the differential equation that now bears his name:

$$\frac{d^2p}{dt^2} + (\theta_0 + \theta_1 \cos 2t + \theta_2 \cos 4t + \cdots)p = 0,$$

where the θ_n are real constants. His approach depended on developing a certain infinite determinant associated with an infinite system of linear equations. He showed that in order for the original equation to have a nontrivial solution of the form

$$p = \sum_{\ell=-\infty}^{\infty} g_\ell e^{it\rho+2it\ell} \quad \text{for } \rho \text{ a constant},$$

this determinant, which is a function of the exponent ρ, must vanish. Since Hill's interest in this method stemmed from its physical implications, his primary concern lay in deriving numerical results, and he accomplished this by calculating the roots ρ of the determinantal equation to fifteen places for certain specified values of the θ_n. His memoir did leave a number of mathematical questions unanswered, however. One of these, the proof that his infinite determinant actually converged, was resolved by Poincaré in 1886.[106]

Although seminal, Hill's work on the three-body problem did not fit well with the fresh directives Newcomb, as the new Superintendent, put in place at the Almanac Office. Hill accommodated himself to the new order by shifting his research focus and his residence in 1882. Moving to Washington, he tackled yet another difficult problem, the determination of the mutual perturbations of Jupiter and Saturn. Hill worked on this and other questions for the next ten years, before retiring to resume his life of relative seclusion in West Nyack. He died there in 1914.[107]

the Mean Motions of the Sun and Moon (Cambridge, MA: privately printed, 1877), 28 pp.; and "Researches in the Lunar Theory," *American Journal of Mathematics* 1 (1878):5–26, 129–147, and 245–260. The first work was eventually reprinted in 1886 in *Acta Mathematica.* See note 129 below. Poincaré lauded Hill's accomplishments in the introduction to George W. Hill, *Collected Mathematical Works*, pp. vii–xviii. Indeed, he opened by stating that "Mr. Hill is one of the most original faces [physionomies] in the American scientific world [p. vii, our translation]." For a mathematical discussion of the Hill equation (mentioned below), consult E. T. Whittaker and G. N. Watson, *A Course of Modern Analysis*: *An Introduction to the General Theory of Infinite Process and of Analytic Functions*; *With an Account of the Principal Transcendental Functions*, 4th ed. (Cambridge: University Press, 1927; reprint ed., Cambridge: University Press, 1963), pp. 413–417.

[106]Henri Poincaré, "Sur les Déterminants d'Ordre infini," *Bulletin de la Société mathématique de France* 14 (1886):77–90, or Henri Poincaré, *Oeuvres de Henri Poincaré*, ed. Paul Appell *et al.*, 11 vols. (Paris: Gauthier-Villars et Cie., 1928–1956), 4:95–107. Poincaré returned to this topic in 1900, giving a less complicated proof of the convergence of Hill's infinite determinant in "Sur le Déterminant de Hill," *Bulletin astronomique* 17 (1900):134–143, or *Oeuvres de Henri Poincaré*, 4:108-116.

[107]Hill did, however, lecture on celestial mechanics at Columbia for four years from 1893–1895 and 1898–1900. He also served as the third President of the American Mathematical Society from 1895–1896.

The succession of names—Bowditch, Benjamin Peirce, Newcomb, Hill—suggests that the United States sustained a line of continued and mutually dependent investigations in mathematical astronomy throughout the nineteenth century. Supported to some extent by the federal government, this work clearly—and perhaps naturally—fell into an *applied* tradition and represented America's closest approach to research-level mathematics prior to the founding of the Johns Hopkins University in 1876. In the last half of the century, developments crucial to the professionalization of American science, together with educational trends both at home and abroad, would shape a very different *pure* tradition in mathematics at the research level. This would find its support within the institutions of higher education and would come to dominate an emergent American mathematical research community.

Mathematics and Scientific Professionalization

The establishment of an enclave within the federal government devoted to applied mathematical research owed largely to the early efforts of two men, Alexander Dallas Bache and Benjamin Peirce. Their influence on American science in general, and on mathematics in particular, reached beyond the bounds of government and academe, however. As founding members of an exclusive, but informal, group of scientific intellectuals known as the Lazzaroni, Bache and Peirce worked to organize and to professionalize science in the United States from the late 1840s into the 1860s.[108]

The Lazzaroni, or scientific "beggars," did not hold their first real meeting until 1857, but Bache had articulated the group's animating ideas at least as early as 1838. Writing to his friend, Bache, in August of 1838, the physicist Joseph Henry admitted that

> I am now more than ever of your opinion that the real working men in the way of science in this country should make common cause and endeavour by every proper means unitedly to raise their own scientific character. [They need] [t]o make science more respected at home to increase the facilities of scientific investigations and the inducements to scientific labours. There is the disposition on the part of our government to advance the cause if this were properly directed. At present however Charlatanism is much more likely to meet with attention and reward than true unpretending merit.[109]

---

[108]For discussions of the Lazzaroni, see, for example, Bruce, pp. 217–226; and Mark Beach, "Was There a Scientific Lazzaroni?," in *Nineteenth-Century American Science: A Reappraisal*, ed. George H. Daniels (Evanston: Northwestern University Press, 1972), pp. 115–132.

[109]Joseph Henry to Alexander Dallas Bache, 9 August 1838, in *Science in Nineteenth-Century America: A Documentary History*, ed. Nathan Reingold (Chicago: University of Chicago Press, 1964):81–90 on p. 85.

Bache and the like-minded Henry believed that the time had come to establish a scientific profession in America, and they had a strong sense of what needed to be done. First, they had to get themselves into positions of power and influence: in 1843, Bache became Superintendent of the Coast Survey, and in 1846, Henry became the first Secretary of the Smithsonian Institution. Second, they needed to enlist into their cause other influential scientists who shared their vision. To accomplish this, they rounded out the charter members of the club with Benjamin Peirce and his Harvard colleague, Louis Agassiz.

Never numbering more than a dozen, the Lazzaroni were more of a loose network of well-placed individuals than a formal organization. The members were all strong personalities with the energy to scheme and to work for science. Although they often disagreed and worked at cross purposes on specific issues, they were instrumental in the establishment of the American Association for the Advancement of Science (AAAS) in 1847 and of the National Academy of Sciences in 1863.[110] Furthermore, they hoped to provide much-needed vehicles for the dissemination of American scientific research by supporting the publication of the *Proceedings* of these two organizations.

The vision of the Lazzaroni reflected well the perception of many American scientific intellectuals at mid-century. In a much expanded United States, American scientists sensed the need for organizations of scope greater than that of local societies like Philadelphia's American Philosophical Society (founded in 1769) or Boston's American Academy of Arts and Sciences (founded in 1784), which, catering to the scientific cultivator, practitioner, and researcher alike, had sufficed in the earlier decades of the 1800s.[111] While societies of a more regional character worked well in their limited venues, national organizations would better serve to set and maintain scientific standards, to encourage and promote research and publication, and to provide a network for communication. In the words of Sally Gregory Kohlstedt, American scientists sensed the need for national organizations "with the potential

[110]For the complete story of the development of the AAAS out of the Association of American Geologists and Naturalists, see Sally Gregory Kohlstedt, *The Formation of the American Scientific Community: The American Association for the Advancement of Science* 1848–60 (Urbana: University of Illinois Press, 1976). On the founding of the National Academy of Sciences, see, for example, Bruce, pp. 301–305; Dupree, pp. 135–141; and Rexmond C. Cochrane, *The National Academy of Sciences: The First Hundred Years,* 1863–1963 (Washington: The Academy, 1978).

[111]Nathan Reingold drew the distinction between scientific cultivators, those who are interested in and engage in scientific activities without remuneration; practitioners, those who work at science full-time but may or may not publish; and researchers, those who work at science full-time and make real contributions. Consult his article, "Definitions and Speculations: The Professionalization of Science in America in the Nineteenth Century," in *The Pursuit of Knowledge in the Early American Republic: American Scientific and Learned Societies from Colonial Times to the Civil War,* ed. Alexandra Oleson and Sanborn C. Brown (Baltimore: Johns Hopkins University Press, 1976), pp. 33–69.

to coordinate scientific inquiry and to establish science as a true and viable profession in the United States."[112]

Previous attempts to found societies in the individual sciences had tended to failure, largely due to the small number of participants in any given field. As early as 1817, the transplanted Englishman William Marrat had organized a mathematical society in New York with a total membership of eight. This venture lasted for several years, apparently collapsing with Marrat's return to England.[113] George Gibbs and Benjamin Silliman of Yale started the American Geological Society in 1819 with a greater number of subscribed members, but distances and difficulties of travel and communication resulted in small meetings and eventual disbanding in 1826.[114] Learning from such failures, the all-encompassing AAAS and the initially less influential National Academy of Sciences allowed American scientists of all stripes to work together to define the scientific profession as a whole. They also provided prototypes for the national societies of the individual sciences, which would develop during the century's closing quarter as a result of increasing numbers and a growing sense of autonomy relative to American science as a whole.[115]

Just as early efforts to found societies in the individual sciences had failed, so endeavors to publish specialized journals had come to naught. In 1810, for example, the *Mineralogical Journal* came and went, while a lone volume of the *Transactions of the Geological Society of Pennsylvania* appeared in 1834–1835.[116] But mathematical journals seem to have had a particularly high mortality rate. (See Table 1.2.) The earliest American journal devoted solely to mathematics, George Baron's *The Mathematical Correspondent*, survived less than two years. Baron, another Englishman on American shores,

---

[112]Kohlstedt, p. 79. On the professionalization of science and the role of the AAAS in that development, see, also George H. Daniels, "The Process of Professionalization in American Science: The Emergent Period, 1820–1860," *Isis* 58 (1967):151–166.

[113]Smith and Ginsburg, p. 88; and Edward R. Hogan, "Robert Adrain: American Mathematician," *Historia Mathematica* 4 (1977):157–172 on pp. 168–169.

[114]Kohlstedt, p. 63. For more on this early society, see George P. Merrill, *The First One Hundred Years of American Geology* (New Haven: Yale University Press, 1924).

[115]For example, the American Chemical Society first met in 1876. The New York Mathematical Society, which began in 1888, held its first meeting as the American Mathematical Society in 1894. The American Physical Society started up in 1899. Regarding numbers of scientists, Robert Bruce found 477 scientists listed in the *Dictionary of American Biography* who were active at some time during the period 1846 to 1876. Of these, only forty-one or 8.6% of all scientists had mathematics as their primary field of investigation. The life sciences claimed the greatest percentage of scientists during the period with 31.5% or 150 working scientists. With such small numbers, particularly in mathematics, it is little wonder that early organizational efforts failed. See Robert V. Bruce, "A Statistical Profile of American Scientists, 1846–1876," in *Nineteenth-Century American Science: A Reappraisal*, ed. George H. Daniels (Evanston: Northwestern University Press, 1972), pp. 63–94.

[116]Daniels, *American Science in the Age of Jackson*, p. 17; and Kohlstedt, p. 64. In Appendix II of his book, Daniels also gives several tables reflecting data gathered on 337 periodicals which published scientific articles. His tables include data on the number of journals in print in any given year between 1805 and 1849, the lengths of their publications, etc. See pp. 229–232.

eventually settled in New York City and set up a school in which he offered instruction in navigation and mathematics. From 1804 to 1806, he edited and published one volume of nine issues of his mathematical journal. He hinted at the motivation behind his new enterprise when he wrote in the volume's preface: "When we consider the great exertions of learned men to disseminate mathematical information in other countries we must be surprised to find that this kind of knowledge is most shamefully neglected in the United States of America."[117] Baron sought to remedy this situation by offering a journal in which problems proposed in one issue hopefully reappeared with readers' solutions in subsequent issues. While he did not assume a very high level of mathematical competence initially among his readership, he intended that the future would bring a gradual ascension "towards the higher regions of those sciences, as far as may be thought consistent with the abilities of our readers."[118] One of the earliest readers and problem-solvers of *The Mathematical Correspondent* already possessed a much higher level of mathematical sophistication than even the editor himself.

Robert Adrain, a self-taught, Irish-born mathematician, sought refuge in the United States after the Irish uprisings of 1798 and promptly found employment as a mathematics teacher.[119] When Baron's journal appeared in 1804, he took an immediate and active interest in the venture, offering both solutions and new problems to his fellow subscribers. Thus, although the problem of showing that $3/(2+\sqrt{3}) = 3(2-\sqrt{3})$ was typical of the journal's level of difficulty, Adrain posed and tackled much more challenging questions. For example, he proposed "[t]o determine the nature of the catenaria volvens, or the figure which a perfectly flexible chain of uniform density and thickness will assume, when it revolves with a constant angular velocity about an axis, to which it is fastened at its extremities, in free and non gravitating spaces."[120] He demonstrated that this problem reduced to that of evaluating a certain elliptic integral, but could take it no further. The fact that it defied evaluation until Alfred Clebsch developed a theory to handle it after mid-century gives some indication of Adrain's mathematical adventurousness.[121] As the *Correspondent*'s most talented participant, Adrain was the natural successor to the journal's editorship when Baron abandoned the project after the ninth issue. Unfortunately, Adrain succeeded in keeping it alive through only

---

[117] Smith and Ginsburg, p. 86.

[118] From the preface of *The Mathematical Correspondent* 1 (1804):iv, as quoted in Edward R. Hogan, "George Baron and the *Mathematical Correspondent*," *Historia Mathematica* 3 (1976):403–415 on p. 404.

[119] On Robert Adrain, see Greene, pp. 132–134; Gillispie, ed., *DSB*, s.v. "Adrain, Robert," by Dirk J. Struik; Julian L. Coolidge, "Robert Adrain, and the Beginnings of American Mathematics," *American Mathematical Monthly* 33 (1926):61–76; and Edward Hogan, "Robert Adrain: American Mathematician," pp. 157–172.

[120] Hogan, "George Baron and the *Mathematical Correspondent*," pp. 404–405.

[121] See Coolidge, p. 65; and Alfred Clebsch, "Ueber die Gleichgewichtsfigur eines biegsamen Fadens," *Journal für die reine und angewandte Mathematik* 57 (1860):93–110.

one more number, and the journal passed from existence in 1806.

Undaunted by this initial editorial failure, however, Adrain started up his own journal, *The Analyst or Mathematical Museum*, in 1808. This journal was also short-lived, lasting through only one volume of four numbers, but it contained respectable mathematical work by Nathaniel Bowditch and Ferdinand Hassler as well as Adrain's proof of the exponential law of errors from which he deduced the method of least squares.[122] Adrain next tried to resuscitate *The Analyst* in 1814 from his new post as Professor of Mathematics at Columbia. After that attempt failed too, he launched his final journal, *The Mathematical Diary*, in 1825. Although the *Diary* maintained publication for seven years and went through thirteen numbers, it did not share the high standards of Adrain's *Analyst*. This may well have been the secret to its relative longevity, however. As a journal aimed at the amateur problem-solver rather that at the mathematician, it naturally appealed to a greater number of subscribers.[123]

The next sustained effort to mount a purely mathematical publication came in 1836 when Charles Gill brought out *The Mathematical Miscellany*, a publication which he "entered into for the advantage of those who are desirous to *progress* in the important study of mathematics."[124] Among those so inclined were Harvard's Benjamin Peirce and Theodore Strong of Rutgers. The work which they and others submitted reflected not only the growing sophistication of American mathematics but also its increasing absorption of Continental ideas. For example, Strong used Abel's theorem on integrals of algebraic functions (see Chapter 5 for a discussion of this theorem) in his solution to one of the *Miscellany*'s problems, and in another, Benjamin Peirce made use of the differential geometry of Gauss's *Disquisitiones generales circa superficies curvas.*[125] *The Mathematical Miscellany* thus bore witness, in the mid-1830s, to a heightened awareness of the Continental mathematics which Ferdinand Hassler had tried to introduce at West Point as early as 1807 and which John Farrar had espoused at Harvard starting in 1818.

Still, the number of contributors able to grasp such mathematics remained too small to assure the journal's success. One of Gill's correspondents sensed

---

[122]Adrain proved this theorem independently of Gauss and Legendre. See Hogan, "Robert Adrain: American Mathematician," pp. 158 and 167 as well as note 2 on p. 170; and Greene, p. 133.

[123]Other journals, such as *The Portico* (begun in 1816) and *The Ladies' and Gentlemen's Diary or United States Almanac* (begun in 1820), contained problem-solving sections even though they were not solely devoted to mathematics. See Smith and Ginsburg, pp. 87–89; and Cajori, pp. 96–97.

[124]On Gill, see Edward R. Hogan, "The Mathematical Miscellany (1836–1839)," *Historia Mathematica* 12 (1985):245–257. This quote from the first volume of the journal appears on p. 246. Gill's emphasis.

[125]Carl Friedrich Gauss, *Disquisitiones generales circa superficies curvas* (Göttingen: Sumtibus Dieterichianis, 1828). See Hogan, "The Mathematical Miscellany (1836–1839)," pp. 248–249.

this problem from the start. Writing in August of 1836, Evans Hollis expressed his "wonder at your obtaining even sixty subscribers for in this country there are but few who can appreciate such a work and people have not yet acquired wealth or spirit enough to patronize works in which they are not personally interested."[126] By 1839, the dearth of paying subscribers had apparently forced the journal to suspend publication.

Other attempts, such as the 1842 revival of the *Miscellany* in the form of *The Cambridge Miscellany of Mathematics, Physics, and Astronomy* by Benjamin Peirce and Joseph Lovering at Harvard and the *Mathematical Monthly* started in 1858 by John Runkle[127] then at the Nautical Almanac Office in Cambridge, failed just as rapidly. This absence of specialized journals compelled those active and productive researchers such as Peirce, Strong, and later Hubert Newton, Willard Gibbs, and Peirce's son, Charles, to seek other publication outlets. Thus, mathematicians like Strong turned in the early years to general science journals such as Benjamin Silliman's *American Journal of Science* for the dissemination of their ideas. Later, after the foundation of societies like the AAAS and the National Academy of Sciences, Benjamin Peirce, among others, presented his work as part of the published *Proceedings*. Local societies also provided means of communication. Newton and Gibbs entrusted their work to the *Transactions of the Connecticut Academy of Arts and Sciences*, Strong published in the *Proceedings* of Philadelphia's American Philosophical Society, and the Peirces relied on the *Proceedings of the American Academy of Arts and Sciences* in Boston.[128]

These types of publications certainly served to get mathematical ideas into print; they were less effective in actually making such ideas known. Their eclectic contents failed both to attract mathematical readers—particularly from abroad—and to provide the sort of concentrated exposure of a specialized journal. Consequently, some American mathematicians had their work printed privately so that they could personally see that it fell into the

[126] Evans Hollis to Charles Gill, 7 August 1836, as quoted in Hogan, "The Mathematical Miscellany (1836–1839)," p. 251. Hollis was a high school mathematics teacher in Westchester County, New York.

[127] Runkle, who had studied under Benjamin Peirce at the Lawrence School, went on to become Professor of Mathematics and President of the Massachusetts Institute of Technology.

[128] Among other papers, see Theodore Strong, "Theory and Variation of the Arbitrary Constants in Elliptic Motion," *American Journal of Science* 30 (1836):248–266; Benjamin Peirce, "Certain Methods of Determining the Number of Real Roots of Equations Applicable to Transcendental As Well As Algebraic Equations," *Proceedings of the American Association for the Advancement of Science* 1 (1848):38–39; Hubert Newton, "On the Transcendental Curves Whose Equation Is $\sin y \sin my = a \sin x \sin nx + b$, With 24 Plates," *Transactions of the Connecticut Academy of Arts and Sciences* 3 (1875):97–107; J. Willard Gibbs, "Graphical Methods in the Thermodynamics of Fluids," *op. cit.* pp. 309–342; Theodore Strong, "On Analytical Trigonometry," *Proceedings of the American Philosophical Society* 1 (1843):49–50; Benjamin Peirce, "On the Uses and Transformations of Linear Algebras," *Proceedings of the American Academy of Arts and Sciences*, n.s., 2 (1875):395–400; and Charles S. Peirce, "On the Application of Logical Analysis to Multiple Algebra," *Proceedings of the American Academy of Arts and Sciences* 10 (1875):392–394.

appropriate hands. In 1877, for example, George Hill had 200 copies of the twenty-eight-page paper containing his work on the Hill equation printed at his own expense. The work only became more widely known, however, through John Couch Adams's review of it for the Royal Astronomical Society and after its eventual publication in Gösta Mittag-Leffler's *Acta Mathematica* in 1886.[129]

Benjamin Peirce's ground-breaking work, entitled "Linear Associative Algebra," suffered an even slower reception. Peirce originally presented this 153-page study in installments before the National Academy of Sciences between 1866 and 1870. In 1870, he had 100 copies of the work lithographed and distributed to his friends. He entrusted to George Bancroft, who was at that time the American Ambassador to Prussia, the task of presenting his brainchild to the Berlin Academy of Sciences. In that way, he hoped to introduce his ideas to an audience which, unlike any at home, was capable of understanding and fully appreciating them.[130] Unfortunately, Bancroft apparently failed to follow through on this request. The work remained essentially unknown in Germany until a decade after Peirce's death in 1880, and many of its results were ultimately duplicated in the work of Georg Scheffers and Eduard Study.[131] In England, however, the treatise fared somewhat better. William Spottiswoode alerted his countrymen to Peirce's research in his retirement address before the London Mathematical Society in 1872.[132] In spite of this favorable press, the work still generated little interest until later in the century.

A highly original and abstract work, Peirce's "Linear Associative Algebra," like Grassmann's *Ausdehnungslehre*, may have seemed more like a philosophical than a mathematical treatise to most of his contemporaries. It opens, for example, with these metaphysical musings on the nature of mathematics:

> Mathematics is the science which draws necessary conclusions.
>
> This definition of mathematics is wider than that which is ordinarily given, and by which its range is limited to quantitative research. The ordinary definition, like those

[129] George W. Hill, "On the Part of the Motion of the Lunar Perigee Which Is a Function of the Mean Motions of the Sun and Moon," *Acta Mathematica* 8 (1886):1–36.

[130] Writing to Bancroft on 18 November 1870, Peirce added: "I also send you a copy for the Academy of Berlin—which I hope that you will do me the honor to present. If it would be referred to a committee of geometers for report, I should be greatly gratified." For the complete transcript of the letter, see Jekuthiel Ginsburg, "A Hitherto Unpublished Letter by Benjamin Peirce," *Scripta Mathematica* 2 (1934):278–282.

[131] See Georg Scheffers, "Zur Theorie der aus n Haupteinheiten gebildeten complexen Grössen," *Berichte über die Verhandlungen der königlichen sächsischen Gesellschaft der Wissenschaften zu Leipzig* 41 (1889):290–307, and "Ueber die Berechnung von Zahlensystemen," *Berichte über die Verhandlungen der königlichen sächsischen Gesellschaft der Wissenschaften zu Leipzig* 41 (1889):400–457, as well as Eduard Study, Über Systeme von complexen Zahlen," *Nachrichten der Gesellschaft der Wissenschaften zu Göttingen, Mathematisch-Physicalische Abteilung* (1889): 237–268.

[132] William Spottiswoode, "Remarks on Some Recent Generalizations of Algebra," *Proceedings of the London Mathematical Society* 4 (1873):147–164.

> of other sciences, is objective; whereas this is subjective. Recent investigations, of which quaternions is the most noteworthy instance, make it manifest that the old definition is too restricted. The sphere of mathematics is here extended, in accordance with the derivation of its name, to all demonstrative research, so as to include all knowledge strictly capable of dogmatic teaching. ...
>
> Mathematics under this definition, belongs to every enquiry, moral as well as physical.[133]

In this metamathematical guise, however, Peirce presented a fundamental extension and generalization of Hamilton's work on the algebra of quaternions and laid much of the theoretical groundwork used in 1907 by Joseph H. M. Wedderburn to set up a general theory of algebras over arbitrary fields.[134]

As mentioned earlier, Peirce embraced Hamilton's work on quaternions almost immediately and incorporated it into his Lawrence School curriculum as early as 1848. Following Hamilton, his notion of algebra extended to systems whose multiplication did not necessarily obey the commutative law. But Peirce went much further. For him, division was not necessarily an operation defined for the system either. As if his algebras did not deviate enough from the usual concept based on the real numbers, they could contain zero divisors, or *nilfactors* as he termed them, that is, nonzero elements a satisfying $ab = 0$ or $ba = 0$ for some nonzero element b in the algebra. They could also have within them *nilpotents*, or elements which vanished when raised to a sufficiently high power, and *idempotents*, or elements e satisfying $e^2 = e$.[135] Peirce used these special elements to analyze the structure of algebras over the complex numbers.

He began by showing that "in every linear associative algebra, there is at least one idempotent or one nilpotent expression."[136] He then used this fact to derive what is now called the *Peirce decomposition* of an algebra with idempotent: if A is an algebra with idempotent e, then

$$A = B \oplus eB_1 \oplus B_2e \oplus eAe$$

where $B_1 = \{x \in A | xe = 0\}$, $B_2 = \{x \in A | ex = 0\}$, and $B = B_1 \cap B_2$.[137] On the other hand, he showed that "in ... an algebra which has no idempotent

[133] As we shall see below, Peirce's work actually was published in the much more accessible *American Journal of Mathematics*. We quote from the published source: Benjamin Peirce, "Linear Associative Algebra With Notes and Addenda, by C. S. Peirce, Son of the Author," *American Journal of Mathematics* 4 (1881):97–229. The quotation here appears in *ibid.*, p. 97.

[134] For the historical development of the theory of algebras from Sir William Rowan Hamilton's discovery of the quaternions in 1843 to Wedderburn's epoch-making work of 1907, see Karen Hunger Parshall, "Joseph H. M. Wedderburn and the Structure Theory of Algebras," *Archive for History of Exact Sciences* 32 (1985):223–349.

[135] Benjamin Peirce, "Linear Associative Algebra," p. 104.

[136] *Ibid.*, p. 109.

[137] While Peirce's notation is fairly modern in spirit, he, like his contemporary Sylvester,

expression, all the expressions are nilpotent."[138] With these theorems and their consequences providing "the key to the research,"[139] Peirce proceeded to determine and dissect the over 150 algebras of dimensions one through six by explicitly exhibiting their multiplication tables. An algebraic *tour de force*, Peirce's classification lay in almost total obscurity during the first decade of its existence due to both its mathematical and its physical inaccessibility.

The reception of Peirce's work on linear associative algebras not only reflected the state of research-level mathematics in the America of the first three-quarters of the nineteenth century but also augured the promise of that century's closing years. As late as 1870, few in America had the mathematical wherewithal to understand a work such as the one Peirce had written. Presented to the National Academy of Sciences, the organization which Peirce had long championed, his "Linear Associative Algebra" fell on deaf ears. Perhaps therein lay the reason why Peirce opted for private printing and distribution of his ideas. Not even that honorific society—composed of those whom Bache, Agassiz, and Peirce himself viewed as the country's best in science—included members competent to appreciate the true import of Peirce's work. Given this state of affairs, why use the Academy's *Proceedings* as the venue for its publication? Why not simply send the paper directly to some of the individuals who could and would appreciate it? Or why not have the paper read before a scientific academy—like the one in Berlin—which really had mathematically sophisticated members?

Whether or not these thoughts actually crossed Peirce's mind, they do serve to draw the contrast between the United States and Europe in mathematics in 1870. The signs of professionalization—specialized journals and societies as well as advanced graduate-level education—manifested themselves in both places in the nineteenth century. In Europe, however, the process was well underway early in the century while, in the United States, it was a late nineteenth-century phenomenon. Furthermore, European standards of higher education had evolved during America's first hundred years and had produced the mathematical standard-bearers of the nineteenth century. By and large, those capable of comprehending seminal work like Peirce's in the theory of algebras lived and worked in Europe. Unfortunately, his ideas failed to reach them in 1870.

By the 1880s, however, the lay of the American mathematical landscape had begun to undergo important changes. A new university, the Johns Hop-

---

relished the naming of new mathematical objects. Because of this, his own development of the Peirce decomposition would require several definitions and explanatory remarks to render it intelligible to the modern reader. Thus, we give Wedderburn's 1907 treatment of the decomposition here. See Joseph H. M. Wedderburn, "On Hypercomplex Numbers," *Proceedings of the London Mathematical Society*, 2d ser., 6 (1907):77–118 on p. 91.

[138] Benjamin Peirce, "Linear Associative Algebra," p. 113.

[139] Benjamin Peirce to George Bancroft, 18 November 1870, in Ginsburg, "A Hitherto Unpublished Letter," p. 282. This was also quoted by Helena Pycior in her article on Peirce, entitled "Benjamin Peirce's 'Linear Associative Algebra'," *Isis* 70 (1979):537–551 on p. 546.

kins in Baltimore, had opened with the training of researchers as part of its stated mission and had imported the British algebraist James Joseph Sylvester to spearhead its program in mathematics. Not long after his coming, Sylvester participated in the establishment of the *American Journal of Mathematics*, a journal aimed at a purely mathematical audience both in the United States and abroad. When Peirce's "Linear Associative Algebra" finally appeared reprinted in its pages, it at least had a chance of reaching appreciative readers. Needless to say, Sylvester's arrival on American shores did more for American mathematics than resuscitate Peirce's work. It signaled the beginnings, however modest, of the entry of the United States into the international mathematical arena, a development which would take place within the context of neither the college nor the federal government nor the general scientific society but within that of the emergent American university.

TABLE 1.1
COURSE OF STUDY IN MATHEMATICS AND ASTRONOMY UNDER BENJAMIN PEIRCE AT THE LAWRENCE SCIENTIFIC SCHOOL (1848)*

(1) Curves and Functions: (Regular Course) Peirce—*Curves and Functions*; Lacroix—*Calcul différentiel et intégral*; Cauchy—*Les Applications du Calcul infinitésimal à la Géométrie*; Monge—*Application de l'Analyse à la Géométrie.*

(Parallel Course) Biot—*Géométrie analytique*; Cauchy—*Cours d'Analyse de l'École royale polytechnique*; Hamilton's researches respecting quaternions from volume twenty-one of the *Transactions of the Royal Irish Academy.*

(2) Analytical and Celestial Mechanics: (Regular Course) Laplace—*Mécanique céleste, translated with a Commentary, by Dr. Bowditch,* volume 1; Bowditch—"On the Computation of the Orbits of a Planet or Comet," Appendix to volume three of his translation; Airy—"Figure of the Earth," from the *Encyclopedia Metropolitana*; Airy—"Tides," from the *Encyclopedia Metropolitana.*

(Parallel Course) Poisson—*Mécanique analytique*; Lagrange—*Mécanique analytique*; Hamilton—"General Method in Dynamics," from the *Philosophical Transactions* from 1834 and 1835; Gauss—*Theoria Motus Corporum Coelestium*; Bessel—*Untersuchungen*; Leverrier—*Développements sur plusieurs Points de la Théorie des Perturbations des Planètes*; Leverrier—*Les Variations séculaires des Elémens des Orbites, pour les sept Planètes principales*; Leverrier—*Théorie des Mouvements de Mercure*; Leverrier—*Recherches sur les Mouvements de la Planète Herschel*; Adams—*Explanation of the Observed Irregularities in the Motion of Uranus, on the Hypothesis of Disturbances Caused By a More Distant Planet.*

(3) Mechanical Theory of Light: (Regular Course) Airy—*Mathematical Essays*; MacCullagh—"On the Laws of Crystalline Reflection and Refraction" in volume eighteen of the *Transactions of the Royal Irish Academy.*

(Parallel Course) Cauchy—*Exercises d'Analyse et de Physique mathématiques*; Neumann—"Theoretische Untersuchungen der Gesetz, nach welchen das Licht reflectirt und gebrochen wird" in the *Transactions of the Berlin Academy* for 1835.

* From Florian Cajori, *The Teaching and History of Mathematics in the United States* (Washington: Government Printing Office, 1890), pp. 137–138.

TABLE 1.2
AMERICAN MATHEMATICS JOURNALS: 1800–1900**

(The dates in parentheses give the years of the journal's existence.)

The Mathematical Correspondent (1804–1806)

Analyst or Mathematical Museum (1808)

The Monthly Scientific Journal (1818)

The Mathematical Diary (1825–1832)

The Ladies' and Gentlemen's Diary or United States Almanac (1820–1822)

The Mathematical Companion (1828–1831)

Mathematical Miscellany (1836–1839)

The Cambridge Miscellany of Mathematics, Physics and Astronomy (1842)

Mathematical Monthly (1858–1860)

The Analyst: A Monthly Journal of Pure and Applied Mathematics (1874–1883)

Mathematical Visitor (1877–1894)

American Journal of Mathematics (1878–present)

Mathematical Magazine (1882–1884)

Mathematical Messenger (1884–1895)

Annals of Mathematics (1884–present)

Bulletin of the New York (later *American*) *Mathematical Society* (1891–present)

The American Mathematical Monthly (1894–present)

The Mathematical Review (1896)

Transactions of the American Mathematical Society (1899–present)

** List compiled from Cajori, pp. 94–97 and 277–286; and David Eugene Smith and Jekuthiel Ginsburg, *A History of Mathematics in America Before 1900*, Carus Mathematical Monographs, no.5 (Chicago: Mathematical Association of America, 1934; reprint ed., New York: Arno Press, 1980).

Chapter 2
A New Departmental Prototype: J. J. Sylvester and the Johns Hopkins University

In 1876, a university opened in Baltimore, Maryland which would set new standards not only for American higher education generally but also for American mathematics particularly. The Johns Hopkins University—under the direction of its first president, Daniel Coit Gilman, and financed through the generous bequest of the Baltimore millionaire, Johns Hopkins—opened as an institution of higher education devoted to the ideal of research. Unlike other, older colleges and universities in the United States, the Johns Hopkins[1] emphasized graduate education, although it did not neglect the important function of the undergraduate college as a sort of feeder into its graduate programs. Having as two of its primary goals the training of future researchers and the maintenance of high levels of research productivity among its faculty, this institution departed radically from the traditional model of the undergraduate teaching college.

Gilman searched both at home and abroad in order to find a faculty equal to these two objectives. He personally interviewed and rejected many possible candidates in all fields, seeking people of either demonstrated or promising research ability. As his Professor of Mathematics, he hired a man decidedly in the former category, the sixty-one-year-old British mathematician, James Joseph Sylvester. With the advent of Sylvester in 1876, research-level mathematics finally gained a significant foothold in the United States. Indeed, Sylvester succeeded in founding America's first school of mathematical research within the institutional framework of the Johns Hopkins University.

A New Experiment in Baltimore: The Johns Hopkins University

In 1867, one of Baltimore's leading citizens, the financier and septuagenarian Johns Hopkins, laid the legal groundwork for a new educational institution to be funded by his private bequest and officially organized only after his

[1] See note 16 in the Preface.

death. Although Hopkins broadly conceived of the school as a sort of educational complex—composed of a university and a teaching hospital—which would satisfy the strong humanitarian convictions engendered by his Quaker beliefs, he ultimately provided little insight into his precise vision. Thus, following his death on Christmas Eve in 1873 and the probation of his will early in 1874, his appointed trustees found themselves in unfettered control of some $7,000,000, the largest single private donation to higher education up to that time.[2] They immediately set to work trying to decide how best to satisfy their deceased friend's wishes.[3]

Although Hopkins had perhaps consciously stacked his Board with men who, unlike himself, had at least attended college, none of its members had had any real experience in educational planning or policy-making. Acutely aware of their deficiency on this score, they not only selected a preparatory reading list of contemporary books on education—its history, issues, and institutions—but they also sought advice directly from the leading college and university presidents of the day.[4] They chose as their chief presidential advisers Charles W. Eliot of Harvard, Andrew D. White of Cornell, and James B. Angell of the University of Michigan, each an acknowledged trend-setter in American higher education and each representative of a different brand of American educational institution. The Hopkins Board hoped to gain some consensus as to precisely what form the Johns Hopkins University should take by consulting each of these men privately.

By no means passive in this data-gathering process, the Trustees posed detailed and informed questions to each of their three advisers—to Eliot and Angell in person in Baltimore and to White by correspondence. The Board sought to apprise itself of the various options open to a new educational institution, as well as of the pitfalls and difficulties inherent in those options, by addressing such issues as admissions standards, coeducation, fixed versus elective-based curricula, professional schools, and graduate programs. In particular, both Eliot and Angell doubted the ability of a brand-new university to launch graduate programs successfully from the outset. As Eliot expressed it: "The post-graduate course is a matter far off for you. Not until

[2]Hopkins did restrict the University's expendable funds to the interest—and not the principle—of the Baltimore and Ohio Railroad stock which formed the bulk of the endowment. In the early years of the institution's existence, a fluctuating stock market would repeatedly cause financial worries. For much more on the history of the Johns Hopkins University, see, for example, John C. French, *History of the University Founded by Johns Hopkins* (Baltimore: Johns Hopkins University Press, 1946); and Hugh Hawkins, *Pioneer: A History of the Johns Hopkins University,* 1874–1889 (Ithaca: Cornell University Press, 1960).

[3]The key role played by the Johns Hopkins University in the history of American higher education has been well documented. Consult, for instance, Richard J. Storr, *The Beginnings of Graduate Education in America* (Chicago: University of Chicago Press, 1953).

[4]For the reading list, see French, pp. 24–25. As both French and Hawkins pointed out, it is not clear how faithfully the Trustees carried out this particular self-appointed task.

you have organized the whole of the College [that is, undergraduate] course, could that question practically present itself."[5] Clearly, Eliot and Angell proposed an organizational strategy which started with preparing a clientele for graduate-level studies.[6] White, on the other hand, suggested that, through a finely crafted system of fellowships, the Johns Hopkins *could* implement graduate training from the start, a position consonant with that of Trustee George William Brown. As early as 1869 in a speech at St. John's College in Annapolis, Brown had argued that, contrary to the prevailing situation, the faculty members of the nation's universities "should be teachers in the largest sense, that is, should have the ability and the leisure too, to add something by their writings and discoveries to the world's stock of literature and science."[7] In his view, America, in all of its egalitarian zeal, had created an educational system like "a temple without a dome, a column without a capital, a spire without a pinnacle."[8] In order to be complete, it needed a graduate school which encouraged and, indeed, expected original research from its faculty.

Their conflicting opinions on graduate training aside, the Board's three earliest advisers, as well as others consulted, all independently agreed on one key point: the man best suited for the new university's presidency was the then President of the University of California, Daniel Coit Gilman.[9] Thus, even as they continued their fact-finding both at home and abroad throughout the fall and winter of 1874 and 1875, the Board sounded Gilman out on his willingness to leave his position in Berkeley and assume the new post in Baltimore.[10] Just what qualities did Gilman possess which elicited such overwhelming unanimity among his peers and inspired such confidence in the Hopkins Board?

A man educated at Yale College in the late 1840s, Gilman had traveled to Europe prior to taking up a position as a fund-raiser for the still-infant Scientific School at his *alma mater*.[11] While abroad, he had observed not

---

[5]Hawkins, *Pioneer*, p. 13.

[6]Angell remembered his stand a bit differently at the University's twenty-fifth anniversary celebration. He recalled arguing in favor of the establishment of a graduate school when the Trustees called on him somewhat later in Ann Arbor. See *Johns Hopkins University Celebration of the 25th Anniversary of the Founding of the University and Inauguration of Ira Remsen LL.D. as President of the University* (Baltimore: Johns Hopkins University Press, 1902), p. 134. This is quoted in French, pp. 23–24.

[7]French, p. 6.

[8]*Ibid.*, pp. 6–7. For more details on Brown's early views, see Hugh Hawkins, "George William Brown and His Influence on The Johns Hopkins University," *Maryland Historical Magazine* 52 (1957):173–186.

[9]In addition to Angell, Eliot, and Gilman's close friend, White, Noah Porter of Yale also suggested Gilman. Hopkins's second President, Ira Remsen, said that, in all, five people independently proposed Gilman's name to the Trustees. See Hawkins, *Pioneer*, pp. 14–20.

[10]Fabian Franklin, *The Life of Daniel Coit Gilman* (New York: Dodd, Mead and Co., 1910), pp. 184–185 for the text of the letter of inquiry. Franklin quoted this, as well as many other archives now held in the collection at the Johns Hopkins University, in full in his biography.

[11]*Ibid.*, pp. 39–109 details Gilman's years in New Haven.

only the social and cultural mores of the various countries he visited but also their different educational systems. In fact, he wrote about the latter after his return to New Haven in several articles in Henry Barnard's *American Journal of Education.* These reflect Gilman's awareness of the institutionalization of higher level teaching and original research on the Continent as well as his emergent ideas on higher education in the American context.[12] In particular, they underscore his conviction in the value of "original inquiries and investigations" and in the effectiveness of the "new" as opposed to the classical education.[13] Thus, although Gilman appreciated the value of the largely classical curriculum he had followed at Yale, he also believed that the sciences, modern languages, and history—elements of the "new" education—merited a prominent place in university studies.

In New Haven, Gilman had occasion to air these and related views as early as 1856 when he drafted a "Proposed Plan for the Complete Organization of the School of Science connected with Yale College." To that short pamphlet, he appended a sketch of his European observations, thereby underscoring the intimate connections between what he had seen of the European scientific and technical schools and the plan of organization he proposed for Yale.[14] Nevertheless, he did not advocate merely the transplantation of a foreign institution to American soil. Rather, he believed "that it is important to gain a thorough knowledge of what is being done in kindred foreign institutions, not in order to copy their methods but to adapt them to local conditions and to the wants of this country as acknowledged by practical men."[15] Through the reflection of ideas like these in his various activities in New Haven from 1856 to 1872, Gilman had secured a reputation as an educator, as an organizer, and as an innovator. Furthermore, his presidency (from 1872) of the University of California had given him firsthand experience in actually running a university. All of these qualities combined to make Gilman a choice catch for the new university in Baltimore, when he accepted its offer on 30 January 1875.[16]

---

[12]Daniel Coit Gilman, "Scientific Schools in Europe," *American Journal of Education* 1 (1855):317–328; "German Universities," *op. cit.*, pp. 402–404; and "The Higher Special Schools of Science and Literature in France," *op. cit.*, 2 (1856):93–102. Francesco Cordasco identified Gilman as the author of the anonymous, second article in his book *Daniel Coit Gilman and the Protean Ph.D.: The Shaping of American Graduate Education* (Leiden: E. J. Brill, 1960), p. 12. See, also, Laurence R. Veysey, *The Emergence of the American University* (Chicago: University of Chicago Press, 1965), pp. 159–160.

[13]Gilman, "Scientific Schools in Europe," p. 328.

[14]The pamphlet was only printed privately. See Cordasco, pp. 25–26 for a discussion of Gilman's plan.

[15]Franklin, *The Life of Daniel Coit Gilman*, p. 41.

[16]Gilman was also moveable. As President of an institution tightly controlled by the partisan interests of the state legislature, Gilman felt continual frustration in California. It is little wonder, then, that when the overture from the Hopkins Board arrived in the fall of 1874, with its promise of an institution free of the "shackles of state or political influence," Gilman wasted little time in contemplating the offer. See Hawkins, *Pioneer*, pp. 19–20; and Cordasco, pp. 45–

Although he officially assumed his post on 1 May 1875, Gilman had begun to formulate his plan for the University as early as January and had faithfully communicated his evolving thoughts to the Board. The final blueprint, which received unanimous approval on 27 May, thus held no surprises while it incorporated significant educational innovations.[17] Gilman called for a postponement of the development of the proposed Hospital and Medical School in order to concentrate first on the departments of literature and science.[18] This would allow the administration to focus on one set of issues and problems at a time and thereby prevent it from spreading itself too thin. Also inherent in this decision was an equal emphasis on the sciences and the more traditional liberal arts courses. Unlike the situations at Yale or at the University of California, where the sciences were grafted onto an originally classical curriculum, the history of the Johns Hopkins University was just beginning. Gilman found himself in the position of creating precedents and traditions, not working against them. His new university would stress the sciences from the start, but not at the expense of the liberal arts curriculum. Furthermore, whereas Yale and California had tried with mixed results to fuse graduate schools onto extant undergraduate institutions, the Johns Hopkins would open as a graduate school buttressed by an undergraduate college.[19] Thus, its permanent faculty would consist of researchers committed to advancing their respective disciplines both through their own work and through the high-level training of students. Gilman fully realized, however, that in order to create an atmosphere conducive to research, he first had to liberate his research faculty from the then standard fifteen- to twenty-hour-a-week teaching load.[20] To this end, he set a precedent of utilizing visiting professors from other institutions, who would supplement teaching at the graduate level, as well as adjuncts and assistants, who would provide

52. In fact, Gilman had felt so embattled earlier in 1874 that he had tendered his resignation at Berkeley. The University's Regents talked him out of making such a drastic move at that point.

[17]Cordasco, p. 65. Cordasco gave the complete plan as published in Gilman's Second Annual Report of 1877 on pp. 65–67.

[18]Johns Hopkins had mandated that his $7,000,000 bequest be divided equally between the University proper and the medical complex.

[19]In his analysis of the history of American higher education, Richard Storr delineated three models of university and graduate school building: (1) the coupling of a graduate school with a preextant undergraduate college, (2) the creation of a complete, full-scale university *de novo*, and (3) the elevation or expansion of an undergraduate college into a graduate school. See Storr, pp. 129–134. The Johns Hopkins University represents a prime example of the second model. As Storr pointed out, however, these models were not necessarily mutually exclusive.

[20]On teaching loads in general, see Veysey, pp. 77 and 358. In 1890, Florian Cajori published a survey of the mathematics programs at 168 American colleges and universities. Of 117 responding to his question about teaching loads, 47% lectured sixteen or more hours a week and 27% taught between eleven and fifteen hours a week. See Florian Cajori, *The Teaching and History of Mathematics in the United States* (Washington: Government Printing Office, 1890), pp. 345–349. Some of his findings have been systematized and analyzed in Karen Hunger Parshall, "A Century-Old Snapshot of American Mathematics," *The Mathematical Intelligencer* 12 (3) (1990):7–11.

undergraduate instruction.[21] Within this sort of tripartite system, the permanent faculty members, relieved of the full teaching burden, could concentrate on that dual goal of training graduate students and pursuing their own research. This research ethic defined the guiding philosophy of the new university from the outset. Gilman next had to assemble a faculty capable of realizing it.

The President seemingly had no preconceived notion of precisely what departments the University would open with. First and foremost, he sought men of proven research ability in their chosen areas of specialization. If initially he could secure a Latinist but no chemist who met his standards, then Hopkins would have a Department of Classics but no Department of Chemistry. Next, he looked for teachers with a gift for sharing their knowledge and for instilling a zest for research in their students. Finally, Gilman needed colleagues who would cooperate fully with him in the organization and running of the University.[22] He began his search for men satisfying these criteria with an American tour that ultimately netted three of his six original faculty members: Basil Gildersleeve in classics, Henry Rowland in physics, and Ira Remsem in chemistry. Finally, after conducting a foreign search which took him to the British Isles, France, Germany, Switzerland, and Austria, he brought his other three initial members from England: Charles D'Urban Morris as undergraduate professor of classics; Henry Newell Martin in biology; and, in perhaps his boldest and riskiest hiring move, an unemployed, sixty-one-year-old mathematician with a reputation as a hot-tempered eccentric, James Joseph Sylvester.[23]

---

[21] Interestingly enough, Gilman envisioned that these adjuncts and assistants might come from the ranks of the graduate students, thereby setting a precedent for the graduate teaching assistant so essential—for better or for worse—to university teaching in twentieth-century America.

[22] Gilman outlined his criteria for faculty appointments to the Board in a position paper entitled "On the Selection of Professors." In keeping with one of the few stipulations of the benefactor, the faculty would also be chosen without consideration of religious or political persuasion. See Hawkins, *Pioneer*, pp. 39–41.

[23] The search for and securing of the Hopkins faculty has been told in many places. For Gilman's own account looking back fondly late in his life, consult Daniel Coit Gilman, *The Launching of a University and Other Papers* (New York: Dodd, Mead, & Co., 1906), pp. 12–22 and 47–85. For secondhand accounts, see Franklin, *The Life of Daniel Coit Gilman*, pp. 196–218; French, pp. 33-39; Hawkins, *Pioneer*, pp. 28–62; and Cordasco, pp. 69–78. Judging by the appointments of Martin, Rowland, and Remsen, Gilman wished to institutionalize the scientific component of the "new" education at the Johns Hopkins from the start by placing the sciences on an equal footing with the more traditional subjects of Latin, Greek, and mathematics. In fact, Hopkins had the first graduate program in biology in the United States. On its importance, see Keith R. Benson, "American Morphology in the Late Nineteenth Century: The Biology Department at Johns Hopkins University," *Journal of the History of Biology* 18 (1985):163–205. The remainder of the present chapter, together with Chapter 3, details the impact that Hopkins had on mathematics in America. The institution also played a major role in physics, chemistry, history, and other fields. See Daniel J. Kevles, *The Physicists: The History of a Scientific Community in Modern America* (New York: Alfred A. Knopf, 1978); Edward A. Beardsley, *The Rise of the American Chemical Profession,* 1850–1900 (Gainesville: University of

THE JOHNS HOPKINS'S FIRST PROFESSOR OF MATHEMATICS

Born in London into the family of a Jewish merchant on 3 September 1814, James Joseph Sylvester received his early education in London, primarily at the boarding school at Highgate. There, he came under the tutelage of Leopold Neumegen, considered "a good mathematician" by at least one of his contemporaries.[24] Recognizing a certain mathematical talent in the young Sylvester, Neumegen arranged to have his charge examined by Olinthus Gregory, the Professor of Mathematics at the Royal Military Academy at Woolwich, who was apparently also impressed by the boy's abilities. Thus, Sylvester had, at the age of eleven, his initial contact with the institution which would provide him his first sustained academic position in mathematics in 1855 and which would deal him one of the most crushing blows of his career fifteen years later.

Leaving Highgate at the end of the 1826–1827 term, Sylvester moved on to Islington for a year and a half before entering the University of London (later University College London) during its first academic year of operation, 1828–1829. Conceived as early as 1825 by a prominent group of men who deplored the fact that England's institutions of higher education effectively excluded all but subscribers to the tenets of the Church of England from earning degrees, the University of London offered educational opportunities to young men regardless of their religious affiliation.[25] It thus presented a viable option to a student in Sylvester's position. Furthermore, its first Professor of Mathematics, the then twenty-two-year-old Augustus DeMorgan, introduced his students to the progressive French analytic approach to algebra through the translation he was preparing of Bourdon's *Éléments d'Algèbre*.[26]

Florida, 1964); John Servos, *Physical Chemistry from Ostwald to Pauling: The Making of a Science in America* (Princeton: University Press, 1990); and Peter Novick, *That Noble Dream: The "Objectivity Question" and the American Historical Profession* (New York: Cambridge University Press, 1988).

[24]David S. Blondheim, "James Joseph Sylvester," *Jewish Comment* 23 (May 25, 1906). For many details pertaining to Sylvester's life, see the biographical notice by H. F. Baker in *The Collected Mathematical Papers of James Joseph Sylvester*, ed. H. F. Baker, 4 vols. (Cambridge: University Press, 1904–1912; reprint ed., New York: Chelsea Publishing Co., 1973), 4:xv–xxxvii (hereinafter cited as *Math. Papers JJS*); Raymond Clare Archibald, "Unpublished Letters of James Joseph Sylvester and Other New Information Concerning His Life and Work," *Osiris* 1 (1936):85–154; Raymond Clare Archibald, "Material Concerning James Joseph Sylvester," in *Studies and Essays in the History of Science and Learning Offered in Homage to George Sarton on the Occasion of His Sixtieth Birthday, 31 August* 1944 (New York: Schuman, n.d.), pp. 209–217; and Charles C. Gillispie, ed., *The Dictionary of Scientific Biography*, 16 vols., 2 supp. (New York: Charles S. Scribner's Sons, 1970–1990), s.v. "Sylvester, James Joseph" by J. D. North (hereinafter cited as *DSB*).

[25]Students who did not subscribe to the tenets of the Church of England could *attend* Cambridge, for example, but they could not take a degree at the end of their course of study. See H. Hale Bellot, *University College London* 1826–1926 (London: University of London Press, Ltd., 1929) on the history of that University.

[26]This appeared in 1828 as *The Elements of Algebra; Translated from the First Three Chapters of the Algebra of M. Bourdon, and Designed for the Use of Students in and Preparing for the University of London*, trans. Augustus DeMorgan (London: John Taylor, 1828). See, also, Louis Pierre Marie Bourdon, *Éléments d'Algèbre*, 4th ed. (Paris: Bachelier, 1825).

JAMES JOSEPH SYLVESTER (1814–1897)

Fresh from Trinity College, Cambridge where he had absorbed the ideals of the Cambridge Analytical Society, DeMorgan shared the belief that England lagged far behind France relative to mathematical education in the early decades of the nineteenth century, and he hoped to help reverse this state of affairs through his translation of Bourdon's work.[27] The young Sylvester may thus have come into contact with these more progressive ideas, if only briefly. Sylvester's student days at DeMorgan's university came abruptly to an end after only five months: his family withdrew him from the institution on account of his apparent immaturity and inability to behave in a manner befitting a university student.[28]

After this forced hiatus in his higher education, Sylvester proceeded to the school attached to Liverpool's Royal Institution, where he studied until gaining admittance to St. John's College, Cambridge on 7 July 1831. Interrupted twice by illness, Sylvester's stay at Cambridge reached its climax in January of 1837 when he passed the Mathematical Tripos as Second Wrangler behind fellow Johnian, William Griffin, but ahead of George Green, the Fourth Wrangler and a student of Gonville and Caius.[29] The Tripos had become increasingly controversial in mathematical circles during the nineteenth century, stressing, its opponents claimed, the mastery of minutiæ rather than active mathematical thought. The Fourth Wrangler in 1827, Augustus DeMorgan, leveled the criticism that "[i]t is impossible in such an examination to propose a matter that would take a competent mathematician two or three hours to solve, and for the consideration of which it would be necessary for him to draw his materials from different sources, and see how he can put together his previous knowledge, so as to bring it to bear most effectually on this particular subject."[30] While DeMorgan clearly faulted the Tripos system for failing to measure adequately a candidate's potential for mathematical research, this was the goal of neither the examination nor mathematical higher education in nineteenth-century Cambridge. Both aimed, in the broadest terms, to provide British gentlemen with a liberal education combined with a certain mental discipline and attention to detail. At a more concrete level, however, the Tripos served as a measure of academic achievement, with Wranglers naturally falling within the most promising lot.[31] In light of

---

[27]For references pertaining to the Cambridge Analytical Society, see note 20 in Chapter 1 above.

[28]In his history of University College, Bellot alleged that Sylvester was expelled from the University "for taking a table-knife from the refectory with the intention of sticking it into a fellow student who had incurred his displeasure [p. 180]." Correspondence between one of Sylvester's elder brothers and University authorities, however, indicates that the family—and not the University—took the final action. See College Correspondence, Sylvester, E. J. 1829: 1614, Manuscripts and Rare Books Room, Bloomsbury Science Library, University College London.

[29]Baker, "Biographical Notice," p. xvii.

[30]*Ibid.*, p. xxi.

[31]For further discussion of the Tripos and its influence in British scientific society, see, for example, P. M. Harmon, ed., *Wranglers and Physicists: Studies on Cambridge Physics in the*

his ranking as Second Wrangler, Sylvester might reasonably have expected to compete well academically for further Cambridge plums—the Smith's Prizes or a college fellowship—had his religious convictions not put an insurmountable obstacle in his path. As a Jew, he could not subscribe to the Thirty-Nine Articles of the Church of England, a necessary formality dating from the seventeenth century. Thus, by law, he was ineligible not only for prizes and fellowships but also for the credential he had rightly earned. Forced to leave Cambridge without a degree, Sylvester's chances for an academic career in England seemed slim at best. Virtually only University College London allowed Jews on its faculty, and there the youthful DeMorgan still held the chair in mathematics.[32]

At University College, however, a not unrelated chair, the professorship of natural philosophy, did fall vacant, and in 1837 Sylvester offered himself as a candidate. Securing testimonials from such influential figures as Olinthus Gregory of Woolwich and from George Peacock and Philip Kelland then both of Cambridge, Sylvester won the appointment effective the 1838–1839 academic year.[33] Unfortunately, he found himself uninterested in and ill-equipped for the duties associated with this position and resigned after only three years to take a post in the United Stated in mathematics, his actual subject of interest. By the time he left London in 1841, Sylvester had published no fewer than fifteen papers on topics such as the mathematics of fluid mechanics and, more importantly, on Charles Sturm's theory for locating the roots of an algebraic equation as well as on the not unrelated theory of elimination. (These last two areas, in particular, would repeatedly attract Sylvester's research attention over the next two decades.)

In elimination theory, for example, Sylvester formulated in 1840 what he would subsequently call the *dialytic* method for determining when two polynomial equations $f(x) = 0$ and $g(x) = 0$ have a common root.[34] Although he described his construction in the fullest generality for f and g polyno-

---

Nineteenth Century (Manchester: University Press, 1985). This source also includes an extensive bibliography of the primary and secondary literature on this topic, including biographies of the various figures involved. For a nineteenth-century assessment of the British university system as a whole, see William Whewell, *On the Principles of English University Education* (London: John W. Parker, 1838).

[32]The Test Acts for university faculty were finally repealed in 1871, a circumstance which, as we shall see in Chapter 3, would prove important later in Sylvester's life.

[33]*Testimonials of J.J. Sylvester, Esq, A.M., F.R.S., etc.*, (privately printed), copy held in Daniel C. Gilman Papers Ms. 1, Special Collections Division, The Milton S. Eisenhower Library, The Johns Hopkins University (hereinafter denoted Gilman Papers).

[34]James Joseph Sylvester, "A Method of Determining by Mere Inspection the Derivatives From Two Equations of Any Degree," *Philosophical Magazine* 16 (1840):132–135, or *Math. Papers JJS*, 1:54–57. (In what follows, we have used α, β, and γ for the coefficients of the second equation instead of Sylvester's l, m, and n in order to avoid confusion with his choice of n and m for the degrees of the two equations.) The dialytic method is one of the results for which Sylvester is still remembered today. For a concise modern treatment of it, see B. L. van der Waerden, *Algebra*, 2 vols. (New York: Springer-Verlag, 1991), 1:102–105.

mials of degree n and m, respectively, Sylvester provided no proof of its validity and illustrated it only in the case of two quadratic polynomials. For $f(x) = ax^2 + bx + c = 0$ and $g(x) = \alpha x^2 + \beta x + \gamma = 0$, he formed the 4×4 (or more generally $(n+m) \times (n+m)$) array

$$\begin{matrix} a, & b, & c, & 0 \\ 0, & a, & b, & c \\ \alpha, & \beta, & \gamma, & 0 \\ 0, & \alpha, & \beta, & \gamma \end{matrix}$$

from the polynomial coefficients. (Here, note that in the general case the first row involving the coefficients of f terminates with $m-1$ zeros, the second row begins with one zero and terminates with $m-2$ zeros, etc., while the first row involving the coefficients of g terminates with $n-1$ zeros, the second begins with one zero and terminates with $n-2$ zeros, etc.) He then calculated, without introducing this terminology, the determinant of this array and noted that it vanished when the two equations $f(x) = 0$ and $g(x) = 0$ had a common root.[35] As Augustin-Louis Cauchy showed later in 1840, the algebraic methods of elimination devised independently by Euler and Bézout in 1764 boiled down to Sylvester's determinant, namely, the modern form of the resultant.[36]

One year prior to completing this early work in elimination theory, Sylvester's first research effort (on Fresnel's optical theory of crystals) had already secured his election to the Royal Society at the age of twenty-five.[37] In 1841 his combined output earned him the B.A. and M.A. from Trinity College, Dublin.[38] It was thus a confident and qualified young mathematician who applied for the vacant chair of mathematics at Thomas Jefferson's nonsectarian University of Virginia in faraway Charlottesville.

With this application, he offered even stronger testimonials based not merely on his promise as a mathematician but actually on his achievements since 1837. Although Sir John Herschel and Charles Babbage, among many others, supported his candidacy strongly, the most prescient endorsement came from his colleague, DeMorgan. Writing on 22 May 1841, DeMorgan

[35] While this is actually an "if and only if" statement, Sylvester did not explicitly recognize it as such in his paper. He first used the term *determinant* in 1841 in James Joseph Sylvester, "Examples of the Dialytic Method of Elimination as Applied to Ternary Systems of Equations," *Cambridge Mathematical Journal* 2 (1841):232–236, or *Math. Papers JJS*, 1:61–65. See Thomas Muir, *The Theory of Determinants in the Historical Order of Development*, 4 vols. (London: Macmillan & Co., Ltd., 1906–1923; reprint ed., New York: Dover Publications, Inc., 1960), 1:243. This work is a veritable gold mine of information on determinants up to 1900.

[36] For a discussion of Cauchy's work, consult Muir, 1:240–243.

[37] James Joseph Sylvester, "Analytical Development of Fresnel's Optical Theory of Crystals," *Philosophical Magazine* 11 (1837):461–469 and 537-541; 12 (1838):73–83 and 341–345, or *Math. Papers JJS*, 1:1–27.

[38] Trinity College, Dublin had received a special exemption from the Test Acts in 1793 when the Roman Catholic Relief Act was passed. Thus, not only could Catholics take their degrees at Trinity but also Jews, like Sylvester, and those of other religious persuasions.

declared that

> [n]o person of his years in this country has more reputation than Mr. Sylvester as an original Mathematician, or bids fairer to extend the exact sciences by his labours. From my own knowledge of what he has done, I can most safely say that he is a Mathematician of great power, well acquainted with the most modern forms of the science, and very zealous in his prosecution of his inquiries. By these qualifications he will certainly make his name well known as an original contributor of Mathematics, to the credit of any institution with which he may be connected.[39]

The strengths of endorsements such as this, together with Sylvester's own record of publication and his acquaintance with the first incumbent in the chair, Thomas Key, undoubtedly all contributed to his appointment at the University of Virginia in 1841.

Arriving in central Virginia somewhat late for the fall term in November of 1841, Sylvester received a warm welcome into the University community, with the students illuminating the academic buildings and lighting bonfires on the University grounds, and he immediately set about the duties of his professorship.[40] With a total of forty-seven students, he taught mathematics ranging in level from arithmetic and elementary algebra to the differential and integral calculus, and he, too, adopted the French approach of authors like Legendre and Lacroix as opposed to the more outmoded British presentation.[41] Standing before his classes rhapsodizing mathematically to an audience which did not exactly share his enthusiasm, the short and burly Englishman with the Cockney accent might well have become the much caricatured, belovedly eccentric professor had he come to Virginia in different times.

In the 1830s and 1840s, however, the University of Virginia suffered from extreme student drunkenness and lawlessness, and tensions often ran high between the faculty and student body. The year before Sylvester's arrival, in fact, a student had murdered a member of the faculty. In this atmosphere, Sylvester, as a foreigner teaching a "foreign" subject, elicited not knowing, understanding smiles but direct, challenging confrontations.[42] In his efforts

---

[39] *Testimonials of J. J. Sylvester*, p. 24.

[40] "Professor Sylvester's Arrival," *Collegian* 4 (1841):62. This source was uncovered by Lewis S. Feuer and used in his paper "America's First Jewish Professor: James Joseph Sylvester at the University of Virginia," *American Jewish Archives* 36 (1984):151–201 on p. 155. He published an abbreviated version of this paper as "Sylvester in Virginia," *The Mathematical Intelligencer* 9 (2) (1897):13–19.

[41] Feuer, "America's First Jewish Professor," p. 154. Consult, also, note 22 in Chapter 1.

[42] Feuer first situated Sylvester's problems at the University of Virginia within the context of the student violence of this period. He also underscored the provinciality, if not xenophobia, of certain segments of the population of Virginia at the time of Sylvester's arrival. Taking this line of thought even further, Feuer linked Sylvester's particular problems to anti-Semitism, which he

to deal with the troublemakers in his courses, Sylvester tried to assert his authority. On 1 February 1842, he reported one of his first-year students to the faculty for insubordination. By 23 February, the situation had worsened, and Sylvester recounted to his colleagues a heated verbal exchange between himself and the same student. Perhaps mindful of the delicate student-faculty relationship at the time, the faculty condoned Sylvester's handling of the affair and reprimanded the student for the abusive language and disrespect he had directed toward his professor, but did not go so far as to bring strong sanctions against the young man and thereby risk further student unrest. The faculty minutes recorded Sylvester's strenuous objections to this decision on 19 March and on 22 March the acceptance of his resignation from the University.[43]

Sylvester left Charlottesville for the refuge of his brother's home in New York City. The young mathematician not only made contacts from this temporary home base with the mathematicians of the Northeast, most notably Benjamin Peirce, but also began searching for another academic job on American shores. After his hopes of job prospects at Harvard, at Columbia College in New York, at the University of South Carolina, and in the Washington, D.C. area came to naught and after his proposal of marriage to a Miss Marston of New York City met with refusal on religious grounds, Sylvester returned home to England, dejected and despondent, late in the fall of 1843.[44]

He realized that he went back to face a most uncertain future. After all, when he left England in 1841, he had given up virtually the only academic job open to him. To return almost certainly meant to change his career plans altogether, but to change them to what? This remained an open question until 9 December 1844 when he took the posts of Secretary and Actuary at the Equity and Law Life Assurance Company in London. By 1846, in

---

argued contributed to what he called the "attempts to suppress the facts" surrounding Sylvester's short tenure in Charlottesville. The argument Feuer presented relative to this last contention lacks the strength and convincingness of the evidence he put forth relative to the other two points.

[43]Faculty Minutes, Rare Books and Manuscripts Room, Alderman Library, University of Virginia. Sylvester's resignation from Virginia sparked stories and rumors from the very beginning. Sylvester allegedly bought a sword-cane for protection sometime in February of 1842 in order to protect himself in the face of possible student violence. Sylvester was also allegedly confronted by a bludgeon-wielding student and defended himself by striking the student with the sword. Feuer argued that this event occurred between 19 and 22 March and precipitated Sylvester's resignation. He further argued that the event *did* occur on the basis of two separate accounts, by contemporaneous individuals, one a faculty member. Since neither of these men was an eye-witness and since, as Feuer himself admitted, the faculty member clearly harbored animosity against Sylvester, the truth of this story still seems open to question.

[44]The actual date and even the year of Sylvester's departure from the United States has never been adequately pinned down in the literature, although Archibald came the closest to settling the issue in "Unpublished Letters of J. J. Sylvester." There, he remarked that "[a]s to where Sylvester was during July 1843–November 1844 I have been unable to determine [p. 101]." A letter in the Joseph Henry Papers, Smithsonian Institution Archives, from Sylvester's brother, Sylvester Joseph Sylvester, to Joseph Henry and dated 3 January 1844 answers this question, at least in part. He told Henry "that my brother James sailed for England on the 20 Nov. [1843]."

addition to these duties, he had also taken on the responsibilities of tutor to private pupils preparing for the university and had entered the Inner Temple to ready himself for the Bar.

Sylvester slowly returned to his mathematical researches as his financial situation increasingly strengthened. After publishing nothing during his American interlude, he produced two short articles in 1844 and three brief pieces in 1847. Following his call to the Bar in 1850, however, he exploded onto the mathematical scene, reaching new heights of productivity and creativity. In particular, Sylvester defined and analyzed various specific forms of determinants arising from systems of equations in an effort to extend his work in elimination theory into what he called a "general theory of associated forms." This resulted in his publication of some twenty papers in 1850 and 1851 alone.[45] (This work, as we shall see below, led Sylvester naturally into the emergent theory of invariants.) While the peace of mind occasioned by his improved personal situation undoubtedly contributed to this turnabout, the single most important influence in Sylvester's life at this time was his new friend, Arthur Cayley.

Six years Sylvester's junior, Cayley had entered Trinity College, Cambridge in the same year Sylvester took the natural philosophy professorship at London. Cayley finished as Senior Wrangler on the Mathematical Tripos of 1842 and stayed on at his College as a Fellow for the next four years.[46] Celibate, able, and unfettered by his religious faith, Cayley could have stayed on indefinitely in this sort of post, but given the shortage of professorships and so the slim chances for real advancement, the overall employment picture in British higher academe looked grim. Thus in 1846, Cayley, too, opted for the surer and more lucrative legal path.[47] While it remains unclear precisely when the two misplaced mathematicians met, by 1849 Sylvester and Cayley had already established both the friendship and the mathematical exchange which only Cayley's death in 1895 would end.

As early as 1845, Cayley had fruitfully explored a new realm of mathematics—invariant theory—which the work of George Boole had opened up to him and which was not unrelated to the work Sylvester had done on the theory of elimination. In a two-part article published in 1841–1842, Boole had laid

---

[45] See, for example, James Joseph Sylvester, "On the General Theory of Associated Algebraical Forms," *Cambridge and Dublin Mathematical Journal* 6 (1851):289–293, or *Math. Papers JJS*, 1:198–202; and "An Essay on Canonical Forms, Supplement to a Sketch of a Memoir on Elimination, Transformation and Canonical Forms," privately printed, or *Math. Papers JJS*, 1:203–216.

[46] For a thorough discussion of Cayley's life and works, see Anthony James Crilly, "The Mathematics of Arthur Cayley with Particular Reference to Linear Algebra," (unpublished Ph.D. Dissertation, Middlesex Polytechnic, June 1981).

[47] Unlike Sylvester, Cayley did actually practice law after his call to the Bar in 1849. He earned his living as a conveyancing lawyer until he took the newly formed Sadlerian Chair of Mathematics at Cambridge in 1863.

the foundation for what would become the British approach to this subject.[48] There, Boole studied the behavior of binary forms under the action of a linear transformation (implicitly of nonzero determinant), pursuing a line of investigation suggested by his reading of Joseph-Louis Lagrange's *Mécanique analytique*. In particular, he wanted "to determine the relations by which the [coefficients of the binary form before and after the transformation is applied] are held in mutual dependence."[49] Like Gauss before him but motivated by different concerns, Boole had observed that the discriminant $b^2 - ac$ of the binary quadratic form $Q = ax^2 + 2bxy + cy^2$ remains invariant under the action of the linear transformation defined by $x \mapsto mx' + ny'$, $y \mapsto m'x' + n'y'$, where $m, n, m', n' \in \mathbb{R}$, and $\Delta = mn' - m'n \neq 0$.[50] In other words, applying the transformation to Q to get a new binary quadratic form $R = A(x')^2 + 2Bx'y' + C(y')^2$, the discriminant $B^2 - AC$ of R equals the discriminant $b^2 - ac$ of Q up to a power of the determinant Δ. (In this case, an easy calculation shows that $B^2 - AC = \Delta^2(b^2 - ac)$.) Furthermore, Boole had devised a technique involving formal partial differentiation which associated to any binary form Q an expression $\theta(Q)$ in the coefficients of Q. Using Boole's notation, the process thus yielded $\theta(Q) = b^2 - ac$, the discriminant, for the binary quadratic form Q. Passing to the most general case in which Q is a homogeneous polynomial of degree n in m unknowns, Boole stated the theorem (which he actually verified only for binary quadratic and binary cubic forms) that if γ denotes the degree of $\theta(Q)$, then[51]

$$\theta(R) = \Delta^{\frac{\gamma n}{m}} \theta(Q). \tag{2.1}$$

A relation like the one expressed in (2.1) would soon come to define the notion of an *invariant*. (See the discussion of equations (3.1), (3.2), and (3.3) in Chapter 3 below.)

The recent Cambridge graduate, Arthur Cayley, read this work some time prior to writing Boole directly on 13 June 1844 to relate not only his enthusiasm for Boole's results and approach but also his own findings in the area. By November, Cayley had developed his ideas completely and had

[48] George Boole, "Exposition of a General Theory of Linear Transformations," *Cambridge Mathematical Journal* 3 (1841–1842):1–20 and 106–119. The British approach to invariant theory has been discussed by Tony Crilly in "The Rise of Cayley's Invariant Theory (1841–1862)," *Historia Mathematica* 13 (1986):241–254; and "The Decline of Cayley's Invariant Theory (1863–1895)," *Historia Mathematica* 15 (1988):332–347; and by Karen Hunger Parshall in "Toward a History of Nineteenth-Century Invariant Theory," *The History of Modern Mathematics*, ed. David E.Rowe and John McCleary, 2 vols. (Boston: Academic Press, Inc., 1989), 1:157–206. The latter paper also discusses the rival Continental methods in some detail.

[49] Boole, p. 3.

[50] Gauss's interest in invariant-theoretic phenomena stemmed from his work in number theory, while Boole came to them from the point of view of differential operators. Later, Cayley would focus, at least partially, on the subject's geometrical implications, while Sylvester's concerns would remain largely algebraic in nature. Boole did not use the term "invariant" in 1841.

[51] Boole, p. 19. Boole used the notation E (for "elimination") for the determinant instead of Δ. It is easily verified that $\frac{\gamma n}{m} \in \mathbb{Z}^+$.

written them up in a paper "On the Theory of Linear Transformations," which appeared in 1845.[52] In this, his first invariant-theoretic effort, Cayley succeeded in isolating a more general technique for generating invariants than Boole's partial differentiation method. Called the hyperdeterminant, Cayley's new construct actually produced invariants of multilinear forms, which, after suitable identification of the variables, reduced to invariants of binary forms. Moreover, this identification did not merely recapitulate the invariants which arose from Boole's method; it produced new and different expressions satisfying the invariantive property inherent in (2.1). This quite naturally suggested the following naive problem: for a given n-ary form, to find all of its invariants.

As Boole rather quickly realized, however, this problem was complicated by the fact that invariants may satisfy algebraic relations. The specific example Boole stumbled upon involved the binary quartic form $U = \alpha x^4 + 4\beta x^3 y + 6\gamma x^2 y^2 + 4\delta x y^3 + \epsilon y^4$. Cayley's hyperdeterminant produced the invariant $I = \alpha\epsilon - 4\beta\delta + 3\gamma^2$ of U, while the technique Boole had put forth in 1841 generated a very complicated invariant K. In studying Cayley's work, Boole hit upon yet another invariant-producing method late in 1844 which yielded the invariant $J = \alpha\gamma\epsilon - \alpha\delta^2 - \epsilon\beta^2 - \gamma^3 + 2\beta\gamma\delta$ of U. He then showed by a brute-force calculation that

$$K = I^3 - 27J^2. \tag{2.2}$$

Cayley generalized this observation, stating that "the two functions on which the linear transformations of the fourth order ultimately depend are the very simple ones [I] and [J], the function [K] of the sixth order being merely a derivative from these."[53] By 1846, Cayley had realized that even more was involved. Although the invariants I and J in (2.2) comprise a minimal generating set of invariants of the binary quartic form, the problem still remained to determine whether I and J are *independent*, that is, to decide if they satisfy no algebraic dependence relation. As Cayley forewarned, working up a theory of these dependence relations (or *syzygies* as Sylvester would later call them) for general binary forms would "present many great difficulties."[54] This and other problems would soon come to shape much of the research that Cayley and Sylvester would produce over the next four decades.

In the meantime, Cayley coupled his study of conveyancing at Lincoln's Inn with further exploration of the ramifications first of the hyperdeterminant defined in 1845 and then of the hyperdeterminant derivative devised

[52] Arthur Cayley, "On the Theory of Linear Transformations," *Cambridge Mathematical Journal* 4 (1845):193–209, or Arthur Cayley and A. R. Forsyth, ed., *The Collected Mathematical Papers of Arthur Cayley*, 14 vols. (Cambridge: University Press, 1889–1898), 1:80–94 (hereinafter cited as *Math. Papers AC*. All subsequent page references refer to this collection).

[53] *Ibid.*, p. 94.

[54] Arthur Cayley, "On Linear Transformations," *Cambridge and Dublin Mathematical Journal* 1 (1846):104–122, or *Math. Papers AC*, 1:95–112 on p. 95.

in 1846. Owing probably to the extreme calculational difficulty of both of these approaches, he shifted from his invariant-theoretic researches into other areas of mathematics (most notably geometry and elliptic function theory) in the years from 1846 to 1850. By then, however, he and his new friend, Sylvester, had finished up their law studies and had established an active mathematical working relationship. On an almost daily basis, they shared their thoughts—both in person and through correspondence—on geometry as well as on elimination theory and the general theory of forms. By 1851, Sylvester had entered the invariant-theoretic fray with still another new generating technique, dubbed compound permutation, which he had worked out within the more general context of his research on forms and which subsumed Cayley's hyperdeterminant as a special case.[55]

The next two years saw Sylvester, the actuary-by-day and mathematician-by-night, plunge headlong into the study of invariants. In his bipartite effort of 1852, entitled "On the Principle of the Calculus of Forms," he made great strides in systematizing the isolated techniques and results on invariants and in molding these into an actual invariant *theory*.[56] He followed this up in 1853 with another gargantuan labor, "On the Theory of Syzygetic Relations of Two Rational Integral Functions ... ," in which he began to come to terms with that thorny problem of algebraic dependence relations among invariants.[57] There, too, in addition to providing a valuable lexicon of the terminology which had been documented primarily on the pages of his private correspondence to Cayley, Sylvester gave the proof of his *law of inertia*: "[i]f a quadratic homogeneous function of any number of variables be (as it may be in an infinite variety of ways) transformed into a function of a new set of variables, linearly connected by real coefficients with the original set, in such a way that only positive and negative squares of the new variables appear in the transformed expression, the number of such positive and negative squares respectively will be constant for a given function whatever be the linear transformations employed."[58] By 1854, Sylvester's successes had

[55] James Joseph Sylvester, "Sketch of a Memoir on Elimination, Transformation, and Canonical Forms," *Cambridge and Dublin Mathematical Journal* 6 (1851):186–200, or *Math. Papers JJS*, 1:184–197.

[56] James Joseph Sylvester, "On the Principles of the Calculus of Forms," *Cambridge and Dublin Mathematical Journal* 7 (1852):52–97 and 179–217, or *Math. Papers JJS*, 1:284–363.

[57] James Joseph Sylvester, "On a Theory of the Syzygetic Relations of Two Rational Integral Functions, Comprising an Application to the Theory of Sturm's Functions, and That of the Greatest Algebraical Common Measure," *Philosophical Transactions of the Royal Society of London* 143 (1853):407–548, or *Math. Papers JJS*, 1:429–586.

[58] *Ibid.*, p. 511. Apparently, Jacobi knew the law of inertia as early as the 1840s and had made it known to others, although not in print. See Leo Königsberger, *Carl Gustav Jacob Jacobi* (Leipzig: B. G. Teubner, 1904), pp. 412–413. Jacobi gave a proof of the theorem in a paper entitled "Ueber einen algebraischen Fundamentalsatz und seine Anwendungen," and published posthumously by Carl Borchardt in the *Journal für die reine und angewandte Mathematik* 53 (1857):275–280. See, also, Carl Borchardt and Karl Weierstrass, ed., *C. G. J. Jacobi's Gesammelte Werke*, 7 vols. (Berlin: Georg Reimer, 1881–1891), 3:590–598.

seemingly lured his friend back into invariant theory, and Cayley published the first of the ten "Memoirs on Quantics" which would issue from his pen over the course of the next quarter-century. From this point on, both men labored to establish invariant theory on a firm footing, and their researches informed a parallel—but quite distinct—development on the Continent, and particularly in Germany.[59]

With his mathematical reputation established by the early 1850s, Sylvester began once more to consider the possibility of academic employment. In 1854, he put himself up for the vacant professorship in mathematics at the Royal Military Academy in Woolwich but lost out on the appointment to a politically connected—if mathematically far inferior—candidate. Not allowing this defeat to daunt him, Sylvester immediately applied for the open chair in geometry at London's Gresham College, but after giving "A Probationary Lecture on Geometry ..." on 4 December 1854, he failed to win that post, too.[60] Just when academe must have looked permanently closed to him, however, the successful Woolwich candidate died prematurely, creating an opening there once again early in 1855. On this second go-round and thanks perhaps to the intercession of his powerful friend Lord Brougham, the Academy finally named Sylvester to the position, and he assumed his duties there in the fall of 1855.[61]

Although once again gainfully employed in higher education, Sylvester did not particularly thrive during the fifteen years he spent at the Military Academy. First, he found the instructional side of his new job less than stimulating intellectually, confined as it was to relatively elementary teaching directed primarily at students of engineering. Second, he clashed on at least two occasions with the military authorities over his teaching load, and these confrontations left him emotionally spent and mathematically despondent.[62] On the positive side, however, Sylvester was able to partake fully of London's intellectual and social life from his vantage point just on the outskirts of the capital city. The Royal Society and his club, the Athenæum, also

---

[59]For more on the Continental developments, see Parshall, "Toward a History of Nineteenth-Century Invariant Theory," pp. 170–180.

[60]James Joseph Sylvester, "A Probationary Lecture on Geometry, Delivered Before the Gresham Committee and the Members of the Common Council of the City of London, 4 December 1854," *Math. Papers JJS*, 2:2–9.

[61]Correspondence suggesting Lord Brougham's involvement may be found in the Brougham Papers, Manuscripts and Rare Books Room, Bloomsbury Science Library, University College, London. As one of the original founders of University College London, Brougham had probably known of Sylvester since the time of his brief stint on the London faculty. Brougham was also an influential, and often vocal, member of the House of Lords.

[62]Correspondence dating from this period and documenting Sylvester's problems during these years may be found in the Brougham Papers and the London Mathematical Society Papers, both housed in the Manuscripts and Rare Books Room, Bloomsbury Science Library, University College London. Selections from this and other of Sylvester's correspondence will appear in Karen Hunger Parshall, ed., *Selected Correspondence of James Joseph Sylvester and His Circle*, Teubner-Archiv zur Mathematik (Stuttgart/Leipzig: B. G. Teubner, forthcoming).

provided him with ample opportunities to interact with other scientists and with diverse brands of Victorian intellectuals. Furthermore, until Cayley's assumption of the Sadlerian Chair at Cambridge in 1863, Sylvester continued to meet and talk often with his friend about mathematical matters of mutual interest. Both effectively outside of Oxbridge mathematics, Cayley and Sylvester shared and reinforced in one another a deep commitment to research which was largely absent from mathematics as it had evolved within the British academic establishment.

The first ten Woolwich years witnessed Sylvester's continued success in the theory of invariants, his new and promising investigations in the related area of partition theory, and his (the first rigorous) proof of Newton's rule for the discovery of imaginary roots of a polynomial equation,[63] while the last five found him groping for mathematical direction. This troubled period nevertheless saw Sylvester win recognition both nationally and internationally in such forms as a Royal Medal from the Royal Society in 1861, the position of Foreign Correspondent to the French *Académie des Sciences* in 1863, the second presidency of the newly formed London Mathematical Society in 1866, and the presidency of the Mathematics and Physics Section of the British Association at its Exeter meeting in 1869. These accolades aside, Sylvester's period of mathematical creativity seemed over by the late 1860s. Even worse, in 1870, he found himself unemployed and financially vulnerable, after changes in the regulations at Woolwich forced his premature retirement and denied him a full pension. At the age of fifty-five, Sylvester appeared to have reached the end of his academic career.

The next five years, from 1870 to 1875, found Sylvester stumbling to regain his footing. Poetry and a bid for the London School Board came to occupy time which, in better days, would have been devoted almost exclusively to mathematics.[64] One of his poems, entitled "Destiny," movingly captured his feelings of dejection and confusion.

> Why should I feed on contumely and wrong?
> To me more lofty destinies belong;
> Why should I lavish on one thankless face
> Thought that might bind the homage of my race?
> Out of ourselves, by Nature's hard decree,

[63] James Joseph Sylvester, "Algebraical Researches, Containing a Disquisition on Newton's Rule for the Discovery of Imaginary Roots, and an Allied Rule Applicable to a Particular Class of Equations, Together with a Complete Invariantive Determination of the Character of the Roots of the General Equation of the Fifth Degree, &c," *Philosophical Transactions of the Royal Society of London* 154 (1864):579–666, or *Math. Papers JJS*, 2:376–479.

[64] Sylvester read and published his flights of poetic fancy throughout his lifetime. He collected not only his thoughts on the writing and proper translation of poetry but also many of his pieces in his only book-length publication, James Joseph Sylvester, *The Laws of Verse: or, Principles of Versification Exemplified in Metrical Translations: Together with an Annotated Reprint of the Inaugural Presidential Address to the Mathematical and Physical Sections of the British Association at Exeter* (London: Longmans, Green, & Co., 1870).

Vain is our hope, our struggle to be free;
Our life is clay spun on the potter's wheel,
Our barren power is only—not to feel.
Why then accuse the inexorable past!
Shall I be angry at the simoon's blast,
Or that the marble-mocking human skill,
Shaped as a statue stands a statue still?[65]

As these lines suggest, Sylvester felt adrift, and this feeling manifested itself in the fact that he wrote only eight short articles during this half-decade. The fall of 1875, however, brought with it new hope and the opening up of quite unforeseen possibilities.

Word of the organization of the Johns Hopkins University in Baltimore traveled rapidly through American intellectual circles following Daniel Gilman's assumption of the presidency in mid-1875. For his University's proposed chair of mathematics, Gilman heard the name of just one candidate repeated in almost every quarter—Sylvester. Writing to Gilman on 25 August 1875 in reaction to correspondence recently received from Sylvester, Smithsonian Secretary Joseph Henry apparently first put the British mathematician's name before the Hopkins authorities. He wrote:

> I take a deep interest in the University over which you have been appointed to preside and I think it offers the best opportunity ever presented for elevating the character of the literature and the science of our country. It has ample means and can afford to employ the best men that can be found and establish chairs which shall be looked up to as the most desirable positions to be filled only by those who have achieved a big reputation: it is not dependent upon public favor and may be the means of forming and directing instead of following popular opinion.
>
> I take the liberty of expressing these sentiments to you since I have been informed that they coincide with your own, and to assure you that it will give me pleasure at all times to enforce and defend them. I am however induced at this time to write to you, by a letter received from my friend Prof. Sylvester of London who intimates a willingness to accept a chair in your University provided one were tendered him: he is one of the very first living mathematicians and his appointment would give a celebrity to the institution which would at once direct to it the attention of the whole scientific world.[66]

Although Henry certainly overstated the British mathematician's ranking in

[65]Taken from James Joseph Sylvester, *Fliegende Blätter*: *Supplement to the Laws of Verse* (privately printed in London by Grant & Co. in 1876), p. 25.

[66]Joseph Henry to Daniel C. Gilman, 25 August 1875, Gilman Papers.

mathematics internationally, he quite rightly pointed out that Sylvester was a "name" that would draw attention to the new university in mathematical as well as British scientific circles.

On his European tour and in England when he received Henry's letter, Gilman clearly had many things on his mind besides a mathematics chair, but on contacting the distinguished botanist and Director of Kew Gardens Joseph Hooker for advice on scientific candidates in Britain, the mathematics post once again came unexpectedly to the fore. Hooker echoed Henry's suggestion in his response to Gilman's inquiries on 11 September 1875, offering "that I can perhaps help you to a Professor of Mathematics of the very highest distinction & order & [a] practised teacher" and adding that "[m]y friend J. J. Sylvester FRS LLD (Correspondent of the Institute of France) who was Professor at Woolwich, is tired of inaction & would gladly accept a sufficiently paid Professorship in America."[67] One week later, yet another endorsement of Sylvester's candidacy for the Hopkins chair was sent on its way to Gilman, this one from the Harvard mathematician Benjamin Peirce.

The strength of Sylvester's allies together with his own research (if not teaching) record won Sylvester several interviews with Gilman during the latter's London sojourn. By all accounts, the mathematician impressed the President favorably during these meetings, but Gilman, ever-cautious, continued to harbor doubts about Sylvester's candidacy. After all, Gilman's enterprise was young, and Sylvester had already passed his sixty-first year. It was thus not at all clear whether Sylvester would remain active long enough to establish firmly the kind of research-oriented department Gilman sought. Furthermore, he had essentially retired from mathematical research in 1870 and had, in fact, not been all that active since the mid-1860s, yet Gilman wanted and needed active researchers to carry out his plans for the new university. There were certainly no guarantees that Sylvester could bounce back mathematically. Finally, the Englishman had never earned kudos for his teaching abilities, while he had deservedly won a certain notoriety for doggedness in defending what he deemed matters of principle. Gilman, however, saw his new venture as dependent on a faculty both able in the lecture room and willing to pursue common goals harmoniously. In his letter of 18 September 1875, Peirce had tried to assure Gilman, particularly about Sylvester's qualities as a teacher:

> If you inquire about him, you will hear his genius universally recognized but his power of teaching will probably be said to be quite deficient. ... [A]s the barn yard fowl cannot understand the flight of the eagle, so it is the eaglet only who will be nourished by his instruction. ... Among your pupils, sooner or later, there must be one, who has a genius for geometry. He will be Sylvester's special pupil—

[67]Joseph Hooker to Daniel C. Gilman, 11 September 1875, Gilman Papers.

> the one pupil who will derive from the master, knowledge and enthusiasm—and that one pupil will give more reputation to your institution than the ten thousand, who will complain of the obscurity of Sylvester, and for whom you will provide another class of teachers.[68]

As mentioned above, Gilman was not actively seeking particular specialists, but rather individuals whom he felt would fit in with his overall plans for the University. So, although he apparently had no other mathematical candidates in the wings, he felt no need to make a hasty judgment in Sylvester's or any other case. If he ultimately concluded that Sylvester was not right for the position, he would continue his search until he found an appropriate choice, even if that meant opening the university with no one in a chair of mathematics.

For his part, Sylvester, still stinging from his premature retirement from Woolwich, demanded absolute guarantees of job and financial security which Gilman could not deliver. Thus, even though the President did finally satisfy himself of Sylvester's fitness for the post, and even though he had extended an offer to Sylvester in the name of the University at the end of November 1875, the mathematician still had reservations. By 1875, he had almost completely resigned himself to his situation and was living reasonably comfortably on his carefully invested savings. Furthermore, although he found the idea of participating in Gilman's new educational undertaking appealing and even flattering, he was well aware of his reputation within the international scientific community and saw no reason to sell himself short in order to make a transatlantic move. He responded in kind to Gilman's overtures with their accompanying set of conditions from the Board:

> I now beg to say that I will accept the offer contained in Mr. Reverdy Johnson's letter if the $5000 salary therein named be understood to mean gold and if a house and the club fees be attached to my appointment.
>
> Highly as I should appreciate my connexion with your University and much as I should rejoice in filling so congenial a sphere of mental activity as it promises to afford, I should not according to the best of my judgement confirmed by the corroborating [?] opinion of my friends here feel justified in accepting less favorable terms.[69]

The Trustees balked at these extravagant demands, particularly at the condition of the full salary payable in gold, and countered by telegram with a

---

[68] Benjamin Peirce to Daniel C. Gilman, 18 September 1875, Gilman Papers.

[69] J. J. Sylvester to Daniel C. Gilman, 17 December 1875, Gilman Papers. The salary conditions were a direct consequence of what he viewed as underhanded treatment earlier by the Woolwich authorities. Since Sylvester felt that he had been cheated out of his rightfully earned pension fund at Woolwich, he wanted firm salary arrangements from the Hopkins officials.

lesser, compromise offer. Sylvester answered (also by telegram) with new conditions of his own on 10 January 1876: "House from 1877 or Equivalent[;] with gold[;] without fees[;] ultimatum answer."[70] By the twenty-sixth, though, he had broken off negotiations, cabling that the Johns Hopkins represented an "[u]ntried institution[;] uncertain tenure[;] favorable home prospects[;] stipend Crowning Career inadequate against risk insured[;] regret[;] thanks[;] decline."[71] Without the intervention of Joseph Hooker late in January, this would have represented the final word. The two sides finally reached an agreement—$5000 in gold annually plus a yearly house allowance of $1000 also in gold—after another round of letters and some hastily sent telegrams. Sylvester accepted the Hopkins offer on 17 February 1876 and was officially appointed by the Board on 6 March.[72]

The Department of Mathematics at the Johns Hopkins University

If tensions had often run high during the negotiations early in 1876, neither side harbored any animosity after their successful resolution in February. Quite the contrary, both Gilman and Sylvester seemed elated at the outcome: Gilman had secured a noted mathematician for his faculty, and Sylvester had won acceptance at an institution fully appreciative of him as a mathematician. Almost immediately each man embraced the other as a fellow in common cause, with Gilman assuring Sylvester "of my high regard & of my hope that a new career of honor & usefulness is opening before you" and with Sylvester anticipating "a new course of usefulness in connexion with your and my University to which I already begin to feel the attachment of a favored son."[73] They began in earnest to organize their new Department of Mathematics after Sylvester's arrival in the United States early in May.

A teaching Associate had to be engaged, the students—both graduate and undergraduate—had to be secured, fellowships had to be meted out, and the course of study had to be settled upon, all before the official opening on 3 October. Although wayward baggage and then the severe heat of the

[70] J. J. Sylvester to Daniel C. Gilman, telegram, 10 January 1876, Gilman Papers.

[71] J. J. Sylvester to Daniel C. Gilman, telegram, 26 January 1876, Gilman Papers.

[72] J. J. Sylvester to Daniel C. Gilman, telegram, 17 February 1876, Gilman Papers. See, also, Daniel C. Gilman to Joseph Hooker, telegram, 15 February 1876, Letter Book 1, Gilman Papers. Gilman tried to give Sylvester an idea of just how generous a $5000 per year salary was for an academic in late nineteenth-century America. In the letter of 29 November 1875 in which he extended the University's offer to Sylvester, Gilman noted that "[i]f I am rightly informed, at Yale College in New Haven the highest professional salaries are $3500,—& at Harvard College in Cambridge $4000,—& there are very few colleges which pay even these figures." He also offered that "the Judge of the highest courts are paid $3500,—the ministers of the largest churches from $3500 to $4000,—cashiers of banks from $3000 to $4000." See Daniel C. Gilman to J. J. Sylvester, 29 November 1875, Gilman Papers.

[73] Daniel C. Gilman to J. J. Sylvester, 29 February 1876, Letter Book 1, Gilman Papers, and J. J. Sylvester to Daniel C. Gilman, 23 March 1876, Gilman Papers.

East Coast summer prevented Sylvester from actually taking up residence in Baltimore, he and Gilman met in Philadelphia and exchanged much correspondence relating to plans for the department. Sylvester retreated northward in his attempts to flee the oppressive weather and enjoyed an extended visit in Cambridge at the home of his long-time friend Benjamin Peirce. While there, he quite naturally cast about for the names of candidates capable of serving as his teaching Associate, and his inquiries turned up two possibilities: William E. Byerly and William E. Story. Both Harvard A.B.'s under the sway of Benjamin Peirce, Byerly had stayed on at his *alma mater* for the Ph.D. in 1873 while Story had gone on to Germany to take a doctorate at Leipzig in 1875.[74] In the spring of 1876, as Hopkins officials went busily about the task of staffing their university, Byerly held an assistant professorship at Cornell, and Story served as one of the Harvard Tutors. Both men seemed eminently qualified and potentially movable, so Sylvester probed a bit more deeply to get senses of the two as mathematicians. Writing to Gilman in mid-June, he reported that "I have ... conversed with Prof[esso]r B. Peirce and his son Prof[esso]r James Peirce concerning Mr. Byerly. Their report of him as a well-read mathematician and good teacher is very favorable—but they think slightly more highly of Mr. Story."[75] As for himself, Sylvester tended to agree on the basis of "[a] little work of his on the Higher Algebra which has been shown to me [and which] is enough to satisfy me that he is not undeserving of their commendations as a mathematician of decided power and originality."[76] Given that Story's thesis research at Leipzig as well as some of his early post-doctoral work centered on Sylvester's own specialty, the algebra of quantics (or forms), it must have appealed to the transplanted Englishman. How better to lessen the shock of the transatlantic move than to have a mathematically kindred spirit close at hand?

The opinions of Sylvester and the Harvard mathematicians did not suffice to clinch Story's appointment, however. The President insisted on personally interviewing the candidate and making his own judgment before putting him before the Board of Trustees. In fact, after Sylvester's sudden departure for England at the end of July (the unusually hot summer temperatures had become unbearable for him), all of the remaining negotiations with Story devolved to Gilman. By early August, the matter had come to a successful conclusion, and the Mathematics Department had its full staff of two.

Prior to and concurrent with the search for the appropriate Associate, Gilman fielded applications for the ten university-wide graduate fellowships of $500 annually, which he had announced and advertised early in 1876. So

---

[74] While it is not clear who served as Story's adviser, Roger Cooke and V. Frederick Rickey conjectured that it was Carl Neumann in their article, "W. E. Story of Hopkins and Clark," in *A Century of Mathematics in America—Part* III, ed. Peter Duren *et al.* (Providence: American Mathematical Society, 1989), pp. 29–76 on p. 33.

[75] J. J. Sylvester to Daniel C. Gilman, 12 June 1876, Gilman Papers.

[76] *Ibid.*

overwhelming was the response that at the time of the June deadline for submission, the University had received over 150 serious applications, and over 100 of the applicants actually met the eligibility requirements.[77] In order to capitalize on this situation, the Board authorized an additional ten fellowships, and Gilman, in concert with his faculty, began the screening process. They looked for students already focused within their respective areas of specialization and capable of supplementing the University's regular course offerings at the undergraduate—and even at the gradaute—levels. Furthermore, they sought students likely to meet the expectation of annual evidence of continued growth and development within their fields. As with his faculty, Gilman wanted Fellows committed both to research and to the ideals of the Johns Hopkins University. In mathematics, once again, two clear choices surfaced fairly early on.

Thomas Craig, a twenty-year-old from the mining region of Pennsylvania, caught Gilman's eye first with his letter to the President of 23 March.[78] An 1875 graduate in civil engineering from his home state's Lafayette College, Craig had continued to pursue his mathematical studies diligently while engaged in high school teaching in Newton, New Jersey. Guided in his work by Benjamin Peirce and the University of Edinburgh's Peter Guthrie Tait, men to whom he had written for advice and counsel, Craig had seemingly adopted the research ethic from the start and sought to prepare himself as best he could for the professorial calling. Thus, he jumped at the chance to become part of the new university in Baltimore, especially when he learned of its proposed fellowships of then unprecedented generosity. For Craig, a young man of modest means who nevertheless aspired to the scholarly life, these fellowships afforded the opportunity to engage in graduate study full-time while still maintaining a comfortable—if humble—existence. Accordingly, he put forth his qualifications, his current course of study, and his plans for the future in a letter to Gilman dated 23 March 1876. The young man's conviction and sincerity so impressed the President that he arranged to meet with him within the week and sent him down to Washington to talk with his principal scientific adviser, the Naval Observatory's professor, Simon Newcomb. Newcomb only confirmed Gilman's own impression when he reported back on 3 April that "I am glad you sent Mr. Craig .... If he digests all he reads he stands a good chance to prove a first rate man, and if he is really a 'sample,' you will start your university full-fledged indeed."[79] That same

[77]For more on the fellowship scheme, see Hawkins, *Pioneer*, pp. 79–90. The figures given here may be found on pp. 80–81. A set number of fellowships per graduate department was not allocated in advance.

[78]*Ibid.*, pp. 85–86. For more on Thomas Craig's life, see F. P. Matz, "Professor Thomas Craig, C.E., Ph.D.," *The American Mathematical Monthly* 8 (1901):183–187; and Clark A. Elliott, *Biographical Dictionary of American Science*: *The Seventeenth through the Nineteenth Centuries* (Westport: Greenwood Press, 1979).

[79]Simon Newcomb to Daniel C. Gilman, 3 April 1876, Gilman Papers.

day, Gilman sent Craig a preliminary program of courses with the tiding that "[w]e shall expect you here when Professor Sylvester arrives, not later than Oct. 1."[80]

With a Mathematics Fellow chosen long before the June closing date for receipt of applications and with Gilman still operating under the assumption that the budget limited the University as a whole to only ten graduate fellowships, subsequent inquiries in mathematics necessarily received more gingerly treatment. When the Princeton A.B., George Bruce Halsted, heard of the Hopkins fellowship scheme in mid-April, he, too, fired off a letter to Gilman, confessing "such overmastering anxiety to be a partaker in your rich feast of learning that I cannot wait a single day, but would this very instant lay before you my humble petition."[81] The young Halsted proceeded to detail his own educational accomplishments in glowing terms, describing his class rankings, his success on an intercollegiate mathematics contest, and his varied reading in the mathematical literature. When Simon Newcomb contacted Gilman to tell him of his own letter from Halsted, Gilman betrayed his impression of the Princetonian's over-enthusiasm in his response: "Halsted wrote so to me, 'and more so.' I told him in reply that I could not promise a fellowship,—but I thought it clear he was one of the young men we are in search of, & I begged him to plan to come here, some-how, next year."[82] Gilman had not allowed the young man's egotism to blind him to the candidate's qualifications and promise.

By the end of May, when the Board seemed sure to expand the number of fellowships, Gilman put the applicants before Sylvester for his judgment and left open the possibility of two and maybe even three graduate fellowships for mathematics. Sylvester came down with a categorical verdict:

> Messers Craig and Halsted appear to me from the evidence they submit to be particularly well fitted to receive each of them an appointment as Fellow.
>
> If a third Fellowship is to be given away Mr. Gore's claims might be considered. He seems to have a fair working acquaintance with the ordinary subjects and analytical principles of Mathematical study—but his claims do not

[80]Daniel C. Gilman to Thomas Craig, 3 April 1876, Gilman Papers. In fact, Gilman took such a personal interest in Craig that he lent him money on several occasions over the spring and summer of 1876 in order to help him make ends meet until the start of school in the fall. See, for example, Daniel C. Gilman to Thomas Craig, 13 June 1876, Gilman Papers.

[81]George B. Halsted to Daniel C. Gilman, 10 April 1876, Gilman Papers. For more on Halsted's life and subsequent career, see Albert C. Lewis's articles, "George Bruce Halsted and the Development of American Mathematics," in *Men and Institutions in American Mathematics*, ed. J. Dalton Tarwater, John T. White, and John D. Miller, Graduate Studies, Texas Tech University, No. 13 (Lubbock: Texas Tech Press, 1976), pp. 123–129; and "The Building of the University of Texas Mathematics Faculty, 1883–1938," in *A Century of American Mathematics-Part* III, ed. Peter Duren *et al.* (Providence: American Mathematical Society, 1989), pp. 205–239.

[82]Daniel C. Gilman to Simon Newcomb, 12 April 1876, Gilman Papers.

> appear to me even approximately comparable with those of Messers Craig and Halsted.
>
> As regards the other Candidates I consider that there is not any evidence of any other of them possessing the necessary qualifications.[83]

The money came through, and Sylvester's call stood: the first Hopkins Fellows in mathematics were Craig, Halsted, and Gore.

Fifteen students—seven undergraduates and eight graduates—attended the Mathematics Department's courses during the University's inaugural year of operation. For the undergraduates, Story taught conic sections and the theory of equations while Craig lectured on differential and integral calculus. At the graduate level, Sylvester discoursed on determinants and modern algebra in addition to running the so-called "Mathematical Seminary"; Story offered the mathematical theory of elasticity; and, as a testament to the level of his prior preparation, Craig treated elementary mechanics and the theory of definite integrals.[84] Necessarily modest in scope initially, the course of study expanded during Sylvester's tenure at Hopkins in response to the demands of a mathematical student body which increased from the fifteen of the opening academic year to thirty-five in 1883–1884.[85] By the time of Sylvester's departure for the Savilian Professorship of Mathematics associated to Oxford's New College in December of 1883, the mathematics staff had doubled in size to meet the growing needs with the additions of two of its own Ph.D.'s: Thomas Craig (Ph.D. 1879) came on board as an Instructor in 1879, and Fabian Franklin (Ph.D. 1880) began as an Assistant in 1879. (Both men stayed on and eventually moved through the ranks to attain full professorships at their *alma mater*.) This increase in faculty allowed for a broadening of the program to include courses on higher plane curves, solid geometry, and differential equations for the undergraduates as well as topical courses on number theory, quaternions, the theory of algebras, higher plane curves, and invariant theory for the graduate students. Arthur Cayley further supplemented the University's mathematical offerings during the Spring Semester of 1882, when he came over from England as one of

---

[83]J. J. Sylvester to Daniel C. Gilman, 29 May 1876, Gilman Papers. Of the three men, Craig and Halsted earned doctoral degrees. Joshua W. Gore, who had taken a degree in civil engineering at the University of Virginia in 1875, held a Hopkins fellowship from 1876 to 1878, but left before earning a Ph.D. He subsequently became Professor of Natural Sciences at Southwestern Baptist University in Jackson, Tennessee. See *The Johns Hopkins University Circulars* 1 (3) (1880):23.

[84]William E. Story to Daniel C. Gilman, "Report on the Department of Mathematics 1876 to 1884," dated 17 December 1883 (and its undated continuation), Gilman Papers. See, also, *The Johns Hopkins University Circulars* for those years; and Cajori, pp. 270–271. With no teaching experience, Halsted and Gore were not initially called upon as instructors. Because of his independent preparation, Craig, however, was entrusted with a graduate course while only a graduate student himself. Somewhat later, Frank Nelson Cole would also teach graduate courses as a graduate student, but at Harvard. See Chapter 5 below.

[85]Story to Gilman, "Report on the Department of Mathematics 1876 to 1884."

Gilman's Visiting Professors to give a course on "Algebraical Geometry and Abelian and Theta Functions." On the whole, the mathematics students at the Johns Hopkins during these early years sampled from a course of study which far surpassed even that put together by Benjamin Peirce in the late 1840s. (Compare Tables 1.1 and 2.1.)

Sylvester generated considerable interest through his courses, which, more often than not, reflected his immediate research interests. Not surprisingly, then, he tended to focus on topics of a purely algebraic nature, although he did offer classes in number theory for five consecutive semesters. The brunt of the instructional burden thus fell to faculty member, Story, and graduate student, Craig, with Story providing the teaching in geometry, Craig working up various courses in analysis, and the two men dividing the more applied offerings between them.[86] By leaving Sylvester free to teach precisely that subject matter which most excited him, the Johns Hopkins University provided him with his first real opportunity to test his mettle as a teacher. By all accounts, Sylvester met the challenge, although he did so in a characteristically idiosyncratic way.

Sylvester was the antithesis of an organized lecturer. When a mathematical thought popped into his mind, it would take control of him. He would forget what he had been doing or saying, and, like a man possessed, he would pursue the thought until diverted by some new idea. This peculiarity wreaked havoc with his classroom presentations. No matter how firm his resolve to follow a text during a given course, when an idea came over him, the text, and sometimes even the whole course, would be forgotten. William Pitt Durfee, a Hopkins Fellow from 1881 to 1883 and later Professor of Mathematics at Hobart College in Geneva, New York, vividly captured the mathematical drama of a typical semester:

> Sylvester began to lecture on the Theory of Numbers, and promised to follow Lejeune Dirichlet's book; he did so for, perhaps, six or eight lectures, when some discussion which came up led him off, and he interpolated lectures on the subject of frequency [of prime numbers], and after some weeks interpolated something else in the midst of these. After some further interpolations he was led to the consideration of his Universal Algebra, and never finished any of the previous subjects. This finished the first year [1881–1882], and, although we never received a systematic course of lectures on any subject, we had been led to take a living interest in several subjects, and, to my mind, were greatly gainers thereby.[87]

[86]First Craig and later other gradaute students like Fabian Franklin served as much more than teaching assistants in the modern sense of the term. They frequently taught advanced courses in their developing areas of expertise.

[87]William P. Durfee to Florian Cajori, as reported in Cajori, p. 267. Undoubtedly, the

Indeed, Sylvester's mathematical verve also rendered him virtually incapable of reading and absorbing the published work of others. Mathematically egocentric, he had little interest in the known results of others unless they were of immediate import for his own work. He thus tended neither to draw inspiration from the current literature nor to encourage his students in this direction. For him, mathematics was much too spontaneous for such calculated, bibliographical study. (Contrast the discussion of Felix Klein in Chapter 4.) Durfee seemed to recognize precisely this aspect of—if not flaw in—his teacher's personality when he concluded that Sylvester simply

> could not lecture on a subject which was not at the same time engaging his attention. His lectures were generally the result of his thought for the preceding day or two, and often were suggested by ideas that came over him while talking. The one great advantage that this method had for his students was that everything was fresh, and we saw, as it were, the very genesis of his ideas. One could not help being inspired by such teaching, and many of us were led to investigate on lines which he touched upon.[88]

Many of Sylvester's other Hopkins students echoed Durfee's remarks.[89] The Englishman was completely incapable of satisfying the student who wished to take away a notebook containing a crystallization of some mathematical topic. He viewed mathematics as a live and growing subject, not a static and fossilized one, and he wanted his students to contribute to its growth.

Sylvester's mathematical teaching and working style perfectly reflected the broader philosophy of mathematics which he had articulated in his Presidential Address at the Exeter meeting of the British Association for the Advancement of Science in 1869. He had used that occasion to counter publicly remarks the biologist Thomas Huxley had made earlier in the year on the nature of mathematical research.[90] In two separate published articles, Huxley had characterized mathematics first as "almost purely deductive"[91] and then as "that study which knows nothing of observation, nothing of experiment,

---

book referred to here was one of Dedekind's editions of Peter Lejeune-Dirichlet's classic text in number theory. See, for example, Peter Lejeune-Dirichlet, *Vorlesungen über Zahlentheorie*, ed., Richard Dedekind, 1st ed. (Braunschweig: F. Vieweg und Sohn, 1863).

[88]Cajori, pp. 267–268.

[89]Cajori also included reminiscences of student days under Sylvester by George Bruce Halsted, Ellery Davis, and Arthur Hathaway on pp. 265–267.

[90]For Sylvester's ideas, see James Joseph Sylvester, "Presidential Address to Section 'A' of the British Association," *Math. Papers JJS*, 2:650–661; and the two appendices "Additional Notes to Prof. Sylvester's Exeter British Association Address," and "On the Incorrect Description of Kant's Doctrine of Space and Time Common in English Writers," *Math. Papers JJS*, 2:714–719 and 719–731, respectively.

[91]Thomas Huxley, "Scientific Education: Notes of an After-Dinner Speech," *Macmillan's Magazine* 20 (1869):177–184 on p. 182.

nothing of induction, nothing of causation!"[92] Sylvester took strong exception to this view in his address and argued that while mathematical results may be published in a deductive form, they are more often than not arrived at inductively through the observation and codification of specific examples. In a characteristic burst of literary hyperbole, Sylvester informed his audience that

> [w]ere it not unbecoming to dilate on one's personal experience, I could tell a story of almost romantic interest about my own latest researches in a field where Geometry, Algebra, and the Theory of Numbers melt in a surprising manner into one another, like sunset tints or the colours of the dying dolphin, ... which would very strikingly illustrate how much observation, divination, induction, experimental trial, and verification, causation, too (if that means, as I suppose it must, mounting from phenomena to their reasons or causes of being), have to do with the work of the mathematician.[93]

For Sylvester, it was the spirit of the inductive hunt and not the more routine composition of the deductive proof which brought the mathematical endeavor alive. Later, he worked to instill precisely this sense of mathematical vitality in his students at Hopkins.

In his lecture room, Sylvester presented numerous examples and worked in concert with his class to draw meaningful conclusions from them. He posed strings of open questions, some of which he pursued on the spot and some of which he left for his students to answer. Through this open-endedness, he continually encouraged his auditors to *do* mathematics actively, not merely to study it passively. He challenged his classes to keep up with him as he proved new theorems, and he urged them to follow his example and to make new discoveries of their own. Consonant with the University's broader institutional philosophy, new and original research came first and foremost in the Department of Mathematics under Sylvester, and as Durfee noted, it was undertaken in a true spirit of community.[94]

---

[92]Thomas Huxley, "The Scientific Aspects of Positivism," *Fortnightly Review* 11 (1869):653–670 on p. 667. Exclamation point in the original. Sylvester's reaction to Huxley's remarks along with a discussion of the mathematician's philosophy of mathematics may be found in Karen Hunger Parshall, "Chemistry through Invariant Theory? James Joseph Sylvester's Mathematization of the Atomic Theory," to appear.

[93]Sylvester, "Presidential Address to Section 'A' of the British Association," *Math. Papers JJS*, 2:656–657. The paper Sylvester referred to here was "Outline Trace of the Theory of Reducible Cyclodes, That Is a Particular Family of Successive Involutes to a Circle Whose Determination Depends on the Solution of an Algebraico-Diophantine Equation, and of the Number and Classification of the Forms of Such a Family for Any Given Order of Succession," *Proceedings of the London Mathematical Society* 2 (1869):137–160, or *Math. Papers JJS*, 2:663–688.

[94]See Karen Hunger Parshall, "America's First School of Mathematical Research: James Joseph Sylvester at the Johns Hopkins University 1876–1883," *Archive for History of Exact*

THE OLD CAMPUS OF THE JOHNS HOPKINS UNIVERSITY CA. 1895. FROM LEFT TO RIGHT, THE ADMINISTRATION BUILDING, HOPKINS HALL, THE CHEMICAL LABORATORY, AND THE BIOLOGICAL LABORATORY.

During the seven-and-a-half years Sylvester spent in Baltimore, some two dozen students entered into this unique mathematical sodality either as Fellows or as independently supported graduate students, and eight of them actually earned their doctoral degrees under Sylvester's guidance. (See Table 2.2.) One student stood out in this group both for her talent and for her gender—Christine Ladd. Her presence at the Johns Hopkins at once symbolized the opening to women of research-level mathematics in America and highlighted the difficulties women encountered in entering this or any other academic field throughout the last quarter of the nineteenth century.[95]

Sciences 38 (1988):153–196 for a detailed description of this community at work on the theory of partitions.

[95]These difficulties persisted well into the twentieth century. For an historical account of women in American science, see Margaret W. Rossiter, *Women Scientists in America: Struggles and Strategies to* 1940 (Baltimore: The Johns Hopkins University Press, 1982). On women in the American mathematical research community at the turn of the twentieth century, consult Della Dumbaugh Fenster and Karen Hunger Parshall, "Women in the American Mathematical Research Community: 1891–1906," in *The History of Modern Mathematics*, vol. 3, ed. Eberhard Knobloch and David E. Rowe (Boston: Academic Press, Inc., 1994), forthcoming. Judy Green

Ladd's connection with the Johns Hopkins University began rather unmomentously on 27 March 1878 when she wrote directly to Sylvester expressing her "desire to listen next year to such of your mathematical lectures as I may be able to comprehend. Will you kindly tell me whether the Johns Hopkins University will refuse to permit it on account of my sex?"[96] An 1869 graduate of Vassar College, Ladd had already sat in on several courses at Harvard as an unmatriculated "special student" and realized that her chances of formally pursuing an advanced degree there were nil. Like Craig and Halsted, however, she saw in the opening of the new university in Baltimore a chance not only to enter into research-level mathematics but also to study under a mathematician of established reputation, a rare commodity in the United States of the late 1870s. She thus hazarded an inquiry, although well aware of the University's official posture as a "males only" institution.

Sylvester promptly took this matter up with Gilman, explaining with complete candor that he had

> written to Miss Ladd saying that I did not personally anticipate that her sex would be an objection when attending lectures at our University and I should rejoice to have her as a fellow worker among us—but that ... I would bring the matter officially before the Authorities of the University and acquaint her with the result. ... My own impression is that her presence among us would be a source of additional strength to the University. I regard her as more than another Miss [sic] Somerville in prospect and I cannot but think that with your fertility of resource you would hit upon some plan of utilizing her for the purposes of the University.[97]

Judging by this note, the mathematician responded to the potential student fully cognizant of his University's position on women students and hopeful that the right candidate might effect a positive change in its essentially exclusionary policy. Although Sylvester may not have intended, through his handling of the Ladd inquiry, to force Gilman's hand relative to the admission of women, his letter of 2 April sent the President scurrying to the Board of Trustees.

The issue of coeducation had haunted the University from the very beginning. The fact that Cornell and some of the state universities already admitted women seemingly forced the Johns Hopkins to take a stand as well.

---

gives the biographical details of Christine Ladd (later Franklin) in her article "Christine Ladd-Franklin (1847–1930)," in *Women of Mathematics: A Biobibliographic Sourcebook*, ed. Louise S. Grinstein and Paul J. Campbell (New York: Greenwood Press, 1987), pp. 121–128.

[96]Christine Ladd to J. J. Sylvester, 27 March 1878, Gilman Papers.

[97]J. J. Sylvester to Daniel C. Gilman, 2 April 1878, Gilman Papers. The reference here is to Mrs. Mary Somerville (1780–1862), a noted British scientific expositor whose scientific attainments were highly regarded by her contemporaries.

In his inaugural address, Gilman had tried to avoid the problem by supporting the possibility of an associated women's college at some later point in time, but as early as November of 1876, Board member James Carey Thomas (the father of then Cornell senior and future Bryn Mawr President M. Carey Thomas) had formally raised the question of coeducation and had vigorously supported its adoption by the Board.[98] No decision came from his initiative, but by November of the following year a resolution did pass which allowed women, in exceptional cases, to attend courses at Hopkins as special students. This resolution provided Gilman and the Board with a precedent for considering Ladd's case and reopened the coeducation issue.

Indeed, Ladd did represent the sort of exceptional case presaged in the 1877 statement. While pursuing a career as a secondary school teacher during most of the nine years after her graduation from Vassar, she had published a surprising amount of mathematical work in *The Analyst*, a recreationally oriented American journal for pure and applied mathematics begun by Joel E. Hendrick in Des Moines, Iowa in 1874 (see Table 1.2), and in the British *Educational Times.*[99] Her contributions to both of these periodicals, although they merely involved posing or solving problems, showed marked originality which favorably impressed Sylvester.[100] After weighing the evidence put forth by the Mathematics Department and after three weeks of debate and discussion over the larger matters at stake, the Trustees authorized Gilman to convey their decision to Ladd: they were unwilling to open their university up to coeducation, but they would allow Ladd to attend Sylvester's classes with a waiver of all usual tuition charges.[101] In essence, the Board succeeded in maintaining its official stance on coeducation, while granting Ladd the exceptional status of "invisible student."[102]

Ladd apparently distinguished herself in Sylvester's courses, for she won the Department's strong backing after her arrival in Baltimore in the fall of 1878. Not only did Sylvester support her further request to attend the courses of both Story and the logician Charles S. Peirce, but he also recom-

---

[98]Hawkins, *Pioneer*, pp. 259–265.

[99]Sylvester had Fabian Franklin do a literature search on Ladd's work, and this was subsequently included in the documentation reviewed by the Trustees. In addition to numerous problem solutions in *The Educational Times*, Franklin found four contributions in volumes four and five of *The Analyst* in 1877 and 1878. Fabian Franklin, "Articles in '*The Analyst*,' by Miss Christine Ladd," dated 1 April 1878 in Gilman Papers.

[100]In an undated letter to Gilman, Sylvester confidently posited that Ladd "was a Mathematician of high attainments and splendid promise. She is favorably known to Dr. Salmon and Prof[esso]r Cayley with each of whom she has corresponded on Mathematical subjects correcting an error or imperfection in a treatise of the one and suggesting an improvement in a memoir or work of the other. She is well known in America and England and in a recent number of the Analyst ... she has written a profound paper on Quaternions." See Gilman Papers. George Salmon taught mathematics at Trinity College, Dublin. He was a noted contemporary and friend of Sylvester and Cayley.

[101]Daniel C. Gilman to Christine Ladd, 26 April 1878, Gilman Papers.

[102]Compare the discussion in Rossiter, pp. 29–33.

mended Ladd for one of the coveted Hopkins fellowships for the 1879–1880 academic year. Gilman and the Trustees went along with the Department's recommendations on both counts, but the fellowship question produced deep divisions. Still officially "invisible" relative to the student roster, Ladd received the stipend, but not the title, associated with the fellowship. As a result of this compromise, Trustee Reverdy Johnson, who had studied in Germany and who wished to follow the German example of his day of excluding women from the university, stepped down from the chair of the Board's Executive Committee and withdrew from the committee altogether the following year.[103]

In spite of the political turmoil occasioned by her presence at Hopkins, Ladd participated as an equal in the intellectual life of the Mathematics Department. Together, she and her fellow graduate students pursued their collective goals of first learning and then creating mathematics. Obviously, the course offerings were designed to help them realize the first of these objectives and, in some cases, even the second. The actual generation of new mathematical ideas, however, clearly represented the much more elusive of the two goals. Since it was not, by and large, a classroom activity, it required a separate and special venue. At Hopkins, as at the German universities from which it was adapted, that place was the Mathematical Seminary.

A book- and journal-lined room, the Mathematical Seminary served as a primary locus for mathematical activities outside the classroom. Students met there casually to read, to study, and to discuss their research, while the full mathematical contingent convened there formally once a month throughout the academic year for the development, presentation, and critical review of new mathematical research. More often than not these monthly meetings—also called the "Mathematical Seminary"—served as an extension of Sylvester's regularly scheduled classes, with the students presenting the fruits of their own researches inspired by the myriad, impromptu questions posed by their professor during the course of his lectures. In effect, the Seminary functioned like a sort of laboratory for the creation of new mathematics. In keeping with his inductive philosophy of mathematics, Sylvester knew, or thought he knew, how a given mathematical "experiment" should go. Like a laboratory director, he guided his student assistants through to its successful completion. They, in turn, presented their "laboratory reports" before the regularly assembled Mathematical Seminary. These then underwent the scrutiny of the assembled participants prior to their publication in a revised and more polished form often in the pages of *The Johns Hopkins University Circulars*.[104] This highly interactive process served to train the developing

[103]Hawkins, *Pioneer*, p. 264. By this time, though, some German universities, like Heidelberg, had fairly liberal policies relative to the admission of *foreign* women. See Chapter 5 on the changing atmosphere at the German universities relative to the admission of women.

[104]This in-house periodical came out several times each semester. In addition to functioning as a university calendar of courses and events, it also served to publicize and encourage original

THE MATHEMATICAL SEMINARY AT THE JOHNS HOPKINS UNIVERSITY CA.1895

mathematicians not only as lecturers but also as expositors.

Story underscored the importance of the Seminary to the department's educational mission and summarized its productivity in 1883 when he reported to Gilman that "[t]he Mathematical Seminary has held monthly meetings since October 1878, and the graduate students, more particularly the fellows, have taken an active part in its proceedings, the papers read by them generally indicating a high order of ability and mastery of the subject under discussion. Since November 1879 abstracts of 35 such communications have been published in the University Circulars, viz. 11 in 1879–80, 7 in 1880–81, 6 in 1881–82, and 11 in 1882–83."[105] More than generating departmental statistics for internal consumption, however, the Seminary sparked a rare *esprit de corps* which Durfee vividly captured in recounting "the work we did while I was at the Johns Hopkins University. I say we, as I always think of the whole staff as working together, so thoroughly did Sylvester inspire us all."[106] (See Chapter 3 for details on the joint work undertaken by the Hopkins school.)

And this inspiration reached beyond the confines of the Seminary Room. Significantly, much of the work presented in the *Circulars* merited and received fuller treatment in refereed research journals both at home and abroad. Quite naturally, the *American Journal of Mathematics* brought most of this research to light.

---

research activities on campus through publication of research announcements, abstracts, and short papers.

[105]William E. Story to Daniel C. Gilman, "Report on the Department of Mathematics," as in note 84 above.

[106]Durfee to Cajori, as reported in Cajori, p. 267.

THE *American Journal of Mathematics*

First issued in 1878 under the auspices of the Johns Hopkins University, the *American Journal of Mathematics* marked the partial realization of an idea which had occurred to Gilman as early as 1875 during his European tour—university sponsorship of scholarly publications.[107] This bold and unprecedented concept fit perfectly with Gilman's overarching research goals for the participants in his university. If research and publication were to be institutionally mandated activities, Gilman reasoned, then the institution needed to provide suitable outlets for this work. Moreover, by establishing research-level journals, the Johns Hopkins University could prove influential in setting, monitoring, and controlling American standards of publication in the various scholarly areas. In the case of mathematics, for example, the last American journal to emphasize the publication of research—as opposed to mere recreations—had been Benjamin Peirce's *The Cambridge Miscellany of Mathematics, Physics, and Astronomy*, and that had come in and gone out of existence in 1842. Unlike their European (and particularly German) colleagues, the mathematicians at Hopkins had neither the proceedings of local scientific societies nor nationally based, research-level mathematics journals for the ready submission of their work.[108]

Gilman rather quickly saw one possible solution to this problem. Sylvester had served as one of the editors of England's *Quarterly Journal of Pure and Applied Mathematics* since its first appearance under this title in 1855. A well-established publication with a loyal readership, steady subscription levels, and regular contributors, the *Quarterly Journal*, if moved to Baltimore, would have presented little or no financial risk to the University, while providing for the Hopkins mathematicians. Unfortunately for Gilman, neither Sylvester nor the journal's publisher greeted this idea with enthusiasm, and the journal remained in Great Britain.[109]

With the failure of this initiative but with Story's independent endorsement,[110] Gilman continued to press the obstinant Sylvester to initiate a jour-

---

[107]Gilman, *The Launching of a University*, p. 115.

[108]This was the case for Hopkins faculty members in other disciplines as well. In fact, the university ultimately supported several other scholarly publications during its early years: the *American Chemical Journal* began in 1879 under the direction of Hopkins chemist, Ira Remsen; *Studies from the Biological Laboratory* also appeared in 1879 but came out only irregularly and was discontinued in 1893; in 1880, Hopkins classicist, Basil Gildersleeve, started the *American Journal of Philology*; Michael Foster's *Journal of Physiology* (begun in 1878) moved to the United States from England and gained Hopkins support in 1881, although it continued under Foster's editorship; and historian, Henry B. Adams, founded the monograph series, entitled *Johns Hopkins University Studies in Historical and Political Science*, in 1882. See French, pp. 53–55.

[109]Gilman, *The Launching of a University*, pp. 115–116. As remarked in the previous note, this strategy of relocating an established journal did work for Gilman several years later relative to the *Journal of Physiology*.

[110]*Ibid.*, p. 116; and Hawkins, *Pioneer*, pp. 44–45. Cooke and Rickey detail Story's associa-

nal in the United States. As Sylvester later recounted: "I said it was useless, there were no materials for it. Again and again he returned to the charge, and again and again I threw all the cold water I could on the scheme."[111] By November of 1876, however, Gilman had apparently worn down the defenses of his senior mathematician by enlisting the help of Simon Newcomb and Hopkins physicist Henry Rowland.

On 8 November, these men, together with Story and Sylvester, circulated a letter to the mathematical scientists of America announcing their intention to found a research-level mathematics journal and soliciting advice and support. As they stated the case:

> It is believed that a periodical of a high class published in America, in which Mathematicians might interchange ideas and impart their investigations and discoveries has been long felt to be a desideratum, and that the want of such a medium of communication operates as a serious impediment to the propagation and advancement of mathematical knowledge in this country.
>
> This want it is proposed to supply by issuing at regular stated periods and probably in a quarto form a journal to be called the American Journal of Pure and Applied Mathematics.[112]

Yet, in both recognizing this need and putting forth this solution, they fully realized that the success of the venture was by no means assured:

> Previous attempts to found such a journal in the United States have been made but without permanent success—a result partly owing it may be supposed to the want of sufficient contributions of a nature to attract subscribers, but still more to the expense and risk unavoidably attendant on an undertaking in which only a limited portion of the public can be expected to take an interest.[113]

To overcome these obstacles, they offered new strategies for securing research contributions and financial backing.

Quite naturally, they expected American mathematicians to provide the bulk of the original work. To supplement what they feared might be an inadequate volume of material from this source, however, they proposed to solicit submissions from abroad, "confidently anticipat[ing] receiving valuable

---

tion with the *American Journal* in their article, "W. E. Story of Hopkins and Clark," particularly on pp. 35–43.

[111] "Remarks of Prof. Sylvester, at the Farewell Reception tendered to him by the Johns Hopkins University, Dec. 20, 1883 (Reported by Arthur S. Hathaway)," 24 typescript pages, Gilman Papers. The quote appears on p. 10.

[112] J. J. Sylvester, Simon Newcomb, H. A. Rowland, and W. E. Story to the mathematical community, 8 November 1876, Gilman Papers.

[113] *Ibid.*

contributions from Mathematicians of eminence in the old world."[114] As for securing the capital necessary to launch and sustain the publication, they appealed to

> the public spirit of the Trustees of the Johns Hopkins University, who there are grounds for believing may be induced to afford to the proposed American Journal the same aid and 'furtherance' as has been for many years afforded to Crelle's Journal of Pure and Applied Mathematics by the enlightened government of Prussia, by bearing for some time at least the expenses of the publication in the event of it's [sic] not proving self-supporting.[115]

With the university underwriting the journal, it, unlike its predecessors, would not have to depend solely on subscriptions or personal fortune to guarantee its continuance.

Although Sylvester, Newcomb, Rowland, and Story cited *Crelle's Journal* as a precedent for their plan, the case of the proposed *American Journal* was hardly parallel. The American enterprise, unlike its German counterpart, was to be underwritten by a university which was not governmentally financed. With only limited funds available, Hopkins nevertheless had several departments all equally wanting in publication outlets.[116] Thus, if the university supported the mathematics journal, it would have to be both willing and able to insure the financial solvency of similar undertakings in other scholarly areas. In short, the Trustees supported Gilman's idea for university-sponsorship of research-level publications in principle, but, with the overall financial welfare of the institution to consider, they would require hard and compelling evidence of broadly based commitment before entering into such a scheme in any particular discipline.[117] (The difficulties in funding the publication of high-level mathematical work would persist throughout the period 1876 to 1900. Compare the discussion in Chapter 9 of similar problems encountered in launching the *Transactions of the American Mathematical Society* and in publishing Felix Klein's Evanston Colloquium lectures.)

Mindful of this, Sylvester and his colleagues closed their letter with an appeal for written testimonials in favor of their plan and for further suggestions for its implementation. Coming from Ann Arbor, Annapolis, Cambridge, Ithaca, New Haven, Princeton, Washington, and elsewhere, over forty let-

[114] *Ibid.*

[115] *Ibid.*

[116] See note 108 above for some of the other departments which received such support in the University's early years.

[117] Here, again, it is important to note that Gilman ran the Johns Hopkins University not autocratically but cooperatively. He had his own educational and institutional agenda, but the Trustees oversaw the operation and held the purse-strings as a consequence of Johns Hopkins's original bequest. Gilman's success as the University's first president thus hinged largely on his ability to convince the Board of the necessity and viablility of the plans he wished to see put into place.

ters arrived in overwhelmingly positive response to the appeal.[118] Benjamin Peirce of Harvard expressed his willingness "to do anything I can to advance the Mathematical Journal"; from the University of Michigan, James Watson opined that "The Johns Hopkins University by fostering an enterprise of this character will confer lasting benefits upon science, and will receive the hearty thanks of all who have at heart the progress of intellectual development"; and Cornell's James Oliver welcomed such a publication, having "long felt it to be needed in this country."[119] The force of the endorsements of the nation's mathematical scientists provided Gilman with the evidence he needed to sway the Hopkins Trustees in their decision. By June of 1877, the organizational machinery for the Johns Hopkins's first scholarly publication, the *American Journal of Mathematics*, was in place.[120]

Sylvester served as the venture's editor-in-chief, a post he defined in purely scientific terms. Willing to help fill the journal's pages both with his own researches and with work solicited from his many contacts in Europe, Sylvester had neither the desire nor the aptitude for running its day-to-day affairs. He would referee papers and make editorial changes in them, but he would deal with neither the general correspondence nor the printers. Furthermore, he wanted nothing to do with the publication's financial aspects. Writing to Gilman on 9 May 1877, he explained that "[m]y desire is to be as useful as I possibly can to the undertaking *quâ* mathematician and to have my mind undisturbed by being mixed up in any way with its financial arrangements."[121]

During the negotiations with the Trustees in the spring of 1877, however, it became clear that, while they were willing to offer monetary support for the proposed journal, they had no intention of accepting full financial responsibility for it. After Gilman went vainly in search of a publisher favorably disposed to taking on this risk, the Trustees naively assumed that Sylvester would agree to shoulder the burden personally.[122] In so doing, they completely betrayed their inexperience in the business of scholarly publication, and the Professor of Mathematics set them straight in no uncertain terms in a lengthy follow-up letter to Gilman on 14 May 1877:

---

[118]List of abstracted quotations entitled "*American Journal of Mathematics*—Approvals by prominent men," Gilman Papers.

[119]*Ibid.* The Smithsonian's Joseph Henry also wrote in support of the mathematical journal, but he suggested a key modification of its purpose. He advocated that, in addition to research-level articles, the journal publish expository papers aimed at the student of mathematics desirous of bridging the gap between ordinary course work and the frontiers of mathematics. As Henry himself noted, this suggestion did not impress Sylvester and was not adopted. See Joseph Henry to J. J. Sylvester, S. Newcomb, H. A. Rowland, and W. E. Story, 12 December 1876, Gilman Papers; and Joseph Henry, diary entry for 12 November 1876 Joseph Henry Papers, Smithsonian Institution Archives.

[120]Hawkins, *Pioneer*, p. 75.

[121]J. J. Sylvester to Daniel C. Gilman, 9 May 1877, Gilman Papers.

[122]Hawkins, *Pioneer*, p. 75.

> I had understood that either a publisher was to be found for the purpose or else that the legal title and responsibility were to be vested in some one representing the interest of the University and for the first time at the interview to which you refer found myself unexpectedly placed in a position and assumed to be ready to undertake a responsibility which I had never contemplated and which (as far as my experience extends) is altogether unusual on the part of a scientific Editor.[123]

Part of the problem was ultimately resolved by enlisting Story's services as "associate editor in charge," that is, in charge of all operational and financial affairs. As to the latter, the University agreed to provide $500 annually, or roughly one-fifth of the production cost, to John Murphy & Co. of Baltimore.[124] When the first number of the new journal appeared late in the spring of 1878, its title page announced that it was "published under the auspices of the Johns Hopkins University" but "printed for the Editors by John Murphy & Co."[125] This arrangement set a precedent for subsequent university co-sponsored publication ventures both at Hopkins and elsewhere: the university provided some financial security, but subscriptions were intended to offset most of the cost.

Although Murphy & Co. took a loss in the first year, subscriptions nevertheless reached more encouraging levels. By 1 September 1878, over 200 subscribers had paid the $5.00 annual fee.[126] College, university, and public libraries both at home and abroad ordered copies; individuals subscribed, including such noted mathematicians as Arthur Cayley in England and Charles Hermite in France; and, what especially gratified Sylvester, the journal had favorably impressed the Germans. In a letter to Gilman written on 7 June 1878, shortly after the appearance of the first number, Sylvester glowingly reported that "[a]ll the world speaks admiringly of the Journal. There appears to be a brisk demand for it in Germany which is a very good symptom of it's [sic] possessing some sterling claim to regard."[127] While he surely overstated the mathematical public's initial reaction, Sylvester seemed genuinely pleased that—despite his own initial reservations about the undertaking—the mathematics journal had gotten off to a solid start.

The editors—Sylvester, Benjamin Peirce, Simon Newcomb, Henry Rowland, and especially Story—had produced a handsome, quarto first number

---

[123]J. J. Sylvester to Daniel C. Gilman, 14 May 1877, Gilman Papers.

[124]Thomas Craig to Daniel C. Gilman, "Financial Report of the *American Journal of Mathematics*," 8 November 1882, Gilman Papers; and Hawkins, *Pioneer*, p. 75. As Craig reported, during its first four years of operation, the journal cost $9,654.83 to produce.

[125]*American Journal of Mathematics* 1 (1878), title page.

[126]Story included lists of the subscribers in the first three numbers of the journal. For the information on subscription rates, see *American Journal of Mathematics* 1 (1878), p. iv.

[127]J. J. Sylvester to Daniel C. Gilman, 7 June, 1878, Gilman Papers.

containing over 100 pages of respectable mathematical research in both the pure and applied veins.[128] George William Hill presented the first installment of his pioneering paper on lunar motion, which, as noted in Chapter 1, would later attract the attention of Henri Poincaré.[129] Arthur Cayley sent a brief note on the abstract theory of groups in which he stated, for perhaps the first time in the American mathematical literature, that "[t]he general problem is to find all the groups of a given order n."[130] Sylvester offered a fascinating memoir linking invariant theory and the atomic theory of matter.[131] Rounded out with additional contributions by Newcomb, Rowland, Henry T. Eddy of the University of Cincinnati, Charles S. Peirce then of the United States Coast and Geodetic Survey, and Guido Weichold, a future student of Felix Klein, the début of the *American Journal* did indeed possess "some sterling claims to regard," and the subsequent numbers and volumes only crystallized this first impression. The *American Journal*, unlike the many failed American attempts at mathematics journals before it, was a consistently serious, strongly supported, research-level mathematics publication.

Under the editorship principally of Sylvester and Story from 1878 to 1884, the journal continued to attract research of high quality. English mathematicians like Cayley, Thomas Muir, and Percy MacMahon published there; from Denmark, Julius Peterson sent a new proof of the law of quadratic reciprocity;[132] other articles and notes came from France, Italy, and Canada. Of the work from abroad, however, A. B. Kempe's fallacious proof of the so-called four color problem in the second volume caused the greatest stir in mathematical circles.[133] From the United States, articles appeared by Yale's

---

[128]In fact, the first volume of the journal ran to almost four hundred pages and included work from American, British, French, and German mathematicians. Peirce, Newcomb, and Rowland served as cooperating editors in mechanics, astronomy, and physics, respectively.

[129]George William Hill, "Researches in the Lunar Theory," *American Journal of Mathematics* 1 (1878):5–26. This paper continued under the same title over several numbers of the journal's first volume. See, also, pp. 129–147 and pp. 245–260.

[130]Arthur Cayley, "The Theory of Groups," *American Journal of Mathematics* 1 (1878):50–52 on p. 51, or *Math. Papers AC*, 10:401–403. As is well-known, the problem of classifying finite groups came to occupy a prominent place in American mathematical research in the twentieth century. After much effort in the 1960s and 1970s, the early 1980s witnessed what now seems to be the complete classification of all finite simple groups.

[131]J. J. Sylvester, "On an Application of the New Atomic Theory to the Graphical Representation of the Invariants and Covariants of Binary Quantics,—With Three Appendices," *American Journal of Mathematics* 1 (1878):64–125, or *Math. Papers JJS*, 3:148–206. This work is discussed in detail in Karen Hunger Parshall, "Chemistry through Invariant Theory?" James Joseph Sylvester's Mathematization of the Atomic Theory," to appear.

[132]Julius Peterson, "A New Proof of the Theorem of Reciprocity," *American Journal of Mathematics* 2 (1879):285–286.

[133]A. B. Kempe, "On the Geographical Problem of the Four Colours," *American Journal of Mathematics* 2 (1879):193–200. As is well known, the problem was to show that four colors suffice to tint a map so that no two adjacent regions are of the same color. Story noticed certain problems with Kempe's proof, which he dealt with in a "Note on the Preceding Paper," *American*

J. Willard Gibbs, Emory McClintock of Northwestern Mutual Life Insurance Company in Milwaukee, Wisconsin, and the Cincinnati Observatory's Ormond Stone.[134] The journal also reprinted Benjamin Peirce's groundbreaking treatise on linear associative algebra in 1881 with additions and addenda by his son, Charles S. Peirce. Among the latter, the younger Peirce gave a proof, independently of Georg Frobenius, of the fact that the only finite-dimensional division algebras over the real numbers are the reals, the complex numbers, and the quaternions.[135]

Although foreigners and other Americans provided slightly more than half of the contributions appearing in the *American Journal* during Sylvester's editorship, all of the remainder came from within his own institution. The *American Journal* thus showcased the mathematical research of the Johns Hopkins University and brought it before an extended mathematical audience, in keeping with Gilman's original conception of university-sponsored publication. With the journal as its publication outlet and guided by its unique institutional mandate and leadership, the Hopkins Department of Mathematics functioned not merely as a training ground for would-be American mathematicians but as America's first center of mathematical research. It is to an examination of the actual research output and interests of the Hopkins group that we now turn.

Journal of Mathematics 2 (1879):201–204. A computer-assisted proof of this result was finally obtained in 1979 by Kenneth Appell and Wolfgang Hacken.

[134]McClintock never held an academic position (although he did spend one year as Tutor of Mathematics prior to earning his Columbia A.M. in 1862). He was, however, a respected member of the emergent American mathematical community, serving as the second President of the New York Mathematical Society from 1891 through 1894. During his regime, the Society assumed the name American Mathematical Society. See Chapters 6 and 9 for more on the Society and its development.

[135]Benjamin Peirce, "Linear Associative Algebra With Notes and Addenda, by C. S. Peirce," *American Journal of Mathematics* 4 (1881):97–229. C. S. Peirce's proof appeared in the third addendum, "On the Algebras In Which Division Is Unambiguous," pp. 225–229. Frobenius published his proof of the theorem in 1878 in "Ueber lineare Substitutionen und bilineare Formen," *Journal für die reine und angewandte Mathematik* 84 (1878):1–63, or Ferdinand Georg Frobenius, *Ferdinand Georg Frobenius: Gesammelte Abhandlungen*, ed. Jean-Pierre Serre, 3 vols. (Berlin: Springer-Verlag, 1968), 1:343–405. Élie Cartan gave yet another independent proof in "Les groupes bilinéaires et les Systèmes de Nombres complexes," *Annales de la Faculté scientifique de Toulouse* 12B (1898):B1–B99, or Élie Cartan, *Oeuvres Complètes*, 3 vols. in 6 pts. (Paris: Gauthier-Villars, 1952–1955),1, pt. 2:7–105.

TABLE 2.1
THE MATHEMATICAL OFFERINGS AT THE JOHNS HOPKINS UNIVERSITY 1876–1884

UNDERGRADUATE COURSES

Year-Long Courses: Conic Sections (WES, FF)
- Differential and Integral Calculus (TC, FF)
- Differential Equations (WES, TC, FF)
- Theory of Equations and Determinants (WES, FF)

Semester Courses: Higher Plane Curves (WES)
- Solid Analytic Geometry (WES, FF)

GRADUATE COURSES

Year-Long Courses: Theory of Numbers (JJS)
- Determinants and Modern Algebra (JJS)
- Quaternions (WES)
- Higher Plane Curves (WES)
- Modern Synthetic Geometry (FF)
- Elliptic and Theta Functions (TC)
- Mathematical Astronomy (WES)
- Mathematical Theory of Elasticity (WES, TC)
- Advanced Logic (CSP)

Semester Courses: Algebra of Multiple Quantity (JJS)
- Theory of Substitutions (JJS)
- Theory of Partitions (JJS)
- Surfaces and Curves in Space (WES)
- Geometric Congruences (WES)
- General Theory of Functions, including Riemann's Theory (TC)
- Partial Differential Equations (TC)
- Calculus of Variations (TC)
- Definite Integrals (TC)
- Elementary Mechanics (TC)
- Theoretical Dynamics (TC)
- Hydrodynamics (TC)
- Mathematical Theory of Sound (TC)
- Algebraical Geometry and Abelian and Theta Functions (AC)
- Theory of Probabilities (CSP)

Short Courses: Theory of Invariants (FF) 10 lectures
Spherical Harmonics (TC) 10 lectures
Cylindric or Bessel Functions (TC) 10 lectures
Logic of Relatives (CSP)

Instructor's initials are in parenthesis: JJS = Sylvester; WES = Story; TC = Craig; FF = Franklin; AC = Arthur Cayley, in residence Spring Semester 1881–1882; CSP = Charles S. Peirce, on the faculty 1879–1884. Compiled from William E. Story's report to Daniel C. Gilman on the Mathematics Department composed partially on 17 December 1883 in Gilman Papers.

Table 2.2
The Mathematics Fellows at the Johns Hopkins University 1876–1884

Thomas Craig (1876–1879), Ph.D. 1878 for "The Representation of One Surface Upon Another, and Some Points in the History of the Curvature of Surfaces."

George Bruce Halsted (1876–1878), Ph.D. 1879 for "Basis for a Dual Logic."

Joshua W. Gore (1876–1878).

Fabian Franklin (1877–1879), Ph.D. 1880 for "Bipunctual Coordinates."

Washington Irving Stringham (1878–1880), Ph.D. 1880 for "Regular Figures in n-Dimensional Space."

Charles A. Van Velser (1878–1881).

Oscar Howard Mitchell (1879–1882), Ph.D. 1882 for "Some Theorems in Numbers."

Robert W. Prentiss (1879–1881).

Christine Ladd (1879–1880), Ph.D. earned 1882 for "On the Algebra of Logic," received 1926.

Herbert M. Perry (1880–1882).

William Pitt Durfee (1881–1883), Ph.D. 1883 for "Symmetric Functions."

George Stetson Ely (1881–1883), Ph.D. 1883 for "Bernoulli's Numbers."

Arthur Safford Hathaway (1881–1883).

Gustav Bissing (1882–1884).

Archibald L. Daniels (1882–1884).

Ellery William Davis (1882–1884), Ph.D. 1884 for "Parametric Representations of Curves."

Chapter 3
Mathematics at Sylvester's Hopkins

While it goes without saying that Gilman fostered a unique environment for the pursuit of graduate training and original research at the Johns Hopkins University, he had no assurances that his new educational experiment would work. All of the inducements, all of the moral and institutional support, all of the best wishes and best laid plans, do not create new knowledge. That requires not only talent but also the wherewithal to forge beyond the boundaries of what is known.

And then there is the matter of how far those boundaries are extended. New results might represent a natural and obvious generalization of known research. They might fill in a gap or correct a mistake in the published literature in an area. They might reflect a new idea which unexpectedly propels an accepted approach beyond its previous limits. Or they might truly represent a radical departure from contemporary thinking. In the first case, the research, new though it may be, rarely evokes great enthusiasm or inspires subsequent work. In the second, the discipline clearly benefits, but while new insights may follow from the isolation and correction of an error, these need not lead to further breakthroughs. The best research tends to fall in the third and fourth categories. Fresh ideas and novel points of view enliven a subject by suggesting unexplored avenues of research and fostering continued activity.

At Hopkins, Sylvester and the other Head Professors continued their own lines of inquiry unfettered by the heavy teaching loads and burdensome duties which encumbered most of their colleagues around the country. Motivated—and often talented—students accepted graduate fellowships and came to Baltimore determined to do original work. In mathematics, that work tended to reflect Sylvester's interests in the theories of invariants, partitions, and algebras, but geometry and mathematical logic also found proponents in William Story and Charles Peirce (who taught part-time at Hopkins from 1879 to 1884), respectively. Not all of the research produced was exciting; not all of it was important; but most of it was solid; and some of it was genuinely remarkable. Of importance here, however, is not so much the *quality* of the research as the evidence of the conviction, shared by the members of the Department of Mathematics (and by the University as a whole for that matter), in the importance and primacy of the production of new knowledge. As what follows will show, the Hopkins mathematical group consistently strove to develop new mathematics.

Sylvester and Invariant Theory

Although his premature retirement from the Royal Military Academy in 1870 had plunged Sylvester into a deep mathematical depression, the challenge of his new duties in the United States so thoroughly revived him that he resumed his researches almost immediately upon his arrival in 1876. As it had since the early 1850s, invariant theory captured his mathematical interest, and, not surprisingly, he led some of his Hopkins students into precisely that area of research.

More than thirty years had passed since Boole had isolated the phenomenon of invariance, and invariant theory had undergone substantial development in Great Britain largely at the hands of Cayley and Sylvester. In particular, Sylvester had succeeded in drawing the various isolated facts and techniques concerning invariants into a *bona fide* mathematical theory in 1852.[1] One year later, he had mounted an assault on the problem of recognizing dependence relations among invariants, and had provided an invaluable glossary of the vocabulary of invariant theory.[2] Building on this foundation in 1854, Cayley had launched what would become the British approach to invariant theory.

In the first in his series of ten "Memoirs on Quantics," Cayley cast the theory in more general terms than those he had employed in his early papers of 1845 and 1846.[3] He considered a binary *quantic* (or *form* in modern terminology)

$$(3.1)\qquad a_0x^m + a_1\binom{m}{1}x^{m-1}y + a_2\binom{m}{2}x^{m-2}y^2 + \cdots + a_m\binom{m}{m}y^m$$

[1]James Joseph Sylvester, "On the Principle of the Calculus of Forms," *Cambridge and Dublin Mathematical Journal* 7 (1852):52–97 and 179–217, or *The Collected Mathematical Papers of James Joseph Sylvester*, ed. H. F. Baker, 4 vols. (Cambridge: University Press, 1904–1912; reprint ed., New York: Chelsea Publishing Co., 1973), 1:284–363 (hereinafter cited *Math. Papers JJS*). For more complete historical discussions of the early British work in invariant theory, see the various works referenced in note 48 in Chapter 2 above.

[2]James Joseph Sylvester, "On a Theory of Syzygetic Relations of Two Rational Integral Functions, Comprising an Application to the Theory of Sturm's Functions, and That of the Greatest Algebraical Common Measure," *Philosophical Transactions of the Royal Society of London* 143 (1853):407–548, or *Math. Papers JJS*, 2:429–586.

[3]Arthur Cayley, "An Introductory Memoir Upon Quantics," *Philosophical Transactions of the Royal Society of London* 144 (1854):244–258, or *The Collected Mathematical Papers of Arthur Cayley*, ed. Arthur Cayley and A. R. Forsyth, 14 vols. (Cambridge: University Press, 1889-1898), 2:221–234 (hereinafter cited *Math. Papers AC*). (In what follows, all subsequent page references to articles cited in the original and in Cayley's collected works will refer to the pagination in the *Math. Papers AC*.)

and a linear transformation of its variables given by $T : x \mapsto aX + bY$, $y \mapsto a'X + b'Y$ of nonzero determinant Δ which sends (3.1) to

$$(3.2) \qquad A_0X^m + A_1\binom{m}{1}X^{m-1}Y + A_2\binom{m}{2}X^{m-2}Y^2 + \cdots + A_m\binom{m}{m}Y^m.$$

In this set-up, a homogeneous polynomial $K(a_0, a_1, \dots, a_m; x, y)$ in the variables and nonbinomial coefficients of (3.1) is a *covariant* if it satisfies the equation

$$(3.3) \qquad K(A_0, A_1, \dots, A_m; X, Y) = \Delta^k K(a_0, a_1, \dots, a_m; x, y),$$

for some nonnegative integer k.[4] An *invariant* is a covariant which does not involve the variables x and y. Furthermore, defining the formal differential operators

$$\mathfrak{X} := a_0\partial_{a_1} + 2a_1\partial_{a_2} + \cdots + ma_{m-1}\partial_{a_m} \text{ and}$$
$$(3.4) \qquad \mathfrak{Y} := a_m\partial_{a_{m-1}} + 2a_{m-1}\partial_{a_{m-2}} + 3a_{m-2}\partial_{a_{m-3}} + \cdots + ma_1\partial_{a_0},$$

Cayley established that any homogeneous polynomial K annihilated by $\mathfrak{X} - y\partial_x$ and $\mathfrak{Y} - x\partial_y$ is a covariant.[5] This linearization technique provided him with an efficacious method for determining covariants. He also announced his solution to one of the key problems surrounding the determination of a maximal collection of *linearly* independent or *asyzygetic* covariants in the postscript he added on 7 October 1854 to his "Introductory Memoir Upon Quantics." When he finally presented that critical result in 1856, it hinged on several of the fundamental, if elementary, concepts of invariant theory.[6]

Given a covariant K of a binary form (3.1), the components of K which are homogeneous in the coefficients $a_0, \dots, a_m$ and in the variables x and y are also covariants. Thus, for simplicity, we may assume that K has constant degree of homogeneity in $a_0, \dots, a_m$—denoted θ and called the *degree* of K—as well as constant degree of homogeneity in x and y—denoted μ and termed the *order* of K. Assigning x weight 1, y weight 0, and a_i weight i, each monomial $a_0^\alpha a_1^\beta \cdots a_m^\gamma x^\sigma y^\tau$ of K has a *weight* given by $0\cdot\alpha + 1\cdot\beta + \cdots + m\cdot\gamma + 1\cdot\sigma + 0\cdot\tau$. It is easily verified that every monomial of K has the same weight, expressible in terms of θ and μ as $\frac{m\theta+\mu}{2}$.[7] In his paper "A Second Memoir Upon Quantics," Cayley exploited the combinatorial aspects of the concept of weight to get not only an explicit construction algorithm

[4]In the expression (3.1), the coefficients $a_0, a_1, \dots, a_m$ are also to be regarded as variables. The coefficients $A_0, A_1, \dots, A_m$ are linear combinations of the $a_0, a_1, \dots, a_m$.

[5]Cayley, "An Introductory Memoir," p. 230.

[6]Arthur Cayley, "A Second Memoir Upon Quantics," *Philosophical Transactions of the Royal Society of London* 146 (1856):101–126, or *Math. Papers AC*, 2:250–275.

[7]For an exposition of these invariant-theoretic facts, see, for example, Edwin Bailey Elliott, *An Introduction to the Algebra of Quantics* (Oxford: University Press, 1895; reprint ed., Bronx, N. Y.: Chelsea Publishing Co., 1964), pp. 40–45. Cayley dealt with this aspect of the theory in "An Introductory Memoir," p. 233.

for covariants (see (3.5) below) but also a combinatorial formula for the (maximal) number of asyzygetic, that is, linearly independent covariants of any given degree and order.

He took a (homogeneous) polynomial A_0 in the coefficients of (3.1) of degree θ and of weight $\frac{m\theta-\mu}{2}$ (for some nonnegative integer μ) which satisfies the additional condition that $\mathfrak{X}A_0 = 0$. Setting $A_j = \frac{\mathfrak{Y}A_{j-1}}{j}$, for $1 \leq j \leq \mu$, determines the coefficients of a homogeneous polynomial

$$K := A_0 x^\mu + A_1 x^{\mu-1} y + \cdots + A_{\mu-1} x y^{\mu-1} + A_\mu y^\mu \tag{3.5}$$

of weight $\frac{m\theta-\mu}{2} + \mu = \frac{m\theta+\mu}{2}$ and order μ. Cayley proved not only that K in (3.5) is always a covariant but also that *every* covariant of order μ and degree θ is expressible in this way.[8] Employing the well-known theory of generating functions,[9] Cayley further noted that the number of monomials in the coefficients of (3.1) "of the degree θ and of the weight q is obviously equal to the number of ways in which q can be made up as a sum of θ terms with the elements $(0, 1, 2, \ldots, m)$, a number which is equal to the coefficient of $x^q z^\theta$ in the development of $\frac{1}{(1-z)(1-xz)(1-x^2z)\cdots(1-x^m z)}$."[10] The assumption that the system of linear equations arising from the equation $\mathfrak{X}A_0 = 0$ is linearly independent[11] readily yields Cayley's combinatorial formula that the maximal number of asyzygetic covariants of degree θ and weight q equals the coefficient of $x^q z^\theta$ minus that of $x^{q-1} z^\theta$.

By putting these various facts and observations together, Cayley attempted to tackle the more difficult problem of determining *irreducible* covariants for a binary quantic of a given degree. (These irreducible covariants form a minimum collection such that every other covariant can be written as a polynomial in them. The irreducible covariants would later be called the *Grundformen* by the German invariant-theorists and the *groundforms* by the British school. The latter would also term a set of groundforms a *fundamental system* of covariants of a given binary quantic.) Cayley correctly calculated

---

[8] Thus, if $\mu = 0$, $K = A_0$ is an invariant. Note that Cayley uses the A_i's in two different senses; those in (3.5) are not the same as those in (3.2).

[9] Leonard Euler explored and analyzed the properties of certain generating functions in Chapter 16, "On the Partitions of Numbers," of his fundamental treatise of 1748, *Introduction to the Analysis of the Infinite*, Book 1, trans. John D. Blanton (New York: Springer Verlag, 1988), pp. 256–279. At its most elementary level, any closed expression for a (formal) power series $\sum_{n=0}^{\infty} h_n x^n$ can be regarded as a *generating function* for the sequence $h_0, h_1, \ldots$. For a readable, modern treatment of these matters, consult Richard A. Brualdi, *Introductory Combinatorics*, 2nd ed. (New York: North-Holland, 1992), pp. 216–231. In the section "From Invariant Theory to the Theory of Partitions" below, we examine the partition-theoretic interpretations of generating functions in the context of the research of Sylvester and his students.

[10] Cayley, "A Second Memoir," p. 260. In modern terms, this number is the dimension of the space of polynomials A_0 of degree θ and weight $q = \frac{m\theta-\mu}{2}$.

[11] Although correct, Cayley tacitly accepts—without providing proof or indicating that a proof might be warranted—this fact. As we shall see later in this chapter, Sylvester only noted and filled this gap some twenty-five years later.

that the binary quadratic form had two irreducible covariants, the binary cubic had four, and the binary quartic had five. Due to a flaw in his reasoning, however, he mistakenly concluded that the binary quintic had an infinite number of irreducible covariants and, undaunted, he proceeded to calculate twelve of them explicitly.[12]

Seemingly entranced by these computations, Cayley immediately published "A Third Memoir Upon Quantics" in which he displayed three new additions to the quintic list and began his catalogues for the binary sextic, septemic, octavic, nonic, and dodecadic as well as for the ternary quadric and ternary cubic forms.[13] Although a tireless calculator, Cayley stopped here, at least for the moment, and returned, in his "Fourth Memoir" of 1858, to the general theory of binary forms.

Cayley, as well as and a number of his contemporaries, had quickly come to recognize the potential utility of invariant theory in projective geometry. Indeed, George Salmon, the Irish friend and correspondent of both Cayley and Sylvester, first exploited this connection in a systematic way.[14] In Germany, Otto Hesse emerged as the area's leading proponent, while Gotthold Eisenstein came to it motivated by more number-theoretic concerns. Picking up on their work, but particularly on Hesse's geometrically inspired results, other researchers like Siegfried Aronhold and Alfred Clebsch succeeded in formulating what came to be known as the *symbolic* method of invariant theory. This approach operated at a somewhat higher level of abstraction and employed a more general notation than that of its British counterpart.[15]

---

[12]Cayley, "A Second Memoir," pp. 270 and 273–275. Although Cayley incorrectly concluded that the number of irreducible covariants is infinite for binary forms of degree five and greater, the number, in fact, gets large quickly. This result follows, for example, from work of V. L. Popov as reported by Jacques Dixmier in "Quelques Résultats de Finitudes en Théorie des Invariants," *Seminaire Bourbaki Volume* 1985/86 *Exposés* 651–668, *Astérisque* 145–146 (1987):163–175. The late Roger Richardson pointed out this reference to us.

[13]Arthur Cayley, "A Third Memoir Upon Quantics," *Philosophical Transactions of the Royal Society of London* 146 (1856):627–647, or *Math. Papers AC*, 2:310–335.

[14]Felix Klein emphasized Salmon's pioneering role in this effort when he wrote: "[a]ll that is needed in the geometry of invariant theory especially is to be found in the textbooks of G. Salmon, which have contributed most to spread the ideas which arise here." Among the textbooks Klein mentioned explicitly was George Salmon, *A Treatise on Higher Plane Curves*, 2nd ed. (Dublin: Hodges, Foster, and Co., 1873). See Felix Klein, *Elementary Mathematics from an Advanced Standpoint. Geometry*, trans. E. R. Hedrick and C. A. Noble (New York: Dover Publications, Inc., 1939), p. 135, and his further discussion of these applications in *ibid.*, pp. 135–148.

[15]Karen Hunger Parshall gave an encapsulated technical discussion of the algebraic side of the German approach to invariant theory in "Toward a History of Nineteenth-Century Invariant Theory," in *The History of Modern Mathematics*, ed. David E. Rowe and John McCleary, 2 vols. (Boston: Academic Press, Inc., 1989), 1:157–206 on pp. 170–180. For a mathematical discussion from a modern point of view, consult Jean Dieudonné and James B. Carrell, *Invariant Theory, Old and New* (New York: Academic Press, Inc., 1971); or T. A. Springer, *Invariant Theory* (Berlin: Springer-Verlag, 1977). A detailed, contemporaneous presentation of the geometrical side of the theory may be found in Alfred Clebsch, *Vorlesungen über Geometrie*, ed. Ferdinand Lindemann (Leipzig: B. G. Teubner, 1876).

As a result of these roughly contemporaneous developments, two distinct approaches to invariant theory had emerged on the two sides of the English Channel by the end of the 1850s. Although they shared the same basic goals, namely, the establishment of a *theory* of invariants and the calculation of invariants and covariants for given forms, the Germans tended to stress theory-building while the British focused more on computational algorithms. Perhaps because of this slight yet important difference in emphasis, the German approach, especially in the masterful hands of Paul Gordan, succeeded in rectifying a crucial error in the British literature.

In 1868, Gordan proved that Cayley's results of 1856 on the number of irreducible covariants of a given form were incorrect. Whereas Cayley had employed a case-by-case argument hinging on an analysis of generating functions to conclude that this number is infinite for the binary quintic and higher forms, Gordan gave a general, constructive proof showing that it is, in fact, *finite* for *all* binary forms. He then used his method to calculate explicitly fundamental systems of the binary quintic and sextic, systems which contained twenty-three and twenty-six covariants, respectively.[16] In the context of projective geometry, Gordan's results provided a complete list of those basic entities which remain unchanged under projective transformations and which are thus independent of the choice of coordinate system.

It was a startled Cayley who scrambled, in "A Ninth Memoir on Quantics," to minimize the damage done to the British theory by Gordan's result. Admitting that his theorem of 1856 had led to false conclusions, Cayley acknowledged that "[t]he theory [was] thus in error, by reason that it omits to take account of the interconnexion of the syzygies."[17] Knowing this, Cayley was still "unable to make this correction in a general manner so as to show from the theory that the number of the irreducible covariants is finite, and so to present the theory in a complete form," when he offered his "Ninth Memoir" to the Royal Society in the spring of 1870.[18] In spite of this gap in the general theory, however, Cayley salvaged his attack at least partially by demonstrating that it could "be made to accord with the facts" in the case of the binary quintic.[19] He closed his paper expressing the "hope that a more simple proof of Professor Gordan's theorem will be obtained—a theorem the importance of which, in reference to the whole theory of forms,

[16]Paul Gordan, "Beweis, dass jede Covariante und Invariante einer binären Form eine ganze Function mit numerische Coefficienten einer endlichen Anzahl solchen Formen ist," *Journal für die reine und angewandte Mathematik* 69 (1868):323–354. In short, Gordan proved a special case of what is now known as the First Main Theorem of Invariants. He enumerated the fundamental systems for the binary quintic and sextic on pp. 343–354. Hermann Weyl popularized the terminology "First Main Theorem" in his book, *Classical Groups: Their Invariants and Representations* (Princeton: University Press, 1939), Chapter 2.

[17]Arthur Cayley, "A Ninth Memoir on Quantics," *Philosophical Transactions of the Royal Society of London* 161 (1871):17–50, or *Math. Papers AC*, 7:334–353 on p. 335.

[18]*Ibid.*, p. 335.

[19]*Ibid.*

it is impossible to estimate too highly."[20] Clearly, Cayley wanted to see the British methods yield not only this simpler proof but also an extension of Gordan's results to an explicit determination of the exact number of irreducible covariants for binary forms of degrees higher than the quintic and sextic. After his embarrassment at the hands of Gordan, this would at least establish the equality, if not the superiority, of the British approach over the German techniques. Not until the winter of 1877 did it appear that this vindication might be imminent.

Having arrived in Baltimore for the start of the Fall Semester in 1876, Sylvester did not take up his teaching duties until the Spring Semester of 1877. At that time, he started a course, entitled "Determinants and Modern Algebra," which he most probably taught from George Salmon's *Higher Algebra*.[21] In the inaugural lecture he gave on 22 February at the University's Commemoration Day ceremonies, he admitted that he had only reluctantly settled on this topic, having been compelled by a student (namely, Halsted) who "would have the New Algebra (Heaven knows where he had heard about it, for it is almost unknown on this continent), that or nothing. I was obliged to yield, and what was the consequence! In trying to throw light upon an obscure explanation in our text-book, my brain took fire, [and] I plunged with re-quickened zeal into a subject which I had for years abandoned."[22] The subject into which his teaching had drawn him was invariant theory, and it almost immediately led him back to the problem of finding a nonsymbolic proof of Gordan's finiteness theorem.

By 23 April 1877, Sylvester had emerged from his teaching-induced reveries with what seemed to be the long-sought result. He wrote to Cayley that "two or three weeks have passed like a shadow whilst I have been engaged in this research. I think I may now announce with moral certainty that my method completely solves the problem of finding the [*G*]*rundformen* for binary forms and systems of binary forms & (without admixture of superfluous forms) in all cases."[23] He added that he intended to put the students in his class to work calculating certain key examples (presumably to change his "moral certainty" to actual certainty), even though one of the computations he wanted to see would "be a slinger to the Calculator."[24] Thus, if his stu-

[20] *Ibid.*, p. 353.

[21] George Salmon, *Lessons Introductory to the Modern Higher Algebra*, 3rd ed. (Dublin: Hodges, Foster, & Co., 1876). In fact, Salmon dedicated this book to Arthur Cayley and Sylvester "in acknowledgement of the obligation I am under, not only to their published writings but also to their instructive correspondence."

[22] James Joseph Sylvester, "Address on Commemoration Day at Johns Hopkins University 22 February, 1877," *Math. Papers JJS*, 3:72–87 on p. 76.

[23] J. J. Sylvester to Arthur Cayley, 23 April 1877, Sylvester Papers, Box 11, The Library of St. John's College, Cambridge (hereinafter denoted Sylvester Papers SJC). Sylvester's emphasis.

[24] *Ibid.* Franklin ultimately carried out these calculations and received due credit for them in James Joseph Sylvester, "Sur les Covariants fondamentaux d'un Système cubo-biquadratique binaire," *Comptes rendus* 87 (1878):242–244 and 287–289, or *Math. Papers JJS*, 3:127–131.

dents had drawn him back into invariant theory, he insisted on pulling them in along with him.

From the outset, Sylvester, who had never before had the opportunity to teach highly motivated advanced students, succeeded in engaging and involving them. Consonant with his inductive philosophy of mathematics, the students in Sylvester's classroom had to do more than listen passively to lectures. They, like their mentor, had to "get their hands dirty," delve into the examples, understand the theorems at more than a logical level. In short, they had to learn through doing, and, in Sylvester, they had an energetic teacher and guide. Yet in giving them their first exposure to mathematical research, Sylvester also unwittingly exposed them to some of the endeavor's pitfalls.

In sketching out his new findings to Cayley in that letter of 23 April, Sylvester sensed the presence of a problem lurking behind his new method for calculating a fundamental system of groundforms of a given binary form.[25] He forewarned Cayley "that *anterior to all* verification this method could not give superfluous forms—but it is metaphysically conceivable that it might give *too few* [G]rundformen."[26]

Characteristic of Sylvester's mathematical exuberance, he triumphantly, but prematurely, announced to his friend a new and important result, based on compelling circumstantial evidence, but without a rigorous proof to clinch it. Sylvester also sent off a short announcement of the result to the *Comptes rendus* of the French Academy and immediately piqued the interest of an incredulous Camille Jordan. On 13 May, Jordan queried Sylvester concerning his most recent claims and urged him not to delay in publishing the details of his new method.[27] Sylvester came up short in his efforts to comply with Jordan's request, however. He could use his technique to reproduce the calculational results of the German method, and he could even use it to uncover errors in the German calculations, but he could not rigorously prove that his method gave *the* finite number of covariants in a fundamental system associated to a given form. Because of its failure to detect dependence relations, the German symbolic algorithm potentially overestimated the number of covariants in a fundamental system, while Sylvester's method potentially underestimated it.[28] Sylvester had nonetheless succeeded in providing the British invariant theorists with a competitive process for calculating irreducible covariants, even though a British-style proof of Gordan's theorem still eluded him.

[25] Furthermore, Sylvester thought that his method applied to more than just a single binary form. In fact, he believed he had hit upon a technique that allowed him to calculate the fundamental system for systems of binary forms.

[26] J. J. Sylvester to Arthur Cayley, 23 April 1877, Sylvester Papers SJC, Box 11. Sylvester's emphasis.

[27] Camille Jordan to J. J. Sylvester, 13 May 1877, Sylvester Papers SJC, Box 2.

[28] Elliott, pp. 173–176. Tony Crilly also makes this point in his unpublished thesis, "The Mathematics of Arthur Cayley with Particular Reference to Linear Algebra," (unpublished Ph.D. dissertation, Middlesex Polytechnic, June 1981), p. 140.

Captivated by his success rather than his failure, Sylvester, together with his student Fabian Franklin, continued to push the new calculational technique. Their work over the next several years yielded a complete enumeration of the groundforms for the binary forms of the first ten orders.[29] If Sylvester's efforts in 1877 relative to a British proof of Gordan's theorem ended with mixed results, he *did* succeed in filling a major gap in one of the key operational theorems of the British approach in the autumn of that same year.

As noted above, Cayley had provided a combinatorial formula for the number of linearly independent covariants of a given degree and order associated with a given binary form in 1856. At a key juncture in his argument, however, he had assumed, without proof, that the system of linear equations resulting from $\mathfrak{X}A_0 = 0$ was linearly independent (using the notation set up in (3.4) and (3.5)).[30] Writing to Sylvester about his discovery, Cayley focused precisely on this gap when he admitted that "there is no reason for doubting that these equations are independant [sic]."[31] Cayley's willingness to leave the argument at this points out perhaps not only the primitive state of linear algebra in 1856 but also the potential danger of this linear algebraic naiveté for the British approach. After the publication of Cayley's paper in 1856, he, Sylvester, George Salmon, and others fully exploited this result in establishing the British approach to invariant theory. Had Cayley's intuition been faulty here, he would have led himself and his compatriots in the wrong mathematical direction. Fortunately, but quite unexpectedly in November of 1877, Sylvester realized that he could supply a rigorous proof of the required linear independence. Once again, his teaching inspired his research.

Having accepted the truth of Cayley's theorem back in 1856, Sylvester had used it unquestioningly in the intervening years. Early in 1877, however, George Bruce Halsted, the student responsible for Sylvester's class on determinants and modern algebra, brought the result back under Sylvester's

---

[29]See, for example, Sylvester, "Sur les Covariants fondamentaux d'un Système cubo-biquadratique binaire," pp. 242–244 and 287–289, or *Math. Papers JJS*, 3:127–131; "Table des Nombres de Dérivées invariantives d'Ordre et de Degré donnés, appartenant à la Forme binaire du dixième Ordre," *Comptes rendus* 89 (1879):395–396, or *Math. Papers JJS*, 3:256; "Sur le vrai Nombre des Covariants fondamentaux d'un Système de deux Cubiques," *Comptes rendus* 89 (1879):828–832, or *Math. Papers JJS*, 3:258–261; "Tables of Generating Functions and Ground-Forms for the Binary Quantics of the First Ten Orders," *American Journal of Mathematics* 2 (1879):223–251, or *Math. Papers JJS*, 3:283–311; "Tables of the Generating Functions and Ground-Forms for Simultaneous Binary Quantics of the First Four Orders Taken Two and Two Together," *American Journal of Mathematics* 2 (1879):293–306 and 324–329, or *Math. Papers JJS*, 3:392–410. See, also, Fabian Franklin, "On the Calculation of the Generating Functions and Tables of Groundforms for Binary Quantics," *American Journal of Mathematics* 3 (1880):128–153. As Tony Crilly recounted, further announcements of proofs of Gordan's theorem continued to appear in Sylvester's correspondence through 1886. See his "The Decline of Cayley's Invariant Theory," *Historia Mathematica* 15 (1988):332–347 on pp. 340–341.

[30]This is precisely equivalent to the difficulty alluded to in note 11 above.

[31]Arthur Cayley to J. J. Sylvester, undated, Sylvester Papers SJC, Box 2.

scrutiny. Halsted noted that Francesco Faà de Bruno had called Cayley's claim of the required linear independence into doubt in his book *Théorie des Formes binaires.* In fact, Faà de Bruno had asserted that the linear independence did not always hold and that Gordan's work of 1868 had actually proven this.[32] In explaining the situation to Halsted, Sylvester held (correctly) that while Gordan's research had uncovered an error in Cayley's result on the number of irreducible covariants for the binary quintic form, the mistake did not lie in the underlying combinatorial theorem. Although still convinced of the truth of the result, Sylvester could not provide Halsted with the proof which would categorically settle the matter. As he put it, "the extrinsic evidence in support of the independence of the equations which had been impugned rendered it to my mind as certain as any fact in nature could be, but that to reduce it to an exact demonstration transcended, I thought, the powers of the human understanding."[33] However, just prior to 6 November 1877, Sylvester hit upon a clear and remarkable proof of Cayley's combinatorial formula.[34] In its published form, Sylvester's demonstration exploited fully the actions of the operators $\mathfrak{X}$ and $\mathfrak{Y}$ defined in (3.4) as well as of the operator $\mathfrak{Y}\mathfrak{X} - \mathfrak{X}\mathfrak{Y}$ (now commonly called the commutator of $\mathfrak{X}$ and $\mathfrak{Y}$ and denoted $[\mathfrak{Y}, \mathfrak{X}]$) on the set of polynomials of degree θ in $a_0, \ldots, a_m$ to achieve the long-overdue result.[35] While this proof certainly vindicated a key aspect of the British approach to invariant theory, neither Sylvester nor Cayley nor any of their followers ever succeeded in winning the big prize, a proof of Gordan's finiteness theorem using the British techniques.[36]

Geometrical Research at Johns Hopkins University

While Sylvester, as the department's Head Professor and mathematical star, concentrated almost exclusively on his own research and related course work, most of the department's teaching burden fell on William Story in his role as Associate. For example, Story supervised the undergraduate teach-

[32]Francesco Faà de Bruno, *Théorie des Formes binaires* (Turin: P. Marietti, 1876), p. 150; and James Joseph Sylvester, "Proof of a Hitherto Undemonstrated Fundamental Theorem of Invariants," *Philosophical Magazine* 5 (1878):178–188, or *Math. Papers JJS*, 3:117–126 on p. 117.

[33]Sylvester, "Proof of a Hitherto Undemonstrated Fundamental Theorem of Invariants," p. 178, or p. 117.

[34]J. J. Sylvester to Arthur Cayley, 6 November 1877, Sylvester Papers SJC, Box 11.

[35]Sylvester and Cayley clearly did not have the notion of a Lie algebra, yet their work led them to derive what would come to be seen as some of the elementary, finite-dimensional representation theory of the simple Lie algebra $\mathfrak{sl}_2$ of 2×2 complex matrices of trace 0.

[36]See below for hints of Sylvester's repeated assaults on this problem. This is also discussed in Parshall, "Toward a History of Nineteenth-Century Invariant Theory." Ironically, both Gordan's theorem and his methods would soon fall victim to the farther-reaching methods of David Hilbert. For a brief account of Hilbert's role in the history of invariant theory, see Karen V. H. Parshall, "The One-Hundredth Anniversary of the Death of Invariant Theory?," *The Mathematical Intelligencer* 12 (4) (1990):10–16.

WILLIAM E. STORY (1850–1930)

ing program and lectured nine hours a week on three distinct topics during the first year of operation, while Sylvester taught one graduate course twice weekly and supervised the as yet irregular Mathematical Seminary. Given that he maintained this schedule throughout Sylvester's years at Hopkins in addition to picking up the associate editorship of the *American Journal of Mathematics* after its founding in 1878, Story played a pivotal role in the success of the overall Hopkins program. While these demands clearly left precious little time for mathematical investigations of his own, Story nevertheless proved miscellaneous results and made various observations arising from his lecture preparations.[37] More importantly, however, he nurtured his interest in algebraic geometry, while branching out into non-Euclidean geometry, and succeeded in contributing at least one note or paper to each of the first six volumes of the *American Journal.*

Story's first major contribution, a paper "On the Theory of Rational Derivation on a Cubic Curve," appeared in the *American Journal* in 1880 and largely formed a sequel to Sylvester's work "On Certain Ternary Cubic-Form Equations."[38] There, spurred by the year-long course in number theory which he

[37] Although Story's contribution to the first volume of the *American Journal* did not involve geometry, it did stem from his lectures. For three consecutive semesters beginning in the spring of 1877, he taught a course on the mathematical theory of elasticity and subsequently produced "On the Elastic Potential of Crystals," *American Journal of Mathematics* 1 (1878):177-183.

[38] William E. Story, "On the Theory of Rational Derivation on a Cubic Curve," *American*

had given for the first time beginning in the fall of 1879, Sylvester had considered the now familiar process of generating points on a cubic curve (in his case) from a finite collection of points on the curve by means of the chord-tangent law of composition.[39] Story sought to recast Sylvester's number-theoretic approach in a more purely geometrical light by drawing from the researches of Cayley, George Salmon, and Alfred Clebsch in the theory of higher plane curves. Thus, in his paper, Story "propose[d] to develop this new *theory of indices* [that is, Sylvester's approach] in a more general and symmetrical form than that originally given to it; and, finally, by combining it with the theory of parameters, [to] solve a number of problems especially relating to the enumeration of points having certain properties analogous to those of singular points or of the contacts of singular tangents."[40] Although the work of both Sylvester and Story on this problem is largely unknown today, Story's paper, in particular, displays a mastery of contemporaneous research on the theory of plane curves. Furthermore, unlike Sylvester, who had little patience when it came to reading and absorbing published results, Story revealed himself in this work as a true mathematical scholar capable of fruitfully synthesizing his own ideas and the work of others in the realization of his research objectives.[41]

Story exposed yet another aspect of his mathematical persona, namely, the careful and systematic expositor at work, in his 1884 piece "On the Absolute Classification of Quadratic Loci, and On Their Intersection With Each Other

Journal of Mathematics 3 (1880):356–387; and James Joseph Sylvester, "On Certain Ternary Cubic-Form Equations," *American Journal of Mathematics* 2 (1879):280–285 and 357–393; and 3(1880):58–88 and 179–189, or *Math. Papers JJS*, 3:312–391. Over twenty years earlier, Sylvester had touched on the same mathematical issues dealt with in this work. See James Joseph Sylvester, "Note on the Algebraical Theory of Derivative Points of Curves of the Third Degree," *Philosophical Magazine* 16 (1858):116–119, or *Math. Papers JJS*, 2:107–109.

[39] For a modern account of this process as well as a discussion of its place in number theory, and especially in the theory of elliptic curves, see, for example, Dale Husemöller, *Elliptic Curves* (New York: Springer-Verlag, 1986), pp. 10–17. For a more geometrically oriented discussion, consult Egbert Breiskorn and Horst Knörrer, *Plane Algebraic Curves*, trans. John Stillwell (Boston: Birkhäuser Verlag, 1986), p. 306. The chord-tangent law of composition is closely related to—but distinct from—what is now referred to as the *group law* on cubics. Norbert Schappacher has given an historical account of the development of the group law on a cubic in "Développement de la Loi des Groupes sur une Cubique," preprint.

[40] William E. Story, "On the Theory of Rational Derivation on a Cubic Curve," p. 356.

[41] Story continued to make notable contributions to invariant theory. In 1893, he published "On the Covariants of a System of Quantics," *Mathematische Annalen* 41 (1893):469–480. David Hilbert described this paper to *Mathematische Annalen* editor Felix Klein saying that "[Story's] paper is by no means to be thrown in the same pot with the numerous English works in this field." After providing a detailed explanation of the significance of Story's principal result, Hilbert continued that "[s]ince the other parts of the paper leave the impression of great care and exactness, I am of the opinion that it is well worthy of publication." See David Hilbert to Felix Klein, 13 April 1892, in *Der Briefwechsel David Hilbert–Felix Klein* (1886–1918), ed. Günther Frei, Arbeiten aus der Niedersächsischen Staats- und Universitätsbibliothek Göttingen, 19 (Göttingen: Vandenhoeck & Ruprecht, 1985), p. 79 (our translation).

and With Linear Loci."[42] At the beginning of this paper, he underscored the importance of thorough exposition when he commented that "[m]any of the results here obtained are well known, but I believe some of them are new, and the collection of the criteria in such a form that they can be applied to any real forms of the equations of quadratic and linear loci will probably not be without a practical interest."[43] This statement clearly illuminates some of the key differences between Story and Sylvester as personalities and as professors of mathematics. Sylvester virtually never measured his words when speaking of his latest mathematical discovery. It was, for him, almost always a "new" and "remarkable" result. As we have seen, this tended to translate into a classroom presentation full of his latest, as yet unsystematized ideas, which the students were to follow and absorb as best they could. Story, however, stressed a careful preliminary reading of the literature, which was reflected both in his research and in his classroom presentations.

That Story's more scholarly and methodical approach to mathematical research also paid off in new results was particularly evidenced in two papers on non-Euclidean geometry published in the *Journal* in 1881 and 1882.[44] In the first, Story pursued a line of investigation initially suggested by Cayley in his "Sixth Memoir Upon Quantics" of 1859.[45] There, Cayley had hit upon the idea of introducing metrics into projective geometry by means of a fixed, embedded conic called the *absolute*. The geometry associated with this absolute figure is determined by considering those projective transformations acting on the complex projective plane $\mathbf{P}^2(\mathbb{C})$ which leave the figure fixed. One of Cayley's main examples involved replacing the conic by the *circular points at infinity* $(1, \pm i, 0)$. (This can be viewed as a kind of degenerate case.) Cayley deduced from this set-up a metric which yields a geometry equivalent to the standard model of Euclidean geometry when restricted to an appropriate set of points in $\mathbf{P}^2(\mathbb{R})$.[46] Cayley thus effectively showed how to view Euclidean geometry as a special construction within the context of plane projective geometry. By 1871, Klein had picked up on Cayley's work and had used it, in conjunction with other absolute figures, to realize models for (the usual) elliptic and hyperbolic non-Euclidean geometries.[47] In

[42]William E. Story, "On the Absolute Classification of Quadratic Loci, and on Their Intersections With Each Other and With Linear Loci," *American Journal of Mathematics* 6 (1884):222–245.

[43]*Ibid.*, p. 222.

[44]William E. Story, "On the Non-Euclidean Trigonometry," *American Journal of Mathematics* 4 (1881):332–340; and "On the Non-Euclidean Geometry," *American Journal of Mathematics* 5 (1882):180–211, respectively. See, also, William E. Story, "On Non-Euclidean Properties of Conics," *American Journal of Mathematics* 5 (1882):358–381.

[45]Arthur Cayley, "A Sixth Memoir Upon Quantics," *Philosophical Transactions of the Royal Society of London* 149 (1859):61–90, or *Math. Papers AC*, 2: 561–592.

[46]*Ibid.*, p. 590.

[47]Felix Klein, "Ueber die sogenannte Nicht-Euklidische Geometrie," *Mathematische Annalen* 4 (1871):573–625, or Felix Klein, *Gesammelte Mathematische Abhandlungen*, 3 vols. (Berlin:

response to techniques employed by Klein, Cayley also utilized his own previous approach to derive a special case of non-Euclidean trigonometry in the plane.[48]

Story continued this line of investigation by formulating non-Euclidean trigonometry in a more general setting. He proved that "in a uniformly curved space of two dimensions, with a projective measurement based upon any conic, whose constants of linear and angular measurement are k and k' respectively, *the trigonometrical formulae will be obtained from those of spherical trigonometry by replacing each side* (*or arc*) *by the corresponding side* (*or straight line*) *divided by* $2ik$, *and each angle by the corresponding angle divided by* $2ik'$ [here $i = \sqrt{-1}$]."[49] In other words, non-Euclidean plane trigonometry follows from Euclidean spherical trigonometry. By 1882, Story had further extended this result to show that, in fact, non-Euclidean spherical trigonometry may also be deduced from the Euclidean setting.[50]

While this research generated a certain amount of mathematical interest in geometrical questions at Hopkins, Story's lecture courses had a much more direct impact on the graduate program there. W. Irving Stringham came to the university as a Fellow in 1878 after earning a Harvard A.B. under Benjamin Peirce the previous year. Pursuing his undergraduate interest in geometry, Stringham took Story's courses on higher plane curves and elliptic functions in 1878–1879 and then again in 1879–1880.[51] By 1880, he had successfully drawn enough from Story's teaching to earn his Hopkins doctorate for a thesis on "Regular Figures in n-Dimensional Space." Although officially a student of Sylvester, Stringham clearly identified his mentor when the published version of his results appeared later that year in the *American Journal.* There, he "[made] grateful acknowledgment to my coworkers at this University, and especially to Dr. Story, for valuable suggestions."[52] The preparation he received under Story well qualified the fresh Ph.D. for the next step in his career, a Parker Fellowship (under the aegis of Harvard

J. Springer, 1921–1923), 1:254–305; and Felix Klein, *Vorlesungen über Nicht-Euklidische Geometrie,* ed. W. Rosemann (Berlin: J. Springer, 1928), pp. 128–144. For further mathematical discussion of the Cayley metric and its connections with Klein's work, see Hans Wussing, *The Origin of the Abstract Group Concept,* trans. Abe Shenitzer (Cambridge: The MIT Press, 1984), pp. 171–177.

[48]Arthur Cayley, "On the Non-Euclidean Geometry," *Mathematische Annalen* 5 (1872):630–634, or *Math. Papers AC,* 8:409–413.

[49]Story, "On the Non-Euclidean Trigonometry," pp. 334–335. Story's emphasis. The constants which Story denotes by k and k' arose in Klein's reformulation of Cayley's work (Klein denoted them by c and c'). See Felix Klein, "Über die sogenannte Nicht-Euklidische Geometrie," p. 279.

[50]Story, "On the Non-Euclidean Geometry," pp. 180–211.

[51]W. Irving Stringham to Daniel C. Gilman, 23 May 1879 and 20 May 1880, Daniel C. Gilman Papers Ms. 1, Special Collections Division, Milton S. Eisenhower Library, The Johns Hopkins University (hereinafter denoted Gilman Papers).

[52]W. Irving Stringham, "Regular Figures in n-Dimensional Space," *American Journal of Mathematics* 3 (1880):1–14 on p. 14.

University) for further mathematical training in Leipzig under Felix Klein. (See Chapter 5 for more on Stringham's experiences abroad.)

Although perhaps less directly influenced by Story than Stringham, the first two Hopkins Fellows in mathematics, George Bruce Halsted and Thomas Craig, also showed marked interest in geometrical topics. As noted above, Halsted had begun his graduate career as a sort of research assistant to Sylvester in the fall of 1876, but apparently did not share his professor's enthusiasm for invariant theory. Instead, as he reported to Gilman in his fellowship application for the 1877–1878 academic year:

> I have been carrying on alone a totally distinct mathematical investigation upon the two subjects Absolute Geometry and Pro-space. On these subjects there seems to be such a woeful ignorance in America that I had to begin by finding out for myself what had been written on the subject in every language. With no one to aid me, you can imagine what a task was the Bibliography I have now completed, and which I enclose with this application.[53]

Never one to understate his own abilities and attainments, Halsted built on this early interest in non-Euclidean geometry, producing numerous expository and historical articles on the subject throughout an itinerant career which saw him at the College of New Jersey (after 1896 called Princeton University) (1878–1884), the University of Texas (1884–1902), St. John's College in Annapolis, Maryland (1903), Kenyon College (1903–1906), and the State Normal School in Greeley, Colorado (1906–1914). In particular, the bibliography he mentioned in his letter to Gilman appeared with several additions in the *American Journal* beginning with the first volume in 1878.[54]

Like Halsted, Thomas Craig also steered a mathematical course different from that of the algebraically inclined Sylvester. In light of his undergraduate degree in civil engineering and his subsequent self-directed studies, Craig focused more on the applications-oriented areas of mathematics: function theory, differential equations, fluid mechanics. These interests tended to reflect the French tradition in mathematics rather than the British strain of algebraic studies that Sylvester embodied. In fact, as a Hopkins graduate student, Craig essentially prepared *himself* mathematically, ultimately writing his 1878 dissertation on a topic in differential geometry. (See Table 2.2.) By 1882, Craig had not only published several papers in the *American Journal* but also incorporated many of the ideas in his thesis into *A Treatise on Projections*, a book designed primarily for the workers at the United States

[53] George Bruce Halsted to Daniel C. Gilman, [?] April 1877, Gilman Papers.

[54] George Bruce Halsted, "Bibliography of Hyper-Space and Non-Euclidean Geometry," *American Journal of Mathematics* 1 (1878):261–276 and 384–385; and 2 (1879):65–70. On Halsted's life and career, see note 81 in Chapter 2.

George Bruce Halsted (1853–1922)

THOMAS CRAIG (1855–1900)

Coast and Geodetic Survey.[55] From there, his work moved into both elliptic function theory and the geometry of the ellipsoid, and he published several papers on these topics after officially joining the Hopkins mathematical faculty in 1879.[56]

Although Craig quite naturally submitted most of his papers to his home-based *American Journal*, he also ventured into the European mathematical arena. In 1882, for example, one of his geometrical papers appeared in the *Journal für die reine und angewandte Mathematik*. There, Craig studied the relationships between an ellipsoid and its parallel surface. In particular, he focused on the relationship between an element of area dS on the parallel surface and the corresponding element $d\Sigma$ on the ellipsoid, working out and analyzing an explicit formula for $\frac{dS}{d\Sigma}$ in terms of the radii of curvature of the ellipsoid.[57] His article formed part of the small trickle of mathematical research which began to flow from the United States to Europe in the 1880s and which would become a steady stream by the turn of the century.

From Invariant Theory to the Theory of Partitions

Because of the combinatorial nature of the post-1876 research stemming from Cayley's method for generating covariants given in (3.5) above, it is not at all surprising that the research of the Hopkins mathematicians ultimately shifted from invariant-theoretic to purely combinatorial questions. In the years from 1878 to 1883, first Franklin, then Sylvester, and finally most of the latter's other mathematics students became embroiled in some very original combinatorial research.[58]

Fabian Franklin came to the University as a Fellow in the fall of 1877 and participated in Sylvester's regular courses as well as in the Mathematical

[55]Thomas Craig, *A Treatise on Projections* (Washington: United States Coast and Geodetic Survey, 1882). As Craig explained in his preface, the book was divided into two parts: "The first part contains the mathematical theory of projections, while the second part contains merely such a sufficient account of the various projections as will enable the draughtsman to construct them." Craig himself held a part-time position at the Coast Survey from 1879 to 1881 in order to supplement his Hopkins income.

[56]See, for example, Thomas Craig, "Orthomorphic Projection of an Ellipsoid on a Sphere," *American Journal of Mathematics* 3 (1880):114–127; "On Certain Metrical Properties of Surfaces," *American Journal of Mathematics* 4 (1881):297–320; "The Counter-Pedal Surface of the Ellipsoid," *American Journal of Mathematics* 4 (1881):358–378; "Some Elliptic Function Formulae," *American Journal of Mathematics* 5 (1882):62–75; and "Note on the Counter-Pedal Surface of an Ellipsoid," *American Journal of Mathematics* 5 (1882):76–78.

[57]Thomas Craig, "On the Parallel Surface to an Ellipsoid," *Journal für die reine und angewandte Mathematik* 93 (1882):251–270; and "Note on Parallel Surfaces," *Journal für die reine und angewandte Mathematik* 94 (1883):162–170. Story, Stringham, Halsted, and Craig did not produce all of the geometrical work at Hopkins, either. Christine Ladd published an article on "The Pascal Hexagram" in the *American Journal of Mathematics* 2 (1879):1–12.

[58]This research was analyzed in historical context in Karen Hunger Parshall, "America's First School of Mathematical Research: James Joseph Sylvester at the Johns Hopkins University," *Archive for History of Exact Sciences* 38 (1988):153–196.

Seminary. He not only gained exposure there to all of his teacher's latest ideas but also partook fully of the unique Hopkins educational experience. Thus, in 1878, when Sylvester was hard at work on his main contribution to the inaugural volume of the *American Journal of Mathematics*, he undoubtedly exposed Franklin and his other listeners to his evolving ideas on the interrelations between invariant theory in algebra and the atomic theory in chemistry by punctuating his lectures, as usual, with open questions and unverified conjectures.[59] Franklin's first partition-theoretic paper was most likely stimulated by one of Sylvester's remarks.

Entitled "On a Problem of Isomerism," Franklin's note in the *American Journal* treated the following problem: "to find the number of different compounds that can be formed by n m-valent atoms and $(m-2)$ $n+2$ univalent atoms; the word 'compound' being understood to mean any arrangement, *whether continuous or not*, in which every atom appears, with exactly the number of bonds to which its valence entitles it; it being understood, moreover, that no two univalent atoms are connected with each other."[60] Under these constraints, Franklin showed that this reduces to the purely partition-theoretic question: "[i]n how many ways can $n-1$ bonds be distributed among n atoms, no atom having *more* than m bonds attached to it?"[61] He answered this query in the special case of $m=2$; his proof hinged on recovering, in a clever way, the number of partitions of $n-1$ from the number of partitions of n which do not contain more than one 1.[62]

In addition to rising to Sylvester's challenges from the floor of the lecture room, Franklin almost immediately assumed the role of his mathematical assistant, carrying out many of the tedious calculations involved in codifying the irreducible covariants for binary forms of the first ten orders.[63] These computations necessarily involved Cayley's theorem on the number of linearly independent covariants of given degree and order associated to a given form as well as its subsequent refinements particularly at the hands

---

[59]Sylvester presented these novel ideas in "On an Application of the New Atomic Theory to the Graphical Representation of the Invariants and Covariants of Binary Quantics,—With Three Appendices," *American Journal of Mathematics* 1 (1878):64–125, or *Math. Papers JJS*, 3:148–206. For an historical discussion of this work, see Karen Hunger Parshall, "Chemistry Through Invariant Theory? James Joseph Sylvester's Mathematization of the Atomic Theory," to appear. The actual titles of talks presented in the Mathematical Seminary were only officially recorded after 1879, however.

[60]Fabian Franklin, "On a Problem of Isomerism," *American Journal of Mathematics* 1 (1878): 365–368. Franklin's emphasis.

[61]*Ibid.*, p. 366. Franklin's emphasis.

[62]*Ibid.*, pp. 365–366. A partition of a positive integer n is a finite nonincreasing sequence of nonnegative integers $a_1, a_2, \ldots, a_r$ such that $a_1+a_2+\cdots+a_r=n$. The a_i's are called the *parts* of the partition. For a cogent introduction to the modern theory of partitions, see George E. Andrews, *The Theory of Partitions*, Encyclopedia of Mathematics and its Applications, vol. 2 (Reading: Addison Wesley Publishing Co., 1976).

[63]Sylvester's student George Bruce Halsted also helped with the calculations relating to the binary decimic. See "On an Application of the New Atomic Theory," p. 196.

of Sylvester. They thus gave Franklin further exposure to combinatorics and partition theory, which quickly led to new results. In 1879 and also in the *American Journal*, he published another "Note on Partitions" in which he gave an alternate proof of a partition-theoretic result Sylvester had presented in the *Messenger of Mathematics* earlier that year.[64]

If $(w : i, j)$ denotes the number of partitions of w into j parts taken from the integers $0, 1, 2, \ldots, i$ with repetitions allowed, then Cayley's theorem on the number of asyzygetic covariants of degree θ and order μ reduces to calculating $(\frac{m\theta-\mu}{2} : m, \theta) - (\frac{m\theta-\mu}{2} - 1 : m, \theta)$, or, more generally, to the problem of determining $\Delta(w : i, j) := (w : i, j) - (w - 1 : i, j)$. Obviously, the calculation of this difference by brute force, that is, by counting up the entries in tabular listings of the partitions comprising $(w : i, j)$ and $(w - 1 : i, j)$ and then subtracting, could be time-consuming depending on the relative sizes of the particular numbers w, i, and j. In his paper, Sylvester gave a more efficient algorithm for computing the desired difference, which turned on the inherent similarities between the partitions in each of the two tables.

Consider, as Sylvester did, the specific example of computing $(7 : 5, 4) - (6 : 5, 4)$. Exhibiting the partitions associated to $(7 : 5, 4)$ and omitting the terminal zeros, we have: $(5, 2)$, $(5, 1, 1)$, $(4, 3)$, $(4, 2, 1)$, $(4, 1, 1, 1)$, $(3, 3, 1)$, $(3, 2, 2)$, $(3, 2, 1, 1)$, $(2, 2, 2, 1)$. Thus, $(7 : 5, 4) = 9$. Now, taking the partitions in this list and subtracting one from each of the leading parts yields the following seven distinct partitions associated to $(6 : 5, 4)$: $(4, 2)$, $(4, 1, 1)$, $(3, 3)$, $(3, 2, 1)$, $(3, 1, 1, 1)$, $(2, 3, 1) = (3, 2, 1)$, $(2, 2, 2)$, $(2, 2, 1, 1)$, and $(1, 2, 2, 1) = (2, 2, 1, 1)$. (Notice that the highest leading part in these partitions is 4 and that all of the partitions of 7 generate distinct partitions of 6 except the two with two equal leading parts.) To complete the list of partitions associated to $(6 : 5, 4)$, we include the partitions involving 5, namely, the single partition $(5, 1)$. From this, it is easy to see that $(7 : 5, 4) - (6 : 5, 4)$ equals the number of partitions of 7 with leading part repeated minus the number of partitions of 6 involving the greatest possible part equals $2 - 1 = 1$.[65] Sylvester proved this fact in general, but, as Franklin showed, he employed an overly complicated argument.

Franklin succeeded in simplifying Sylvester's proof by using the well-known fact, which Sylvester actually mentioned at the beginning of his paper, that $(w : i, j) = (w : j, i)$, or, the number of partitions of w into j parts taken from 0 to i inclusive equals the number of partitions of w into i parts taken from 0 to j inclusive. In unpublished work of 1853, Norman M.

[64] Fabian Franklin, "Note on Partitions," *American Journal of Mathematics* 2 (1879):187–188; and James Joseph Sylvester, "On a Rule for Abbreviating the Calculation of the Number of In- and Covariants of a Given Order and Weight in the Coefficients of a Binary Quantic of a Given Degree," *Messenger of Mathematics* 8 (1879):1-8, or *Math. Papers JJS*, 3:241–248.

[65] Sylvester, "On a Rule for Abbreviating the Calculation of the Number of In- and Covariants," pp. 1–2, or pp. 241–242.

Ferrers, a British mathematician and later the head of Gonville and Caius College, Cambridge, had given a strikingly elementary, graphical proof of this theorem which now bears his name.[66] An example best illustrates his argument.

Take the partition $(5, 3, 2)$ of 10 into 3 parts none of which exceeds 5 and write it graphically as

$$\begin{matrix} * & * & * & * & * \\ * & * & * & & \\ * & * & & & \end{matrix} \tag{3.6}$$

Notice that by interchanging the resulting rows and columns, we get

$$\begin{matrix} * & * & * \\ * & * & * \\ * & * & \\ * & & \\ * & & \end{matrix}$$

a partition of 10 into 5 parts none of which exceeds 3. The equality of $(w : i, j)$ and $(w : j, i)$ falls easily from this one-to-one correspondence.

To verify Sylvester's algorithm, Franklin argued simply that those partitions associated to $(w : i, j)$ with highest part *not* repeated equaled those partitions associated to $(w - 1 : i, j)$ *not* containing i. His proof was elegant; it began by invoking Ferrers's theorem $(w : i, j) = (w : j, i)$. "Now, if we do this," he argued, "it is plain that those partitions whose highest number was not repeated become partitions containing 1; and that those partitions which did not contain i become partitions ... containing 0 ... . And it is plain that the number of partitions of w which contain 1 is equal to the number of partitions of $w - 1$ which contain 0."[67]

As this argument shows, by 1879, Franklin had already become comfortable with partition-theoretic argumentation and with at least one of the graphical devices—Ferrers's theorem—which would come to play extremely important roles in subsequent work of Franklin and most of the Hopkins school. The "Note on Partitions" also revealed a familiarity with another key notion in partition theory, namely, the interpretation of numbers such as $(w : i, j)$ in terms of generating functions.[68] Franklin developed this idea in an invariant-theoretic context in 1880.[69]

---

[66]Ferrers entrusted the proof of his theorem to Sylvester. See James Joseph Sylvester, "Note on the Graphical Method in Partitions," *Johns Hopkins University Circulars* 2 (1883):70–71, or *Math. Papers JJS*, 3:683–684; and "On Mr. Cayley's Impromptu Demonstration of the Rule for Determining at Sight the Degree of Any Symmetrical Function of the Roots of an Equation Expressed in Terms of the Coefficients," *Philosophical Magazine* 5 (1853):199–202, or *Math. Papers JJS*, 1:595–598 on p. 597.

[67]Franklin, "Note on Partitions," p. 187.

[68]*Ibid.*, p. 188. The concept of a generating function, recall, is defined in note 9 above.

[69]Fabian Franklin, "On the Calculation of the Generating Functions and Tables of Groundforms for Binary Quantics," p. 128.

An expository piece growing out of his work with Sylvester on calculating the groundforms for binary forms of degree up to and including ten, Franklin's 1880 article represented a much needed encapsulation of the British methods for effecting these computations. In particular, it focused on generating functions and their relationship to the values $(w : i, j)$ arising in Cayley's theorem on the number of linearly independent covariants of a given degree and order associated to a given form. Echoing Cayley in the 1856 "Second Memoir Upon Quantics," Franklin stated the fact, well-known since the time of Euler, that " $(w : i, j)$ is the coefficient of $c^j z^w$ in the development of $\frac{1}{(1-c)(1-cz)(1-cz^2)\cdots(1-cz^i)}$ .... "[70] As he went on to elaborate,[71]

> [f]rom [this] is also deduced Euler's theorem that $(w : i, j)$ is the coefficient of z^w in the development of
> $$\frac{(1-z^{j+1})(1-z^{j+2})\cdots(1-z^{j+i})}{(1-z)(1-z^2)\cdots(1-z^i)}$$
> so that $\Delta(w : i, j)$ is the coefficient of z^w in the development of
> $$\frac{(1-z^{j+1})(1-z^{j+2})\cdots(1-z^{j+i})}{(1-z^2)\cdots(1-z^i)}.$$

However, if, as we have seen, $(w : i, j)$ did not yield easily to brute force computations, it hardly appeared more accessible through a calculation of the coefficients of these generating functions. Franklin admitted that, in these forms, generating functions were "far from being well adapted for calculation, and moreover furnishe[d] no means of determining the complete system of groundforms of the quantic."[72] Nevertheless, Cayley and Sylvester had developed techniques for successfully manipulating these functions, and Franklin presented reasonably cogent descriptions of these methods in his paper.

By the time he earned his Ph.D. later in 1880, Franklin had clearly mastered not only much of the British approach to invariant theory but also

[70] *Ibid.*, p. 129.

[71] *Ibid.* Thus, the Gaussian polynomial $\left[{i+j \atop i}\right] := \frac{(1-z^{j+1})(1-z^{j+2})\cdots(1-z^{j+i})}{(1-z)(1-z^2)\cdots(1-z^i)}$ is equal to $\Sigma(w : i, j)z^w$, an important combinatorial identity. Sylvester had already hit upon a key property of these coefficients $(w : i, j)$ in his 1879 "Proof of a Hitherto Undemonstrated Fundamental Theorem of Invariants," discussed above: the sequence $(w : i, j)$, $w = 0, \ldots, i+j$ is *unimodal*; that is, it increases up to a point and then decreases. (Sylvester did not use this terminology, however.) Such counting sequences have attracted considerable attention in combinatorics. In particular, there has been interest in finding a *constructive* proof of the unimodality of the $(w : i, j)$. See, for example, Doron Zeilberger, "Kathy O'Hara's Constructive Proof of the Unimodality of the Gaussian Polynomials," *The American Mathematical Monthly* 96 (1989):590–602; and Robert A. Proctor, "Solution of Two Difficult Combinatorial Problems with Linear Algebra," *The American Mathematical Monthly* 89 (1982):721–734. Each of these papers also contains pertinent references to the modern literature.

[72] Franklin, "On the Calculation of the Generating Functions," p. 129.

most of what was then known about partitions. The University rewarded his expertise with a continuation of his assistantship in the Mathematics Department for the 1880–1881 academic year, and Sylvester showed his support by putting the course on "Determinants and Modern Algebra" in Franklin's capable hands. Although he now served officially as his teacher's colleague, Franklin continued to participate in Sylvester's courses. In particular, he sat in on the number theory course, which Sylvester had begun teaching in the fall of 1879 to a class including Arthur Hathaway, Christine Ladd, and Oscar Mitchell.[73] As before, Franklin had produced new, course-related results by midterm, and, once again, they were partition-theoretic. Sometime prior to March of 1881, Franklin had come up with a surprisingly elementary, purely graphical proof of Euler's theorem on pentagonal numbers.

Dating from the mid-eighteenth century, Euler's pentagonal number theorem equated the following infinite product and sum:

$$\prod_{m=1}^{\infty}(1-x^m) = \sum_{n=-\infty}^{\infty}(-1)^n x^{\frac{n(3n-1)}{2}}. \tag{3.7}$$

For positive n, the exponents $m = \frac{n(3n-1)}{2}$ form the infinite sequence $1, 5, 12, 22, \ldots$ of pentagonal numbers, thereby giving the theorem its name. (Note that, for $n \geq 0$, $\frac{n(3n-1)}{2} = \frac{(-n)(3(-n)+1)}{2}$, so we might call numbers of the form $\frac{n(3n\pm1)}{2} (n \geq 0)$ *generalized* pentagonal numbers.) When Leonard Euler presented this result in 1748, he aimed to link partition theory with generating functions such as the one on the left-hand side of (3.7). However, in this work, he did not include a properly partition-theoretic interpretation of (3.7).[74] By 1830, Adrien-Marie Legendre had made this connection in his *Théorie des Nombres*. There, Legendre employed an argument by generating functions (in the spirit of Euler) to show that, for any positive integer m, the difference between the number of partitions of m into an even number of parts and the number of partitions of m into an odd number of parts equaled

[73]"Enumeration of Classes, First Half-Year, 1880–1881," *Johns Hopkins University Circulars* 1 (1880):74. (This internal publication provides a wealth of information about the early years of the Johns Hopkins University.) Growing out of this course, the Hopkins group also produced results in number theory. See, for example, Oscar H. Mitchell, "On Binomial Congruences; Comprising an Extension of Fermat's and Wilson's Theorems, and a Theorem of Which Both Are Special Cases," *American Journal of Mathematics* 3 (1880):294–315, and "Some Theorems in Numbers," *American Journal of Mathematics* 4 (1881):25–38; James Joseph Sylvester, "On Tchebycheff's Theory of the Totality of the Prime Numbers Comprised Within Given Limits," *American Journal of Mathematics* 4 (1881):230–247; George S. Ely, "Bibliography of Bernoulli Numbers," *American Journal of Mathematics* 5 (1882):228–235; and Arthur S. Hathaway, "Some Papers on the Theory of Numbers," *American Journal of Mathematics* 6 (1884):316–330.

[74]Leonard Euler, *Introduction to the Analysis of the Infinite*, Book 1, pp. 273–274. For a discussion of Euler's work on the pentagonal number theorem, see André Weil, *Number Theory: An Approach through History from Hammurapi to Legendre*, (Boston: Birkhäuser, 1984), pp. 279–281. Also, see Komaravolu Chandrasekharan, *Elliptic Functions* (New York: Springer-Verlag, 1985), Ch. 8, which contains a discussion (with references) of Jacobi's work in this direction.

$(-1)^n$ if $m = \frac{n(3n\pm1)}{2}$ and zero otherwise.[75] In particular, this difference was nonzero if and only if m was a generalized pentagonal number. In his 1881 article in the *Comptes rendus* of the French *Académie des Sciences*, Fabian Franklin proved Legendre's version of the pentagonal number theorem without recourse to generating functions.[76] He had hit upon a clever graphical justification of the result. In order to convey the flavor of Franklin's argument without setting up too much notation, consider the following example.[77]

Consider the partitions $\lambda = (8, 7, 5, 3, 2)$ and $\lambda' = (9, 8, 5, 3)$ of 25 and display them graphically to get

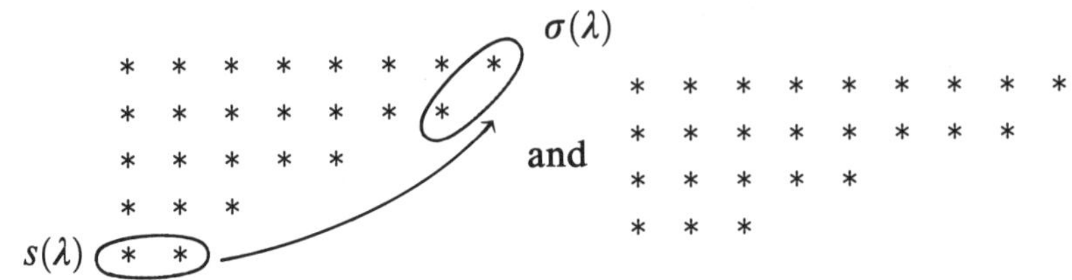

respectively. Notice that $(8, 7, 5, 3, 2)$ is a partition of 25 into an odd number of parts while $(9, 8, 5, 3)$ is a partition of 25 into an even number of parts. Let $s(\lambda)$ be defined as the smallest part in the given partition λ, and let $\sigma(\lambda)$ denote the length of the longest diagonal drawable from the uppermost right-hand node of λ. As shown in the diagram, $\sigma(\lambda) = 2$ and $s(\lambda) = 2$. For an arbitrary partition λ of an integer m, whenever $s(\lambda) \leq \sigma(\lambda)$, we can always transform λ into a new partition λ' by removing $s(\lambda)$ and attaching it alongside $\sigma(\lambda)$ as indicated above, except when m is of the form $\frac{n(3n-1)}{2}$, for $1 \leq n \in \mathbb{Z}$. (Note that the transformation cannot be carried out when the number of parts in the partition λ of m is n and $\sigma(\lambda) = s(\lambda) = n$. It is easy to see that these restrictions imply that $m = n + (n+1) + \cdots + (2n-1) = \frac{n(3n-1)}{2}$.) When λ has an odd (respectively, even) number of parts, the λ' which results necessarily has an even (respectively, odd) number of parts. In his proof, Franklin also showed that a perfectly analogous dissection applied if $s(\lambda) > \sigma(\lambda)$, except when m is of the form $\frac{n(3n+1)}{2}$, for $1 \leq n \in \mathbb{Z}$. (Note that in this case the transformation cannot be carried out precisely when the partition λ of m has n parts, $\sigma(\lambda) = n$, and $s(\lambda) = n + 1$. These restrictions force $m = (n+1) + (n+2) + \cdots + 2n = \frac{n(3n+1)}{2}$.) Franklin thus concluded that all partitions λ of m satisfy one or the other of these conditions on $s(\lambda)$ and $\sigma(\lambda)$, except certain partitions of pentagonal numbers. Furthermore, his method showed that, when $m = \frac{n(3n\pm1)}{2}$, the number of λ having an even

[75] Adrien-Marie Legendre, *Théorie des Nombres*, 2 vols. (Paris: Firmin Didot, 1830; reprint ed., Paris: Albert Blanchard, 1955), 2:128–133.

[76] Fabian Franklin, "Sur le Développement du Produit infini $(1-x)(1-x^2)(1-x^3) \cdot (1-x^4)\cdots$," *Comptes rendus* 82 (1881):448–450.

[77] In what follows, we use the notation set up in Andrews, pp. 10–11. The example presented here, like the one given in Andrews, comes from Franklin's paper. See Franklin, "Sur le Développement du Produit infini," p. 450.

number of parts exceeds the number of partitions λ having an odd number of parts by $(-1)^n$.

Presented to the *Académie* by Charles Hermite, Sylvester's longtime friend and the *grand homme* of French mathematics, Franklin's note and the technique he used in it raised eyebrows in France. Writing to Sylvester on 29 April 1881, Hermite described the interest generated in French mathematical circles by the young American's ideas:

> It certainly will not be unpleasant for you to hear that I was not the only person to be very interested in Mr. Franklin's very original and ingenious proof .... Mr. Halphen, one of our most eminent young mathematicians, who has just won the Academy's Grand Prix in mathematics, found Franklin's method so remarkable that he lectured on it in one of the most recent sessions of the Philomathic Society. Please tell Mr. Franklin that his talent is appreciated, as it deserves to be, by the mathematicians of the old world.[78]

This praise for his student's accomplishment, coming as it did from Hermite, must have especially gratified Sylvester.

Three years earlier, this same Hermite had questioned the ability of the American people to contribute to research-level mathematics and, by implication, the ultimate wisdom of Sylvester's decision to move to the United States. In a letter to his transplanted British friend dated 16 August 1878, Hermite had wondered: "[i]s there really a mathematical future for the New World, and, the extraordinary genius of the American people which has revealed itself in the physical discoveries of Edison with an unexpected flash, could it also be called upon to open new paths into analysis [i.e., mathematics]?"[79] By April of 1881, however, Franklin's graphical proof of the pentagonal number theorem along with many results published in the first three volumes of the *American Journal of Mathematics* gave Hermite and other skeptical Europeans little reason to doubt the mathematical potential of the Americans.

As noted, Franklin had made his breakthrough by March of 1881, most probably within the context of the course on number theory Sylvester gave during the 1880–1881 academic year. That spring found Sylvester mathematically torn between two research agendas: results in classical number theory suggested by his teaching[80] and the ever-elusive proof of Gordan's theorem.

[78]Charles Hermite to J. J, Sylvester, 29 April 1881, Gilman Papers (our translation).

[79]Charles Hermite to J. J. Sylvester, 16 August 1878, Sylvester Papers SJC, Box 2 (our translation).

[80]Sylvester wrote various papers on number theory in the early 1880s. In addition to his work "On Certain Ternary Cubic-Form Equations," Sylvester published such articles as "Sur les Diviseurs des Fonctions cyclotomiques," *Comptes rendus* 90 (1880):287–289 and 345–347, or *Math. Papers JJS*, 3: 428–432; "Sur la Loi de Réciprocité dans la Théorie des Nombres," *Comptes rendus* 90 (1880):1053–1057 and 1104–1106, or *Math. Papers JJS*, 3:433–437; and "Sur

In a letter to Cayley, Sylvester betrayed this divided mathematical loyalty, describing not only Franklin's "beautiful proof" of the pentagonal number theorem but also announcing that "I believe that I have proved Gordan's theorem and moreover can assign a superior limit to the number of fundamental invariants."[81] When his efforts to write up the complete argument failed three days later, he had to admit that "[t]he theorem which I sent you the day before yesterday is wrong—it proceeded from a certain new principle which is correct—but the application made of the principle was erroneous."[82]

Sylvester continued to pursue the proof of Gordan's theorem via related invariant-theoretic researches throughout the remainder of 1881. His quest was interrupted first in January of 1882 by Cayley's much-hoped-for and long-awaited visit to Baltimore and then by the depression which followed his friend's return to England, but Sylvester returned to invariant theory afresh in the fall of 1882. The specter of Gordan's theorem materialized in his work almost immediately. In his attempts to analyze the groundforms further, Sylvester had reduced the problem to one of showing that a certain sequence increased without bound. He explained in a letter to Cayley on 6 September that

> Of this law of progression holding I have little or no doubt and the proof (however difficult it may be) belongs to the province of ordinary algebra or the theory of partitions.
>
> Supposing it established, I have *proved* Gordan's theorem.[83]

Although by 6 October Sylvester had realized that his "supposed proof of Gordan's theorem was a *Delusion*," his latest work toward its proof had focused his thoughts once again on partition-theoretic arguments.[84] In fact, this new line of research had so captivated him by November of 1882 that in December he abandoned his regularly scheduled, year-long course on number theory to teach a specialized course on the theory of partitions in the spring of 1883. Working in concert with the ten students in the class, Sylvester developed the graphical approach to partition theory suggested by Franklin's note and linked it to the traditional generating function techniques. The results of their combined labors testify to the vitality of the mathematical laboratory at the Johns Hopkins.

The researches of Sylvester and his students, playfully entitled "A Con-

---

les Diviseurs des Fonctions des Périodes des Racines primitives de l'Unité," *Comptes rendus* 92 (1881):1084–1086, or *Math. Papers JJS*, 3:479–480.

[81] J. J. Sylvester to Arthur Cayley, 23 March 1881, Sylvester Papers SJC, Box 11.

[82] J. J. Sylvester to Arthur Cayley, 26 March 1881, Sylvester Papers SJC, Box 11.

[83] J. J. Sylvester to Arthur Cayley, 6 September 1882, Sylvester Papers SJC, Box 11. Sylvester's emphasis.

[84] J. J. Sylvester to Arthur Cayley, 6 October 1882, Sylvester Papers SJC, Box 11. Sylvester's emphasis.

structive Theory of Partitions, Arranged in Three Acts, an Interact and an Exodion," appeared in 1883 in the 1882 volume of the *American Journal.*[85] They took a novel approach to questions involving partitions: instead of viewing them analytically as generating functions, they treated them more abstractly as mathematical entities in and of themselves. In short, they established a *partition theory*, which they then interpreted in terms of the extant theory of generating functions.[86] Sylvester opened the paper by laying down the premise upon which their new point of view depended. "In the new method of partitions," he explained,

> it is essential to consider a partition as a *definite thing*, which end is attained by regularization of the succession of its parts according to some prescribed law. The simplest law for the purpose is that the arrangement of the parts shall be according to their order of magnitude. A leading idea of the method is that of correspondence between different complete systems of partitions regularized in the manner aforesaid. The perception of the correspondence is in many cases facilitated by means of a graphical method of representation, which also serves *per se* as an instrument of transformation.[87]

Thus, Sylvester advocated displaying partitions graphically as in (3.6) and then proving theorems through clever reorganizations of the graphs. Once the theorems had been determined in this way, they could be translated into analogous theorems about generating functions, thereby circumventing the often otherwise complicated argumentation. In short, he sought to fashion a theory around the technique Franklin had used to advantage for the pentagonal number theorem. Once again, an example conveys best the spirit of the work.[88]

Consider the partition $(9, 6, 5, 4, 3, 2, 1, 1, 1)$ of 32 represented

---

[85] James Joseph Sylvester, "A Constructive Theory of Partitions, Arranged in Three Acts, an Interact, and an Exodion," *American Journal of Mathematics* 5 (1882):251–330, or *Math. Papers JJS*, 4:1–83.

[86] The work of the Sylvester school generated contemporary interest and further research, particularly in Great Britain. See, for example, the numerous papers on partition theory by Percy Alexander MacMahon in the first volume of *Collected Papers*, ed. George E. Andrews, 2 vols. (Cambridge, MA: MIT Press, 1978–1986). The work continues to retain its interest and importance today. George Andrews pointed out in his book, *The Theory of Partitions*, that this graphical approach, while elementary, is potent and "effective [p. 6]." He further characterized the paper, "A Constructive Theory of Partitions," by Sylvester and his students as "monumental [p. 14]."

[87] *Ibid.*, p. 251, or p. 1. Sylvester's emphasis.

[88] For a discussion of the content of this paper, see Parshall, "America's First School of Mathematical Research," pp. 184–188.

graphically:

* * * * * * * * *
* * * * * *
* * * * *
* * * *
* * *
* *
*
*
*

Notice that this graph has the property that interchanging its rows and columns results in exactly the same graph, namely, one corresponding to the original partition (9, 6, 5, 4, 3, 2, 1, 1, 1) of 32. A graph having this property is termed *self-conjugate* or *symmetric* as is its associated partition.

As Sylvester and others had realized, self-conjugate partitions arose in the context of tabulating symmetric functions, a problem not unrelated to that of efficiently displaying complete tables of linearly independent covariants of given degrees and orders associated to a given binary form. By the spring of 1882, Sylvester had set first-year graduate student, William Pitt Durfee, on this problem, and Durfee had made some important observations relative to the graphical method.[89]

Again, referring to the self-conjugate partition (9, 6, 5, 4, 3, 2, 1, 1, 1) of 32 above, Durfee noticed that its graph decomposes into a maximal (now called *Durfee*) square and two conjugate appendages as follows:

(3.8)

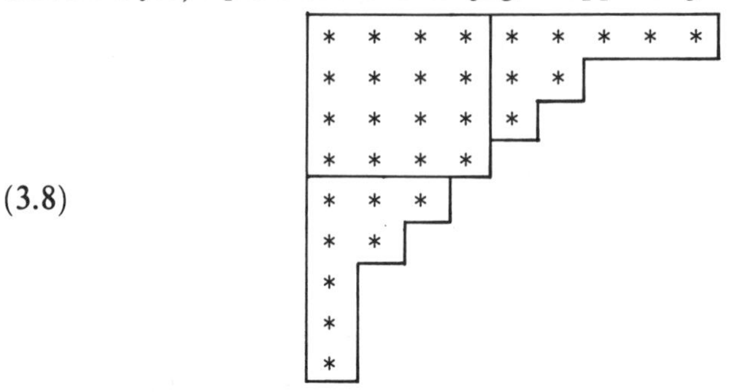

Thus, this graph of 32 has maximal square of side 4 containing $4^2 = 16$ nodes, while the two appendages each contain $\frac{32-4^2}{2}$ nodes. Sylvester incorporated Durfee's decomposition into the evolving, abstract graphical theory by stating the general fact that any self-conjugate graph of a partition of n "may be dissected into a square (which may contain one or any greater square number) of say i^2 nodes, and two perfectly similar appended graphs, each having the content $\frac{n-i^2}{2}$, and subject to the sole condition that the number

[89] William Pitt Durfee, "Tables of the Symmetric Functions of the Twelfthic," *American Journal of Mathematics* 5 (1882):45–61; and "The Tabulation of Symmetric Functions," *American Journal of Mathematics* 5 (1882):348–349.

of its lines (or columns), that is that the number (or magnitude) of the parts in the partition which it [i.e., the appendage] represents, shall be i or less."[90] As this formulation made clear, the collection of self-conjugate graphs (with maximal square of side i) of some number n was equivalent to the collection of partitions of $\frac{n-i^2}{2}$ into parts none of which exceeded i. With this identity established graphically, Sylvester next presented the well-known interpretation of this latter collection of partitions within the classical theory of generating functions.

Consider the product

$$\frac{1}{(1-x)(1-x^2)(1-x^3)\cdots(1-x^i)}. \tag{3.9}$$

One approach to evaluating this expression involves explicitly writing each of the i factors $\frac{1}{1-x^j}$ in its power series expansion. This transforms the product in (3.9) into this product of infinite sums:

$$(1+x^1+x^{1+1}+x^{1+1+1}+\cdots)(1+x^2+x^{2+2}+x^{2+2+2}+\cdots) \\ \cdots(1+x^i+x^{i+i}+x^{i+i+i}+\cdots).$$

Finally, formal multiplication termwise yields a representation of this product as an infinite series of the form $\sum_{k=1}^{\infty} a_k x^k$, where a_k is precisely the number of partitions of k with no parts exceeding i. Thus, the generating function (3.10)

$$\frac{1}{(1-x)(1-x^2)(1-x^3)\cdots(1-x^i)} = \sum_{k=0}^{\infty} \begin{pmatrix} \text{number of partitions of} \\ k \text{ into parts} \leq i \end{pmatrix} x^k,$$

which gives the desired product-sum identity.[91] Sylvester next applied this result to the graphical observation made above to deduce effortlessly yet another product-sum identity.

In (3.10), he let $k = \frac{n-i^2}{2}$ to get that the number of partitions of $\frac{n-i^2}{2}$ into parts none of which exceed i equals the coefficient of $x^{\frac{n-i^2}{2}}$ in

$$\frac{1}{(1-x)(1-x^2)(1-x^3)\cdots(1-x^i)},$$

or, as a simple algebraic manipulation shows, to the coefficient of x^n in

$$\frac{x^{i^2}}{(1-x^2)(1-x^4)\cdots(1-x^{2i})}.$$

[90] Sylvester, "A Constructive Theory of Partitions," p. 276, or p. 26.

[91] Euler carried out these same calculations in 1748. He, however, did not explicitly express the coefficients on the right-hand side of (3.10) in partition-theoretic terms. For the reference, see note 9 above.

This immediately yielded the product-sum identity[92]

$$\frac{x^{i^2}}{(1-x^2)(1-x^4)\cdots(1-x^{2i})}$$
$$= \sum_{n=0}^{\infty} \begin{pmatrix} \text{number of self-conjugate partitions} \\ \text{of } n \text{ with Durfee square of side } i \end{pmatrix} x^n.$$

While elementary, this argument highlighted the efficacy of the new graphical methods in interpreting generating functions and only hinted at the full capabilities of the graphical techniques in establishing partition-theoretic identities. Since the so-called constructive method hinged on clever decompositions and rearrangements of given graphs, at least one natural question arose: given a class of partitions, to what other distinct class or classes of partitions is it equivalent? Put another way, what classes of partitions are isomorphic? To get an idea of the ease with which the constructive method handled this question, consider once more the case of the self-conjugate graphs.

As noted above, the Durfee decomposition of a self-conjugate graph involves dividing it up into its maximal square and its two isomorphic appendages. Clearly, however, other decompositions are possible. Returning to the partition $(9, 6, 5, 4, \ 3, 2, 1, 1, 1)$ of 32, suppose that instead of decomposing it as in (3.8), we dissect it into what Sylvester called "bends" or "angles" in the following way:

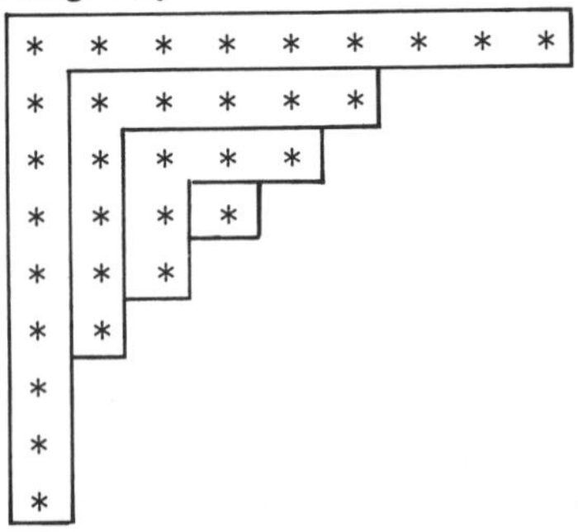

```
* * * * * * * * *
* * * * * *
* * * * *
* * * *
* * *
* *
*
*
*
```

Counting up the nodes in each of the four angles yields the new partition $(17, 9, 5, 1)$ of 32 into unrepeated, odd parts. Sylvester immediately recognized the generalization from this example: the number of self-conjugate graphs of n equals the number of partitions of n into unrepeated, odd parts, and so

$$\frac{x^{i^2}}{(1-x^2)(1-x^4)\cdots(1-x^{2i})} = \sum_{n=0}^{\infty} \begin{pmatrix} \text{number of partitions of } n \text{ into} \\ \text{exactly } i \text{ unrepeated odd parts} \end{pmatrix} x^n.$$

This time, the new product-sum identity fell out of the graphical method as a bonus; it required absolutely no additional finagling.[93] Following the

[92] *Ibid.*, p. 276, or pp. 26–27.

[93] *Ibid.*, pp. 277–278, or pp. 28–29. Similarly, Sylvester and his school proved results involving the Gaussian polynomials, as defined in note 71 above.

simple prescript of treating graphical entities graphically, the Sylvester school had demonstrated the power of its new constructive techniques; moreover, their results showed the fruitfulness of Sylvester's approach to teaching at the graduate level.[94]

Indeed, this research had emerged from a Sylvester *school*, that is, from the united and concerted effort of Sylvester and his students. Although Sylvester's name alone appeared as author of this mathematical play in three acts, an interact, and an exodion, inside it the contributing students in his class received their credit due. Of the student actors, Fabian Franklin, having the most experience, earned the starring role and dominated Act One, Scenes 6, 12, and 20 as well as the First Interact, Scenes 24 and 25. William Durfee appeared prominently in Act Two, Scene 27 with his special decomposition of self-conjugate graphs. Among the lesser actors, second-year graduate student, George Ely, played in Scenes 50 and 51 of the Second Interact, while third-year student, Arthur Hathaway, walked on in the Exodion. Working behind the scenes, others also competed for parts, whether large or small. Sylvester conveyed the sense of common cause, if not outright competition, on the Hopkins mathematical stage in a letter to Cayley written during production. He proudly declared that "I sent to the Comptes Rendus two or three days ago my proof of the wonderful theorem (discovered by observation) [on] partitions of n into odd numbers and its partitions into unequal numbers. Franklin, Mrs. Franklin [the former Christine Ladd], Story, Hathaway, Ely, and Durfee were all at work trying to find the proof—but I was fortunately beforehand with the theory and the only one in at the death."[95] Directed by Sylvester, the Hopkins players sharpened their mathematical talent on the research stage and aspired to take starring roles themselves.

[94]Using the constructive method, Sylvester also gave alternate proofs of the pentagonal number theorem in "An Instantaneous Graphical Proof of Euler's Theorem on Partitions of Pentagonal and Non-Pentagonal Numbers," *Johns Hopkins University Circulars* 2 (1882):71, or *Math. Papers JJS* 3:685–686; and Euler's theorem on the number of partitions of a number into odd and into distinct parts in "Démonstration graphique d'un Théorème d'Euler concernant les Partitions des Nombres," *Comptes rendues* 96 (1883):1110–1112, or *Math Papers JJS* 4:95–96. In their paper's "final act" and "exodion," Sylvester and his students demonstrated that their method also worked in establishing more complicated product-sum identities like Jacobi's identity: $(1 \pm q^{n-m})(1 \pm q^{n+m})(1-q^{2n})(1 \pm q^{3n-m})(1-q^{3n+m})(1-q^{4n}) \cdots = \sum_{i=-\infty}^{+\infty} (\pm 1)^i q^{ni^2+mi}$. Their specific results have also generated interest in modern mathematical circles (see, for instance, George F. Andrews, "On a Partition Problem of J. J. Sylvester," *Journal of the London Mathematical Society* 2 (1970):571–576, and "Generalizations of the Durfee Square," *Journal of the London Mathematical Society* 3 (1971):563–570). Needless to say, many of the constructs they studied, like the Gaussian polynomials, have become crucially important in areas as diverse as statistics and quantum groups, but any analysis of this vast literature would take us much too far afield of our present historical objectives.

[95]J. J. Sylvester to Arthur Cayley, 16 March 1883, Sylvester Papers SJC, Box 11. Sylvester's emphasis.

Research in Mathematical Logic at the Johns Hopkins University

As evidenced by the foray into partition theory, the mathematics graduate student had unique opportunities at Hopkins. Thanks to Gilman's tripartite instructional system consisting of Professors, Visiting Lecturers, and student Fellows, the graduate program had the flexibility to cover not only the basics but also various special topics. When Gilman hired the geodesist-mathematician-logician-philosopher Charles Sanders Peirce as Lecturer in Logic, one of those special topics thus became mathematical logic, taught by a man whom the twentieth century would recognize as one of America's premier philosophers.

Charles Peirce, son of Harvard mathematician Benjamin Peirce, joined the United States Coast Survey in 1859 in his twentieth year. Serving in the post of Assistant from 1861 until his forced retirement from government service in 1891, he worked on a wide variety of geodetic projects—including eclipse observations, pendulum studies, and mathematical cartography—while he privately pursued his studies in philosophy, logic, and pure mathematics.[96] By the 1870s, his Survey work had won Peirce recognition as a geodesist, although he was as yet little known in mathematical logic. This fact may go a long way toward explaining Gilman's initial response to Peirce's inquiry regarding a permanent faculty position in logic at Hopkins. Replying on 23 January 1878 to Peirce's letter of 18 December 1877, Gilman explained that, after consultation with several specialists within the University, he could not see his way clear to making Peirce an official offer. Nevertheless, he did not close the door on further negotiations. He promised that "I shall continue however to reflect upon all the possibilities. One thought I have had is this;—that retaining your place in the Coast Survey, you might like to be a Professor of Logic, & give to the University a certain portion of your time, what might be considered on half-service; or that you might like to be included among our non-resident lecturers, coming here for the next two or three years, at an appointed time."[97]

This was clearly not the sort of response that Peirce had hoped for from the University. The disappointment turned to anger when he heard it rumored that one of the specialists consulted, his family's long-time friend Sylvester, had not spoken highly of his work in logic. George Bruce Halsted, who, as we

---

[96]Much has been written on the life, works, and thought of Charles Sanders Peirce. For some idea of the vast literature, we direct the interested reader to Joseph Brent, *Charles Sanders Peirce: A Life* (Bloomington: Indiana University Press, 1993) and Charles C. Gillispie, ed., *Dictionary of Scientific Biography*, 16 vols. 2 supps. (New York: Charles Scribner's Sons, 1970–1990), s.v. "Peirce, Charles Sanders," by Carolyn Eisele. On Peirce's properly mathematical work, see the definitive, multivolume work, Charles S. Peirce, *The New Elements of Mathematics*, ed. Carolyn Eisele, 4 vols. in 5 pts. (The Hague: Mouton Publishers; Atlantic Highlands, N. J.: Humanities Press, 1976).

[97]Daniel C. Gilman to Charles S. Peirce, 23 January 1878, Gilman Papers.

have seen, did not share Sylvester's interest in invariant theory, had shifted his mathematical focal point to non-Euclidean geometry while maintaining the interest in logic he had nurtured as an undergraduate under the influence of his Princeton professor, James McCosh. During the course of his preparations for a series of lectures on modern logic in the spring of 1878, Halsted wrote an injudicious letter to Peirce recounting some of the opinions he claimed to have run up against in his efforts to include Peirce's ideas in his presentation. In Halsted's words,

> I desired to devote a lecture to you by name, and to attempt to give the people here an account of your contributions to Modern Logic. But the opinion was expressed here to me that I had better not do it and that you had exhibited in your writings a tendency to undervalue everybody and everything you mentioned. Besides this I was particularly discouraged by Prof. Sylvester's adding that your articles in the Popular Science [M]onthly were pretentious without being at all profound and that any body [sic] could have written them.[98]

Not surprisingly, this report, coming as it did from such an unexpected source, occasioned a bitter letter from Peirce to Sylvester. "I was surprized [sic] to learn from the enclosed letter," Peirce wrote,

> that you are acting against my being invited to the Johns Hopkins University. I thought you had given me to understand that you would be friendly to me in this matter.
>
> In regard to your opinion, in itself, I have nothing to say. I am satisfied with the reception which my writings met with in the logical world, and the opinion of outsiders does not greatly concern me.[99]

Distraught by what he viewed as a terrible misunderstanding, Sylvester went immediately to Gilman, who willingly and promptly intervened on the Professor's behalf. Gilman wrote to Peirce assuring him "that in his conversations with me, so far from appearing unfriendly to you [Sylvester] has constantly referred to you in terms of high appreciation and has left upon me the abiding impression that in his opinion you would bring to the J. H. University all means of intellectual power [?] which would be most [desirable in the same body?] in their influence upon other departments of university work."[100] This assurance from Gilman apparently assuaged Peirce and smoothed relations between him and Sylvester, for on 29 March, Sylvester informed Gilman that he was leaving Baltimore to spend a week in New York City and that he could

[98] George B. Halsted to Charles S. Peirce, 14 March [1878], Gilman Papers.

[99] Charles S. Peirce to J. J. Sylvester, 19 March 1878, Gilman Papers.

[100] Daniel C. Gilman to Charles S. Peirce, 20 March 1878, Gilman Papers. Words in [ ] are smeared in the original.

be reached there in "care of C. S. Peirce Esq 42 7^{h} St."[101]

Although Peirce and Sylvester emerged from this episode amicably, the same might not be said for Sylvester and his young student, Halsted. Whether because Peirce was not named to the faculty for the fall of 1878 and so would be unavailable to advise him in his dissertation research or because he felt he could no longer work under Sylvester given the circumstances, Halsted left the University to spend the 1878–1879 academic year back at Princeton. While there he wrote the dissertation on a "Basis for a Dual Logic" which earned him the Hopkins Ph.D. *in absentia* in 1879.

When Peirce did join the Hopkins faculty part-time in the fall of 1879, the unpleasantness of the spring of 1878 seemed completely forgotten. He enjoyed close ties with the Mathematics Department throughout his five-year-long association with the University and drew a majority of his students from the mathematical ranks. Among the faithful in his classes, the talented and controversial Christine Ladd developed into his closest disciple.

Ladd and her fellow graduate students learned firsthand of Peirce's work on what he termed "the logic of relatives" through his courses as well as through his lectures in both the University's Metaphysical and Mathematical Clubs. Initially formulated as early as 1870, Peirce's version of symbolic logic received thorough treatment in the first number of the *American Journal's* third volume in 1880. There, Peirce drew from and reacted to the work primarily of Augustus DeMorgan and George Boole to systematize an algebra which governed his notion of relatives.[102]

For Peirce, a relative

> is a term whose definition describes what sort of a system of objects that is whose first member (which is termed the *relate*) is denoted by the term; and names for the other members of the system (which are termed the *correlates*) are usually appended to limit the denotation still further.
> ...
> A relative of only one correlate, so that the system it supposes is a pair, may be called a *dual* relative.[103]

Thus, if $A, B, C, \ldots$ represent mutually disjoint sets of individuals, Peirce designated the set of dual relatives formed from these sets by

$$\begin{array}{cccc} A:A & A:B & A:C & \ldots \\ B:A & B:B & B:C & \ldots \\ C:A & C:B & C:C & \ldots \\ \vdots & \vdots & \vdots & \end{array}$$

By way of example, consider a family, and suppose that A denotes the set

[101] J. J. Sylvester to Daniel C. Gilman, 29 March 1878, Gilman Papers.

[102] Charles S. Peirce, "On the Algebra of Logic," *American Journal of Mathematics* 3 (1880): 15–57.

[103] *Ibid.*, p. 43. Peirce's emphasis.

of parents while B stands for the set of all children, then the set of dual relatives corresponding to A and B is

$$\begin{matrix} A:A & A:B \\ B:A & B:B \end{matrix}$$

where $A : A =$ spouse, $A : B =$ parent, $B : A =$ child, and $B : B =$ sibling. In his 1880 paper, Peirce gave a full algebraic treatment of this new symbolism, which included such observations as $(A : B)(B : A) = (A : A)$ and $(A : B)(B : C) = (A : C)$.[104]

By 1882, Christine Ladd had extended her professor's symbolic study even further. She submitted her findings in the form of a doctoral dissertation to the Hopkins authorities in 1882 and published them a year later in a volume of essays edited by Peirce.[105] Given the as yet unaltered policy on women students at the University, however, Ladd's degree petition was denied over the protestations of the Mathematics Department. The University sought to redress this wrong by awarding Christine Ladd-Franklin her long overdue degree at its semicentennial celebrations more than forty years later in 1926.[106]

Although perhaps not immediately obvious, the ideas which Charles Peirce expounded on the symbolic manipulation of dual relatives paralleled certain aspects of the algebra of matrices, which had been developing since at least the 1850s in the hands of such mathematicians as Arthur Cayley. (See the next section below.) Furthermore, as noted in Chapter 1, Benjamin Peirce had classified all linear algebras of dimension one through six in 1870, an accomplishment well understood and appreciated by his son, Charles. Not surprisingly, then, the interrelation of the symbolic algebra of dual relatives (also developed in 1870) and linear associative algebra did not escape Charles Peirce's notice.

In a paper communicated by his father and published in the *Proceedings of the American Academy of Arts and Sciences* in 1875, Charles showed, using his algebra of dual relatives, that his father's linear algebras actually corresponded to matrix algebras. Benjamin underscored the importance of

[104]For these properties, see *ibid.*, pp. 45–46 and 50, respectively. Compare Thomas Hawkins, "Hypercomplex Numbers, Lie Groups, and the Creation of Group Representation Theory," *Archive for History of Exact Sciences* 8 (1972):243–287 on p. 246.

[105]Christine Franklin, "On the Algebra of Logic," in *Studies in Logic by Members of the Johns Hopkins University*, ed. Charles S. Peirce (Boston: Little and Co., 1883), pp. 17-71.

[106]Ladd-Franklin eventually held lectureships at the Johns Hopkins (1904–1909) and at Columbia (1910–1930), continuing her work in logic as well as in the theory of color vision. She was also a notable figure in the early women's movement. For more on Christine Ladd-Franklin, see, for example, Judy Green, "Christine Ladd-Franklin (1847–1930)," in *Women of Mathematics: A Biobibliographic Sourcebook*, ed. Louise S. Grinstein and Paul J. Campbell (New York: Greenwood Press, 1987), pp. 121–128. In particular, this article contains a brief discussion of Ladd-Franklin's work in symbolic logic and a list of her scientific publications. Among the latter, her paper "On De Morgan's Extension of the Algebraic Processes," *American Journal of Mathematics* 3 (1880):210–225 might also be mentioned in the context of the work in mathematical logic of the Hopkins group.

Charles's discovery in these words in a paper which followed his son's in the same issue of the *Proceedings*:

> Mr. [C. S.] Peirce has shown by a simple logical argument that the quadrate [matrix algebra] is the legitimate form of a complete linear algebra, and that all the forms of the algebras given by me must be imperfect quadrates, and has confirmed this conclusion by actual investigation and reduction. His investigations do not however dispense with the analysis, by which the independent forms [basis elements] have been deduced in my treatise, but they seem to throw much light upon their probable use.[107]

In 1881 when Sylvester agreed to have Charles edit his father's treatise for posthumous publication in the *American Journal*, the younger Peirce used this as an opportunity to present his father's work in a broader context. By systematically adding footnotes which gave the matrix algebra representation of each of the abstract algebras presented in the text, Charles securely linked the two mathematical theories. Clark University's Henry Taber put Charles Peirce's achievement in perspective in his 1889 "Sketch of the History of the Development of the Theory of Matrices." As he expressed it, the younger Peirce

> has shown that every linear associative algebra has a relative form, i.e., its units may be expressed linearly in terms of the vids [basis elements] (denoted in his notation by $(A : A)$, $(A : B)$, etc.) of a linear transformation; and consequently, that any expression in the algebra can be represented by a matrix. Whence the theory of all possible sets of linear associative algebras is only the theory of all possible sets of matrices constituting a group in Benjamin Peirce's sense, i.e. which are such that the product of any two members of the set can be expressed linearly in terms of itself and the other members of the set alone.[108]

Although he did not explicitly trace the intellectual lineage of his own research in the theory of linear associative algebra through the work of Charles Peirce, Sylvester certainly had known of his colleague's seminal ideas by the time his thoughts turned to such matters in 1883.

---

[107]Benjamin Peirce, "On the Uses and Transformations of Linear Algebra," *Proceedings of the American Academy of Arts and Sciences*, n.s., 2 (1875):395–400 on p. 397. C. S. Peirce and his place in the history of the theory of algebras is discussed in Karen Hunger Parshall, "Joseph H. M. Wedderburn and the Structure Theory of Algebras," *Archive for History of Exact Sciences* 32 (1985):223–349 on pp. 257–261. Consult, also, the papers by Thomas Hawkins cited in notes 104 and 112.

[108]Henry Taber, "On the Theory of Matrices," *American Journal of Mathematics* 11 (1889): 337–395 on p. 353.

Sylvester and the Theory of Matrix Algebra

With the completion of the paper on partition theory in the spring of 1883 and with the onset of summer, Sylvester left Baltimore and his teaching behind for his annual restorative sojourn in England. Generally, he spent his summer vacations doing mathematics at his club, the Athenæum, and partaking of London's scientific and intellectual milieu. The summer of 1883, however, found Sylvester floundering after the close of a totally consuming, major project; he was unsure of what mathematical inquiry to pursue next. Furthermore, the death of the British geometer Henry Smith had opened the Savilian Professorship of Mathematics at Oxford, and Sylvester's name figured prominently in the list of contenders. These circumstances so distracted the sixty-eight-year-old mathematician that after two months abroad, he felt he had nothing to show for his time. He lamented to Gilman on 8 August that "I have been able to do scarcely any—indeed no writing—since my return to these leaden and inclement skies, which seem to freeze all the sunshine out of one's sail. I have tried Cambridge and am trying London with equal want of success. It makes me very sad and despondent—having such large materials at my disposal and [being] unable to turn them to my account."[109] With the start of classes back in Baltimore looming, Sylvester tried to concentrate on the topic of his scheduled course, what he called the "theory of multiple algebra," and this ultimately focused his mathematical energies. By 22 August, he was able to write Cayley that "I have been recovering my theory of Multiple Algebra—by slow degrees and have made a good deal out of it already and hope to recover it nearly in it's [sic] entirety before very long."[110]

Sylvester had first taught a course on this subject (what we would today term the "theory of algebras" or more particularly "matrix theory") during Cayley's visit in the spring of 1882. Although, true to form, he had published a couple of course-inspired papers on the subject, Sylvester had spent most of that semester reveling in Cayley's company and not engaged in writing up new mathematical ideas.[111] As we saw, Sylvester had slipped into a period of

[109] J. J. Sylvester to Daniel C. Gilman, 8 August 1883, Gilman Papers. The situation was not quite as grim as Sylvester painted it to Gilman. The mathematician had most probably penned several of his short notes on matrix theory prior to writing this letter in August of 1883. See James Joseph Sylvester, "On the Equation to the Secular Inequalities in the Planetary Theory," *Philosophical Magazine* 16 (1883):267–269, or *Math. Papers JJS*, 4:110–111; "On the Involution and Evolution of Quaternions," *Philosophical Magazine* 16 (1883):394–396, or *Math. Papers JJS*, 4:112–114; "On the Involution of Two Matrices of the Second Order," *British Association Report, Southport* (1883):430–432, or *Math. Papers JJS*, 4:115–117; "Sur les Quantités formant un Groupe de Nonions analogues aux Quaternions de Hamilton," *Comptes rendus* 97 (1883):1336–1340, or *Math. Papers JJS*, 4:118–121.

[110] J. J. Sylvester to Arthur Cayley, 22 August 1883, Sylvester Papers SJC, Box 11.

[111] See, however, James Joseph Sylvester, "Sur les Racines des Matrices unitaires," *Comptes rendus* 94 (1882):396–399, or *Math. Papers JJS*, 3:565–567; "On the Properties of a Split Matrix," *Johns Hopkins University Circulars* 1 (1882):210 and 211, or *Math. Papers JJS*, 3:645–646; "A Word on Nonions," *Johns Hopkins University Circulars* 1 (1882):241 and 242, or *Math.*

depression after his friend's departure early in July and had emerged from it in hot pursuit of partition-theoretic results, the theory of algebras put aside. Thus, his return to matrix algebras in the summer of 1883 marked the beginning of his first sustained assault on the subject.[112] By the winter of 1883, his studies had yielded not only the various observations contained in a series of short notes but also the partial systematization of these new mathematical insights which appeared in his more substantial "Lectures on the Principles of Universal Algebra."[113]

In particular, Sylvester investigated the algebra of quaternions discovered in 1843 by the Irish mathematician Sir William Rowan Hamilton and generalized it to algebras of higher dimensions. In so doing, he not only explored the properties of the characteristic equation of a matrix but also hit upon the importance of what we now call the minimum polynomial.[114] Sylvester's treatment of the quaternions (first formulated during the spring of 1882) neatly captured his slant on matrix-theoretic questions, although he only alluded to it at the close of the first installment of the "Lectures on the Principles of Universal Algebra."

As Hamilton had determined, the quaternions form a (real) algebra with basis 1, i, j, and k satisfying the relations given in (1.1) and (1.2). Clearly, the relations in (1.1) may also be expressed as $uv + vu = 0$ for $u \neq v \in \{i, j, k\}$. Motivated by Hamilton's work and his own growing interest in matrix algebra *per se*, Sylvester sought a matrix-theoretic representation of the quaternions; that is, he wanted to find a collection of matrices that satisfied (1.1) and (1.2).

Consider two nonsingular 2×2 matrices u and v with complex entries, and form a new 2×2 matrix $z+yv+xu$. (Here, x, y, and z are variables, and we interpret the z in the sum as zI, where I is the 2×2 identity matrix.) The determinant of this new matrix is thus $z^2 + 2bxz + 2cyz + dx^2 + 2exy + fy^2$, where the coefficients may be expressed alternatively as $2b = \operatorname{tr} u$, $2c = \operatorname{tr} v$, $d = \det u$, $2e = \operatorname{tr}(u \operatorname{adj} v)$, and $f = \det v$.

Papers JJS, 3:647–649; and "On Mechanical Involution," *Johns Hopkins University Circulars* 1 (1882):242 and 243, or *Math. Papers JJS*, 3:651–652.

[112] For an historical treatment of the development of matrix theory, see Thomas Hawkins, "The Theory of Matrices in the 19th Century," *Proceedings of the International Congress of Mathematicians: Vancouver*, 1974, 2 vols. (n.p.: Canadian Mathematical Congress, 1975), 2:56–70; "Weierstrass and the Theory of Matrices," *Archive for History of Exact Sciences* 17 (1977):119–163; and "Another Look at Cayley and the Theory of Matrices," *Archives internationales d'Histoire des Sciences* 26 (1977):82–112.

[113] James Joseph Sylvester, "Lectures on the Principle of Universal Algebra," *American Journal of Mathematics* 6 (1884):270–286, or *Math. Papers JJS*, 4:208–224. For a discussion of Sylvester's place in the history of algebras, consult Parshall, "Wedderburn and the Structure Theory of Algebras," pp. 241–250

[114] Sylvester used different terminology. For him, the "characteristic equation" was the "latent equation" and so the "characteristic roots" were "latent roots." He also defined the concept of a "derogatory matrix," that is, a matrix having the property that its characteristic polynomial does not equal its minimal polynomial.

Relative to this set-up, Sylvester asserted that[115]

> the necessary and sufficient conditions for the equation $uv + vu = 0$ are the following, namely,
>
> $$b = 0, \quad c = 0, \quad e = 0.$$
>
> If to these conditions we superadd $d = 1$, $f = 1$, and write $uv = w$, then
>
> $$u^2 = -1, \quad v^2 = -1, \quad w^2 = -1, \quad uv = -vu = w, \quad vw = -wv = u,$$
>
> $$wu = -uw = v.$$

Thus, the formation of a matrix representation of the quaternions hinged on finding a pair of 2×2 matrices u and v such that

$$\det(z + yv + xu) = z^2 + y^2 + x^2. \tag{3.11}$$

In the 2×2 case, it is easy to see that $u = \begin{bmatrix} 0 & 1 \\ -1 & 0 \end{bmatrix}$ and $v = \begin{bmatrix} 0 & \theta \\ \theta & 0 \end{bmatrix}$ (where $\theta = \sqrt{-1}$) satisfy (3.11), and that therefore the matrices

$$\begin{bmatrix} 1 & 0 \\ 0 & 1 \end{bmatrix}, \begin{bmatrix} 0 & 1 \\ -1 & 0 \end{bmatrix}, \begin{bmatrix} 0 & \theta \\ \theta & 0 \end{bmatrix}, \text{ and } \begin{bmatrix} \theta & 0 \\ 0 & -\theta \end{bmatrix}$$

form a quaternion system. Sylvester's argument, however, generalized to the case of 3×3 matrices and produced a nine-dimensional analogue of the quaternions which he dubbed the nonions.[116] (While an analogue of sorts, the nonions do not form a division algebra.)

From this research on matrix representations of quaternion-like algebras, that is, matrix algebras generated by a pair of $n \times n$ matrices u and v satisfying relations like $u^n = v^n = \pm 1$, and $uv = \rho vu$, for ρ a primitive nth root of unity, Sylvester came very naturally to questions about commuting matrices. Among other things, he asked, what conditions must two $n \times n$ matrices u and v satisfy in order for $uv = vu$? Using the characteristic polynomials $\det(uv - Ix)$ and $\det(vu - Ix)$ of the matrix products as his primary tools, Sylvester ultimately focused on and studied the properties of the characteristic roots of such equations. He recognized that the characteristic roots of a given characteristic equation need not be distinct—and so distinguished between the characteristic and the minimum polynomial—but he did not pursue the ramifications of this distinction in a more general setting.[117]

[115] Sylvester, "A Word on Nonions," p. 241, or p. 647.

[116] *Ibid.*, p. 242, or p. 649.

[117] Sylvester analyzed commuting matrices in "Sur les Quantités formant un Groupe de Nonions analogues aux Quaternions de Hamilton," pp. 1336–1340, or pp. 118–121; "On Quaternions, Nonions, Sedenions, etc.," *Johns Hopkins University Circulars* 2 (1883):7–9, or *Math. Papers JJS*, 4:122–132; and "Lectures on the Principles of Universal Algebra," pp. 279–282, or pp. 216–220.

As noted, Sylvester's attempts to systematize these and other "arguments in progress" resulted in the unfinished "Lectures on the Principles of Universal Algebra." These lectures, which were ostensibly the published version of the course he gave in the fall of 1883, reflected a classroom atmosphere very different from that of the previous spring's course on partition theory. This time, his six students seemingly did not partake as fully of the creative mathematical experience. As his correspondence to Cayley during the summer of 1883 shows, Sylvester had proven most of the new results he presented in his fall series of lectures during the months of August and September. Furthermore, by the time he was forced to return to Baltimore to resume his classes, he had made a momentous career decision: in a letter to Gilman dated 12 September 1883, he resigned his Hopkins position effective 1 January 1884—with an eye toward the vacant Savilian Chair—even though the successor to that post remained unnamed.[118]

Sylvester had passed his sixty-ninth birthday on 3 September 1883; he felt tired and longed for security in his homeland. As he confided to Cayley on 5 September, "[i]f I am not appointed to Oxford I shall buy an annuity and study abroad or go into business with my small acquired capital the product of my seven years savings."[119] Obviously, a Sylvester different from the one who had directed the partition-theoretic research in the spring had returned to teach his final course at the Johns Hopkins in the fall. With his heart no longer in his teaching and his mind on the impending decision at Oxford, Sylvester merely presented his own completed work to the students in his course rather than engaging them in its development. The spirit of interaction no longer sparked his classroom.

Dénouement

With the departure of his Professor of Mathematics imminent, Gilman found himself in search of a suitable successor, someone who could continue to build upon the foundation Sylvester had laid both as a researcher and as a teacher. Sylvester had already formulated a distinct opinion on the question of succession even as he deliberated his possible resignation. He named Felix Klein as the man he would most like to see as the next Hopkins chair in the same letter of 5 September in which he first wrote Cayley of his evolving thoughts on the future. There, too, Sylvester asked Cayley if he would unofficially sound Klein out on the feasibility of a move from Leipzig to Baltimore before he approached the President with a definite suggestion. According to Sylvester,

> Gilman warned by experience and confirmed by the information which recently he has acquired in Germany is

[118] J. J. Sylvester to Daniel C. Gilman, 12 September 1883, Gilman Papers.

[119] J. J. Sylvester to Arthur Cayley, 5 September 1883, Sylvester Papers SJC, Box 11.

> determined to make no application to any German Professor in any department until he has ascertained that there is a *bona fide* intention of entering into the negotiation with a view to acceptance.
>
> The too usual practice of certain professors in Germany seems now to be to encourage an application to be made with the sole object and intention of making use of it as a *Call* to better their position and obtain more favorable conditions in their own country.
>
> Such at least is Gilman's consideration.[120]

Cayley dutifully followed through on his friend's request, although he expressed his own opinion that Klein was comfortable in Leipzig and would probably not be tempted by an offer from Hopkins.[121]

In a letter dated 5 October 1883, Cayley presented the situation to Klein and wrote positively of the impressions he had formed of the University and its students during his six-month stay there in the winter and spring of 1882.[122] Indeed, Cayley apparently considered the Baltimore institution a more promising center for mathematics than Oxford for, when the Savilian Chair had fallen open at Henry Smith's death, he had informed Klein about that vacancy in decidedly unenthusiastic terms.[123] Not surprisingly, Klein showed little inclination toward Oxford, whereas the Hopkins situation genuinely piqued his interest.[124] The prospect of heading the Mathematics Department at Hopkins seems, in fact, to have shaken him from his research-related melancholy. (See Chapter 4 on Klein's mathematical competition with Henri Poincaré.) Klein approached the possibility of the American offer with a renewed desire of becoming a leading international scientific figure, yet he recognized that his acceptance might mean scaling back his own research activity.

Over two months passed before Klein received any official communication from Gilman, and then it came in the form of a telegram offering him a professorship. In the interim, Sylvester had won (on 7 December) and accepted the Savilian Chair, thereby ending all of Gilman's hopes of coaxing

---

[120] *Ibid.*

[121] J. J. Sylvester to Daniel C. Gilman, 15 September 1883, Gilman Papers.

[122] Arthur Cayley to Felix Klein, 5 October 1883, Klein Nachlass XXII L, Niedersächsische Staats- und Universitätsbibliothek, Göttingen (hereinafter abbreviated NSUB, Göttingen).

[123] Arthur Cayley to Felix Klein, 4 April 1883, Klein Nachlass VIII, NSUB, Göttingen. Clearly, Cayley may have had ulterior motives for discouraging Klein from considering the vacant Savilian Chair. He may already have been thinking of Sylvester as a possible candidate and might not have wished to weaken his sixty-eight-year-old friend's chances by putting him in competition with the much younger Klein. Also, Cayley, never having succeeded in animating a school around his own work at Cambridge, may simply have believed that Klein's talents would have been wasted in the Britian of the 1880s.

[124] Felix Klein to the Saxon Ministry of Education, 16 April 1883, Klein Nachlass XXII L, NSUB, Göttingen. See, also, Constance Reid, "The Road Not Taken: A Footnote in the History of Mathematics," *The Mathematical Intelligencer* 1 (1978):21–23.

him to remain in Baltimore. Waiting a week before mailing his reply, Klein carefully considered the matter before him. As he put it to Gilman: "I am inclined to accept the position. I am attracted by the novelty of the task and the grandeur of the perspective that it offers; I am even young enough [he was then thirty-four] to find something enticing about the change itself."[125] Still, he also had understandable reservations: he stressed the commitment and effort it had taken to make Leipzig one of the world centers for mathematics; he pointed to his many personal and family connections in Germany; and he worried for his health in an academic environment which required too much diverting administrative work. Most importantly, however, he took exception to several of the financial aspects of Gilman's offer.

With Sylvester's permission, Cayley had informed Klein that the professorship of mathematics at Hopkins currently entailed a salary of $6,000.[126] Unfortunately, Gilman could only convince his Trustees to offer Sylvester's successor $5,000 a year plus $1,000 for moving expenses.[127] Furthermore, unlike the German universities, the Johns Hopkins had no pension provisions for the widows and dependents of university professors. These shortcomings prompted Klein to set two conditions for continued negotiations: (1) financial parity with his predecessor and (2) the creation of an annuity to provide for his family in the event of his death. Setting 31 January 1884 as the deadline for receipt of Gilman's next communication, Klein closed his reply with the promise to respond to it by return post.[128] Obviously, he wished to see these negotiations to a swift, and happy, conclusion.

As Klein waited for Gilman's counteroffer, the University made preparations to honor its first Professor of Mathematics. On 20 December, the eve of Sylvester's departure from Baltimore, his colleagues hosted a gala farewell celebration in Hopkins Hall. True to form, Sylvester took the podium and gave a lengthy and effervescent speech in which he lauded the University, its President, its Trustees, its faculty, and its students. As to his successor, Sylvester told his audience that

> [y]ou want one to succeed me who possesses a breadth of culture and a systematic knowledge of all branches of this prodigiously fruitful science which I cannot lay claim to. I think, however, I may venture to say that I have done, to the best of my power, a good turn to this University in suggesting the name of one who can do all that I can do, and a great deal more besides. I would be flattered to think that there is considerable probability that we may be able to attract him to these shores to take part in the work

[125]Felix Klein to Daniel C. Gilman, 18 December 1883, Klein Nachlass XXII L, NSUB, Göttingen (our translation).

[126]Cayley to Klein, 5 October 1883.

[127]J. J. Sylvester to Arthur Cayley, 13 December 1883, Sylvester Papers SJC, Box 11.

[128]Klein to Gilman, 18 December 1883.

> which devolves upon the Professors of this University. ... If so, I feel that I shall have done more good in leaving this University, ... , if my suggestion be the means of having brought him in connection with this University, than all the work I have done, or may have been supposed to have done in the last seven years.[129]

While Sylvester pushed for Klein in Hopkins Hall, Klein wrote in search of advice to many of his friends and colleagues as well as to several members of the Hopkins faculty. Clearly interested in the job, he must have found the reconnaissance and advice he received encouraging. Paul Haupt, who had just left a professorship at Göttingen to assume the newly created chair in Semitic languages at Hopkins, told Klein that his new institution was "fully capable of competing with the German universities" and that its student body seemed "even more diligent and enthusiastic."[130] As for the faculty, Haupt described it as very collegial and suffering from none of the cliquishness so often characteristic of academe. He suggested that Klein approach the Saxon Ministry of Education about the possibility of taking a leave during the Winter Semester of 1884–1885 with the idea of testing the waters in Baltimore. In that way, he argued,

> you could come here and see the place yourself, and if you like it stay, if not—which I doubt—you could then return to your old position. You would then be the richer for having had this interesting experience and having seen something more of the world. It is certainly not that easy to get to America otherwise, and to cross the ocean as Sylvester's successor is something that does not happen to every mathematician.[131]

Klein also heard from Sylvester's Associate, the obviously not disinterested William Story, who had become increasingly disenchanted with Sylvester in the wake particularly of editorial and managerial disagreements concerning the *American Journal*.[132] Story described a very loose mathematics program

---

[129] "Remarks of Prof. Sylvester at a Farewell Reception tendered to him by the Johns Hopkins University, Dec. 20, 1883 (Reported by Arthur S. Hathaway)," 24 typescript pages, Gilman Papers. The quote appears on p. 19.

[130] Paul Haupt to Felix Klein, 4 January 1884, Klein Nachlass XXII L, NSUB, Göttingen (our translation).

[131] *Ibid.*

[132] The first sign of friction between the two men appeared in the summer of 1880, when Story, in Sylvester's absence and in his capacity as Associate Editor in Charge, allowed Henry Rowland to exceed the page limits Sylvester had set. See J. J. Sylvester to Daniel C. Gilman, 22 July 1880, Gilman Papers; and William E. Story to Daniel C. Gilman, 26 July 1880, Gilman Papers. The two mathematicians resumed cordial working relations after this incident, but Story continued to feel that his exertions on behalf of the University were insufficiently recognized. In June of the following year, he asked for and was denied a promotion. See Daniel C. Gilman to William E. Story, 6 June 1881, Gilman Papers.

at Sylvester's Hopkins with no fixed lecture courses in the curriculum and no set area requirements for the graduate students. Betraying his displeasure with this lack of structure, he explained that as it then stood a student could actually earn a Hopkins doctorate with no more background in a subject like geometry than the usual course on analytic geometry required for the Bachelor's Degree. He also offered the opinion (later shared by Klein) that would-be American mathematicians favored those areas that placed a premium on calculations, while hastening to add that he had nevertheless enjoyed some success in awakening an interest in geometry in his students. As noted above, Story had carried virtually all of the teaching burden in geometry during the Sylvester years, and he welcomed a Klein regime as one in total sympathy with and in full support of his mathematical interests. In fact, he hoped for a curricular shift away from Sylvesterian mathematics and toward the intuition-oriented, Kleinian approach.[133]

For his own part, Sylvester must have assuaged any of Klein's professional doubts about the move by his frank and open responses to the German's inquiries. Relative to the program itself, Sylvester assured Klein that "[t]he *complete control* of the mathematical studies of the University rests with the person who may succeed to my position there."[134] He also enumerated the reasons behind his decision to leave Baltimore in a characteristically Sylvesterian way:

> 1° because I was anxious to return to my native country
> 2° because I had reasons of a strictly individual and personal nature for wishing to quit Baltimore
> 3° (and paramountly) because I did not consider that my mathematical erudition was sufficiently extensive nor the vigor of my mental constitution adequate to keep me abreast of the continually advancing tide of mathematical progress to that extent which ought to be expected from one on whom practically rests the responsibility of directing and molding the mathematical education of 55 million of one of the most intellectual races of men upon the face of the earth.[135]

As singular as this reply was, the most remarkable of the several letters Klein received regarding the Hopkins offer came from Sophus Lie, who strongly counseled him to take the position in spite of the Norwegian's apprehensions as to the fate of Klein's journal, the *Mathematische Annalen*. Obviously, since Klein would have to assume the editorship of the *American Journal* as part of the duties of the new professorship, he would have to

[133] William E. Story to Felix Klein, 10 January 1884, Klein Nachlass XXII L, NSUB, Göttingen.

[134] J. J. Sylvester to Felix Klein, 17 January 1884, Klein Nachlass XXII L, NSUB, Göttingen. Sylvester's emphasis.

[135] *Ibid.*

turn over the editorial control of the *Annalen.* Understandably, Lie fretted that "[y]ou were the focal point there, and I have no idea who could replace you."[136] Still, he contended that a five- to ten-year stay in the United States would give Klein the opportunity to build the sort of international prestige necessary to achieve ultimate victory over the controlling mathematical powers in Berlin. Lie saw Klein's political situation as follows:

> between you and Berlin, as before between Clebsch and Berlin, there stands a rivalry. You, on the one hand, are just with respect to the Berliners, whom you understand and value. The Berlin school, on the other hand, has tried for a long time, if not actually to ignore your activity, then at least to belittle it. ... Although I ... consider you the victor in this battle with the Berliners, I believe it would be good for you to leave this provocative affair for a few years ... . I feel convinced that the Berlin school would follow you more closely once you were in Baltimore.[137]

Although this analysis perhaps reveals the most about Lie's vision of the mathematical world from his 1883 vantage point in Norway, the affinities between the outlooks of Lie and Klein in matters of fame and power cannot be denied. Thus, Klein probably valued Lie's blunt assessment of the offer most highly of all. Perhaps his career *would* ultimately benefit from such a bold move.

Unfortunately for the mathematics program at the Johns Hopkins, however, when Klein finally received Gilman's reply, dated 12 January 1884 and curiously sent not as a telegram but rather as surface mail, his terms had not been met. First, Gilman explained that, like all professors, Klein could purchase a life insurance policy for himself, but this did not fall under the University's purview. Second, he reiterated the salary figure of $5,000, stating that it was the highest amount currently being paid any professor at the University and citing "peculiar reasons" for the augmentation Sylvester had received.[138] As promised, Klein responded to the counteroffer immediately: he would not succeed Sylvester at Hopkins.

With these negotiations closed, Gilman sounded out Arthur Cayley, also without success, and finally focused on the mathematical astronomer Simon Newcomb. The Superintendent of the Nautical Almanac Office in Washington, Newcomb, as noted, had served as a scientific adviser to Gilman since the earliest organizational days of the University. Furthermore, he had also

---

[136] Sophus Lie to Felix Klein, January, 1884, Klein Nachlass X, NSUB, Göttingen (our translation).

[137] *Ibid.*

[138] Daniel C. Gilman to Felix Klein, 12 January 1884, Klein Nachlass XXII L, NSUB, Göttingen. Gilman's fiscal conservatism in dealing with Klein may well have reflected the increasingly precarious state of the Baltimore and Ohio Railroad stock on which the University's financial security depended.

been formally attached to the University both as a Visiting Lecturer and as an Associate Editor of the *American Journal* in spite of his obligations in the nation's capital. An internationally renowned astronomer and a personality well known to Gilman, Newcomb seemed a reasonable third choice.

As a successor to Sylvester, however, Newcomb fell far short of the mark.[139] The astronomer accepted only a half-time position at Hopkins, and that as Professor of Mathematics *and* Astronomy, while maintaining his post at the Naval Almanac. With his loyalties thus divided, Newcomb failed to seize the initiative begun by Sylvester of animating a mathematical group in Baltimore.[140] He favored courses in astronomy over those in mathematics when he taught, and even then he hardly proved gifted in the classroom. This left the mathematical teaching almost solely in the hands of Story, Craig, and Franklin, none of whom had the research talent or mathematical exuberance of Sylvester even if they most certainly had greater overall classroom skills.

As the University's senior mathematician in effect, Story seized the opportunity to reform the department's program along the lines he had outlined in his letter to Klein. In particular, he set up a required introductory course for all graduate students which surveyed the entire mathematical landscape and thereby insured their better-rounded mathematical education. When Story left in 1889 to assume the first chair of mathematics at the newly formed Clark University, Hopkins mathematics also lost its pedagogical and organizational focus. Finally, none of that promising, early department remained after Franklin resigned to go into newspaper editing in 1895 and Craig died in 1900. As Hugh Hawkins put it in his history of the University, "the inspiration of an ecstatic creator and living link with the mathematical past had departed with the stocky, absent-minded Victorian gentleman who was so poor an organizer."[141]

After Sylvester's return to England and the subsequent decline of the Hopkins Mathematics Department, prospective American mathematicians suf-

---

[139]Eduard Study, who taught at Hopkins in the fall of 1893, made his opinion on this clear in a letter to Sophus Lie. As he put it, "unfortunately, everything here depends on Newcomb, a man who is narrower than one would think and with few civilized manners as well. The fundamental maxim for the mathematics instruction here is 'a little of everything' and to avoid 'one-sidedness,' that is to say, concentration. For that reason, it is impossible to attain anything advanced, despite the splendid organization, so long as this man remains in charge. Professor Franklin, who knows what is necessary, has little influence." See Eduard Study to Sophus Lie, 12 November 1893, Brevsamling, Nr. 289, Universitetsbibliothek, Oslo (our translation).

[140]Newcomb did take over the editorship of the *American Journal*, and he did make strides in improving the University's astronomy department, however. See Albert E. Moyer, *A Scientist's Voice in American Culture: Simon Newcomb and the Rhetoric of Scientific Method* (Berkeley: University of California Press, 1992), p. 79.

[141]Hugh Hawkins, p. 138. On the Hopkins programs in mathematics and mathematical physics after Sylvester's departure, consult Clifford Truesdell, "Genius and the Establishment at a Polite Standstill in the Modern University: Bateman," in *An Idiot's Fugitive Essays on Science: Methods, Criticisms, Training, Circumstances* (New York: Springer-Verlag, 1984), pp. 402–438 and particularly on pp. 414–418.

fered from the absence, on American soil, of mathematical instruction even remotely comparable to that available in Europe. A dozen or so schools—among them, Harvard, Yale, Princeton, Cornell, and the Universities of Virginia, North Carolina, Texas, Michigan, and Wisconsin—did offer programs ostensibly at the graduate level, but none of these institutions yet had a staff of researchers teaching or working at the research level.[142] The Hopkins-trained students emerged from their mathematically charged environment only to find themselves unable to maintain their research momentum in other home institutions. Sylvester himself recognized and commented on this deplorable state of affairs in his farewell address to the Hopkins community. He rhetorically questioned his audience:

> What happens to them? They are absorbed by inferior though valuable colleges and institutions, and their work droops. They write to me or to their friends, "We miss the stimulus of the Johns Hopkins." What a great thing it would be if means were found for providing traveling scholarships or Fellowships for a year or two, that they might prolong their studies, and come in contact with scientific men and science in England and on the Continent of Europe.[143]

Of course, Sylvester offered only a stopgap solution to the problem. His students would only have returned from their trips abroad to face the same mathematical isolation. William Durfee, for example, took a professorship of mathematics at teaching-intensive Hobart College in Geneva, New York in 1884, became Dean in 1888, and dropped from the research ranks. A similar fate befell Sylvester's two number-theoretically oriented students. Oscar Mitchell accepted a position at another small college, Marietta College in Ohio, and George Ely became an examiner at the United States Patent Office. As late as 1888, Sylvester's last student, Ellery William Davis, *was* the Mathematics Department at the University of South Carolina and tried, but failed, to institute a graduate program there before moving on to somewhat greener pastures at the University of Nebraska in 1893.[144]

Part of the problem was surely the overall lack of support for research-level work at the institutions where these students took positions, but part of it was the narrow, very directed training they had received from their mentor. Sylvester had spurred his students to do mathematical research by involving them in his own personal research agendas. He had led them to well-defined open problems and had directed their approaches, but he had

[142]See Cajori, *The Teaching and History of Mathematics in the United States*, (Washington, D.C.: Government Printing Office, 1890) for descriptions of these various programs, and Parshall, "A Century-Old Snapshot of American Mathematics," *The Mathematical Intelligencer* 12 (3) (1990):7–11.

[143]"Remarks of Prof. Sylvester," p. 12, Gilman Papers.

[144]Parshall, "A Century-Old Snapshot of American Mathematics," pp. 9–10.

provided them with precious little indication as to how a mathematician actually *finds* a research problem. He had also failed to give them a sense of the full mathematical research horizon, focusing their view on what he knew best, namely, invariant theory and closely related algebraic areas. On their own and in basically unsupportive academic atmospheres, these students floundered in their attempts both to set their own research goals and to pass on their research-level training to others. With unencouraging educational prospects at home, those Americans desirous of mathematical training thus turned to Europe, and particularly to the halls of Felix Klein. Given this cruel twist, one can only wonder how much more deeply the Hopkins mathematics program might have affected the emergence of an American mathematical research community had Klein succeeded Sylvester there.

Sylvester's success at the Johns Hopkins University was thus relative. He had founded and animated America's first school of mathematical research there in the years from 1876 to 1883, and he had established the *American Journal of Mathematics* for research-level contributions to the field. Although the United States had witnessed virtually nothing comparable to these achievements prior to Sylvester's tenure at Hopkins, it was still not ready to sustain a specialized research community of mathematicians in the early 1880s.

Chapter 4
German Mathematics and the Early Mathematical Career of Felix Klein

Although he declined first the chair in mathematics at the Johns Hopkins University and later other opportunities to teach in the United States, Felix Klein nevertheless exerted a far more pervasive and longer lasting influence on American mathematics than J. J. Sylvester, even given the fact that Sylvester worked directly on American soil. Part of the explanation for this ironic turn of events traces to improvements in mathematics education at leading American colleges and universities, particularly at some of the older schools on the eastern seaboard. Klein's best American students—coming from institutions like Harvard, Princeton, and Wesleyan—had better training than the available pool of talent from which Sylvester had drawn a decade or so earlier. These educational improvements, although of undeniable significance, account only partially for Klein's success, however. They certainly fail to explain both Klein's magnetic attraction for young American mathematicians and his phenomenal success as their teacher. Why did Klein serve as the main conduit for the sudden transfusion of abstract mathematics in the German style that so decisively enlivened the fledgling community of American mathematicians? To understand this, we must go beyond domestic factors and external causes to the man himself and the unusually rich sources that defined and shaped his career.

Klein's mathematics embodied many of the ideals characteristic of German scholarship in the nineteenth century. Even from his youth, he sought to attain a unified conception of mathematical knowledge that embraced the achievements of his predecessors. He thus strove for and attained an extraordinary breadth of knowledge, much of which he acquired in discussions with colleagues and friends. A master of give-and-take, he cultivated scientific relations with many of the leading mathematicians of his day and then imparted the ideas so gained to the students in his lecture courses. Furthermore, he freely shared the hard-won insights which allowed him to capture the essence of a mathematical theory. Klein's approach clearly met with success, for by the age of thirty, he had already begun to attract talented students from outside of Germany, many of whom went on to prominent positions as scholars and leaders within their respective scientific communities.

In contrast to Sylvester, whose mathematical style was dominated by complex computations and directed towards fairly restricted problems within specialized branches of algebra, Klein tended to soar above the terrain that occupied ordinary workaday mathematicians, taking in vast expanses of mathematical knowledge. His one glaring weakness, as Richard Courant once put it, was that he often found it difficult to land his plane.[1] Klein had no patience for thorny problems that required abstruse technical arguments. For him, what counted was the big picture, drawn from the work of his forerunners—Gauss, Plücker, Clebsch, Riemann, and Weierstrass—as well as from that of leading contemporary figures, including Lie, Schwarz, and Dedekind. Throughout the course of his career, Klein made important contributions to geometry, group theory, Riemannian function theory, Galois theory, rigid-body mechanics, and even the general theory of relativity. In his own mind at least, all of this seemingly disparate work was of a piece, and, what is more, he ultimately viewed it as largely embedded within a mathematical tradition associated with a single institution, Göttingen University.

While Klein's achievements have come to be associated primarily with Göttingen mathematics, this must not obscure the wide variety of ideas from other traditions that shaped his thought, especially during his youth. Indeed, a grasp of the full range of Klein's impact on American mathematicians and an appreciation of the deeper historical roots that animated his teaching and subsequently found expression in the work of his many students requires some familiarity with the sources upon which he drew. This chapter thus interrupts the account of mathematical events in the United States proper in order to describe some of the salient features of Klein's early career and of the mathematical ideas, people, and traditions that informed it.

As we shall see, the years from 1869 to 1882 marked the period in which Klein produced his most brilliant and influential work, research in the areas of algebraic curve theory, geometric Galois theory, and the theory of Riemann surfaces. In December of 1883, however, when he received President Gilman's offer to fill the mathematical vacuum left at Hopkins after Sylvester's departure, Klein's life as a creative mathematician was for the most part behind him, although few, if any, of his contemporaries seemed aware of it at the time. Without fully realizing it himself, Klein had slipped into a transitional phase in his professorial career that would last until the middle of the 1890s and would center largely around his lecture courses and seminar activities. The event that marked the turning point from this second, teaching-dominated period to the third, culminating phase in his career was

[1] Richard Courant, "Felix Klein," *Die Naturwissenschaften* 37 (1925):765–772 on p. 772.

FELIX KLEIN (1849–1925)

the appointment at Göttingen in 1895 of David Hilbert. Following Hilbert's call to the faculty, Klein channeled most of his energy into building up and enhancing Göttingen's reputation as one of the world's leading centers for mathematical and scientific research, while his younger colleague emerged as the most influential teacher of the pre-World War I era. In view of the fact that the small, isolated university town of Göttingen played such a key role as a training ground and "home away from home" for American mathematicians, we open this chapter with a brief discussion of its overall place in the history of mathematics before turning to an examination of the early career of one of its principal figures, Felix Klein.

The Göttingen Mathematical Tradition

Göttingen's *Georgia Augusta* has come to be regarded as a prototype for the modern university in the annals of higher education.[2] Founded in 1737 by George II, King of England and Elector of Hanover, Göttingen University initially served as Hanover's solution to an institutional dilemma posed by the rival Hohenzollern monarchy. Prospective employees of the Hanoverian civil service who wished to pursue a higher education prior to its founding were forced to take up studies outside the Electorate, most notably at the Prussian university in Halle. By mid-century, however, Göttingen had surpassed Halle not only in scholarly reputation but also in popularity among German students, many of whom came from an aristocratic background.

European universities had historically been dominated by interests vested in their theological faculties. Göttingen, however, managed to shake off this sectarian yoke under the beneficent leadership of the Hanoverian Minister, Gerlach Adolf von Münchhausen. Its philosophical faculty, which traditionally offered elementary training in the liberal arts to students desirous of studying in one of the three higher faculties (theology, medicine, or law), rapidly developed a reputation for serious scholarship based on a fertile combination of teaching and research. Initially, the vaunted ideals of *Lehr- und Lernfreiheit*—the freedom to teach and to learn without interference from political and religious authorities—flourished to a remarkable degree particularly in the philology seminars of Johann Matthias Gesner and Christian Gottlob Heyne. Spreading from the humanities, the reform spirit soon took hold in the natural sciences after Göttingen acquired the services of such prominent scholars as the anatomist and botanist Albrecht von Haller, the physicist Georg Christoph Lichtenberg, the astronomer Tobias Mayer, and the mathematician Abraham Kaestner.

[2] See, for example, Charles McClelland, *State, Society, and University in Germany,* 1700–1914 (Cambridge: University Press, 1980). On Göttingen University, see Günther Meinhardt, *Die Universität Göttingen, Ihre Entwicklung und Geschichte von* 1734–1974 (Göttingen: Musterschmidt, 1977). On noteworthy figures associated with it, consult Heinz Motel, *Berühmte Persönlichkeiten und ihre Verbindung zu Göttingen* (Göttingen: Verlag Göttinger Tageblatt, 1990).

Although scholars who studied or taught in Göttingen made lasting contributions to a number of fields, in mathematics their influence has become legendary. The famed Göttingen mathematical tradition commenced with the career of Carl Friedrich Gauss—dubbed the *Mathematicorum princeps*—who studied, worked, and taught there for over fifty years. In 1807, six years after he had correctly predicted the location of the asteroid Ceres and published his monumental *Disquisitiones Arithmeticæ*, Gauss was appointed Professor of Astronomy at Göttingen.[3] During the 1830s, under the influence of his younger colleague Wilhelm Weber, Gauss redirected his interests to electromagnetic phenomena. The Gauss-Weber collaboration ended abruptly in 1837, however, when Weber and six other Göttingen professors were removed from their positions for protesting the annulment of the Hanoverian constitution by the new King, Ernst August. This so-called "Göttingen Seven" incident marked a serious setback for the forces of democracy in Hanover. It sent the chilling message to Göttingen scholars that their cherished academic freedoms had strictly defined limits.

Even though his mathematical ideas broke fresh ground and gave a strong impulse to several branches of the discipline, outwardly Gauss's career and personality bore many traits typical of an eighteenth-century scholar. He often wrote in Latin; he never felt compelled to rush into print; and he never published some of his most striking results, choosing to remain true to his motto: "*pauca sed matura*—few, but ripe." Even more characteristic of his eighteenth-century heritage, Gauss showed no inclination to incorporate his own research or that of other mathematicians into his teaching, preferring instead to carry on an extensive scientific correspondence with a handful of friends and peers. Gauss, like his contemporaries Goethe and Alexander von Humboldt, aspired to the ideal of universal knowledge. He ignored distinctions between fields such as mathematics, astronomy, and mechanics, and his research ran the gamut from number theory and algebra to geodesy and electromagnetism. Felix Klein regarded Gauss as representative of both the culmination of the classical period and the dawn of a new age; he likened him to the crowning peak in a gradually ascending chain of mountains that drops off precipitously, leading to a broad expanse of smaller hills nourished by a steady stream flowing down from on high.[4]

In another sense, Gauss appears as the initial link in a chain of important mathematicians associated with Göttingen. His immediate successor, Peter Gustav Lejeune Dirichlet, carried on Gauss's number-theoretic researches, being the first to penetrate his *Disquisitiones Arithmeticæ*. At the outset of

---

[3]Carl Friedrich Gauss, *Disquisitiones Arithmeticæ*, trans. Arthur A. Clarke (New Haven: Yale University Press, 1966). On Gauss's life, see G. Waldo Dunnington, *Carl Friedrich Gauss, Titan of Science* (New York: Exposition Press, 1955); and Walter Kaufmann-Bühler, *Gauss: A Biographical Study* (New York: Springer-Verlag, 1981).

[4]Felix Klein, *Vorlesungen über die Entwicklung der Mathematik im* 19. *Jahrhundert*, 2 vols. (Berlin: Springer-Verlag, 1926–1927), 1:62.

his career, Dirichlet had traveled to Paris to study mathematics with, among others, Joseph Fourier, whose work on mathematical physics and trigonometric series expansions proved crucial for the young German's subsequent research.[5] Returning first to Breslau, Dirichlet moved to Berlin in 1829 and taught there for over twenty-five years, including six with his brilliant friend Jacobi. Intellectually, Dirichlet developed into a worthy heir to Gauss, owing particularly to his penetrating ideas in algebraic and analytic number theory. Dirichlet's lectures and published work, in fact, exerted a profound influence on leading figures of the next generation like Gotthold Eisenstein, Leopold Kronecker, Bernhard Riemann, and Richard Dedekind. The latter, in turn, developed into a key figure linking the nineteenth-century Göttingen tradition with David Hilbert and Emmy Noether, two of the principal architects of modern algebra in the twentieth century. Bernhard Riemann, who assumed Dirichlet's position after the latter's untimely death, was also strongly influenced by Gauss's work, especially his contributions to analysis and differential geometry. Beginning with Riemann's investigations in function theory, one can trace a line of intellectual development that runs through Klein and his students—principally Adolf Hurwitz and Robert Fricke—and on to Hermann Weyl, who gave the theory of Riemann surfaces its first rigorous treatment in his Göttingen lectures of 1911–1912.[6]

These conceptual links might suggest that the Göttingen mathematical tradition, inaugurated by Gauss and perpetuated by Dirichlet and Riemann, enjoyed a continuous pattern of unbroken achievements that carried well into the next century. Nothing, however, could be further from the truth. Obviously, sustaining a mathematical research school requires more than a strong faculty—it also takes talented students. It was precisely the fertile combination of research *and* teaching that differentiated Jacobi's Königsberg and Dirichlet's Berlin from mid-nineteenth-century Göttingen.

Although he dominated the German mathematical scene until his death in 1855, Gauss took little interest in teaching even advanced students, and he confined his Göttingen lectures to rather rudimentary aspects of mathematical astronomy. Throughout this period, and for some years afterward, the lion's share of Göttingen's mathematics courses, including those designed for future *Gymnasium* teachers, was taught by Bernhard Thibaut (from 1802 to 1832), Georg Ulrich (1821 to 1879), and Moritz Abraham Stern (1848 to 1884).[7] These three professors—unlike their colleagues Gauss, Dirichlet, and Riemann—belonged to the commission founded in 1831 for the purpose

[5]For biographical information, consult Kurt-R. Biermann, "Johann Peter Gustav Lejeune Dirichlet, Dokumente für sein Leben und Wirken," *Abhandlungen der deutschen Akademie der Wissenschaften zu Berlin, Klasse für Mathematik, Physik, und Technik* 2 (1959):2–88.

[6]Hermann Weyl, *Die Idee der Riemannschen Fläche* (Leipzig: B. G. Teubner, 1913).

[7]Felix Klein and Eduard Riecke give a brief history of institutional developments in mathematics at Göttingen in their *Neue Beiträge zur Frage des mathematischen und physikalischen Unterrichts an den höheren Schulen* (Leipzig: B. G. Teubner, 1904), pp. 158–174.

of examining teaching candidates. Thus, at this time, Göttingen employed what amounted to a two-tiered mathematics faculty, one devoted to the highest levels of research and the other to the dissemination of knowledge deemed useful for future teachers. Given the obstacles that stood in the way of pursuing a university career, most students gravitated to the more elementary, *Gymnasium*-oriented classes (although, in many cases, the scientific ideals of German *Gymnasium* teachers were remarkably high).

Gauss's successors at Göttingen, Dirichlet and Riemann, represented two gifted proponents of the marriage between teaching and research, but personal misfortune intervened and prevented them ultimately from realizing much in this direction. Consequently, the University failed to realize its promising potential as a research center at mid-century. Dirichlet occupied his chair only three years before suffering a fatal heart attack in 1859, and Riemann took extended leaves in order to nurture his fragile health in the sunny environs of northern Italy. His death in 1866 left Richard Dedekind as the principal mathematical heir.

Dedekind had studied under Gauss, Dirichlet, and Riemann in Göttingen and had taught there himself as a *Privatdozent* during the late 1850s. After lecturing for three years at the *Eidgenössische Technische Hochschule* (ETH) in Zurich, he accepted, in 1862, the position at the Polytechnical Institute in his native Brunswick he would hold for the remainder of his career. This final career move brought with it Dedekind's publication of the first of several editions of Dirichlet's *Vorlesungen über Zahlentheorie.*[8] The new concepts and results with which he supplemented Dirichlet's lectures eventually proved fundamental for the development of algebraic number theory. Dedekind also later joined Heinrich Weber in preparing the first edition of Riemann's *Werke*, which appeared in 1876.[9] Yet, despite the redoubtable mathematical stature these and other works earned for him, Dedekind proved ill-suited to the task of promoting and sustaining the tradition in which he had been trained, owing to his quiet, withdrawn personality and to his isolated academic environment.

---

[8]Peter Lejeune Dirichlet, *Vorlesungen über Zahlentheorie*, ed. Richard Dedekind, 1st ed. (Brunswick: Vieweg, 1863). On Dedekind, see Pierre Dugac, *Richard Dedekind et les Fondements des Mathématiques*, Collection des Travaux de l'Académie internationale d'Histoire des Sciences, 24 (Paris: Librairie philosophique J. Vrin, 1976); and Winfried Scharlau, ed., *Richard Dedekind*: 1831-1981. *Eine Würdigung zu seinem* 150. *Geburtstag* (Wiesbaden: Vieweg, 1981).

[9]A second edition, prepared by Weber, appeared in 1892, and a third, containing additional material from Riemann's *Nachlass* (the "Nachträge" edited by Max Noether and Wilhelm Wirtinger), appeared in 1902. The most recent edition is Bernhard Riemann, *Bernhard Riemann*: *Gesammelte Mathematische Werke, Wissenschaftlicher Nachlass und Nachträge/ Collected Papers*, ed. Raghavan Narasimhan (Leipzig: BSB B. G. Teubner Verlagsgesellschaft; New York: Springer-Verlag, 1990), which contains essays on individual works of Riemann by W. Wirtinger, H. Weyl, C. L. Siegel, P. Lax, S. Chandrasekhar, and N. Lebovitz, plus additional source material and bibliographical information compiled by Erwin Neuenschwander and Walter Purkert. In what follows, we refer to the Weber edition (Leipzig: B. G. Teubner, 1892) of the collected works, which we abbreviate *Riemann*: *GMW*.

This rupture in Göttingen's mathematical heritage came at a key turning point in the University's history; the year of Riemann's death coincided with the dissolution of the Kingdom of Hanover. After the annexation of Hanover in 1866, Göttingen University became a Prussian institution. Shortly afterward, Königsberg-trained Alfred Clebsch moved to Göttingen and established a major school in algebraic geometry during his tenure there from 1868 to 1872. His sudden and premature death, however, reduced Göttingen to just one more flicker in the mathematical firmament, and so it would remain until the late 1880s.

The man largely responsible for restoring Göttingen's former luster and who, in so doing, initiated a process that transformed the whole structure of mathematics at the German universities was Felix Klein.[10] The most dynamic mathematical figure of the last quarter of the nineteenth century, Klein could captivate an audience with the sweep and polish of his lectures. In retrospect, it is clear that as a creative mathematician he never attained a level of achievement that would justify including him in the pantheon of Göttingen immortals—Gauss, Riemann, Hilbert, Weyl—although his contemporaries often placed him there. Nevertheless, his activities and influence far transcended the usual bounds constraining the mathematician; in effect, he redefined these conventional limits through his lifelong crusade to ensure that mathematics received its rightful due as one of the central pillars of human civilization and culture.

Klein's Educational Journey

Christian Felix Klein's multifaceted career coincided with one of the more fateful periods in modern European history.[11] Shortly after his birth in Düsseldorf on 25 April 1849, the city and surrounding region became the battleground for what proved to be the last echo of the 1848 Revolution.

[10]Given the scope of the present study, we can only touch lightly on the more global questions of how this came about and what was at stake. Here, we focus tightly on Klein's mathematical development in order to appreciate fully the depth and extent of his influence on American mathematics. Aspects of Klein's Göttingen career appear in Herbert Mehrtens, *Moderne—Sprache—Mathematik* (Frankfurt am Main: Suhrkamp Verlag, 1990), pp. 327–401; Karl-Heinz Manegold, *Universität, Technische Hochschule und Industrie: Ein Beitrag zur Emanzipation der Technik im 19. Jahrhundert unter besonderer Berücksichtigung der Bestrebungen Felix Kleins* (Berlin: Duncker & Humblot, 1970); Gert Schubring, "Pure and Applied Mathematics in Divergent Institutional Settings in Germany: The Role and Impact of Felix Klein," in *The History of Modern Mathematics*, ed. David E. Rowe and John McCleary, 2 vols. (Boston: Academic Press, Inc., 1989), 2:171–220; David E. Rowe, "'Jewish Mathematics' at Göttingen in the Era of Felix Klein," *Isis* 77 (1986): 422–449; and David E. Rowe, "Klein, Hilbert, and the Göttingen Mathematical Tradition," in *Science in Germany: The Intersection of Institutional and Intellectual Issues*, ed. Kathryn M. Olesko, *Osiris*, 2nd ser., 5 (1989):186–213.

[11]For biographical information on Klein, see Renate Tobies, *Felix Klein*, Biographien hervorragender Naturwissenschaftler, Techniker und Mediziner, 50 (Leipzig: BSB B. G. Teubner Verlagsgesellschaft, 1981); and Felix Klein, "Göttinger Professoren. Lebensbilder von eigener Hand," *Mitteilungen des Universitätsbundes Göttingen* 5 (1923):11–36.

Over the course of the next two decades, Prussia staked its claim as a major military and economic power in Europe. The path to German unification under Prussian hegemony culminated with Austria's defeat at Königgrätz in 1866 and the crushing victory over France five years later. As a member of a voluntary corps of emergency workers, Klein witnessed firsthand the battle sites at Metz and Sedan, where the Empire of Louis Napoléon finally collapsed only to be replaced by another empire, the Second *Reich* as fashioned by Otto von Bismarck. These monumental political events molded Felix Klein's physical and psychological world. His academic career practically coincided with the rise and fall of the Second *Reich* and, like nearly all German academics of this same generation, his allegiance to the Kaiser was practically unconditional.[12] The Hohenzollerns may not have ruled by virtue of divine right, but for Klein and many of his contemporaries, any other system of government was simply unthinkable. In Klein's case, this strongly nationalistic mentality was coupled with a decidedly internationalistic scientific outlook, an ambiguity that during the demise of the monarchy became difficult to maintain and to reconcile with external events.

Klein spent his youth in the Rhineland, where he attended the *Gymnasium* in Düsseldorf before entering Bonn University in 1865 at the age of sixteen. After a somewhat dreary experience imbibing Latin and Greek classics as a *Gymnasium* student, Klein found Bonn's curriculum, with its strong emphasis on the natural sciences, ideally suited to his temperament. In fact, his experience at Bonn contributed significantly to shaping his universalist outlook as well as his specific scientific interests. There, he studied a wide variety of subjects—mathematics, physics, botany, chemistry, zoology, and mineralogy—and he participated in all five sections of the Bonn Natural Sciences Seminar, reporting on such diverse subjects as crystal systems, the structure of plant vessels, and the morphology of butterflies. When he left the seminar after the Summer Semester of 1867 to concentrate on mathematics, the seminar heads expressed regret over his departure.[13]

In mathematics, Klein took a number of courses with the distinguished analyst, Rudolf Lipschitz, including analytic geometry, number theory, differential equations, mechanics, and potential theory. At this early juncture, however, Klein aspired to become a physicist and so gravitated quite naturally to Julius Plücker, a gifted geometer *and* experimental physicist who deeply inspired him during this formative stage in his career.[14] Klein was

[12]On the political views of the German professoriate, consult Fritz Ringer, *The Decline of the German Mandarins: The German Academic Community* (Cambridge, MA: Harvard University Press, 1969).

[13]Schubring, "Pure and Applied Mathematics," p. 210. The history of this institution is examined in Gert Schubring, "The Rise and Decline of the Bonn Naturwissenschaften Seminar," in *Science in Germany*, pp. 56–93.

[14]For biographical information on Plücker, consult Adolf Dronke, *Julius Plücker, Professor der Mathematik und Physik an der Rheinischen Friedrich Wilhelms-Universität in Bonn* (Bonn: Adolf Marcus, 1871); and Wilhelm Ernst, *Julius Plücker: Eine zusammenfassende Darstellung*

only in his second semester when Plücker chose him as an assistant for the laboratory courses in physics, an incident that, in a somewhat unlikely way, affected the young man's future course.

An outsider to both the physics and mathematics communities in Germany, Plücker had found that the leading members of the Berlin establishment tended to view his work with suspicion. His chief rival among the geometers, the influential Jacob Steiner, considered Plücker's analytic methods sheer anathema and a threat to the Steinerian program for placing projective geometry on a purely synthetic footing.[15] The leading physicists, on the other hand, largely ignored Plücker's work, including his discovery of the first three lines of the hydrogen spectrum, a finding that antedated the highly acclaimed spectroscopic researches of Bunsen and Kirchhoff in Heidelberg.[16]

By the time Klein met Plücker in Bonn in 1866, the latter's research interests had swung back to geometry after a nearly twenty-year hiatus spent exclusively in physics. Klein quickly familiarized himself with both his mentor's earlier works, which had played a central role in the development of algebraic geometry, and with the new two-volume *Neue Geometrie des Raumes* which Plücker was busily preparing in the mid-1860s.[17] The novel theory put forth in the latter work was based on the idea of regarding lines in complex projective three-space $\mathbf{P}^3(\mathbb{C})$—rather than points—as the basic geometric entities. Given a line ℓ in $\mathbf{P}^3(\mathbb{C})$ defined by two points (x_1, x_2, x_3, x_4) and (y_1, y_2, y_3, y_4), Plücker represented it by six homogeneous coordinates p_{ij}, $1 \leq i < j \leq 4$, defined by $p_{ij} := x_i y_j - x_j y_i$. The coordinates (p_{ij}) represent a line ℓ in $\mathbf{P}^3(\mathbb{C})$ if and only if they satisfy the *Plücker relation* $p_{12}p_{34} + p_{13}p_{42} + p_{14}p_{23} = 0$. The fundamental objects in this *line geometry*—line complexes—are defined by homogeneous equations in the line coordinates p_{ij}.[18]

Unfortunately, Plücker did not succeed in bringing the work containing

seines Lebens und Wirkens als Mathematiker und Physiker auf Grund unveröffentlicher Briefe und Urkunden (Bonn: Universitäts-Buchdruckerei, 1933).

[15]See Klein, *Vorlesungen über die Entwicklung der Mathematik im* 19. *Jahrhundert*, 1:116.

[16]However, Plücker did enjoy the esteem of scientists outside Germany, particularly Michael Faraday in England. See L. Pearce Williams, *Michael Faraday*: *A Biography* (New York: Basic Books, 1965), pp. 414–416.

[17]Julius Plücker, *Analytisch-geometrische Entwicklungen*, 2 vols. (Essen: G. D. Baedeker, 1828 and 1831); *System der analytischen Geometrie, auf neue Betrachtungsweise gegründet und insbesondere eine ausführliche Theorie der Curven dritter Ordnung enthaltend* (Berlin: Duncker und Humblot, 1834); *Theorie der algebraischen Curven: gegründet auf eine neue Betrachtungsweise der analytischen Geometrie* (Bonn: Adolf Marcus, 1839); *System der analytischen Geometrie des Raumes in neuer analytischer Behandlungsweise, insbesondere die Theorie der Flächen zweiter Ordnung und Classe enthaltend* (Düsseldorf: Schaub'sche Buchhandlung W. H. Scheller, 1846); and *Neue Geometrie des Raumes, gegründet auf die Betrachtung der geraden Linie als Raumelement*, 2 vols. (Leipzig: B. G. Teubner, 1868–1869).

[18]David E. Rowe gives a brief discussion of Plücker's line geometry in "Klein, Lie, and the 'Erlanger Programm,'" in 1830–1930: *A Century of Geometry*, ed. Luciano Boi, D. Flament, J.-M. Salanski (Berlin: Springer-Verlag, 1992), pp. 45–54 on pp. 48–51.

these notions to fruition, owing to his death in May of 1868. Responsibility for completing the second part of the *Neue Geometrie* fell to Göttingen University's geometer, Alfred Clebsch, who delegated the task to Plücker's young assistant, Klein. This chance circumstance thus brought Klein into closer contact with the man who became the second formative influence on his life as a mathematician.

A highly versatile mathematician, Clebsch enjoyed considerable success employing algebraic and analytical methods in geometry. Although only thirty-five years of age when Klein first met him, he had already emerged as a distinguished teacher and leader of a new school in German mathematics at Giessen.[19] During his six years of activity there, Clebsch combined teaching and mathematical research in a particularly fruitful way, attracting such stalwart talents as Paul Gordan, Alexander Brill, Max Noether, and Jacob Lüroth. In fact, just months after arriving in Giessen in 1863, Clebsch founded the University's first mathematics seminar and, together with Gordan, began creating a new theory of algebraic invariants based on the symbolic notation first developed by Siegfried Aronhold.[20] The following year, he also published an article, "Über die Anwendung der Abel'schen Functionen in der Geometrie," which has since been characterized as the "birth certificate" of modern algebraic geometry.[21] There, Clebsch proved the converse of Abel's theorem and derived many fundamental properties of algebraic curves by utilizing function-theoretic methods developed by Jacobi and Riemann. These techniques enabled him to rederive with relative ease a number of results that had cost geometers like Jacob Steiner and Otto Hesse a great deal more effort.[22] (See the section "Stepping to the Podium: Two Seminar Lectures by Henry White" in Chapter 5 for a brief discussion of Abel's Theorem and Clebsch's work.)

Following Riemann, Clebsch defined the *genus* p of an (irreducible) algebraic curve $C_n \subseteq \mathbf{P}^2(\mathbb{C})$ of degree n to be the number of linearly independent, holomorphic differentials that exist on the corresponding Riemann surface. From this, he derived the formula $p = \frac{1}{2}(n-1)(n-2) - d - r$, when C_n has only elementary singularities (d double points and r cusps). He proceeded to show, using transcendental methods, that p is invariant under birational transformations. This study, one of the first to exploit Riemann's function theory, laid the groundwork for many subsequent investigations. In

[19] On Clebsch's mathematical work, see Felix Klein *et al.*, "Rudolph Friedrich Alfred Clebsch, Versuch einer Darlegung und Würdigung seiner wissenschaftlichen Leistungen," *Mathematische Annalen* 7 (1874):1–55.

[20] See Karen Hunger Parshall, "Toward a History of Nineteenth-Century Invariant Theory," in *The History of Modern Mathematics*, 1:157–206.

[21] Alfred Clebsch, "Über die Anwendung der Abel'schen Functionen in der Geometrie," *Journal für die reine und angewandte Mathematik* 63 (1864):189–243; and I. R. Shafarevich, "Zum 150. Geburtstag von Alfred Clebsch," *Mathematische Annalen* 266 (1983):135–140 on p. 136.

[22] A detailed treatment of these ideas can be found in Alfred Clebsch, *Vorlesungen über Geometrie*, ed. Ferdinand Lindemann (Leipzig: B. G. Teubner, 1876).

ALFRED CLEBSCH (1833–1872)

particular, Clebsch and Gordan's *Theorie der Abelschen Functionen* of 1866 gave an *algebraic* proof of the invariance of genus under birational transformations.[23] This signaled a methodological trend later pursued by two of Clebsch's leading students, Alexander Brill and Max Noether, namely, the replacement of the transcendental techniques originally used by Riemann and Clebsch by purely algebraic arguments.[24]

By 1868, when Clebsch accepted the chair formerly held by Riemann at Göttingen, he and his entourage of students were turning out so much new material on algebraic geometry and invariant theory that they began making plans for the inauguration of a new journal designed to give their work more visibility. The following year saw the publication of the first volume of *Die Mathematische Annalen*, which later, under Felix Klein's leadership, displaced Crelle's *Journal für die reine und angewandte Mathematik* as the leading mathematics periodical in Germany.[25]

While Klein busied himself in 1868 with the completion of Plücker's *Neue Geometrie des Raumes*, Clebsch kept him abreast of related literature on line geometry, notably a recent paper by Guiseppe Battaglini that dealt with the representation of second-degree line complexes by canonical algebraic forms.[26] Klein soon discovered that, contrary to appearances, the quadratic form Battaglini had exhibited was not completely general. This gave him the idea that he eventually turned into his doctoral dissertation—but not before his thesis adviser, Rudolf Lipschitz, made him aware that he might use Weierstrass' newly developed theory of elementary divisors to deal with the other degenerate cases.[27] (The latter theory would play a key role in the dissertation research of Klein's outstanding American student, Maxime Bôcher. See Chapter 5.) The impulse provided by Weierstrass' ideas proved important not only for Klein's *Doktorarbeit*, where he combined it with a masterful feel for the theory of linear and quadratic line complexes,

---

[23]Alfred Clebsch and Paul Gordan, *Theorie der Abelschen Functionen* (Leipzig: B. G. Teubner, 1866). See the discussion in Jeremy Gray, "Algebraic Geometry in the late Nineteenth Century," in *The History of Modern Mathematics*, 1:361–385 on pp. 363–370.

[24]Alexander Brill and Max Noether, "Die Entwicklung der Theorie der algebraischen Functionen in alterer und neuerer Zeit," *Jahresbericht der deutschen Mathematiker-Vereinigung* 3 (1892):107–566 on pp. 347–366.

[25]See Renate Tobies and David E. Rowe, ed., *Korrespondenz Felix Klein–Adolf Mayer, Auswahl aus den Jahren* 1871–1907, Teubner-Archiv zur Mathematik, vol. 14 (Leipzig: BSB B. G. Teubner Verlagsgesellschaft, 1990), pp. 28–45.

[26]Guiseppe Battaglini, "Intorno ai Sistemi di Rette di secondo Grado," *Giornale di Matematiche* 5 (1866):217–231. Renatus Ziegler discusses this work in *Die Geschichte der geometrischen Mechanik im* 19. *Jahrhundert* (Stuttgart: Franz Steiner Verlag, 1985), pp. 79–82.

[27]Felix Klein, *Gesammelte Mathematische Abhandlungen*, 3 vols. (Berlin: Springer-Verlag, 1921–1923), 1:4 (hereinafter abbreviated *Klein*: *GMA*). In what follows, page references in dual citations involving Klein's collected works will refer to *Klein*: *GMA*. Characteristically enough, Klein avoided going into a detailed analysis of these degenerate cases and left the task for one of his doctoral students, Adolf Weiler. This work, which was by no means free of errors, appeared as Adolf Weiler, "Ueber die verschiedenen Gattungen der Complexe zweiten Grades," *Mathematische Annalen* 7 (1874):145–207.

JULIUS PLÜCKER (1801–1868)

but also for many of his subsequent geometrical investigations. Fittingly enough, he dedicated this very first work "to his unforgettable teacher, Julius Plücker," [28] the man who essentially founded the subject of line geometry.

Shortly after passing his doctoral examination in December of 1868, Klein moved to Göttingen to study under Clebsch. During this initial stay, he quickly made friends with many of Clebsch's disciples, particularly Max Noether. Whereas Clebsch had opened the door to a new birational geometry of curves based on Riemannian function theory, Noether was beginning to extend the scope of this theory to surfaces.[29] The significance of these new developments for Klein lay in the exposure they gave him to the function-theoretic ideas of Riemann. Thus, although he closely followed the work of the Clebsch school, Klein soon struck out in a new direction, developing his own distinctive approach to the theory of Riemann surfaces. This would ultimately serve as the source of some of his deepest mathematical achievements.

Following eight months in Göttingen in 1869, Klein yearned to see a little more of the mathematical world, including France. Against the advice of Clebsch, he decided to spend the Winter Semester of 1869–1870 in Berlin, for it was from there that the triumvirate of Karl Weierstrass, Ernst Eduard Kummer, and Leopold Kronecker dominated the stage of German mathematics.[30] In Berlin, Klein met the unorthodox, yet brilliant Norwegian mathematician Sophus Lie, the man who would become the third formative influence during this critical stage of his career. Nearly six years older than Klein, Lie had entered mathematics at the research level only after his exposure to the writings of Victor Poncelet, Michel Chasles, and Plücker. Aside from his knowledge of the work of these three geometers, he had relatively little formal training, scant acquaintance with other mathematical literature, and no doctoral degree when he arrived in Berlin in October 1869.[31] This last circumstance apparently proved bothersome for him, since he consciously misled those he met in Berlin by introducing himself as Dr. Lie.[32]

[28]Felix Klein, "Über die Transformation der allgemeinen Gleichung des zweiten Grades zwischen Linien-Koordinaten auf eine kanonische Form," (doctoral dissertation, Bonn University, 1868), or *Klein: GMA*, 1:9 (our translation).

[29]See Alexander Brill, "Max Noether," *Jahresbericht der deutschen Mathematiker-Vereinigung* 32 (1923):211–233 on pp. 216–218. Brill and Noether gave a detailed survey of work in this field in "Die Entwicklung der Theorie der algebraischen Functionen."

[30]In his autobiographical sketch, Klein wrote: "Despite the favorable surroundings in Göttingen, the desire to broaden my horizons and to transcend the boundaries of the scientific 'schools' prompted me to leave. Thus, against the wishes of Clebsch, who, just as Plücker had done earlier, advised me not to pursue this plan, I left for Berlin." Klein, "Göttinger Professoren," p. 15 (our translation). Kurt-R. Biermann documents the Berlin school in his book *Die Mathematik und ihre Dozenten an der Berliner Universität*, 1810–1933 (Berlin: Akademie-Verlag, 1988).

[31]See Friedrich Engel, "Zur Erinnerung an Sophus Lie," *Berichte über die Verhandlungen der königlichen sächsischen Gesellschaft der Wissenschaften zu Leipzig* 51 (1899):11–61 on pp. 15–16.

[32]Klein, for example, referred to him as Dr. Lie in a letter to Max Noether dated 17 December 1869. An excerpt can be found in David E. Rowe, "The Early Geometrical Works of Sophus Lie and Felix Klein," in *The History of Modern Mathematics*, 1:209–273 on p. 231.

SOPHUS LIE (1842–1899)

Klein and Lie quickly gravitated to one another. Deeply immersed in Plückerian line geometry, they felt like fish out of water in the Prussian capital, where geometry had all but disappeared following Steiner's death in 1863.[33] Indeed, under the sway of Weierstrass and Kronecker, everything that smacked of geometrical reasoning or physical intuition was considered taboo and so was systematically shunned. Moreover, Klein had clearly inherited the sense of estrangement that both Plücker and Clebsch had felt toward the Berlin scientific establishment.[34]

It would be wrong to infer from this, however, that Klein took no serious interest in the work of the Berlin school. On the contrary, he attended Kronecker's lecture courses on quadratic forms and number theory, in addition to meeting informally with Kronecker and Weierstrass for discussions of, among other things, recent work in algebraic geometry undertaken by Clebsch and Noether. Writing to the latter, though, he described Weierstrass as a "very imposing" personality, even in private conversations.[35] In Klein's opinion, this aspect of Weierstrass' nature gave his lectures a distinctly uncongenial, authoritarian quality and resulted in a classroom atmosphere very different from that in Clebsch's courses.[36] Choosing, for these reasons, to absent himself from Weierstrass' lectures, Klein still managed to learn something of the Weierstrassian approach to complex analysis through the reports of Ludwig Kiepert.[37] Klein's Berlin sojourn also brought him into contact with the Austrian mathematician Otto Stolz, who exposed him to the ideas of Lobachevsky and Bolyai on non-Euclidean geometry as well as to Christian von Staudt's new approach to synthetic projective geometry. (The latter first introduced the projective concept of cross-ratio without any reliance on an underlying metric for the space.) Around this time, too, Klein learned about the Cayley metric, probably through the German version of one of George Salmon's classic texts.[38] It did not take long before Klein began to speculate on the interrelations among all of these ideas.

Following his winter in Berlin, Klein spent a brief, but intense, spring together with Lie in Paris. They arrived there just after Camille Jordan's landmark *Traité des Substitutions et des Équations algébriques* had come off the press, and during the course of their stay, they got to know both the

---

[33] *Ibid.*, p. 209.

[34] Klein described the tensions between the Clebsch school and the powerful Berlin mathematicians in *Vorlesungen über die Entwicklung der Mathematik im* 19-*Jahrhundert*, 1:297–298.

[35] Felix Klein to Max Noether, 17 December 1869, Klein Nachlass XII, Niedersächsische Staats- und Universitätsbibliothek, Göttingen (hereinafter abbreviated NSUB, Göttingen).

[36] See Felix Klein, *Vorlesungen über die Entwicklung der Mathematik im* 19. *Jahrhundert*, 1:297.

[37] Klein later came to regret his missed opportunity to study under the master himself. See *ibid.*, 1:284.

[38] Wilhelm Fiedler, *Analytische Geometrie der Kegelschnitte von George Salmon* (Leipzig: B. G. Teubner, 1860). Klein refers to his early studies of the Salmon-Fiedler texts in *Klein: GMA*, 1:3 and 52.

man and his imposing book.[39] Jordan's work may very well have suggested a connection between group theory and the theory of what Klein and Lie called W-curves, objects which arose in connection with the theory of tetrahedral line complexes.[40] These curves have the property that they are left invariant by a one-parameter group of commutative projective transformations. In fact, Klein and Lie hinted at the heuristic role they envisioned for the group concept in geometry in two brief notes presented to the Paris Academy by the aged Chasles in 1870.[41] Later the following year, they gave a more explicit reference to this idea in a systematic study of the various types of W-curves that arise in the plane.[42]

Even more significant than their contact with Jordan, however, was the stimulation Klein and Lie received from the geometer Gaston Darboux. While in Paris, Klein became aware of the work of Darboux and Théodore Moutard on confocal cyclides. The latter subsume various surfaces—among them the Dupin cyclides, that is, surfaces enveloped by a two-parameter family of spheres—as special cases.[43] Klein later recorded the impression this work made on him, recalling "how one day Darboux showed me a manuscript giving a detailed treatment of the theory of cyclides, and how he added the remark that he had obtained the same formula as I had earlier only with five variables instead of six. There was no doubt that here there must exist a transfer principle [*Übertragungsprincip*] connecting line geometry with metric geometry."[44]

---

[39] Camille Jordan, *Traité des Substitutions et des Équations algébriques* (Paris: Gauthier-Villars, 1870). Klein, who greatly admired French texts, later called Jordan's *Traité* "remarkably un-French, ponderous, almost German." Felix Klein, *Vorlesungen über die Entwicklung der Mathematik im* 19. *Jahrhundert*, 1:338 (our translation).

[40] The terminology W-curve derived from von Staudt's theory, in which a "Wurf" (or "throw") replaces the usual metric definition of cross-ratio. See Thomas Hawkins, "Line Geometry, Differential Equations, and the Birth of Lie's Theory of Groups," in *The History of Modern Mathematics*, 1:275–329 on pp. 284–289. A *tetrahedral* line complex has the property that its lines meet the faces of a tetrahedron in four points having a fixed cross ratio.

[41] Felix Klein and Sophus Lie, "Deux Notes sur une certaine Famille de Courbes et de Surfaces," *Comptes rendus* 70 (1870):1222–1226, or *Klein*: *GMA*, 1:415–423.

[42] Felix Klein and Sophus Lie, "Über diejenigen ebenen Kurven, welche durch ein geschlossenes System von einfach unendlich vielen vertauschbaren linearen Transformationen in sich übergehen," *Mathematische Annalen* 4 (1871):50–54, or *Klein*: *GMA*, 1:424–459. See, also, Klein, *Vorlesungen über höhere Geometrie*, ed. Wilhelm Blaschke (New York: Chelsea Publishing Co., 1949), pp. 166–173.

[43] See Rowe, "Early Geometrical Works of Lie and Klein," pp. 252–256, and Felix Klein, *Vorlesungen über höhere Geometrie*, pp. 49–52.

[44] Felix Klein, "Über Lie's und meine Arbeiten aus den Jahren 1870–72," unpublished manuscript, Lie Nachlass, Ms. fol. 3839, LXVII: 11, Universitetsbiblioteket, Oslo (our translation). Klein refers here to the formula in six variables for a one-parameter family of second-degree line complexes with a given singularity surface, which he had studied in his paper, "Zur Theorie der Linienkomplexe des ersten und zweiten Grades," *Mathematische Annalen* 2 (1870):198–226, or *Klein*: *GMA*, 1:53–80 on p. 79. Darboux's analogous formula in five variables represented a one-parameter family of confocal cyclides, and was written in so-called pentaspheric rather than line coordinates. For more details, see *ibid.*.

Since the confocal cyclides were intimately tied to Dupin's Theorem,[45] Klein naturally sought a suitable analogy for line geometry. Before he succeeded in doing this, however, Lie made the seminal discovery of a certain contact transformation that maps lines to spheres and that simultaneously sends the asymptotic curves of one surface into lines of curvature on the image surface. The immediate payoff came when Lie found that he could use the results of Darboux and Moutard, who had found the lines of curvature for their generalized cyclides, in order to determine the asymptotic curves of the Kummer surface, an algebraic surface of degree four with sixteen (the maximum number of) double points.[46]

This discovery came just before the outbreak of the Franco-Prussian War in 1870. With Klein's return to the Rhineland to enlist in the German cause and Lie's adventurous attempt to reach Italy on foot, the collaboration necessarily broke off.[47] By January of 1871, the Germans had crowned Wilhelm I as their emperor in Versailles, Lie and Klein had resumed their collaboration, and Klein had been made a *Privatdozent* at Göttingen.[48] He quickly resumed his search for a suitable line-geometric formulation of Dupin's Theorem. Not only did he find it, but along the way he also gained the valuable insight that "line geometry is equivalent to metric geometry in four variables."[49] He spelled out this idea in one of his most important early papers, "Über Liniengeometrie und metrische Geometrie." Although written in October of

---

[45]Dupin's Theorem states that the surfaces in orthogonal families intersect along lines of curvature. The classic example of such surfaces is the one-parameter family of confocal quadric surfaces, special types of confocal cyclides. For a modern treatment of Dupin's Theorem, see, for example, Michael Spivak, *Differential Geometry*, 5 vols. (Boston: Publish or Perish Press, 1970–1975), 3:300.

[46]Sophus Lie, "Über Komplexe, insbesondere Linien- und Kugelkomplexe, mit Anwendung auf die Theorie partieller Differentialgleichungen," *Mathematische Annalen* 5 (1872):145–256, or Sophus Lie, *Gesammelte Abhandlungen*, ed. Friedrich Engel and Poul Heegaard, 7 vols. (Oslo: H. Aschehoug & Co. and Leipzig: B. G. Teubner, 1922–1960), 2:1–121 (hereinafter abbreviated *Lie*: *GA*).

[47]Klein recounted his wartime experiences in Klein Nachlass XXII L.2, NSUB, Göttingen. Suspected as a German spy, Lie actually spent a month in a French jail. See I. M. Yaglom, *Felix Klein and Sophus Lie*: *Evolution of the Idea of Symmetry in the Nineteenth Century*, trans. Sergei Sossinsky (Boston: Birkhäuser, 1988), p. 24. For an account of the military events, consult Michael Howard, *The Franco-Prussian War. The German Invasion of France, 1870–1871* (London: Rupert Hart-Davis, 1961).

[48]Klein, "Über Lie's und meine Arbeiten aus den Jahren 1870–72." During the oral examination required of candidates for the position of *Privatdozent*, Klein defended a new approach to geometry in which the constructs of projective geometry are no longer considered fundamental, thereby hinting at the ideas that would soon emerge full blown in his *Erlanger Programm*. In view of Klein's publication record, his mentor, Clebsch, apparently had no difficulty persuading his colleagues on the Göttingen faculty to waive the usual requirement that called for the candidate to submit a *Habilitationsschrift*.

[49]Felix Klein, "Über Liniengeometrie und metrische Geometrie," *Mathematische Annalen* 5 (1872):257–277, or *Klein*: *GMA*, 1:106–126 on p. 112. For an account of the connection Klein made between line geometry and the geometry of a quadric hypersurface in $\mathbb{R}^5$, see Rowe, "Early Geometrical Works of Lie and Klein," pp. 258–262.

1871, this work already contained many of the fundamental *leitmotifs* and examples that Klein took up one year later in the *Erlanger Programm*.[50]

Around this same time, Klein also returned to the problem of finding a link between the Cayley metric and non-Euclidean geometries. Although Cayley had been well aware that his approach made sense for an arbitrary conic Ω in the projective plane, he had overlooked the possibility of a geometric interpretation involving such general "absolute" figures Ω.[51] Klein took this decisive step in "Ueber die sogenannte Nicht-Euklidische Geometrie," showing that different types of conics lead to different geometries: a real conic to a *hyperbolic* geometry, an imaginary conic to an *elliptic* geometry, and (as a special limiting, *parabolic* case) the degenerate conic Cayley had used to a Euclidean geometry.[52] (This terminology is due to Klein and was introduced in this paper.) Klein employed formulas similar to Cayley's for the projective distances but recast Cayley's formulation by using von Staudt's treatment of cross-ratios, which are purely projective invariants.[53] Unfortunately, the significance of this step was lost upon many readers, perhaps in part because Klein postponed discussion of the projective basis for his theory until the very end of his long paper, but also in part because subtle arguments in the foundations of geometry were not his forte.[54]

As this work exemplified, Klein generally aimed to illuminate the larger picture rather than focusing on cleaning up the details of an argument. By the end of 1871, the twenty-two-year-old Rhinelander had begun to envision his biggest picture yet, a grandiose view of geometry that went far beyond anything his contemporaries had ever imagined. In effect, it ran almost the entire gamut of geometrical research: Plücker's line geometry and French sphere geometry (as well as the variety developed by Lie), the non-Euclidean geometries of Lobachevsky and Riemann, projective geometry in the tradi-

---

[50]This, along with other key works of Klein's, will soon be reprinted as Felix Klein, *The Erlangen Program, Evanston Colloquium Lectures, and Other Selected Works*, ed. Jeremy Gray and David E. Rowe (New York: Springer-Verlag), to appear.

[51]Arthur Cayley, "A Sixth Memoir on Quantics," *Philosophical Transactions of the Royal Society of London* 149 (1859):61–90, and Arthur Cayley and A.R. Forsyth, ed., *The Collected Mathematical Papers of Arthur Cayley*, 14 vols. (Cambridge: University Press, 1889–1898), 2:561–592. Cayley did, however, indicate how his approach related to spherical geometry, making the important observation that the duality of its theorems results from the fact that the absolute is a nondegenerate conic as opposed to the case of Euclidean geometry. In the former, the formulas for distance between points and angles between lines enter symmetrically, whereas in the latter they do not. See the discussion in Boris A. Rosenfeld, *A History of Non-Euclidean Geometry*, trans. Abe Shenitzer (New York: Springer-Verlag, 1988), pp. 233–236.

[52]Felix Klein, "Über die sogenannte Nicht-Euklidische Geometrie," *Mathematische Annalen* 4 (1871):573–625, or *Klein*: *GMA*, 1:254–305.

[53]Edmond Laguerre had used a formula similar to Klein's as early as 1853, but Klein apparently only became aware of this later. See Edmond Laguerre, "Note sur la Théorie des Foyers," *Nouvelles Annales de Mathématiques* 12 (1853):64–68.

[54]Arthur Cayley apparently never understood the purely projective theory of cross-ratios. He later expressed skepticism regarding the correctness of Klein's approach. See *Klein*: *GMA*, 1:242 and 354.

tion of Cayley and Salmon, the birational geometry of Riemann and Clebsch, the *Ausdehnungslehre* of Hermann Grassmann,[55] the synthetic approach to foundations of von Staudt, and even Lie's new insights into contact transformations and their applications to systems of partial differential equations. Klein wanted to extract the common core from this research and, in so doing, provide a definitive answer to the elusive question: just what constitutes geometrical research? To him, nothing less than the very integrity of his chosen discipline was at stake, and he hoped to restore some order to a notoriously disjointed field that had begun to resemble a veritable Tower of Babel.[56]

An opportunity to articulate this new vision of geometry in the broadest possible terms came in the fall of 1872 when Klein left Göttingen to take up a professorship in Erlangen. Custom there called for the new faculty member to submit a written program to accompany his inaugural lecture before the university community.[57] Klein's *Programmschrift*, entitled "Vergleichende Betrachungen über neueren geometrischen Forschungen," so overshadowed the efforts of other Erlangen professors that it has since become known as *the Erlanger Programm.*[58] Its key novelty lay in Klein's insight that geometries could be classified by means of their associated transformation groups, each of which determines a characteristic collection of invariants. For example, the various types of geometries (elliptic, hyperbolic, and parabolic) Klein had derived by means of the Cayley metric could each be characterized by the respective subgroups of projective transformations that leave a certain type of conic (Cayley's absolute figure) or quadric surface fixed.[59] By thus bringing the group concept to bear on the structural features that determine a geometry, Klein anticipated one of the central trends of twentieth-century mathematics.

---

[55]Hermann Grassmann, *Die lineale Ausdehnungslehre, ein neuer Zweig der Mathematik dargestellt und durch Anwendungen auf die übrigen Zweigen der Mathematik, wie auch auf die Statik, Mechanik, die Lehre vom Magnetismus und die Krystallonomie erläutert* (Leipzig: O. Wigand, 1844).

[56]Klein addressed this issue in some detail in the first two notes he appended to the *Erlanger Programm.* See Felix Klein, "Vergleichende Betrachtungen über neuere geometrische Forschungen," (Erlangen: A. Deichert, 1872), or *Klein*: *GMA*, 1:460–497 on pp. 490–491. The first of these dealt with the antagonism between the synthetic and analytic approaches to geometry, a tension he regarded as no longer holding much significance, whereas the second criticized the separation of geometry into subdisciplines.

[57]Klein's inaugural lecture has often been mistakenly identified with his famous *Erlanger Programm.* See David E. Rowe, "A Forgotten Chapter in the History of Felix Klein's *Erlanger Programm,*" *Historia Mathematica* 10 (1983):448–454; and "Felix Klein's 'Erlanger Antrittsrede.' A Transcription with English Translation and Commentary," *Historia Mathematica* 12 (1985):123–141.

[58]As pointed out by Konrad Jacobs and Heinrich Utz in "Erlangen Programs," *The Mathematical Intelligencer* 6 (1) (1984):79, five mathematicians submitted *Programmschriften* at the time they joined the Erlangen faculty. The other four were Heinrich August Rothe (1814), Johann Pfaff (1825), Christian von Staudt (1845), and Paul Gordan (1875).

[59]Hans Wussing discusses the influence of the Cayley metric on Klein's *Erlanger Programm* in his book, *The Origin of the Abstract Group Concept,* trans. Abe Shenitzer (Cambridge: MIT Press, 1984), pp. 175–177.

Just before Klein presented his *Erlanger Programm*, Alfred Clebsch suffered the attack of diphtheria that took his life at age thirty-nine. Clebsch had been a model teacher, combining mathematical genius with an ability to inspire and encourage gifted students. Had he lived longer, he might easily have become the dominant figure of his generation, and the whole framework for mathematics in Germany might have evolved differently. Under his leadership, concrete plans had already been made for the creation of a German mathematical society, but after his death, this effort gradually lost momentum. Not until 1890 did Georg Cantor succeed in reviving the idea in the form of the *Deutsche Mathematiker-Vereinigung*.[60] As the youngest and most dynamic member of the Clebsch circle, Klein assumed much of the burden for furthering his mentor's research ideals.

Anschauung, Riemann Surfaces, and Geometric Galois Theory in Klein's Early Work

Curiously enough, during his tenure in Erlangen, Klein failed not only to publicize the ideas outlined in his *Erlanger Programm* but also to follow up on them in his own research. Most of his relatively little known work from this brief period centered on a constellation of problems in algebraic geometry closely tied to prior investigations of Clebsch. Although he greatly admired his mentor's ideas, Klein favored concrete geometrical structures in contradistinction to Clebsch's more formal algebraic attack. As a true disciple of Plücker, he sought a connection between the analytical results and the geometrical *Gestalt* they described. The inspiration for much of Klein's most original geometrical research thus came from visual elements or intuitive apprehensions, and while Klein championed such a nonformal, *anschauliche* approach with considerable skill and conviction, he did not see it survive very far into the next century. Even during his lifetime, he strongly sensed that the spirit of the age threatened to cast his achievements into the shadows of oblivion.[61]

In his efforts to recast the Clebschian formulation of algebraic geometry in a more intuitive light, Klein developed tools which he believed closer in spirit to Riemann's own way of thinking. One of the more original ideas he hit upon involved what he called a "new type" of (projective) Riemann surface.[62]

---

[60]See Tobies and Rowe, pp. 20–23; and Friedrich Hirzebruch, "The 100th Anniversary of the *Deutsche Mathematiker-Vereinigung*," *The Mathematical Intelligencer* 13 (2) (1991):8–11.

[61]This foreboding helps to account for the passion behind the propaganda campaigns he launched. In these, he consistently extolled the vitality of *anschauliche* approaches while deprecating the sterility of hairsplitting logicism. See, for example, Felix Klein, "Ueber Arithmetisierung der Mathematik," *Zeitschrift für den mathematischen und naturwissenschaftlichen Unterrricht* 27 (1896), or *Klein: GMA*, 2:232–240.

[62]Klein introduced and applied the "new type of Riemann surface" in four papers from the mid-1870s: "Ueber eine neue Art der Riemannschen Flächen (Erste Mitteilung)," *Mathematische Annalen* 7 (1874):558–566, or *Klein*: *GMA*, 2:89–98; "Ueber den Verlauf der Abelsche Integrale

Following standard geometrical procedures, the locus of points satisfying a real polynomial equation $f(z, w) = 0$ (that is, a polynomial equation with $z, w \in \mathbb{C}$ but with the essential restriction that the coefficients of f be *real*) could be studied either by viewing it as the curve given by the real points that satisfy the equation (that is, as a curve in the real (z, w)-plane), or by taking z to be a complex variable and referring to the associated Riemann surface over the z-plane. In the first case, Klein noted that the image obtained, while visually satisfying, was incomplete since it contained nothing corresponding to the imaginary points on the curve (namely, those with at least one nonreal coordinate). In the second case, however, he found the image complete but the visualization unsatisfactory for two reasons: first, such Riemann surfaces cannot be embedded in three-dimensional space due to their branch points; and second, and most important, it is impossible both to visualize the geometric properties exhibited by the real curve (its singularities, inflection points, etc.) and to study the relationships between it and the imaginary points that satisfy the equation. Klein saw merits in each of these models, but he wished to combine their advantages in order to obviate the need for flip-flopping back and forth between them.

To this end, he built his projective Riemann surface on a skeletal framework formed by a suitably adapted image of the real plane curve C inside real projective space $\mathbf{P}^2(\mathbb{R})$.[63] This had the advantage of preserving the particular features of the curve by actually embedding it in the projective Riemann surface. The construction of this surface could, in principal, be carried out in a few easy steps. Far less simple were the arguments Klein concocted in attempting to show that the genus of his new surfaces always agreed with the genus of the conventional Riemann surfaces associated with the curves in question.

Although Klein pursued these ideas in a novel manner, the general thrust of this work was by no means new. Nineteenth-century geometers had routinely interpreted imaginary elements by means of real constructs. The best-known theory of this kind had been set forth by von Staudt, who utilized the two involutions on the *oriented* real line ℓ joining an imaginary point P with its conjugate P' to give a real interpretation for P and P'. Klein later showed how, by appealing to the Cayley metric, the seemingly arbitrary identification in von Staudt's theory of a given point P with one of the two possible

bei den Kurven vierten Grades (Erste Aufsatz)," *Mathematische Annalen* 10 (1876):365–397, or *Klein*: *GMA*, 2:99–135; "Ueber eine neue Art der Riemannschen Flächen (Zweite Mitteilung)," *Mathematische Annalen* 10 (1876):398–416, or *Klein*: *GMA*, 2:136–155; and "Ueber den Verlauf der Abelsche Integrale bei den Kurven vierten Grades (Zweiter Aufsatz)," *Mathematische Annalen* 11 (1876–1877):293–305, or *Klein*: *GMA*, 2:156–169.

[63]He utilized the dual curve C^*, which, by duality, provides an equivalent version of C when C^* is interpreted as a *class* curve, that is, as the locus of tangent lines to C.

oriented involutions of ℓ could be understood from what Klein later called a "higher standpoint."[64]

Klein's research interests during the period 1872–1875 clearly reflected the influence of his former teacher, Alfred Clebsch. Moreover, Klein assumed the principal responsibility for furthering his mentor's teaching legacy, but although a number of talented students gravitated to Erlangen from Göttingen following Clebsch's death, the opportunity to develop a geometrical school in Erlangen did not really present itself. Klein thus had little difficulty deciding whether to accept the offer to succeed Otto Hesse at the *Technische Hochschule* in Munich in 1875, given that during the three years he had taught at Erlangen he had never had more than seven students in any of his courses. While clearly eager for a change, Klein failed to anticipate the difficulties he would encounter teaching mathematics to large classes of prospective engineers. When he entered the lecture hall to teach his first semester course on analytic geometry, he found it filled to the rafters with more than 200 students.[65]

Fortunately, Klein could count on the support of his colleague Alexander Brill, another leading descendant of the Clebsch school.[66] Brill had also just come to Munich from a teaching post at the *Technische Hochschule* in Darmstadt, where he had gained some familiarity with standard subjects in the engineering curriculum such as descriptive geometry, graphical statics, and kinematics. Besides their mutual concern for the mathematical training of engineering students, Klein and Brill shared an interest in *anschauliche* geometry which prompted them to found a laboratory for the construction of mathematical models at the Munich *Technische Hochschule.* There, the two mathematicians and their students built a number of the models that ultimately became part of the collection marketed by Brill's brother Ludwig, the owner of a publishing firm in Darmstadt. Models from the collection of

---

[64]See Felix Klein, "Zur Interpretation der komplexen Elemente in der Geometrie," *Mathematische Annalen* 22 (1883):242–245, or *Klein*: *GMA*, 1:402–405; and Felix Klein, *Elementary Mathematics from an Advanced Standpoint. Geometry*, vol. 2, trans. E. R. Hedrick and C. A. Noble (New York: Dover Publications, Inc., 1939), pp. 117–129. Interestingly enough, Klein's early collaborator, Lie, had developed his own interpretation of imaginary elements in geometry by employing constructs from Plücker's line geometry. In fact, Lie's discovery of this convoluted construction had effectively launched his mathematical career, since his work on this topic earned him the grant that had enabled him to pursue his studies on the Continent. See Sophus Lie, "Repräsentation der Imaginären der Plangeometrie," *Journal für die reine und angewandte Mathematik* 70 (1869):346-353, or *Lie*: *GA*, 1:1–11, and the notes on pp. 536–537. Lie's approach and various other methods for interpreting imaginary quantities in geometry would later attract the attention of several American mathematicians. Consult, for example, Percey F. Smith, "On Sophus Lie's Representation of Imaginaries in Plane Geometry," *American Journal of Mathematics* 25 (1903):165–179, and compare the discussion of the dissertations of J. L. Coolidge and William Graustein in Chapter 10.

[65]The attendance figures can be found in Hörerverzeichnis, Klein Nachlass VII E, NSUB, Göttingen.

[66]See Gerhard Betsch, "Alexander von Brill (1842–1935)," *Bausteine zur Tübingen Universitätsgeschichte* 3 (1989):71–90 on pp. 77–78.

AS COLLEAGUES AT THE TECHNISCHE HOCHSCHULE IN MUNICH DURING THE LATE 1870S, KLEIN AND ALEXANDER BRILL ESTABLISHED A LABORATORY FOR THE CONSTRUCTION OF MATHEMATICAL MODELS. THE STUDENTS' FINISHED WORK ENHANCED THE IMPRESSIVE COLLECTION OF MODELS ON DISPLAY AT THE TECHNISCHE HOCHSCHULE IN MUNICH, SHOWN HERE DURING A SOMEWHAT LATER PERIOD.

L. Brill were later purchased by many mathematics departments throughout Europe and the United States.[67]

Such interests aside, the focus of Klein's activities during his five highly productive years in Munich centered on his research in pure mathematics. His work, however, involved neither topics in projective and algebraic geometry nor the broad scope of ideas he had outlined in the *Erlanger Programm*. By the mid-1870s, Klein's attention had shifted to problems at the inter-

[67]The Verlag L. Brill monopolized this field until 1899 when it was bought out by the firm of Martin Schilling in Halle. A contemporary account of many of the models is presented in Walther von Dyck, *Katalog mathematischer und mathematische-physikalischer Modelle, Apparate und Instrumente* (Munich: C. Wolf, 1892). An excellent modern discussion can be found in Gerd Fischer, ed., *Mathematische Modelle*, 2 vols. (Berlin: Akademie-Verlag, 1986). It was also in Munich that Klein began to take some interest in technological problems. See Tobies, p. 42; and Manegold, pp. 93–94.

face of algebra and the theory of complex variables, an area he enriched by developing both a geometric approach to Galois theory and a unified theory of elliptic modular functions.[68] This work also paved the way for his contributions to an entirely new field of research, the theory of automorphic functions.

One of Klein's early results in this direction concerned the classification of all finite groups of Möbius transformations, that is, linear fractional transformations $z \mapsto \frac{az+b}{cz+d}$ where $z \in \mathbb{C} \cup \{\infty\}$, $a, b, c, d \in \mathbb{C}$, and $ad - bc \neq 0$. (By stereographic projection, these transformations define transformations on the Riemann sphere S^2.) Klein studied these groups by means of the five Platonic solids, which he inscribed in S^2. A projection of the faces of the respective solids onto the sphere yields a tessellation of its surface, and the group of motions of the solid into itself induces an isomorphic group of linear fractional transformations on S^2. The tetrahedron thus corresponds to a group with 12 rotations; the cube (like its dual figure, the octahedron) has a rotation group with 24 elements; and the icosahedron (like its dual, the dodecahedron) admits 60 rotations. In his "Mémoire sur les Groupes de Mouvements" of 1869, Camille Jordan had given a complete enumeration of the finite rotation groups in three-space, showing that they consist of those just mentioned together with the cyclic groups.[69] Besides the Platonic solids, Klein considered *dihedra*, regular polygons with n edges inscribed in the equator. These admit n rotations and as many reflections and so determine the dihedral groups with $2n$ elements. Utilizing Jordan's result, Klein easily showed in 1875 that the finite rotation groups and the dihedral groups comprise the finite groups of linear fractional transformations.[70]

For Klein, the significance of this new finding lay in its connection with his geometric approach to Galois theory and, in particular, with *Galois resolvents* or polynomial equations whose roots are permuted simply transitively by a Galois group. (The role of geometry in this work is central, since Klein essentially identified a one-parameter algebraic equation $f(z, w) = 0$ with its Riemann surface over the z-plane. He then defined the genus of the equation as the genus of the corresponding surface.) In a major study written in 1878, Klein stressed that the only known Galois resolvents of genus zero containing a single parameter were those associated with finite groups of Möbius transformations.[71] He identified these with four types of Galois

[68] For a detailed discussion of Klein's contributions to these subjects, see Jeremy Gray, *Linear Differential Equations and Group Theory from Riemann to Poincaré* (Boston: Birkhäuser, 1986), pp. 196–203, 225–238.

[69] Camille Jordan, "Mémoire sur les Groupes de Mouvements," *Annali di Matematiche* 2 (1869):167–215.

[70] Felix Klein, "Über binäre Formen mit linearen Transformationen in sich," *Mathematische Annalen* 9 (1875–1876):183–208, or *Klein*: *GMA*, 2:275–301. Thus, in modern notation, the finite subgroups of $PGL_2(\mathbb{C})$ are the cyclic and dihedral groups, the alternating groups A_4, A_5, and the symmetric group S_4.

[71] Felix Klein, "Ueber die Transformation der elliptischen Funktionen und die Auflösung

resolvents: the dihedral, tetrahedral, octahedral, and icosahedral equations. His overriding observation was compelling and simple: these four types of resolvents alone admitted Galois groups that could act on the sphere. Here, as so often, Klein's achievement had less to do with the novelty of the results he obtained than with the fresh insight provided by the overall conceptual framework he employed in deriving them.

In much of Klein's work on geometric Galois theory, the rotation group of the icosahedron, the *icosahedral group*, played a particularly important role. He gave an elegant geometric argument showing that this group is isomorphic to the alternating group A_5, by noting that the midpoints of the 30 edges of the icosahedron consist of the vertices of five octahedra. This suggested a direct connection with the Galois theory of the quintic equation, which he attacked by merging group-, function-, and invariant-theoretic ideas in the development of a theory of the icosahedral equation, an equation of degree sixty.[72] Suitably interpreted, this equation provides a Galois resolvent of the quintic equation. The theory surrounding the icosahedral equation would later receive considerable attention in the United States thanks in part to the efforts of Klein's student, Frank Nelson Cole. (See Chapter 5.)

The techniques that Klein brought to bear in developing a geometric Galois theory for the quintic equation in the last half of the 1870s gradually led him to develop a general theory of *(elliptic) modular functions*, that is, meromorphic functions on the upper half-plane $\mathscr{H}^+$ left invariant under the modular group $\Gamma = SL_2(\mathbb{Z})$. Such functions had arisen in the work of Gauss, Riemann, and Weierstrass, among others, but always in the context of elliptic function theory. The latter area involves meromorphic functions on a complex torus $\mathbb{C}/\Omega$, where $\Omega := \mathbb{Z}\omega_1 \oplus \mathbb{Z}\omega_2$ is a lattice in $\mathbb{C}$. Elliptic functions are thus meromorphic functions which are doubly periodic with (fundamental) periods ω_1, ω_2.[73]

der Gleichungen fünften Grades," *Mathematische Annalen* 14 (1879):111–172, or *Klein*: *GMA*, 3:13–75 on p. 55. Klein's first contribution to geometric Galois theory had come seven years earlier in "Ueber eine geometrische Repräsentation der Resolventen algebraisches Gleichungen," *Mathematische Annalen* 4 (1871): 346–358, or *Klein*: *GMA* 2:262-274.

[72]Besides the two works cited in the above two footnotes, Klein also published a third major paper on this subject: Felix Klein, "Weitere Untersuchungen über das Ikosaeder," *Mathematische Annalen* 12 (1877):503–560, or *Klein*: *GMA*, 2:321–380. He subsequently refined and elaborated on this at length in Felix Klein, *Vorlesungen über das Ikosaeder und die Auflösung der Gleichungen vom fünften Grade*, ed. Peter Slodowy (Basel: Birkhäuser and Leipzig: B. G. Teubner Verlagsgesellschaft, 1993), based on the original Teubner edition of 1884; or *Lectures on the Icosahedron and the Solution of Equations of the Fifth Degree*, trans. George Gavin Morrice (London: Kegan Paul, Trench, Truber & Co., 1888). Slodowy's new edition contains an excellent modern introduction and commentary.

[73]A key example of an elliptic function is the Jacobi function $z(u) = \mathrm{sn}(u, k)$ (or more simply $\mathrm{sn}(u)$) obtained by inversion of the elliptic integral $u(z) = \int_0^z \frac{dz}{\sqrt{(1-z^2)(1-k^2z^2)}}$ ($k \in \mathbb{C}$). The Weierstrass $\wp$-function is another famous example of an elliptic function: $\wp(z|\Omega) := \frac{1}{z^2} + \sum_{0\neq\omega\in\Omega}(\frac{1}{(z+\omega)^2} - \frac{1}{\omega^2})$. It provides a connection between function theory and algebraic geometry. In modern terminology, let E be the field of meromorphic functions $f(z)$

Modular functions, on the other hand, arise naturally in the consideration of the conformal equivalence classes of complex tori. Given two lattices Ω as above and $\Omega' := \mathbb{Z}\omega_1' \oplus \mathbb{Z}\omega_2'$ satisfying $\omega := \frac{\omega_1}{\omega_2}$, $\omega' := \frac{\omega_1'}{\omega_2'} \in \mathscr{H}^+$, the tori $\mathbb{C}/\Omega$ and $\mathbb{C}/\Omega'$ are conformally isomorphic if and only if $\omega' = g \cdot \omega$ for some $g \in \Gamma$. The famous Dedekind J-function is an important example of a modular function whose values actually parametrize the conformal equivalence classes of complex tori.[74] Based on this, Klein developed a theory of *modular equations*, that is, equations relating J and its image J' under a transformation $J \mapsto J' = \begin{pmatrix} a & b \\ c & d \end{pmatrix} J$ with $ad - bc = n$ for a prime n.[75]

By late 1879, Klein had developed a general research program designed to deal with these and related matters that had engaged mathematicians from Abel to Weierstrass. He based his approach—known as *Stufentheorie* or level theory[76] —on certain subgroups $\Gamma(n)$ of the modular group Γ. The principal congruence subgroup $\Gamma(n)$ of level n is comprised of those elements in $g \in \Gamma$ satisfying $g \equiv \begin{pmatrix} 1 & 0 \\ 0 & 1 \end{pmatrix} \pmod{n}$. Each such $\Gamma(n)$ is a normal subgroup of finite index. Klein realized that one must also consider general subgroups of level n, that is, subgroups Λ of Γ such that n is minimal with $\Gamma(n) \subset \Lambda \subset \Gamma(1) = \Gamma$. A function f is called a *level n modular function* if f remains invariant under a subgroup of level n but not one of level less than n.

From Klein's point of view, this group-theoretic framework proved particularly advantageous. It provided a view of the theory of modular functions

---

satisfying $f(z+w) = f(z)$, $z \in \mathbb{C}$, $\omega \in \Omega$. Then E is generated by $\wp$ and its derivative $\wp'$. Also, $y = \wp$ satisfies the differential equation $y'^2 = 4y^3 - g_2 y - g_3$ (g_2, $g_3 \in \mathbb{C}$, $g_2^3 - 27g_3^2 \neq 0$), so that E is isomorphic to the field of rational functions on a complex elliptic curve. For further information on elliptic and modular functions, the reader may consult, for example, Robert Fricke, *Die elliptische Funktionen und ihre Anwendungen*, Erster Teil (Leipzig and Berlin: B. G. Teubner, 1930), pp. 194–227; Komaravolu Chandrasekharan, *Elliptic Functions* (New York: Springer-Verlag, 1985); and Joseph Lehner, *Discontinuous Groups and Automorphic Functions*, Mathematical Surveys, vol. 8 (Providence: American Mathematical Society, 1964).

[74]See Egbert Brieskorn and Horst Knörrer, *Plane Algebraic Curves*, trans. John Stillwell (Boston: Birkhäuser Verlag, 1986), pp. 302–303 and 312–322 for a discussion of these ideas. Dedekind introduced the J-function in "Schreiben an Herrn Borchardt über die Theorie der elliptischen Modulfunktionen," *Journal für die reine und angewandte Mathematik* 83 (1877):265–292. He called this function *die Valenz*, but since the appearance of Klein's work it has been known as the *J-function*. In Weierstrass' theory of elliptic integrals, there are two basic invariants g_2 and g_3. In the notation of note 73, $J = \frac{g_2^3}{\Delta}$, where the discriminant $\Delta = g_2^3 - 27g_3^2$. See Klein, "Über die Transformation der elliptischen Funktionen und die Auflösung der Gleichungen fünften Grades," pp. 14–15.

[75]This work was inspired by the theory of modular equations for elliptic integrals as developed by Jacobi, Galois, and others. See Carl G. J. Jacobi, *Fundamenta Nova Theoriae Functionum Ellipticarum* (Königsberg: Bortraeger Brothers, 1829), reprinted in *C. G. J. Jacobi's Gesammelte Werke*, ed. Carl W. Borchardt and Karl Weierstrass, 8 vols. (Berlin: Verlag von G. Reimer, 1881–1891), 1:49–239; and Évariste Galois, "Lettre à M. A. Chevalier sur la Théorie des Équations et les Fonctions intégrales," *Journal des Mathématiques pures et appliquées* 11 (1846):406–418.

[76]Klein set forth his *Stufentheorie* in "Zur Theorie der elliptischen Modulfunktionen," *Mathematische Annalen* 17 (1880):62–70, or *Klein*: *GMA*, 3:169–178.

and their invariants from a "higher standpoint," since any two modular functions stabilized by the same group are algebraically dependent. In connection with these groups, one can follow the images of a fundamental region to generate a polygonal figure, which, after identifying corresponding sides via the group action, yields a closed Riemann surface of genus p. Klein called p the *genus* of the group.

Within a span of only five years, Klein had apparently found a field of research well suited to his mathematical tastes. The mixture of group theory, algebraic equations, and function theory, all served up in a striking visual display full of *anschauliche* elements, clearly whetted his voracious appetite for work. By the end of his tenure in Munich, the thirty-year-old Klein had already published some seventy papers, including a few which would become classics. His contemporaries, friend and foe alike, clearly saw him as *the* rising star in German mathematics. Thus, when the Saxon Ministry of Culture created a new chair for a geometer at Leipzig University, there was little doubt who would receive the call. Klein accepted the offer with alacrity in May of 1880, and in the fall, he took up a new challenge that would mark both the zenith and the nadir of his career as a research mathematician.

PROFESSOR OF GEOMETRY IN LEIPZIG

Over the course of its nearly five-hundred-year history, Leipzig University had attracted considerable mathematical talent, men like Regiomontanus (Johannes Müller), Joachim Rheticus, Gottfried Wilhelm Leibniz, and August Ferdinand Möbius.[77] Although certainly not continuous, no other German university could point to a comparable mathematical lineage, stretching back to the dawn of the Scientific Revolution. Nevertheless, prior to Klein's arrival in 1880, Leipzig lagged behind such universities as Königsberg, Berlin, and Göttingen, where the marriage of research and teaching had been flourishing, at least sporadically, for some time. Klein's six years in Saxony clearly marked a decisive turning point, however, since he, more than anyone else, laid the foundations for the emergence of Leipzig as one of the leading mathematical centers in Germany.

From an institutional standpoint, the situation certainly cried out for vigorous new ideas. Klein's colleagues—Carl Neumann, Wilhelm Scheibner, Adolf Mayer, and Karl von der Mühll—were all considerably older and had shown no real talent for building a strong mathematics program. Moreover, the faculty's strength lay in analysis and mathematical physics, although Scheibner also taught courses in algebra and number theory. Before Klein's

[77]See Walter Purkert, "Die Mathematik an der Universität Leipzig von ihrer Gründung bis zum zweiten Drittel des 19. Jahrhunderts," in 100 *Jahre Mathematisches Seminar der Karl-Marx-Universität Leipzig*, ed. Herbert Beckert and Horst Schumann (Berlin: VEB Deutscher Verlag der Wissenschaften, 1981), pp. 9–40. On Leibniz, see Joseph Ehrenfried Hoffmann, *Leibniz in Paris, 1672–1676, His Growth to Mathematical Maturity* (Cambridge: University Press, 1974).

appointment, Leipzig had no one qualified to teach courses in geometry, a glaring gap for a major university.[78]

In his inaugural lecture, delivered on 25 October 1880, Klein stressed certainly not geometry *per se* but his standard theme: the need for a comprehensive and unified approach to mathematics unencumbered by narrow specialization and free from the insularity of one-sided schools. He further emphasized the importance of the applied side in mathematics instruction, recommending courses in drawing and the construction of geometric models. These types of courses, which he had cultivated with considerable success at the *Technische Hochschule* in Munich, had never found a niche at any of the leading German universities. Klein's suggestion thus amounted to a proposal for transplanting certain curricular innovations of the technical schools into the universities, those older, more tradition-bound bastions of pure learning.[79]

Klein wasted no time in putting his ideas into action at Leipzig. He reinstated the antiquated post of the *Famulus* (a position Goethe had made famous through Faust's Wagner). By November of 1880, the Saxon Ministry had approved Klein's proposal for funding a collection of mathematical models, and one month later he submitted another petition calling for the creation of a Mathematics Seminar to be housed in the Czermakian Spectatorium, a building constructed in the early 1870s and financed by the Leipzig physiologist, J. N. Czermak. At Klein's request, this structure was renovated for the purposes of mathematics instruction and reopened for its new use by the beginning of the second semester. The Spectatorium contained one large lecture hall with a 150-person capacity, a seminar room, mathematics library, and a room for informal meetings. Adolf Mayer and the renowned psychologist Wilhelm Wundt were put in charge of these physical facilities, and Klein, Mayer, and von der Mühll were named co-directors of the associated Mathematics Seminar.[80]

These swift reforms quickly made an impact on mathematics at Leipzig. Whereas in the decade preceding Klein's arrival, the mathematics program there had graduated only nine doctoral students (compared with twenty-nine in Berlin and a high of sixty in Göttingen), it produced thirty-six doctorates in mathematics during the course of his six-year tenure. More than half of these students took their degrees under Klein's supervision.[81] Even more telling, the Leipzig faculty accepted five postdoctoral theses or *Habilitationsschriften*

---

[78]Fritz König, "Die Gründung des 'Mathematischen Seminars' der Universität Leipzig," in 100 *Jahre Mathematisches Seminar der Karl-Marx-Universität Leipzig*, pp. 41–72 on pp. 56–61.

[79]Felix Klein, "Ueber die Beziehungen der neueren Mathematik zu den Anwendungen," *Zeitschrift für den mathematischen und naturwissenschaftlichen Unterricht* 26 (1895):535–540. (This work was not included in *Klein*: *GMA*.)

[80]König, pp. 61–70.

[81]See the statistics presented in König, pp. 50–51.

during this same period, far more than any other university.[82]

As positive and dramatic as these changes were, however, not everyone appreciated Klein's sudden influence in Leipzig. In particular, Carl Neumann, who clearly regarded himself as *the* mathematical physicist and function-theorist on the Leipzig faculty, resented not only his younger colleague's ambitious plans but also his encroachment upon what Neumann apparently regarded as his own private mathematical domain. Thus, when Klein began a year-long course on function theory from a geometric standpoint during his very first semester, Neumann felt that his territory had been violated.[83]

Klein's highly successful lectures drew eighty-nine auditors during the first semester and forty-five the second (including the former Hopkins student Irving Stringham, who attended both semesters). On completing the course, Klein transformed the last half of the material into one of his most influential works, a booklet entitled *Über Riemanns Theorie der algebraischen Funktionen und ihrer Integrale.*[84] In both his 1881 lectures and the resulting booklet, Klein attempted to explicate some of the main results presented in Riemann's two-part paper of 1857 on Abelian function theory.[85] While one of the most profoundly original works in the history of mathematics, Riemann's study had met with a less than smooth reception. In particular, Karl Weierstrass had carefully scrutinized Riemann's methods and had found them wanting. Riemann had invoked an intuitively compelling argument to deduce the existence of certain harmonic functions that assume given boundary conditions, an idea he had learned from Dirichlet and which, in honor of his teacher, he had called the Dirichlet Principle. Weierstrass apparently communicated at least his misgivings about the validity of this principle to Riemann directly before the latter's death in 1866.[86] Not until 1870, however, did he

[82]Five of Leipzig's six *Privatdozenten* in mathematics—Walther von Dyck, Friedrich Schur, Karl Rohn, Eduard Study, and Friedrich Engel—had ties with Klein.

[83]Neumann had published his own well-known textbook, *Vorlesungen über Riemanns Theorie der Abelschen Integrale* (Leipzig: B. G. Teubner, 1884), originally in 1865, and, in all likelihood, was already at work on the substantially expanded second edition that appeared in 1884. Klein remarked on Neumann's resistance to his lectures on function theory in his autobiographical sketch in "Göttinger Professoren," p. 20.

[84]Felix Klein, *Über Riemanns Theorie der algebraischen Funktionen und ihrer Integrale* (Leipzig: B. G. Teubner, 1882), or *Klein*: *GMA*, 3:499–573. The lectures from the first semester have been recently published as Felix Klein, *Funktionentheorie in geometrischer Behandlungsweise*, ed. Fritz König, Teubner-Archiv zur Mathematik, vol. 7 (Leipzig: BSB B. G. Teubner Verlagsgesellschaft, 1987).

[85]Klein's lectures, however, bypassed one of the central aspects of Riemann's theory, namely, the inversion of Abelian integrals by means of Θ-functions. Riemann introduced his Θ-functions, generalizing those of Jacobi, in the second part of his paper, "Theorie der Abel'schen Funktionen," *Journal für die reine und angewandte Mathematik* 54 (1857):115–155, or *Riemann*: *GMW*, pp. 88-142.

[86]See Felix Klein, *Vorlesungen über die Entwicklung der Mathematik im* 19. *Jahrhundert*, 1:264. Klein further noted, however, that, according to what Weierstrass related to him in a conversation, Riemann had reacted with little concern to this finding, apparently in the belief that his existence theorems were correct, independent of the soundness of the Dirichlet Principle.

present his devastating critique of Riemann's argument in a short paper that appeared to bring down the entire Riemannian edifice.[87]

As we have seen, the theory of Riemann surfaces had come to occupy a crucial place in nearly all of Klein's research during his years in Erlangen and Munich. Then, as later in Leipzig and Göttingen, his interests lay in the global implications of the theory rather than in the evidently sticky foundational questions it raised. In his Leipzig lectures, Klein broadened his purview on the theory of Riemann surfaces even further by interpreting it in terms of physics. Through an appeal to models of current flows on closed surfaces, Klein sought not only to bypass the troubled foundations of Riemann's theory but also to offer immediate *physical* evidence for the validity of the basic existence theorems for algebraic functions and their integrals. While Klein's longstanding interest in physics certainly played a role here, the decisive motivation behind this new approach came from a deep sympathy for and even inner-identification with the thought processes that Klein believed had guided Riemann in his own work.

Klein emphasized his *Büchlein*'s novel point of view when he republished it forty years later in the third volume of his collected works. "In modern mathematical literature," he wrote,

> it is altogether unusual to present, as occurs in my [booklet], general physical and geometrical deliberations in naive *anschaulicher* form that later find their firm support in exact mathematical proofs. ... I consider it unjustifiable that most mathematicians suppress their intuitive thoughts and only publish the necessary, strict (and mostly arithmetical) proofs ... . I wrote my work on Riemann precisely as a physicist, unconcerned with all the careful considerations that are usual in a detailed mathematical treatment, and, precisely because of this, I have also received the approval of various physicists.[88]

---

[87] In brief, Weierstrass showed that Riemann had invoked a standard variational principle to deduce that a function u can be found which attains the greatest lower bound of $\int\int \nabla u \, dx \, dy$. Unfortunately, his argument only proved that one could find functions u_n which approach this bound from above. See Karl Weierstrass, "Über das sogenannte Dirichlet'schen Princip," read before the *Königlicher Akademie der Wissenschaften* on 14 July 1870 and published in Karl Weierstrass, *Mathematische Werke*, 7 vols. (Berlin: Mayer & Müller, 1894–1927; reprint ed., Hildesheim: Georg Ohms Verlagsbuchhandlung and New York: Johnson Reprint Corporation, 1967), 2:49–54. Hilbert finally resolved this problem in David Hilbert, "Über das Dirichletsche Prinzip," *Mathematische Annalen* 59 (1904):161–186; "Über das Dirichletsche Prinzip," *Journal für die reine und angewandte Mathematik* 129 (1905):63–67, or David Hilbert, *Gesammelte Abhandlungen*, 3 vols. (Berlin: Julius Springer, 1932–1935), 3:15–37 (hereinafter denoted *Hilbert: GA*). For some of Klein's remarks on the Dirichlet Principle, see Felix Klein, *Vorlesungen über die Entwicklung der Mathematik im* 19. *Jahrhundert*, 2 vols. (Berlin: Springer-Verlag, 1926–1927): 1:262–267. The many historical twists and turns surrounding this theorem have been discussed in A. F. Monna, *Dirichlet's Principle: A Mathematical Comedy of Errors and Its Influences on the Development of Analysis* (Utrecht: Oosthoeck, Scheltema, and Holkema, 1975).

[88] Felix Klein, "Vorbemerkungen zu den Arbeiten über Riemannsche Funktionentheorie,"

By presenting the theory intuitively, Klein tried to convey the source of Riemann's inspiration as he understood it. Always a romantic, he passionately objected to the compression of mathematical content to fit a preconceived form in which all traces of the thought processes that led to a discovery disappear. He truly thought that he had tapped the source of Riemann's original inspiration, namely, potential theory and current flows on surfaces.[89] Klein tried to document this conjecture by asking Riemann's former students and friends if he had ever expressed similar ideas in their presence. The evidence against this turned out to be so compelling that Klein later published a partial retraction of his claim.[90] Klein's booklet may thus be regarded as an inspired, but historically inaccurate, interpretation of Riemann's original theory.

Besides Riemann, Klein also drew from other potent sources. Citing Maxwell's *Treatise on Electricity and Magnetism*,[91] he regarded the current flows on a surface as the fundamental objects of his study. In the special case of a stationary flow in the xy-plane, Klein interpreted the conjugate harmonic functions u, v associated with a holomorphic complex function $w(z) = u(z) + iv(z)$ as the potential functions associated with two conjugate stationary flows. The level curves u thereby correspond to the flow lines determined by the potential function v and *vice versa*.[92]

Examining the behavior of w at a stationary point z_0 (that is, at a crossing point of the flow), adjacent curves of the two families $u = c_1$ and $v = c_2$ meet in z_0 at an angle of $\frac{\pi}{2k}$ if z_0 occurs as a zero of $w - w(z_0)$ of order k. The flow lines at such a crossing point (which corresponds to a branch point of the usual Riemann surface for w) appear in Figure 4.1. To study the singularities of w, Klein introduced singular points for complex potential functions by a direct appeal to physics. This led him to distinguish between two types of logarithmic singularities: *sources* and *sinks*.

Klein used this new approach to illustrate the intimate connection between the genus p of a surface and the number of crossing points that arise from the flows on it. Beginning with the case $p = 0$, he showed that a function defined on the Riemann sphere with μ logarithmic singularities (or the equivalent thereof) will, in general, have $\mu - 2$ crossing points. He then indicated what happens to the flow on a surface if the genus drops from p to $p - 1$, by means of a schematic diagram illustrating the deformation process for $p = 2$ as shown in Figure 4.2. In passing from a surface of genus two to a

---

Klein: *GMA*, 3:478 (our translation).

[89]See his remarks in the "Vorrede," *Klein*: *GMA*, 3:502.

[90]Klein, "Vorbemerkungen," *Klein*: *GMA*, 3:479. Klein's painstaking efforts to substantiate this conjecture are documented in Umberto Bottazzini, "Ursprünge der Riemannschen Theorie der Funktionen einer veränderlichen komplexen Grösse," unpublished manuscript.

[91]James Clerk Maxwell, *Treatise on Electricity and Magnetism*, 2 vols. (Oxford: Clarendon Press, 1873).

[92]See, also, George Springer, *Introduction to Riemann Surfaces* (Reading, MA: Addison-Wesley Publishing Co., 1957), pp. 1–41.

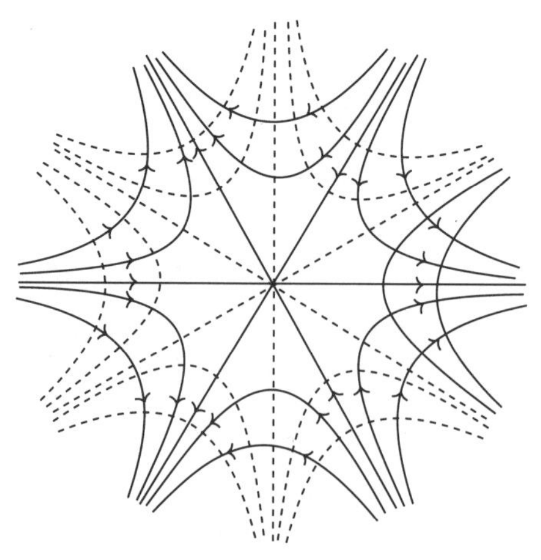

FIGURE 4.1

surface of genus one, two new logarithmic singularities arise (two sinks for the flow on the left, two sources for the conjugate flow on the right, which in both cases are of equal intensity, but opposite sign). This typically Kleinian argument thus suggested that, in the general case of a flow with μ logarithmic singularities on a closed Riemann surface of genus p, there must be exactly $\mu + 2p - 2$ crossing points.[93]

Klein's attack, in which the behavior of algebraic functions and their integrals is derived from potential theory, differs from the then standard approach to this subject. Rather than starting, as Klein did, with a closed surface S and studying the complex functions that live on it, the customary procedure began with an algebraic function $w = w(z)$ satisfying a polynomial equation $f(z, w) = 0$ and then investigated the field of rational functions $R(z, w)$ and their integrals $F(w) = \int R(z, w)\,dz$. For example, suppose w satisfies the equation $w^2 = (z - r_1)(z - r_2)\cdots(z - r_n)$, where the $r_i \in \mathbb{C}$ are distinct. The cases $i = 3, 4$ lead to integrals of elliptic type and Riemann surfaces of genus 1. If $n > 4$, the corresponding integrals are of *hyperelliptic* type and the Riemann surfaces have genus $p = \frac{n}{2} - 1$ (respectively, $\frac{n+1}{2} - 1$) for n even (respectively, odd). More generally, an integral of the type $\int R(z, w)\,dz$ has come to be known as an *Abelian integral.* As we shall see in Chapter 5, the theory of hyperelliptic and Abelian integrals (and their associated functions obtained by inversion) formed one of the main staples in the mathematical diet of the Americans who gravitated to Klein in Göttingen.

In certain respects, Klein's booklet on Riemann's theory of algebraic functions and their integrals bears a striking resemblance to his *Erlanger Programm* of 1872. Neither work was intended as a conventional contribution to mathematical research; rather, each aimed to give a new *interpretation* of

[93]Klein, *Über Riemanns Theorie der algebraischen Funktionen und ihrer Integrale, Klein: GMA*, 3:536–537.

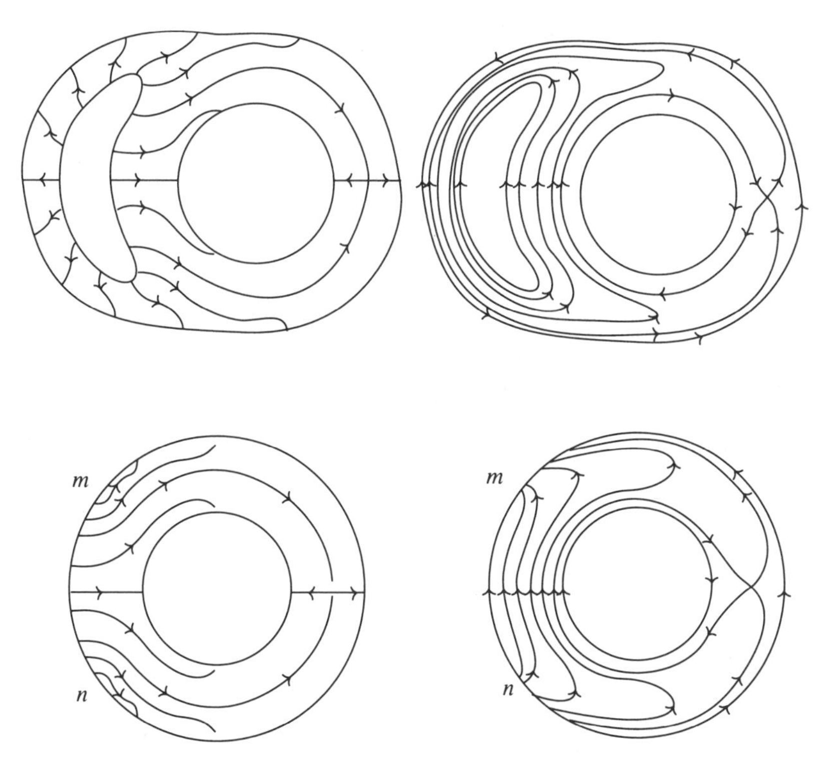

FIGURE 4.2

an already established body of knowledge. Thus, to appreciate the significance of his booklet, it must be seen within the context of the reception and further advancement of Riemannian ideas.[94] For example, Hermann Weyl drew on prior work of L. E. J. Brouwer , Paul Koebe, and David Hilbert to set the theory of Riemann surfaces on a firm foundation in the classic book *Die Idee der Riemannschen Fläche*, which he dedicated to Felix Klein.[95] In his book's introduction, Weyl emphasized the importance of Klein's notion that a Riemann surface does not merely provide a convenient way to describe the behavior of a particular complex-valued function but rather serves as the natural source from which such functions spring. He claimed further that "Klein's approach first allowed the natural simplicity and vital, telling power of Riemann's fundamental conceptions to be fully realized."[96]

While, by today's standards, Klein's mathematics may seem old-fashioned and hopelessly muddled, the fact remains that many of the ideas he intro-

[94]For an authoritative discussion of the fertility of Riemann's conceptions for the development of algebraic geometry, see Brill and Noether, "Die Entwicklung der Theorie der algebraischen Functionen in älteren und neueren Zeit."

[95]Hermann Weyl, *Die Idee der Riemannschen Fläche* (Leipzig: B. G. Teubner, 1913).

[96]*Ibid.*, p. vii. Weyl later recalled Klein's deep influence on his own approach to the subject in the foreword to the third edition, *Die Riemannschen Fläche* (Stuttgart: B. G. Teubner Verlagsgesellschaft, 1955), p. v.

duced have become indispensable for the modern mathematician. Erhard Scholz has carefully traced how the modern concept of a manifold emerged from Riemann's work, and how Klein helped to champion its utility in his work on the foundations of geometry and the theory of modular functions.[97] Along with manifold theory went the study of invariant forms associated with an algebraic curve and the geometric realization of this curve as a Riemann surface. The refinement of the topological structures implicit in this theory led to the rudiments of homology and homotopy theory, and the primitive notions of a covering space, orbit space, and group actions on a manifold. This last-named aspect, in turn, provided the initial impetus for work in combinatorial group theory.[98] These are but a few of the new possibilities which this research tradition would soon open up. Yet, while Klein tilled the ground for many of these developments, it was left for others to reap the harvest that grew from this fertile Riemannian soil. One of the first to do so in a major way was Henri Poincaré.

THE ASCENSION OF A NEW STAR

As Klein brought his summer lectures to a close in 1881, he came upon three notes published earlier that year by Poincaré under the title "Sur les Fonctions fuchsiennes."[99] With them, Poincaré had begun to lay the groundwork for a comprehensive theory of complex-valued functions which remain invariant under an infinite, discontinuous group of linear transformations. Klein immediately wrote the Frenchman on 12 June to inform him of his own previous work on this topic, a letter that marked the beginning of a fascinating correspondence which lasted until September of the following year.[100] At the time Klein initially contacted him, Poincaré was a little-known, twenty-seven-year-old mathematician living in Caen. Although he had studied Lazarus Fuchs's work on the problem of determining when a linear differential equation possesses algebraic solutions, Poincaré had relatively little knowledge of Klein's research or of the approaches of other German function-theorists. Through Klein, however, he quickly became aware

[97] Erhard Scholz, *Geschichte des Mannigfaltigkeitsbegriffs von Riemann bis Poincaré* (Basel: Birkhäuser, 1980).

[98] See Wilhelm Magnus and Bruce Chandler, *The History of Combinatorial Group Theory* (New York: Springer-Verlag, 1982).

[99] This section is based in part on David E. Rowe, "Klein, Mittag-Leffler, and the Klein-Poincaré Correspondence of 1881–1882," in *Amphora: Festschrift für Hans Wussing zu seinem 65. Geburtstag*, ed. Sergei S. Demidov, Menso Folkerts, David E. Rowe, and Christoph Scriba (Basel: Birkhäuser Verlag, 1992):597–618 on pp. 601–608. The pertinent notes by Poincaré are: Henri Poincaré, "Sur les Fonctions fuchsiennes," *Comptes rendus* 92 (1881):333–335, 395–398, and 859–861, or Paul Appell *et al.*, ed. *Oeuvres de Henri Poincaré*, 11 vols. (Paris: Gauthier-Villars, 1916–1956), 2:1–10 (hereinafter *Oeuvres Poincaré*).

[100] The letters were first published in *Klein: GMA*, 3:587–621. They were then published immediately afterward in "Correspondance d'Henri Poincaré et de Felix Klein," *Acta Mathematica* 39 (1923):94–132.

of the fundamental contributions of Hermann Amandus Schwarz, of Klein's own work on modular functions, and even of some of the deeper aspects of the theory of Riemann surfaces. Poincaré managed to assimilate all of this without any apparent effort.[101]

In late 1881, Klein invited Poincaré to write a summary of his results for publication in the *Mathematische Annalen.* The latter agreed and soon made his submission.[102] Before the article went to press, Klein forewarned Poincaré that he had appended a note to it in which he registered his objections to the terminology employed therein. In particular, Klein disputed Poincaré's decision to name the important class of functions possessing a natural boundary circle after Fuchs, a leading exponent of the Berlin school.[103] The importance he attached to this matter, however, went far beyond the bounds of a conventional priority dispute. True, Klein was concerned that his own work receive sufficient acclaim, but the overriding issue hinged on whether the mathematical community would regard the burgeoning research in this field as an outgrowth of Weierstrassian analysis or the Riemannian tradition.[104] As we have seen, Klein had by now become deeply committed to the notion that Riemann's legacy represented the most vital source of that peculiar brand of *anschauliche* mathematics Klein had made his own. Fuchs, on the other hand, stood in the mainstream of the Berlin tradition, and Klein could not afford to stand idly by while the schools of Hermite and Weierstrass captured the limelight by heaping unearned plaudits on each other's work.[105] As the foremost representative of the Clebschian school, he knew that recognition in the world of mathematics seldom came easily, particularly if one's field of research or style of exposition ran counter to current fashion.

Following this early phase of the interchange with Poincaré, Klein planned to spend part of his spring vacation of 1882 working on a new presentation of the existence theorems for algebraic functions on Riemann surfaces as a sequel to his earlier booklet. Suffering from asthma, he traveled to the North Sea island of Nordeney. From there, he wrote to Adolf Hurwitz in a despondent mood, complaining that he seemed to lack a definite goal for his work. He concluded by saying that "actually the Poincaré material should

---

[101]For a detailed discussion of this work and the Klein-Poincaré correspondence, see Gray, *Linear Differential Equations and Group Theory*, pp. 275–315.

[102]Henri Poincaré, "Sur les Fonctions uniformes qui se reproduisent par des Substitutions linéaires," *Mathematische Annalen* 19 (1882):553–564, or *Oeuvres Poincaré*, 2:92–104.

[103]Felix Klein to Henri Poincaré, 13 January 1882, *Klein*: *GMA*, 3:605–606.

[104]See Klein's remarks at the close of his appended note in *Mathematische Annalen* 19 (1882):564–565.

[105]Not surprisingly, Klein's printed remarks left Fuchs far from pleased. See, for example, Lazarus Fuchs to Gösta Mittag-Leffler, 27 December 1882, Mittag-Leffler Correspondence, Institut Mittag-Leffler quoted in Rowe, "Klein, Mittag-Leffler, and the Klein-Poincaré Correspondence of 1881–1882," p. 604. The fallout from this affair had significant consequences for Klein's future relations with mathematicians allied with the Berlin school.

give me enough to do, but I don't have the slightest desire to go into it."[106] Perhaps even worse for Klein's vacation plans, he found himself trapped on the island in stormy weather that brought on violent asthmatic attacks. After eight days of misery, he decided to leave for Düsseldorf. Unable to fall asleep the night before his departure, however, he contemplated the famous 14-sided figure he had discovered three years earlier in connection with his work on the modular equation for the transformations of order seven.[107] Suddenly, he recognized the broader significance of this earlier work—it provided the basic ingredients of the so-called *Grenzkreistheorem*.[108]

This fundamental result asserts that for a Riemann surface of genus $p > 1$ there is, up to conformal automorphism, a uniquely determined uniformizing function η defined on the interior of a *Grenzkreis* or limiting circle in $\mathbb{C}$ and invariant under a given discrete group G of linear transformations. Klein's breakthrough drew heavily upon Poincaré's achievements, especially his demonstration of how very general fundamental domains F could be studied that would produce a domain with a limiting circle under the action of a discontinuous group G. Poincaré had even used these results to prove an early version of the *Grenzkreistheorem* for a closed Riemann surface of genus zero.

As with Klein, Poincaré's key insight in connection with this theory had come in what he later described as a sudden flash of inspiration.[109] What Poincaré saw as he boarded a bus in Caen was the connection between the motions of certain quadrilateral figures inside a disc and the figures that arise in Beltrami's model of a hyperbolic geometry. This discovery had taken place in June of 1880. In his very first letter to Klein, written about one year later, Poincaré referred to non-Euclidean geometry as "the veritable key to the problem which occupies us."[110] Considering that Klein had done pioneering work showing the connection between non-Euclidean geometry and projective geometry, it seems more than a little surprising that he had failed to notice this link himself.[111] On the other hand, Poincaré only published his now-famous model for non-Euclidean geometry more than a year later. In fact, it first appeared (in a two-dimensional form) in his paper of July 1881, entitled "Sur les Groupes kleinéens."[112]

---

[106]Felix Klein to Adolf Hurwitz, 14 March 1882, Mathematiker-Archiv, NSUB, Göttingen. (our translation).

[107]Felix Klein, "Ueber die Transformation siebenter Ordnung der elliptischen Funktionen," *Mathematische Annalen* 14 (1878–1879):428–471, or *Klein*: *GMA*, 3:90–134.

[108]Klein recounted these events years later in *Klein*: *GMA*, 3:584, and in Klein, *Vorlesungen über die Entwicklung der Mathematik im* 19. *Jahrhundert*, 1:372–379.

[109]Henri Poincaré, *Science et Méthode* (Paris: Flammarion, 1908), pp. 51–52.

[110]Henri Poincaré to Felix Klein, 15 June 1881, *Klein: GMA*, 3:590.

[111]Gray offers one possible explanation for this in *Linear Differential Equations and Group Theory*, pp. 299–300.

[112]Henri Poincaré, "Sur les Groupes kleinéens," *Comptes rendus* 93 (1881):44–46, or *Oeuvres*

By March of 1882, when Klein discovered the general *Grenzkreistheorem*, Poincaré's work had already made clear the central importance of non-Euclidean geometry for the theory of automorphic functions. A few days after the fateful sleepless night on Nordeney, Klein wrote a short note announcing the result and had proofsheets sent to Poincaré, Schwarz, and Hurwitz . Little more than one month later, he formulated an even more general result, which he outlined to Poincaré in a letter of 7 May. With it, he hoped to move a step beyond Poincaré by developing a framework broad enough to embrace both Fuchsian and so-called Schottky groups. Klein's more general theorem would thus subsume the two uniformization theorems he had announced earlier—the *Rückkehrschnitttheorem* and the *Grenzkreistheorem*—as special cases.[113] A week later, Klein wrote Poincaré again, this time giving more information about his "proof" and emphasizing that the correspondence he had constructed between the two higher-dimensional manifolds representing the space of Riemann surfaces and the space of uniformizing functions, respectively, had to be analytic as well.[114]

By mid-May, then, Klein was struggling to thrash out some of the difficulties associated with proving this general uniformization theorem. Over the next several months, he delivered a series of lectures on the subject in the Leipzig Mathematical Seminar, which were transcribed and edited by Eduard Study and then reworked by Klein during his fall vacation. The finished product was completed in October of 1882, and Klein circulated offprints in late November, just before the appearance of the first of Poincaré's five articles in *Acta Mathematica.*[115]

The central idea behind Klein's argument in this work, "Neue Beiträge zur Riemannschen Funktionentheorie," turned out to be reasonably sound, namely, to establish a bijection between two abstract manifolds. Unfortunately, it could only be carried through with the help of new topological techniques that would only become available much later. Klein's approach

---

Poincaré, 2:23–25. In the two-dimensional model, one studies the reflections of circular-arc polygons whose edges, when extended, intersect the *Grenzkreis* at right angles. These motions may then be regarded as isometries in a hyperbolic space.

[113]Klein had announced the *Rückkehrschnitttheorem* (which deals with uniformizing Riemann surfaces that are cut open along so-called *Rückkehrschnitte*) and the *Grenzkreistheorem* in two brief notes from January and March of 1882, respectively. Both bore the title "Ueber eindeutige Funktionen mit linearen Transformationen in sich." See *Mathematische Annalen* 19 (1882):565–568, and 20 (1882):49–51, or *Klein*: *GMA*, 3:622–629.

[114]Klein to Poincaré, 14 May 1882, *Klein*: *GMA*, 3:615. Klein had shown that the two manifolds were of equal dimension. For details, see Scholz, *Geschichte des Mannigfaltigkeitsbegriffs*, pp. 205–222.

[115]See Felix Klein, "Neue Beiträge zur Riemannschen Funktionentheorie," *Mathematische Annalen* 21 (1882–1883):141–218, or *Klein*: *GMA*, 3:630–710; and Henri Poincaré, "Sur les Fonctions fuchsiennes," *Acta Mathematica* 1 (1882):193–294, or *Oeuvres Poincaré*, 2:169–257. Consult, also, Henri Poincaré, *Papers on Fuchsian Functions*, trans. John Stillwell (New York: Springer-Verlag, 1985).

hinged on showing that the correspondence between his two manifolds was analytic in nature. As Scholz has pointed out, however, this portion of his argument proved a dead end.[116] The key ideas came nearly thirty years later in the work of L. E. J. Brouwer, particularly in his proof of the topological invariance of dimension. This laid the groundwork for the final breakthrough by Poincaré and Paul Koebe.[117] These same results of Brouwer, incidentally, also provided essential new tools for Hermann Weyl's work on Riemann surfaces, which effectively placed the ideas in Klein's booklet of 1881 on a sound foundation.[118]

Although his "Neue Beiträge" was surely one of the most impressive papers Klein ever wrote, he paid a very heavy price for his exertions. He had repeatedly failed to heed the warning signs of incessant overwork, and this time he had pushed his delicate constitution to the edge of severe physical and mental collapse. When Klein returned from his vacation in the autumn of 1882, the geometer Friedrich Schur found him so exhausted he was surprised to learn that he had not applied for additional time off.[119] Klein insisted on teaching a course on applications of calculus to geometry, but his health problems finally forced him to allow his assistant, Walther von Dyck, to take over the class after about a month. Klein did continue to conduct meetings of his seminar on Abelian functions, assuming all the while that his condition would improve. It did, but only very slowly. In the meantime, he was forced to sit back and watch as Poincaré filled the pages of *Acta Mathematica* with his remarkable memoirs.[120]

The year 1883 marked a decisive turning point in Klein's career, which now entered a new phase of transition. Fighting lethargy and depression, the Leipzig geometer saw his mathematical production fall off sharply over the next two years when, aside from his book *Vorlesungen über das Ikosaeder*, he published only two short papers.[121] He also felt increasingly estranged from his other Leipzig colleagues, although he continued an amiable relationship with Adolf Mayer, coeditor of the *Mathematische Annalen*. Yearning for a fresh start and a new challenge, Klein began to recover a sense of purpose and direction while contemplating President Gilman's offer to succeed Sylvester

---

[116] See the discussion in Scholz, *Geschichte des Mannigfaltigkeitsbegriffs*, pp. 205–222.

[117] L. E. J. Brouwer, "Beweis der Invarianz der Dimensionenzahl," *Mathematische Annalen* 70 (1911):161–165; and "Beweis der Invarianz des n-dimensionalen Gebiets," *Mathematische Annalen* 71 (1912):305–313. See, also, Henri Poincaré, "Sur l'Uniformisation des Fonctions analytiques," *Acta Mathematica* 31 (1907):1–63, or *Oeuvres Poincaré*, 4:70–139; and Paul Koebe, "Ueber die Uniformisierung beliebigen analytischen Kurven," *Journal für die reine und angewandte Mathematik* 138 (1910):192–253.

[118] Hermann Weyl, *Die Idee der Riemannschen Fläche*.

[119] Friedrich Schur to Adolf Hurwitz, 16 November 1882, Mathematiker-Archiv, 79, NSUB Göttingen. See, also, Klein's remarks from his notes, "Vorläufiges über Leipzig," p. 2, Klein Nachlass XXII L, NSUB, Göttingen.

[120] See *Oeuvres Poincaré*, 2:108–462.

[121] See note 72 for the bibliographical reference to this book.

at the Johns Hopkins.[122] As we have seen, he ultimately decided not to accept the position, although the temptation had been great. In fact, Klein emerged from these negotiations with fresh enthusiasm and a clearer vision of the kind of activity he hoped to pursue in the future. Recognizing that he could not keep up with Poincaré's blistering pace, he began to contemplate other ways of retaining his position as head of an important mathematical school. Indeed, by the time he arrived at his final academic destination, Göttingen, in 1886, he had already adapted to the principal role he would fashion for himself over the course of the next decade, that of the "master teacher." From this point onward, Klein focused on cultivating the many ideas he had generated during his youth. He would strive to reap their fruits in his lecture courses by sketching the broad outlines of a theory and by offering suggestions on how to develop it. At the same time he would leave the task of actually working out the details to his students.

As it happened, talented mathematical apprentices were at a premium during the course of this middle period of Klein's career. By the early 1880s, a dismal job market for mathematics teachers had led to plummeting enrollments at the German universities. This decline set in just as a handful of promising young American students—many of whom had studied mathematics at the Johns Hopkins, Harvard, Princeton, or some other East Coast institution—decided to go abroad in order to extend their mathematical horizons and to test their potential as researchers. Rather than opting to study in the more linguistically congenial environs of an English university, however, most of these Americans preferred Germany, where they had a number of first-rate universities—Berlin, Göttingen, Heidelberg, and Leipzig, among others—from which to choose. By the mid-1880s, anyone who wanted to be "somebody" in American mathematics would have been well advised to obtain at least some training at one or more of the German universities.[123] Remarkably enough, the Americans gravitated not just to a single German university, Göttingen, but to one particular German professor, Felix Klein. Throughout the late 1880s and early 1890s, Klein would replace Sylvester as the most influential mentor of America's mathematical aspirants.

---

[122] Klein mentions this, for example, in "Göttingen Professoren," pp. 21–22.

[123] In 1904, it was estimated that about twenty percent of the members of the American Mathematical Society had done either doctoral or postdoctoral study in Germany. See Thomas Fiske, "Mathematical Progress in America," *Bulletin of the American Mathematical Society* 11 (1905):238–246; reprinted in *A Century of Mathematics in America—Part* I, ed. Peter Duren *et al.* (Providence: American Mathematical Society, 1988), pp. 3–11 on p. 5.

Chapter 5
America's *Wanderlust* Generation

As Sylvester had learned at the Johns Hopkins, the younger generation of American mathematicians had a burning desire to contribute to research-level mathematics, and the laboratory environment he created there provided dramatic proof of what could be accomplished in this direction. Even so, it must have been apparent to many of his staunchest American admirers that by the time he left Baltimore in 1883, Sylvester was past his prime and that he, Cayley, and the whole British school had largely fallen out of touch with the then current mathematical developments on the Continent. Sylvester's departure did nevertheless substantially weaken the only viable graduate mathematics program in the United States, and this weakness became ever more apparent as the ranks of those who wished to study advanced mathematics began to grow. Over the next ten years, the slow trickle of students going abroad to study at foreign universities gradually swelled into a steady stream as the "Kleinian era" in American mathematics reached its climax.

Felix Klein and J. J. Sylvester were separated by more than language, culture, and the thirty-five-year difference in their ages. Indeed, their personalities and temperaments contrasted strikingly. Klein appeared to the Americans as the very embodiment of German learning: serious and disciplined with broad, nearly universal interests. He had nothing of Sylvester's flamboyance about him. In his classes at Hopkins, Sylvester had tended to present his students with a conception of mathematical research circumscribed by the well-defined set of unresolved problems that happened to occupy his attention at any given time. This kind of focused activity, and the novel manner in which Sylvester conducted it, had clearly given his students a sense of the excitement that he himself felt in the midst of a mathematical investigation. However, these same Hopkins graduates had received little substantial impression of larger mainstream developments in nineteenth-century analysis and geometry connected with such major figures as Gauss, Abel, Riemann, Clebsch, and Weierstrass. For Klein, on the other hand, nothing proved more rewarding than the exploration of the rich achievements of his immediate forerunners. To learn mathematics from Felix Klein thus meant gaining an overview of a substantial part of nineteenth-century mathematics.

Klein's lectures may have been the single greatest attraction for this generation of aspiring young American mathematicians, but most of those who came to Germany would probably have chosen to do so even if there had been no possibility of attending his classes. By the last quarter of the nineteenth century, the prestige of the German universities drew large numbers of mathematics students from many different foreign countries, including the United States. These students came not only because of the luminaries who taught there—Weierstrass, Kronecker, Klein, Lie, and later Hilbert, among others—but also because the German universities had developed a unique system of seminars for training young talent. (Recall the earlier discussion in Chapter 1.) One of the more important institutional innovations, these seminars brought professors and students together, face to face, in an era when the gulf separating the two groups was far wider than we can easily imagine today. For mathematics students, this meant more than just glimpsing what the creation of new mathematics involved. They learned how to grapple with mathematical ideas by discussing them with others, fielding questions, and addressing critical feedback. In most cases, students took turns presenting recently published research results, although they generally also had ample opportunity to unveil their own work. Occasionally, too, depending on the interests or inclinations of the professor in charge, the seminars were conducted as a forum covering a broad variety of topics, but usually they centered on recent work in a specific field.[1]

Perhaps the greatest advantage of the seminar system at the German universities was that it took students to the threshold of research activity much more quickly than would otherwise have been possible. In traditional lecture courses, students simply sat back and passively took notes while the *Dozent* went through a well-established theory from the ground up; seminars aimed to expose advanced students to the more active side of mathematics so that they could begin to appreciate its creative dimension. Whereas the principle of *Lernfreiheit* entitled students to pick and choose lecture courses at their whim, access to the seminars was generally restricted. (Kummer, for instance, required potential participants to submit a mathematical manuscript *before* they could gain admission to his seminar in Berlin.) For this reason and others, the seminars often provided a challenging setting for would-be mathematicians, enabling the best of an already select group to distinguish themselves. Clearly, only the more talented and highly motivated students reaped the full benefits of this highly competitive and unashamedly élitist approach to mathematics education.[2]

---

[1]On the seminars and mathematics education in general at the German universities, see Wilhelm Lorey, *Das Studium der Mathematik an den deutschen Universitäten seit Anfang des 19. Jahrhunderts*, Abhandlungen über den mathematischen Unterricht in Deutschland, Band III, Heft 9 (Leipzig: B. G. Teubner, 1916).

[2]During the first decade of the twentieth century, for example, when Hilbert and Minkowski

Klein generally organized his seminars by choosing a subject closely related to the content of his lecture courses. In sharp contrast to Sylvester, he orchestrated every detail in advance and unveiled his overall plan during the seminar's initial session. At that time, too, he discussed his goals and assigned the specific topics that he wanted his individual students to research and present in a series of formal lectures. Before the student could step to the podium, however, he or she had to meet with him in order to discuss the content of the lecture. Afterwards, the normal procedure required that the student record a synopsis of its contents in one of the official protocol books that Klein kept throughout his more than forty years of teaching.[3] Although himself a born pedagogue, Klein believed that good teaching was an acquired characteristic that could be inculcated in properly receptive minds. Motivated by this belief as well as by his deeply ingrained sympathy for holistic solutions, he conducted his seminars in accordance with the principle that the activities of teaching and learning must be viewed as integrally related. Whether they liked Klein's tightly controlled approach or not, his students never doubted the earnestness of his commitment to high standards in teaching at every level of instruction. As we shall see in Chapters 6 and 9, this same kind of dedication to and interest in mathematics education ultimately proved to be a characteristic feature within the early American mathematical community.

Klein's Leipzig Students

Klein attracted some gifted foreign students[4] during his years at the *Technische Hochschule* in Munich, including the Italians Luigi Bianchi and Gregorio Ricci-Curbastro, but he did not find Americans among his auditors until he went to Leipzig. In fact, Irving Stringham, fresh from Hopkins with his Ph.D. in hand, arrived in Leipzig for Klein's very first semester in 1880. Stringham attended both semesters of Klein's lectures on geometric function theory, the second half of which served as the basis for Klein's novel booklet on Riemann's theory of algebraic functions.[5] In Klein's seminar, Stringham also gave two presentations based on the results in his Hopkins thesis, which dealt with regular polyhedra in four-dimensional Euclidean space and the

taught at Göttingen, Felix Klein estimated that roughly ten to fifteen percent of the mathematics students there fell into this élite category. See *ibid.*, p. 9.

[3] *Ibid.*, p. 128. These protocol books are presently housed in the library of the Mathematisches Institut, Göttingen University.

[4] Since our concern is, first and foremost, to examine the *overall* influence of Sylvester, Klein, and Moore on American mathematics, we use the term "student" in the broader sense of one who heard the lectures of one of our principals, rather than confining it merely to the category of those who took their doctoral degree under one of them. Fond as mathematicians are of tracing their academic lineage, such information often tells very little about the actual formative influences on their work, a point amply illustrated in this book.

[5] Felix Klein, *Über Riemanns Theorie der algebraischen Funktionen und ihrer Integrale* (Leipzig: B. G. Teubner, 1882).

isometry groups associated with them. (Compare Tables 2.2 and 5.1.) This work led to a publication in the *American Journal of Mathematics* on finite quaternion groups, a paper closely connected with some of Klein's earlier investigations involving the rotation groups of the regular solids in three-space.[6]

In the summer of 1884, two more Americans went to Leipzig to attend Klein's lectures: Frank Nelson Cole and Henry Burchard Fine.[7] At the time, Klein was teaching the second half of a two-semester course on elliptic function theory, so the newcomers found themselves in a difficult situation. Presumably they studied and took notes from the *Vorlesungsheft*[8] prepared by Klein's *Famulus* (or assistant), Otto Fischer, and while neither became an expert in the subject, they both turned in respectable performances. Klein personally advised Fine to try the course, even if he found it difficult to follow at first. The young American took this to heart and, after only a few weeks, felt reasonably comfortable with the material.[9] Given that the "real" graduate training took place in the seminars, however, it was somewhat surprising that neither Fine nor Cole chose to attend Klein's concurrent seminar, which dealt with rather advanced topics related to the same field. Among those who did participate, Otto Hölder lectured on elliptic modular functions, Friedrich Engel presented the Weierstrass $\wp$-function, and Georg Pick spoke on complex multiplication.[10] Many of the other participants, for example, Robert Fricke and Theodor Molien, would also soon make names for themselves.

The following semester, Klein's last in Leipzig, Cole and Fine took his course on higher curves and surfaces, but, again, their names do not appear among those listed as enrolled in the concurrent seminar. Given that the seminar continued with topics from the previous course on elliptic functions, this absence is notable.[11] It thus seems clear that neither had been smitten by

[6]Washington Irving Stringham, "Determination of the Finite Quaternion Groups," *American Journal of Mathematics* 4 (1881):345–357; and Felix Klein, "Ueber binäre Formen mit linearen Transformationen in sich selbst," *Mathematische Annalen* 9 (1875–1876):183–208, or Felix Klein, *Gesammelte Mathematische Abhandlungen*, 3 vols. (Berlin: Springer-Verlag, 1921–1923), 2:275–301 (hereinafter abbreviated *Klein: GMA*).

[7]On Fine, see Oswald Veblen, "Henry Burchard Fine—In Memorium," *Bulletin of the American Mathematical Society* 35 (1929):726–730; and Raymond Clare Archibald, *A Semicentennial History of the American Mathematical Society,* 1888–1938 (New York: American Mathematical Society, 1938), pp. 167–170. On Cole, see Archibald, *Semicentennial History*, pp. 100–103.

[8]In Leipzig, Klein began the practice of having his assistants prepare *Vorlesungshefte*—sometimes called *Ausarbeitungen* or lecture notes—that covered the contents of his lectures and often elaborated points that he chose to skirt in his presentation. After receiving Klein's approval, the *Hefte* were then placed in the Mathematics Library so that students could study them or compare the contents with their own notes.

[9]Fine's notes from these lectures are preserved in the Seeley Mudd Library, Princeton University.

[10]Protokollbuch für das Seminar über elliptische Funktionen, Sommer Semester 1884, Mathematisches Institut der Universität Göttingen. For the definition of the $\wp$-function, see note 73 in Chapter 4.

[11]In fact, during the first three months that the seminar met, Klein gave a series of lectures on elliptic normal curves (that is, nth-order curves of genus one lying in an $(n-1)$-dimensional space with coordinates given by products of Weierstrass σ functions). Protokollbuch für das Seminar über elliptische Funktionen, Winter Semester 1884–1885, Mathematisches Institut der

FRANK NELSON COLE (1861–1926)

this subject—their tastes inclined to algebra rather than geometric function theory—and, with or without Klein's advice, they decided to pursue other lines of study.

Fine, who was three years Cole's senior, had initially been drawn to mathematics during his undergraduate days in Princeton by George Bruce Halsted, that outspoken American champion of non-Euclidean geometry. After studying as an undergraduate at the College of New Jersey (later Princeton University), Fine had stayed on at his *alma mater*, first as a Fellow in experimental science during the academic year 1880–1881, and then as a Tutor in mathematics, before proceeding to Leipzig in 1884. Since he returned to Princeton as an Assistant Professor in 1885, it would seem that Fine went to Germany with one paramount objective in mind: to obtain his doctorate. He managed to achieve this goal in just one year, faster than any of Klein's other American students, despite his less than solid command of the German language and his only modest mathematical talent. Fine's swift success probably owed more to Eduard Study (then a *Privatdozent* in Leipzig) than to Klein, however. Originally, Klein had given Fine a thesis topic in enumerative geometry, which had evidently proved too difficult, because Fine turned to Study for advice and assistance.[12] Study actually posed an alternative research problem; Klein approved it; and by May of 1885, the American had earned his degree.[13] Interestingly enough, this was the only time that Klein allowed one of his students to write the thesis in a language other than German. (See Table 5.2.) After taking his doctorate, Fine moved on to Berlin, where he attended Kronecker's lectures on the theory of elimination. These seem to have made a much stronger impression on him than anything he had experienced in Leipzig, for six years later, he returned to hear some of Kronecker's last lectures. As a tribute to him, Fine wrote a brief article on "Kronecker and his Arithmetical Theory of the Algebraic Equation," which appeared in the first issue of the *Bulletin of the New York Mathematical Society*.[14]

Universität Göttingen. Out of these lectures arose the lengthy study, Felix Klein, "Ueber die elliptischen Normalkurven der n-ten Ordnung," *Abhandlungen der mathematisch-physicalischen Klasse der sächsischen königlichen Gesellschaft der Wissenschaften* 13 (1885), or *Klein*: *GMA*, 3:198–254.

[12]Following a suggestion made by Klein, Study had also worked on enumerative geometry in preparing his *Habilitationsschrift*. Apparently, Klein had hoped he would be able to provide a rigorous foundation for Schubert's calculus. Klein's optimism proved premature, however, for in 1900 David Hilbert called attention to this in the fifteenth—and perhaps murkiest—of his celebrated twenty-three problems. B. L. van der Waerden first provided the sort of rigorous foundation Klein and Hilbert had called for in his paper "Topologische Begründung des Kalküls der abzählende Geometrie," *Mathematische Annalen* 102 (1930):337–362. See, also, W. L. Chow and B. L. van der Waerden, "Zur algebraischen Geometrie IX: Ueber zugeordnete Formen und algebraische Systeme von algebraischen Mannigfaltigkeiten," *Mathematische Annalen* 113 (1937):672–704.

[13]Fine's dissertation "On the Singularities of Curves of Double Curvature" appeared in the *American Journal of Mathematics* 8 (1886):156–177. (Curves of double curvature are usually just called space curves; the terminology derived from the fact that such curves have nonzero curvature and torsion, whereas the torsion vanishes for plane curves.)

[14]Henry B. Fine, "Kronecker and his Arithmetical Theory of Algebraic Equations," *Bulletin of the New York Mathematical Society* 1 (1892):173–184.

HENRY BURCHARD FINE (1858–1928)

Frank Nelson Cole was far less focused than Fine when he came to Leipzig as a Parker Fellow from Harvard. The third of ten children in a family of Massachusetts farmers, Cole had graduated second in his class at Harvard with a strong determination to make a career for himself in mathematics. He apparently learned a good deal from Klein during his year in Leipzig, for, on returning to Harvard in the fall of 1885 to complete his dissertation, he also offered two courses that sparked considerable interest. One of those who attended Cole's lectures that year was William Fogg Osgood. Writing some four decades later from his privileged position as the co-founder (with Maxime Bôcher) of Harvard's strong tradition in analysis, Osgood provided this rather romanticized account of Cole's earlier impact on mathematics there: "[Cole] had just returned from Germany and was aglow with the enthusiasm which Felix Klein inspired in his students. Cole was not the first to give formal lectures at Harvard on the theory of functions of a complex variable, Professor James Mills Peirce having lectured on the subject in the seventies. That presentation was, however, solely from the Cauchy standpoint ... [whereas] Cole brought home with him the geometric treatment [of complex functions] which Klein had given in his noted Leipsic [sic] lectures of the winter of 1881–82."[15] In Osgood's view, Peirce's lectures thus "stood as the Old over and against the New and of the latter Cole was the apostle. The students felt that he had seen a great light. Nearly all the members of the Department attended his lectures. It was the beginning of a new era in graduate education at Harvard, and mathematics has been taught here in that spirit ever since."[16]

Osgood's recollections may pay a fitting tribute to Cole's pioneering role, but they convey little about the actual mathematical working conditions Cole encountered in Cambridge. Roughly thirty auditors had attended his course on substitution groups initially, but by semester's end their numbers had dwindled to only six. In a letter written to Klein almost a year after his return, Cole described his situation in very bleak terms.[17] He lamented that he had been forced to break off his studies in Leipzig just when he had begun to make real progress. At Harvard, Cole could find no one with whom to discuss his dissertation research, a topic on sixth-degree equations suggested to him by Klein. After six months during which he often worked ten hours a day on calculations, he still saw no end in sight. Cole therefore asked Klein to suggest an easier topic, expressing his willingness to work on a problem in

[15]William Fogg Osgood on Cole, as quoted in Thomas S. Fiske, "Frank Nelson Cole," *Bulletin of the American Mathematical Society* 33 (1927):773–777 on pp. 773–774.

[16]*Ibid.*

[17]Frank Nelson Cole to Felix Klein, 26 May 1886, Klein Nachlass VIII, Niedersächsische Staats- und Universitätsbibliothek, Göttingen (hereinafter abbreviated NSUB, Göttingen). Here, Cole mentioned both his lectures on function theory and those on substitution groups. Klein was apparently already familiar with the latter, as Mellen Haskell, then studying in Göttingen, had brought a copy of some lecture notes to show him.

either substitution groups or function theory. Since he had fairly thoroughly studied Klein's *Ikosaeder*, a work he called an "eternal source of new ideas and methods,"[18] Cole probably felt more comfortable with ideas directly related to the theory of quintic equations.

Soon after writing this letter, Cole decided to interrupt his work and submit the preliminary results as his doctoral dissertation. He completed the manuscript in June, and it was published soon afterward in the *American Journal of Mathematics* as "A Contribution to the Theory of the General Equation of the Sixth Degree."[19] In the opening paragraph, Cole expressed his indebtedness to Klein for his "valuable advice and suggestion, which have been of the greatest use to me" and acknowledged that "[t]he fundamental idea of the entire treatment of the subject is due to him, as I have indicated below, and he might claim many of the particular methods involved as his own, if he should consider them worthy of such recognition."[20]

Two German Emigrés

After Klein left Leipzig for Göttingen , the number of Americans attending his classes gradually increased. The first arrival, Mellen Woodman Haskell, had taken his Bachelor's and Master's degrees at Harvard. Arriving in Göttingen in the summer of 1886, he enrolled in both of Klein's offerings that semester: an elementary class on algebra with twenty-three students and a more advanced course on elliptic modular functions that attracted just nine. The latter topic also served as the focus for Klein's seminar, in which Haskell gave a presentation first on fourth-degree resolvents of algebraic equations and then on linear differential equations. (See Table 5.1.) These must have gone well, for Haskell subsequently became something of a fixture in Göttingen, attending nearly all of Klein's courses and seminars over the next four years.[21]

While no new Americans appeared on the scene during Haskell's second semester, he was joined by two young Germans who would soon play a decisive role in guiding American mathematics: Oskar Bolza and Heinrich Maschke.[22] The careers of these two men had already crossed in Berlin in the 1870s, when they had studied together before Maschke left to take his doctorate at Göttingen in 1880. Afterward, Maschke proceeded to a teaching

[18] *Ibid.* The work in question here is Felix Klein, *Vorlesungen über das Ikosaeder und die Auflösung der Gleichungen vom fünften Grade* (Leipzig: B. G. Teubner, 1884). Klein apparently advised Cole to visit Göttingen at the earliest possible opportunity. He returned in 1888.

[19] Frank Nelson Cole, "A Contribution to the Theory of the General Equation of the Sixth Degree," *American Journal of Mathematics* 8 (1886):265–286.

[20] *Ibid.*, p. 265.

[21] Hörerverzeichnis der Vorlesungen Kleins, 1871–1920, Klein Nachlass VII E, NSUB, Göttingen; and Protokollbücher, 1886–1890, Mathematisches Institut der Universität Göttingen.

[22] See Bolza's autobiography, *Aus meinem Leben* (Munich: Verlag Ernst Reinhardt, 1936). On Maschke, see Oskar Bolza, "Heinrich Maschke: His Life and Work," *Bulletin of the American Mathematical Society* 15 (1908):85–95.

job at a *Gymnasium*, whereas Bolza's career followed a rockier course.

Bolza had spent most of his three years in Berlin studying physics under Gustav Kirchhoff and Hermann von Helmholtz. Moving on next to Heidelberg and then to Strassburg, he began work on a dissertation under the physicist August Kundt. Bolza, like many mathematicians before and since, quickly found laboratory work loathsome and finally abandoned physics for pure mathematics. Over the next three years, he studied under Elwin Bruno Christoffel and Theodor Reye in Strassburg, returned to Berlin to hear Weierstrass' lectures, studied with Schwarz in Göttingen, and finally went back to Weierstrass. Although such itinerancy was commonplace at the German universities, it proved detrimental for Bolza, who still found himself unprepared to write a dissertation as late as 1884.

In the meantime, Bolza had resolved to prepare himself for the *Staatsexamen* so that he could at least gain employment as a *Gymnasium* teacher. Beginning in the fall of 1880, it took him another three years to pass the examination and complete the student teaching requirement at the *Gymnasium* in his adopted hometown of Freiburg. Like Maschke, however, he really aspired to a career in higher education. Both men knew full well how difficult it was to make the transition to that level, yet both ultimately decided to try despite the odds against them.

Indeed, an examination of the fragile institutional structures that supported research-level mathematics at the German universities during the nineteenth century can only evoke wonderment that such a system could have attracted any new blood at all. Faculties continued to operate in an atmosphere much like that of a medieval guild, and, to gain entry, one had to overcome truly daunting obstacles. After obtaining a doctorate, the prospective academic had to produce a finished masterwork, the *Habilitationsschrift*, before the faculty granted the *venia legendi*, which allowed the candidate to teach as a *Privatdozent*. This position entitled him (women were barred from pursuing academic careers until the 1920s) to collect the usual, minimal course fees from students, but brought with it no salary from the university. Thus, a *Privatdozent* without private means was presumably a creature with a very short life expectancy. Moreover, as the nineteenth century wore on, it became increasingly difficult to break out of this entry-level stage and so to climb the German academic ladder. (Fortunately for future generations of American scholars, universities in the United States developed a very different system of academic promotion.) Bolza and Maschke's chances for success were thus slim at best, in view of the dearth of positions open to mathematicians in Germany throughout the 1880s, and considering the large pool of highly qualified candidates available to fill them.

Prior to making his move, Bolza, who still had no doctoral degree, had continued to work on mathematics privately. In 1885, his efforts seemingly paid off when he determined which hyperelliptic integrals can be reduced to elliptic integrals by a third-degree transformation. Unfortunately, he found his solution just as Édouard Goursat published an even better result in the *Comptes rendus.*[23] Undaunted, Bolza proceeded to solve the analogous problem in the next highest case of fourth-degree transformations.[24] Following this personal breakthrough, he returned to Berlin to study for a year with Kronecker and Lazarus Fuchs, and he communicated some further, related results to Schwarz and Klein in Göttingen.

In this and in his subsequent research on Abelian functions and on the calculus of variations, Bolza reflected the strong influence of Weierstrass' teaching. Klein, who himself had just begun a period of intense work on an alternative presentation of Weierstrass' theory of hyperelliptic and Abelian functions, therefore took considerable interest in Bolza's research of the mid-1880s. He was keen to acquire the services of a resident expert on the subject, particularly since the Weierstrassian approach remained unpublished.[25] When Klein accepted Bolza's dissertation work in 1886, the not-so-young man was finally able to take his doctoral degree at the age of twenty-nine, and Klein secured his *Fachmann.*[26]

The following semester, Drs. Bolza and Maschke studied together under Klein in Göttingen. Along with Haskell, they worked on the official lecture notes for Klein's course on algebraic equations.[27] Besides attending these lectures and participating in the seminar, both visited Klein once a week for a two-hour-long evening session in his home, then located just north of the *Auditorium* building at Weenderchaussée 6. For these meetings, Klein assigned Bolza the task of investigating the invariants of sixth-order binary forms determined by the roots of their associated Θ-functions, research that led to publications in both the *Mathematische Annalen* and the *American Journal of Mathematics.*[28]

Klein occasionally conducted similar private seminars with other advanced students, usually at the postdoctoral stage, when he felt it profitable to do

---

[23]Édouard Goursat, "Sur un Cas de Réduction des Intégrales hyperelliptiques du second Order," *Comptes rendus* 100 (1885):622–624.

[24]Oskar Bolza, "Zur Reduction hyperelliptischer Integrale auf elliptische," *Berichte über die Verhandlungen der naturforschenden Gesellschaft zu Freiburg im Baden* (1885):330–335.

[25]Weierstrass' work in this field only came out in 1902 with the publication of the fourth volume of his *Mathematische Werke*, 7 vols. (Berlin: Mayer & Müller, 1894–1927; reprint ed., Hildesheim: G. Olms Verlagsbuchhandlung and New York: Johnson Reprint Corporation, 1967).

[26]Bolza sketched the results he had presented in his dissertation in "Ueber die Reduction hyperelliptische Integrale erster Ordnung und erster Gattung auf elliptische durch eine Transformation vierten Grades," *Mathematische Annalen* 28 (1887):447–456.

[27]Hörerverzeichnis der Vorlesungen Kleins, 1871–1920, Klein Nachlass VII E, NSUB, Göttingen.

[28]See Oskar Bolza, "Darstellung der rationalen ganzen Invarianten der Binärform sechsten Grades durch die Nullwerthe der zugehörigen Θ-Functionen," *Mathematische Annalen* 30 (1887):478–495; and "On Binary Sextics with Linear Transformations into Themselves," *American Journal of Mathematics* 10 (1888):47–70.

so, and these students nearly always felt enriched by the experience. The physicist Arnold Sommerfeld, for example, valued his private sessions with Klein as one of the great educational opportunities of his life, and Bolza recalled that "Maschke ... won great and lasting rewards from this year with Klein."[29] For Bolza, however, the experience proved nearly catastrophic. He felt an "immense distance between Klein's brilliant genius, supported by a wonderful capability for geometric visualization ..., [and his own] ... purely analytic gifts, deficient of all fantasy and lying in an entirely different direction."[30] This perceived disparity resulted in an almost total breakdown in Bolza's self-confidence, an odd twist given that Klein had gone through a somewhat similar crisis in the early 1880s when he found that his own relatively modest gifts for hard analysis had left him stranded in the wake of Poincaré's achievements.

Despite this trying experience, Bolza nevertheless soon went on to become an effective and versatile propagator of Klein's mathematics. Since he realized that his hopes of escaping a career as a *Gymnasium* teacher hinged on testing the academic waters abroad, Bolza consulted with and received encouragement from Cole and Haskell before laying plans for a trip to the United States. His first stop, however, was England, where, during the fall and winter months of 1887–1888, he worked both on improving his language skills and on acquainting himself with the British teaching system. His preparations complete, he boarded a ship bound for New York in April 1888.

Armed with nothing more than a letter of introduction from Klein, he arrived in Baltimore, where he visited Simon Newcomb, Story, Craig, and Franklin at the Johns Hopkins. (By this time, Newcomb had joined the Hopkins faculty, while maintaining his position as Superintendent of the Nautical Almanac Office in Washington, D. C.) Newcomb, who shared little of Sylvester's optimism about the future of mathematics in the United States, informed Bolza in no uncertain terms that, had he been consulted first, he would have advised against his coming. After their meeting, Newcomb wrote to Klein offering this assessment of Bolza's prospects:

> I never advise a foreign scientific investigator to come to this country, but always tell him that the difficulties in the way of immediate success are the same that a foreigner would encounter in any other country. ... We have indeed several hundred so-called colleges; but I doubt that if one half of the professors of mathematics in them could tell

---

[29]Bolza, *Aus meinem Leben*, p. 18 (our translation). Sommerfeld wrote of Klein: "The impression I received of F. Klein's imposing personality through his lectures and in conferences with him was overpowering ... . I have always regarded Klein as my teacher, not only in mathematical, but also in mathematical-physical matters, and in the conceptual interpretation of mechanics." Quoted in English translation from Sommerfeld's "Autobiographische Skizze" in Steven H. Schot, "Eighty Years of Sommerfeld's Radiation Condition," *Historia Mathematica* 19 (1992):385–401 on p. 390.

[30]Bolza, *Aus meinem Leben*, p. 18 (our translation).

> what a determinant is. All they want in their professors is an elementary knowledge of the branches they teach and the practical ability to manage a class of boys, among whom many will be unruly. Considerations of religion, personal influence, and former connection with the college also come into play. Of course these drawbacks do not apply to our universities. But, even here the places are usually filled by former students or graduates, and there is little opportunity even for private instruction.[31]

As a foreigner, Bolza thus counted himself lucky when Newcomb managed to arrange an appointment as a Reader in Mathematics (without salary) at the Johns Hopkins. Newcomb requested that he offer a series of twenty lectures on substitution groups and their applications to algebraic equations beginning in January of 1889. During the intervening six-month period, Bolza prepared by studying Jordan's *Traité des Substitutions* and Klein's *Vorlesungen über das Ikosaeder*, together with the notes he had available from Klein's course in the summer of 1886.[32] He ultimately lectured to ten auditors for one hour each day over four weeks and came away quite pleased with the results. As he related in a letter to Klein, he found the University similar in spirit to its English counterparts, but it "makes a much more scientific impression" and has a "more international outlook than at Cambridge."[33] Bolza later recalled how Craig had been the only faculty member to befriend him, and it was he who suggested that Bolza write up his lectures for publication in the *American Journal.*[34] As Eric Temple Bell reported in his study of "Fifty Years of Algebra in America, 1888–1938," Bolza's lectures and the article based on them "did much to popularize the subject and to acquaint young Americans with the outlines of the Galois theory of equations."[35]

In June of 1889, Bolza's long-term career prospects took a decided turn for the better when President G. Stanley Hall of the newly founded Clark University in Worcester, Massachusetts offered him a three-year appointment at an annual salary of $1,750. Bolza accepted immediately, expressing his

[31] Simon Newcomb to Felix Klein, 23 April 1888, Klein Nachlass XI, NSUB, Göttingen.

[32] Camille Jordan, *Traité des Substitutions et des Équations algébriques* (Paris: Gauthier-Villars, 1870); and see note 18 above.

[33] Oskar Bolza to Felix Klein, 27 January 1889, Klein Nachlass VIII, NSUB, Göttingen (our translation). The information regarding Bolza's preparations and the number of listeners comes from this letter.

[34] Oskar Bolza, "On the Theory of Substitution Groups and its Applications to Algebraic Equations," *American Journal of Mathematics* 13 (1891):59–144. In an introductory note to this paper, Simon Newcomb, editor of the *American Journal of Mathematics*, indicated the purpose of publishing these lectures: "This subject being one on which no separate work is found in the English language, Dr. Bolza's development of it is published here, in the belief that it will prove extremely helpful to all students of the subject, especially by supplementing and illustrating the more extensive works of Jordan and Netto." *Ibid.*, p. 59.

[35] Eric Temple Bell, "Fifty Years of Algebra in America, 1888-1938," in *Semicentennial Addresses of the American Mathematical Society*, ed. Raymond C. Archibald (New York: American Mathematical Society, 1938), pp. 1–34 on p. 9.

relief as well as his thanks to Klein for having stuck by him throughout his venture abroad.[36]

Maschke, in the meantime, had been teaching at a *Gymnasium* in Berlin, and trying to extend the work he had begun during his sabbatical with Klein in 1886–1887. Unfortunately, he found the mathematical atmosphere in Berlin generally stifling to his pursuits. As he described the situation to Klein, "everyone here works in isolation and can hardly be moved to talk about his [research]."[37] Clearly desperate to advance his career, Maschke considered applying for a position at Clark alongside his friend. He abandoned this idea on learning from both Klein and Bolza that this could only have negative consequences, but he did resign his position in Berlin in 1890 in order to take up electrical engineering.[38] Intent on leaving Germany, he presumed this field would serve him (just as it had another German immigrant from this period, Charles Steinmetz) should he fail to gain employment in the United States as a mathematician. Indeed, Maschke had no trouble landing a job with an electrical firm in New Jersey.

In spite of this successful career transition, Maschke still hoped for an opportunity to return to academic life. Prior to his departure in 1891, Klein had playfully predicted that, like Odysseus, his wanderings would eventually lead him to Ithaca, that is, to Cornell University in Ithaca, New York![39] Instead, one year later, he and Bolza accepted jobs alongside Eliakim Hastings Moore at the dynamic new University of Chicago, which, as we shall see in Chapter 9 below, became the leading center for mathematics in the nation.

The Kleinian Connection in American Mathematics

Back in Göttingen, Klein continued with his cycle of lectures on hyperelliptic and Abelian functions, which together covered five semesters (from the summer of 1887 to the summer of 1889) and which were inspired in part by Karl Weierstrass' regularly repeated cycle of courses on complex function theory. In the absence of an official assistant to prepare the *Ausarbeitung* of his lectures, however, Klein was forced to rely on his students to write up the material he presented in class, perfecting it in both form and substance.[40] With mathematics enrollments at the German universities plummeting due primarily to poor employment prospects for mathematics teachers at secondary schools, the kind of talented students Klein required for this task represented

[36]Oskar Bolza to Felix Klein, 21 June 1889, Klein Nachlass VIII, NSUB, Göttingen.

[37]Heinrich Maschke to Felix Klein, 8 December 1888, Klein Nachlass X, NSUB, Göttingen (our translation).

[38]Heinrich Maschke to Felix Klein, 21 December 1889, Klein Nachlass X, NSUB, Göttingen.

[39]Heinrich Maschke to Felix Klein, 5 July 1892, Klein Nachlass X, NSUB, Göttingen.

[40]Beginning in 1892, the year in which Klein largely gained control of the mathematics program in Göttingen, he regularly employed an assistant to assume this responsibility. However, between 1886 and 1892, Klein's first eight years in Göttingen, he went without an assistant. A list of Klein's assistants is given in "Anhang," *Klein*: *GMA*, 3:13. On their duties, recall note 8.

a rare commodity. These circumstances help to explain the kind of symbiotic relationship Klein seemed to develop with his American students, who could not have come to study with him at a more propitious point in time.

Klein's first course on hyperelliptic functions in the summer of 1887 managed to attract just nine auditors, among them Maschke, Bolza, the returning Cole, and a newcomer, Henry Dallas Thompson.[41] Thompson hailed from Princeton, where he had studied mathematics under Henry Fine and benefited from the sometimes painful lessons Fine had learned as a graduate student in Leipzig. Wishing to study abroad, Thompson followed Fine's advice and wrote to Klein a year before his departure to obtain information about the courses he intended to teach. As Fine related to Klein: "I recommended Mr. Thompson to take this step, for I often had reason to feel during my stay in Leipzig that I might have come to Germany much better prepared had I only been properly directed in my previous studies."[42] Vaguely recalling his own difficulties in Germany, he added that "[y]ou will find him much better prepared, I think, than the Americans who have hitherto presented themselves to you."[43] Thompson remained in Göttingen throughout the 1887–1888 academic year, attending Klein's lectures on hyperelliptic functions and potential theory. In 1892, he took his doctorate under Klein and, like his American mentor, returned to Princeton, never to move from its bucolic surroundings.

Frank Nelson Cole, on the other hand, came home from his second trip abroad only to face continued hard times and uncertainty. Although 1887 had witnessed the publication of his detailed report on Klein's *Ikosaeder*, the next year marked a creative low in his as yet short career.[44] Cole showed the strains of intense and largely frustrating intellectual activity and, with his future unclear, gave up teaching at Harvard for work laying a railroad as an engineer's assistant.[45] The fresh air and exercise may well have done him good, for he accepted a position as a Lecturer at the University of Michigan for the fall of 1888 and was promoted to Assistant Professor a year later. As at Harvard, Cole enjoyed considerable success in the lecture hall at Michigan, attracting the entire mathematics faculty to one of his advanced courses on substitution groups. Moreover, one of his students, George Abram Miller, went on to become one of the most prolific—although certainly not one of

---

[41] Hörerverzeichnis, Klein Nachlass VII E, NSUB, Göttingen.

[42] Henry B. Fine to Felix Klein, 31 August 1886, Klein Nachlass IX, NSUB, Göttingen. Fine also mentioned that Thompson had won a fellowship from the Johns Hopkins for a short paper on a topic in geometry.

[43] *Ibid.*

[44] Frank Nelson Cole, "Klein's Ikosaeder," *American Journal of Mathematics* 9 (1887):45–61. In this paper, Cole emphasized the connection between the *Ikosaeder* and his teacher's *Erlanger Programm*. See, *ibid.*, pp. 46–47; and note 18.

[45] Frank Nelson Cole to Felix Klein, 22 October 1887, Klein Nachlass VIII, NSUB, Göttingen.

the most original—American contributors to the theory of finite groups.[46]

Through his earlier lectures at Harvard, Cole had already influenced two young undergraduates who went on to become Klein's star American students at Göttingen: William Fogg Osgood and Maxime Bôcher. Osgood arrived in Göttingen first, in the fall of 1887, after completing his A.M. and obtaining a Parker Fellowship.[47] Unlike some of his compatriots, he already had an impressive command of the German language before he went abroad. This, coupled with a profound respect for German learning that bordered on cultural idolatry, made him especially receptive to the experience of studying with Felix Klein. It is safe to say that no American mathematician during this period (and certainly none since) bore the mark of German academic training as visibly and as permanently as Osgood. He married a young Göttingen woman, Therese Ruprecht, whose family owned the local publishing firm of Vandenhoeck & Ruprecht, and the couple's children had names like Rudolf Ruprecht and Frieda Bertha. He even sympathized with the German cause during World War I. Later in life, Norbert Wiener thought he cut a rather ridiculous figure, self-consciously emulating even the mannerisms of a German professor.[48] Quirky though he may have been, Osgood did not merely engage in empty posturing. He wrote his most important works, including the highly influential *Lehrbuch der Funktionentheorie*, in German.[49]

Like Osgood, Maxime Bôcher was born and raised in Boston, attended Harvard, graduated *summa cum laude*, went to Göttingen as a Parker Fellow (arriving there just one year later in 1888), and ultimately married a young Göttingen woman.[50] Finally, both returned to join the Harvard faculty following three years of study in Germany, and both moved through the

---

[46]See George A. Miller, *The Collected Works of George Abram Miller*, 3 vols. (Urbana, Ill.: University of Illinois Press, 1935–1946). Although he did not mention Miller by name, E. T. Bell had him in mind when he wrote the following passage in his book *The Development of the History of Mathematics* (New York: McGraw-Hill, 1945): "it was possible for industrious laborers little advanced beyond mathematical illiteracy to obtain almost any finite number of permutation groups by obstinate grubbing, and to find by the same dull means all the finite abstract groups in certain narrow categories which they themselves had defined, apparently with the express purpose of dignifying their calculations with an air of pseudo generality [p. 445]." That Bell directed this slap at Miller can be inferred from his index, where on p. 627 he cites "Miller, G. A. (1863–), 241, 445,..." although, as noted, Miller's name fails to appear on p. 445 of the text.

[47]On Osgood, see Joseph L. Walsh, "William Fogg Osgood," in *A Century of Mathematics in America—Part* II, ed. Peter Duren *et al.* (Providence: American Mathematical Society, 1989), pp. 79–85; and Archibald, pp. 153–158.

[48]Norbert Wiener, *I Am a Mathematician* (Garden City, NY: Doubleday, 1956), p. 30.

[49]William Fogg Osgood, *Lehrbuch der Funktionentheorie*, 1st ed. (Leipzig: B. G. Teubner, 1907).

[50]On Bôcher, see William Fogg Osgood, "The Life and Services of Maxime Bôcher," *Bulletin of the American Mathematical Society* 25 (1919):337–350; Archibald, pp. 161–166; and George D. Birkhoff, "The Scientific Work of Maxime Bôcher," *Bulletin of the American Mathematical Society* 25 (1919):197–215, reprinted in *A Century of Mathematics in America—Part* II, pp. 59–78.

WILLIAM FOGG OSGOOD (1864–1943)

MAXIME BÔCHER (1867–1918)

academic ranks at their *alma mater* in lockstep fashion. Their single year's separation at Göttingen, however, had significant mathematical consequences for their respective careers, for by the time Bôcher arrived in Germany, Klein was winding down his lecture cycle on hyperelliptic and Abelian functions and preparing to launch a new series of lectures on potential theory and partial differential equations in physics. Thus, while Osgood, Haskell, and also Henry Seely White focused their attention on function theory, Bôcher was the only American to enroll in Klein's new course.

Although Bôcher's career clearly displayed many parallels with Osgood's, it is fair to say that as personalities they had little in common. Osgood had a reputation for being aloof, and his affectations may well have been a mechanism developed to overcome his inherent shyness. Aside from mathematics, his only known interests were smoking cigars and driving automobiles. Bôcher, on the other hand, grew up in a highly cultured family with strong roots in both France and New England. On his mother's side, the family traced its ancestry back to the early days of the Plymouth Colony, even prior to the founding of Harvard College in 1636. His paternal grandfather lived in Caen but visited his son's family often on business trips. Bôcher's father, known for his expertise in the literature, art, and history of France, taught for many years as a professor of foreign languages at Harvard. Coming from this scholarly background, Bôcher could discourse on art, music, or history just as comfortably as he could discuss a recalcitrant problem in higher mathematics.

With the presence of both Osgood and Bôcher, Harvard clearly formed one important focal point for Klein's influence on American mathematics. Another was in the southern New England town of Middletown, Connecticut, where Professor John Monroe Van Vleck taught mathematics and astronomy at Wesleyan University. Himself an 1850 graduate of the institution, Van Vleck had first worked as an assistant in the Nautical Almanac Office (then located in Cambridge), but had returned in 1853 to Wesleyan where he taught until his death in 1912. An active participant in the early life of the American mathematical community, Van Vleck faithfully attended numerous meetings around the country and served as Vice President of the American Mathematical Society (AMS) in 1904.[51] Perhaps more importantly, though, he directed three of his undergraduates to pursue their mathematical studies in Europe: Henry Seely White, Frederick Shenstone Woods, and his own son, Edward Burr Van Vleck (father of the Nobel laureate in physics, John Van Vleck). All three eventually wrote doctoral dissertations under Felix Klein with White and the younger Van Vleck going on to serve as the ninth and twelfth Presidents of the AMS, respectively. (See Tables 5.2 and 5.3.)

Henry White grew up in Cazenovia, New York and attended the local seminary (where his father taught mathematics and surveying) before going

---

[51] Robert A. Rosenbaum, "There were Giants in those Days: Van Vleck and his Boys," *Wesleyan University Alumnus* (November 1956):2–3.

on to study at Wesleyan. After his graduation in 1882, White first spent one year as Van Vleck's assistant in the astronomical observatory, then one year as a teacher of mathematics and chemistry at the Centenary Collegiate Institute in Hackettstown, New Jersey, and finally three years as Registrar and Tutor in mathematics back at Wesleyan. Early in 1887, he traveled to Leipzig in the hopes of studying under the geometers Eduard Study and especially the recently arrived Sophus Lie. White spent only one semester in Leipzig, however, preoccupied mostly with the task of improving his language skills. While there, he obtained some "very valuable advice"[52] from his friend William J. James, who later became a mathematics instructor and librarian at Wesleyan. James had studied in Germany for four years, attending lecture courses given by Klein in Leipzig and by Kronecker and Fuchs in Berlin. As White recalled, James "held up Klein as not only the leading research mathematician but also as a magazine of driving power, whose students received personal attention and stimulus, and in most cases became themselves productive investigators."[53] With this endorsement, White decided to head for Göttingen.

The semester he arrived, the winter of 1887–1888, was the first in which a fairly substantial number of Americans attended Klein's lectures. Besides Haskell and Thompson, White joined three other newcomers, Osgood, B. W. Snow from Cornell, and Harry Walter Tyler from the Massachusetts Institute of Technology. By today's standards six students may not seem like many, but owing to the drastic downturn in mathematics enrollments throughout Germany, they actually represented a substantial portion of the advanced students. At this particular time, Klein was just completing the second half of the two-semester course on hyperelliptic functions mentioned above. This was attended by only four students: Haskell, Thompson, Heinrich Burkhardt, and Joseph Schroeder.[54] Since White, Osgood, and Tyler had missed out on the first half of this course, they opted to wait until the following summer before sitting in on one of Klein's advanced lectures. Along with Thompson and Snow, they enrolled instead in his course on potential theory intended for intermediate-level students.[55]

This interlude gave White a welcome opportunity to prepare for the initial leg of the intensive three-semester course on Abelian functions that Klein had announced for the following semester. White made a copy of a manuscript in Klein's possession, an *Ausarbeitung* of the Leipzig lectures on elliptic function theory which Fine and Cole had heard four years earlier.[56] This prepa-

[52]Henry Seely White, "Autobiographical Memoir of Henry Seely White (1861–1943)," *Biographical Memoirs of the National Academy of Sciences* 25 (1944):16–33 on pp. 22–23.

[53]*Ibid.*

[54]Hörerverzeichnis, Klein Nachlass VII E, NSUB, Göttingen.

[55]*Ibid.*

[56]White's copy of Klein's lectures was recently uncovered at Vassar College by John McCleary. David Rowe is presently preparing an edition based on it together with Fine's lecture

ration evidently stood him in good stead during the cycle of lectures that followed. As White later described the experience, "Klein received me kindly and admitted me to his seminar course... in Abelian functions [in the summer semester of 1888].... Klein expected hard work, and soon had in succession Haskell, Tyler, Osgood, and myself working up the official *Heft* or record of his lectures, always kept for reference in the mathematical *Lesezimmer*. This gave the fortunate student extra tuition, since what the Göttingen geometer gave in one day's lecture (two hours) must be edited and elaborated and submitted for Klein's own correction and revision within 48 hours."[57] This was clearly a task of major proportions, since Klein often left holes in his arguments large enough to trouble even world-class mathematicians today.[58] In the course of working on these lectures, three of the Americans—White, Osgood, and Thompson—gained not only an intimate knowledge of Klein's treatment of the theory of hyperelliptic and Abelian integrals and their related inverse functions but also the skills necessary to tackle the special topics he subsequently gave them for their dissertation research.[59]

The same five Americans—Osgood, White, Thompson, Tyler, Haskell—who took Klein's first semester course on Abelian functions also took part in his parallel seminar on hyperelliptic and Abelian integrals and their role in Clebsch's version of curve theory.[60] In particular, White discussed some of the results Clebsch had derived from Abel's Theorem, including the Clebschian treatment of the 28 double tangents associated with a quartic curve.[61] (We shall examine White's seminar presentations in the following section.)

For his part, Haskell stayed in Göttingen through the summer of 1889, attending all three semesters of Klein's lectures on Abelian functions. His thesis topic, however, had apparently been decided upon long before this, since he spoke about it on two separate occasions in Klein's seminars. (See Table 5.1.) The dissertation Haskell ultimately produced and subsequently published contained an elaborate construction of the projective Riemann surface associated with a certain normal curve that Klein had studied in conjunction with his work on the geometric Galois theory of the modular equation for

notes at Princeton to be published in a forthcoming volume of *Dokumente zur Geschichte der Mathematik* (Wiesbaden: Deutsche Mathematiker-Vereinigung, Friedrich Vieweg & Sohn).

[57]White, "Autobiographical Memoir," pp. 23–24.

[58]See Jean-Pierre Serre, "Extensions icosaédriques," *Seminaire de Théories des Nombres de Bordeaux, Année* 1979–1980 19 (1980):1–7.

[59]Henry Seely White, "Abel'sche Integrale auf singularitätenfreien, einfach überdeckten, vollständigen Schnittcurven eines beliebig ausgedehnten Raumes," *Nova Acta der Kaiserlichen Leopoldinisch-Carolinischen deutschen Academie der Naturforscher* 57 (1891):41–128; William Fogg Osgood, "Zur Theorie der zum algebraischen Gebilde $y^m = R(x)$ gehörigen Abelschen Functionen," (Doctoral Dissertation, University of Erlangen, 1890); and Henry Dallas Thompson, "Hyperelliptische Schnittsysteme und Zusammenordnung der algebraischen und transcendentalen Thetacharakteristiken," *American Journal of Mathematics* 15 (1893):91–123.

[60]Hörerverzeichnis, Klein Nachlass VII E, NSUB, Göttingen.

[61]Protokollbuch für das Seminar über hyperelliptische Funktionen, Sommer Semester 1888, Mathematisches Institut der Universität Göttingen.

the prime $n = 7$.[62] Thus, Haskell's work represented a blend of some of Klein's fondest ideas and interests.[63] It came closer than the work of any of Klein's American students to capturing that peculiar mix of ideas from group theory, algebraic geometry, and complex function theory that lay at the heart of Klein's own mathematical research. In fact, Haskell's effort ultimately represented one of the last concerted attempts to push through this particular set of ideas.

Haskell returned to the United States in the fall of 1889 to accept a position at the University of Michigan alongside Frank Nelson Cole and Alexander Ziwet. As he made his way home on board the "Aller," he wrote to Klein thanking him "for the extraordinary kindness [*Liebenswürdigkeit*] and patience you have always shown me. Be assured," he added, "that I will never forget it, and that I will try to accomplish something worthwhile and useful for science. I have finally learned what it means to *work*. It is a hard thing to learn, but I hope that it sticks with me."[64] Soon after returning to the United States, Haskell began work on an English translation of Klein's *Erlanger Programm*, which he hoped to publish in the *American Journal of Mathematics*.[65] With a teaching load of seventeen to eighteen hours per semester, however, he did not find it easy to carry out this project; nor was his colleague, Cole, prepared to provide anything more than moral support. Against these odds, Haskell managed to complete the project during the summer of 1890, and when it finally appeared three years later, Klein was obviously pleased.[66] He expressed his approval in some brief prefatory remarks, emphasizing the "absolutely literal" nature of Haskell's translation.[67] In the meantime, Haskell had joined Irving Stringham on the faculty at the University of California

[62]See Felix Klein, "Ueber die Transformation siebenter Ordnung der elliptischen Funktionen," *Mathematische Annalen* 14 (1878–1879):428–471, or *Klein*: *GMA*, 3:90–134. He exhibited the modular equations for $n = 2, 3, 4, 5, 7, 13$ in Klein, "Über die Transformation der elliptischen Funktionen und die Auflösung der Gleichungen fünften Grades," *Mathematische Annalen* 14 (1879):111–172, or *Klein*: *GMA*, 3:13–75 on pp. 45–46.

[63]Mellen Woodman Haskell, "Über die zu der Kurve $\lambda^3\mu + \mu^3\nu + \nu^3\lambda = 0$ im projektiven Sinne gehörende mehrfache Überdeckung der Ebene," *American Journal of Mathematics* 13 (1890):1–52.

[64]Mellen W. Haskell to Felix Klein, 27 June 1889, Klein Nachlass IX, NSUB, Göttingen (Haskell's emphasis; our translation).

[65]Mellen W. Haskell to F. Klein, 5 February 1890, Klein Nachlass IX, NSUB, Göttingen. The translation did not ultimately appear in the *American Journal*, however. See note 66.

[66]Felix Klein, "A Comparative Review of Recent Researches in Geometry," trans. M. W. Haskell, *Bulletin of the New York Mathematical Society* 2 (1893):215–249.

[67]*Ibid.*, p. 215. In this preface, Klein took pains to explain why the work had not been translated earlier: "My 1872 Programme, appearing as a separate publication ..., had but a limited circulation at first. With this I could be satisfied more easily, as the views developed in the Programme could not be expected at first to receive much attention. But now that the general development of mathematics has taken, in the meanwhile, the direction corresponding precisely to these views, and particularly since *Lie* has begun the publication in extended form of his *Theorie der Transformationsgruppen* ... it seems proper to give a wider circulation to the expositions in my Programme [p. 215]."

in Berkeley, where he spent the remainder of his career. He later served as Vice President of the AMS, and on three separate occasions he was elected to chair the Society's West Coast Section.[68]

Klein saw relatively few Americans in his courses the following semester, the winter of 1889–1890, when he lectured on non-Euclidean geometry and on the theory of so-called Lamé functions. In terms of his influence on subsequent American research, however, this interlude proved to be one of lasting importance. As his interests shifted from function theory to mathematical physics, Klein took up a geometrical approach to potential theory that had briefly occupied his attention during his first years in Leipzig. Inspired by his reading of Thomson and Tait's *Treatise on Natural Philosophy*, he had contemplated, as early as 1881, the development of a broad framework that would contain all the usual coordinate systems of potential theory as special cases.[69] He returned to this idea nearly ten years later in his 1889–1890 lecture course on Lamé functions.[70] Only five students attended this course, but they included the Americans White, Bôcher, and James E. Oliver, a professor on leave of absence from Cornell.[71]

Once again, the task of working up the *Ausarbeitungen* fell to one of the students, in this case to Bôcher. As an added inducement, Klein arranged for the Göttingen Philosophical Faculty to offer a special prize for the resolution of the following problem: "One can derive the majority of the series expansions and integral representations that appear in potential theory by viewing the various orthogonal systems that thereby come into consideration as different forms of the system of confocal cyclides and, using this as a foundation, by exhibiting suitable series expansions for a body bounded by six confocal cyclides. The faculty wishes that this idea be carried out in detail, and also that the entire theory be given a unified treatment."[72] As noted in Chapter 4, Klein had learned about the theory of confocal cyclides back in 1870 from its principal architect, Gaston Darboux.[73] Since these elaborate figures form systems of orthogonal surfaces in three-space—the usual systems of

---

[68]Archibald, p. 9. For more on the sections of the Society, see Chapter 9.

[69]William Thomson and Peter Guthrie Tait, *Treatise on Natural Philosophy* (Oxford: University Press, 1867). See, also, Felix Klein, "Ueber Körper, welche von konfokalen Flächen zweiten Grades begrenzt sind," *Mathematische Annalen* 18 (1881):410-427, or *Klein: GMA*, 2:521-539 on p. 521.

[70]Klein described the philosophy behind and sketched some of the ideas in these lectures in "Zur Theorie der allgemeinen Laméschen Funktionen," *Nachrichten der Gesellschaft der Wissenschaften zu Göttingen, Mathematisch-Physicalische Abteilung* (1890):85–95, or *Klein*: *GMA*, 2:540–549.

[71]Hörerverzeichnis, Klein Nachlass VII E, NSUB, Göttingen.

[72]Quoted in Maxime Bôcher, *Ueber die Reihenentwicklungen der Potentialtheorie* (Leipzig: B. G. Teubner, 1894), p. 6 (our translation).

[73]On this, see David E. Rowe, "The Early Geometrical Works of Sophus Lie and Felix Klein," in *The History of Modern Mathematics*, ed. David E. Rowe and John McCleary, 2 vols. (Boston: Academic Press, Inc., 1989), 1:209–273 on pp. 254–258.

confocal surfaces and the Dupin cyclides being but two special cases—they had attracted considerable interest among nineteenth-century geometers.[74] Moreover, Gabriel Lamé had shown how to use related systems involving elliptic and hyperelliptic coordinates to advantage in potential theory. This had led to the consideration of certain second-order partial differential equations which generalize the Laplace equation. In the mid-1870s, Alfred Wangerin had demonstrated how this type of equation could sometimes be solved by separation of variables to yield a so-called Lamé triple product.[75] This involved finding solutions to a related two-parameter, second-order differential equation, called a *Lamé equation* by Bôcher. In 1881, Klein had shown that the solutions of Lamé equations can be characterized in terms of their oscillatory behavior.[76] While ingenious, his simple geometrical argument contained major gaps, and he left the task of filling these gaps as well as other difficult matters to his young American student.

Bôcher had already studied potential theory under W. E. Byerly and B. O. Peirce at Harvard, but the challenge he now confronted in developing this theory required the mastery of several branches of higher mathematics, including the theory of elementary divisors, boundary-value problems in partial differential equations, and Lamé polynomials and products.[77] In fact, the task proved far too involved for an exhaustive treatment in the space of the sixty-six pages Bôcher submitted as his prize-winning dissertation of 1891. He thus continued to chip away at this difficult topic after his return to Harvard in the fall. Three years later his efforts resulted in the classic volume, *Ueber die Reihenentwicklungen der Potentialtheorie*, the most substantial piece of research produced by any of Klein's American doctoral students.[78]

---

[74]See Julian Lowell Coolidge, *A Treatise on the Circle and the Sphere* (Oxford: Clarendon Press, 1916).

[75]Alfred Wangerin, "Ueber ein dreifach orthogonales Flächensystem, gebildet aus gewissen Flächen vierter Ordnung," *Journal für die reine und angewandte Mathematik* 82 (1876):145–157.

[76]Felix Klein, "Über Körper, welche von konfokalen Fläschen zweiten Grades begrenzt sind," *Mathematische Annalen* 18 (1881):410–427, or *Klein*: *GMA*, 2:521–539. Earlier work in this direction had been done by Charles Sturm, beginning with a paper read before the Paris Academy in 1833 and published three years later as Charles Sturm, "Mémoire sur les Équations différentielles linéaires du second Ordre," *Journal de Mathématiques pures et appliquées* 1 (1836):106–186. Sturm's work, however, was concerned with one-parameter families of differential equations of, for example, the type $\frac{d^2y}{dx^2} = \phi(x, \lambda)y$, whereas Klein assumed the presence of at least two real parameters. For background on this subject, see Jesper Lützen, "The Solution of Partial Differential Equations by Separation of Variables," in *Studies in the History of Mathematics*, ed. Esther Phillips, MAA Studies in Mathematics, vol. 26 (n.p.: Mathematical Association of America, 1987), pp. 242–277.

[77]See Osgood, "The Life and Services of Maxime Bôcher," p. 342.

[78]The full reference is given in note 72 above. See, also, G. D. Birkhoff's account of this book in "The Scientific Work of Maxime Bôcher," pp. 199–203. Bôcher's work provided part of the classical background to the modern field of multiparameter spectral theory. For a discussion of the connection between Bôcher's work and more recent investigations, see E. G. Kalnins and Willard Miller, Jr., "*R*-Separation of Variables for the Four-Dimensional Flat Space Laplace and Hamilton-Jacobi Equations," *Transactions of the American Mathematical Society*

Bôcher's older compatriot James Oliver also went to Europe with a strong interest in mathematical physics. During the autumn of 1889, he attended lectures in Cambridge offered by Cayley, James Glaisher, and George Gabriel Stokes. Having received "enthusiastic accounts" of Klein's lectures from his former pupil B. W. Snow, Oliver wrote to Klein from Cambridge to ask whether it would be advisable for him to spend some time in Göttingen. [79] As a senior member of the American contingent, Oliver had a viewpoint different from that of Klein's younger students. As he himself expressed it, he mainly sought two things: " (α) to get pretty fully your own ideas as to methods of teaching, topics and courses of study, and promising directions for original research by my young men; (β) to see something of your Seminary work and other University works in Mathematics."[80] Klein and Oliver hit it off very well, as did their wives, and thenceforth Cornell emerged as a prime sphere of Klein's influence in the United States. (For more on Oliver's role in building Cornell's mathematics program, see Chapter 6.)

A third American who came to Göttingen with a solid background in physics, E. B. Van Vleck, arrived in time for the Winter Semester of 1890–1891. Born and raised in the quiet environs of Middletown, Connecticut, Van Vleck attended the local schools before entering hometown Wesleyan University in 1880.[81] After earning his Wesleyan A.B. in 1884, Van Vleck took a job as an assistant in the physics laboratory at his *alma mater* for the 1884–1885 academic year, and then journeyed to Baltimore to study mathematics, physics, and astronomy at the Johns Hopkins University in the fall of 1885. For the next two years, he attended courses given by Story, Craig, Newcomb, and the physicist, Henry Rowland. This preparation allowed him next to take full advantage of the intellectual resources that Göttingen had to offer. In his five semesters abroad, he studied under nearly every mathematician and physical scientist on the Göttingen faculty before completing his dissertation on analytic expansions in continued fractions under Klein in 1893.[82] After graduation, he taught briefly at Wisconsin, returned to Wesleyan from 1895 to 1906, and moved back to Wisconsin for the remainder of his career. A versatile mathematician, Van Vleck won high regard from

244 (1978):241–261. We thank Willard Miller for pointing this out to us.

[79]James E. Oliver to Felix Klein, 11 November 1889, Klein Nachlass XI, NSUB, Göttingen. Snow ultimately took a doctorate in physics at Berlin and later taught at Indiana State University.

[80]*Ibid.*

[81]Rudolf E. Langer and Mark H. Ingraham, "Edward Burr Van Vleck, 1863–1943," *Biographical Memoirs of the National Academy of Sciences* 30 (1957), pp. 399–409; and Archibald, pp. 170–173.

[82]In Göttingen, Van Vleck also attended the lectures of Heinrich Burkhardt, Robert Fricke, Wilhelm Schur, Hermann Amandus Schwarz, Woldemar Voigt, and Heinrich Weber, besides those of Klein. See Archibald, p. 171. Van Vleck's dissertation appeared as "Zur Kettenbruchentwicklung hyperelliptischer und ähnlicher Integrale," *American Journal of Mathematics* 16 (1894):1–91.

his peers for his thesis research as well as for his investigations in other areas of analysis.[83]

As evidenced by Bôcher, Oliver, and Van Vleck, the Americans who came to Göttingen in the early 1890s were exposed to a variety of topics related to mathematical physics. Nevertheless, most of those who studied under Klein during this period learned a good deal more than just potential theory and differential equations. Klein had also rekindled his old interest in geometry around this time and presented a series of retrospective lectures which hearkened back to his research of twenty years earlier. In three year-long courses offered between 1889 and 1893, he covered non-Euclidean geometry (1889–1890), Riemann surfaces (1891–1892), and higher geometry (1892–1893).[84] His lectures on Riemann surfaces, for example, represented an ambitious attempt to cover the major developments in the field over the preceding twenty years. In particular, Klein returned once again to explore the (to his mind) as yet untapped possibilities of the projective Riemann surfaces he had introduced in the mid-1870s.

Three Americans heard Klein present his strong and idiosyncratic views on this subject over the course of two semesters: Van Vleck, F. S. Woods, and Fabian Franklin. (As we shall see shortly, Franklin's wife, Christine Ladd Franklin, also sought to attend, but encountered resistance from University officials.) Woods had taken his Bachelor's degree at Wesleyan in 1885, a fellow undergraduate with White and E. B. Van Vleck. Following the established Wesleyan pattern, he stayed an additional year as an assistant in physics and astronomy before moving on, in his case, to a teaching post at Genesee Wesleyan Seminary in Lima, New York. Woods stayed in this position for four years, earning a Master's degree from his *alma mater* in 1888. In 1890, he accepted an offer to join the faculty at the Massachusetts Institute of Technology, where Harry W. Tyler had just been appointed as an Instructor. These two friends would remain fixtures in the MIT department for the next four decades.

Unlike Klein's former student Henry White, who favored classical algebraic geometry, Woods inclined to the differential side of the subject. His dissertation, submitted in 1894, dealt with so-called pseudominimal surfaces,

---

[83] See George D. Birkhoff, "Fifty Years of American Mathematics," *Semicentennial Addresses of the American Mathematical Society*, pp. 270–315 on p. 295.

[84] All three lecture courses circulated widely in lithograph form during Klein's lifetime, and since his death in 1925 each has appeared in printed editions. See Felix Klein, *Vorlesungen über Nicht-Euklidische Geometrie* (Berlin: Springer Verlag, 1928); Felix Klein, *Riemannsche Flächen*, ed. Günther Eisenreich and Walter Purkert, Teubner-Archiv zur Mathematik, vol. 5 (Leipzig: BSB B. G. Teubner Verlagsgesellschaft, 1986); and Felix Klein, *Vorlesungen über höhere Geometrie*, ed. Wilhelm Blaschke, 3rd ed. (Berlin: Springer Verlag, 1926). Wilhelm Blaschke's edition of Klein's lectures on higher geometry won high praise from Harvard's Julian Lowell Coolidge, who called it "the most stimulating book ever written on geometry." See Julian L. Coolidge, *A History of the Conic Sections and Quadric Surfaces* (Oxford: Clarendon Press, 1945), p. 119.

EDWARD BURR VAN VLECK (1863–1943)

VIRGIL SNYDER (1869–1950)

a topic he surveyed in two seminar lectures delivered during the Winter Semester of 1892–1893. Such surfaces arise in connection with the theory of ordinary minimal surfaces, but satisfy a slightly different type of differential equation. They had been studied earlier by Luigi Bianchi and, in a more general form, by Sophus Lie.[85] In his dissertation, Woods employed techniques first developed by Riemann and Schwarz to deal with conformal representations of minimal surfaces in the plane; he also addressed various special cases satisfying specified boundary conditions.[86] After his return to MIT as an Assistant Professor in 1895, Woods's interest grew in the not unrelated theory of space forms. He gave an overview of how Riemann's ideas on non-Euclidean geometry had been developed by later researchers like Klein, Friedrich Schur, and Wilhelm Killing in a paper of 1901, entitled "Spaces of Constant Curvature."[87] Woods coupled his active research interests with sincere pedagogical concerns; along with Harry Tyler, he played an important role in upgrading the quality of mathematics instruction at MIT. The Woods and Bailey calculus text, the first in the United States to incorporate analytic geometry and algebra into the subject, exerted considerable influence nationwide.[88]

Staying on in 1892–1893 for the third sequence of lectures on higher geometry, Woods was joined by a new student from Cornell, Virgil Snyder. Klein received advance notice of Snyder's arrival from the young man's mentor, James Oliver. "Remembering your kindness to myself, and the inspiration and enlarged outlook that I had got from your lectures, I could think of no one else whose advice and instruction was likely to help him so much," Oliver wrote to Klein. "You will find Mr. Snyder an earnest student and a thoroughly honorable and loyal man. He has made a good record in his works here, and I value him highly, both as a student and as a friend."[89] Snyder, studying on a Brooks Fellowship from Cornell, more than fulfilled the hopes that Oliver and the University had in him. In 1895, he completed his Göttingen dissertation on linear complexes in Lie's sphere geometry (a

[85]On minimal surfaces, see Dirk Struik, *Lectures on Classical Differential Geometry*, 2nd ed. (Reading, MA: Addison-Wesley Publishing Co., Inc., 1961; reprint ed., New York: Dover Publications, Inc., 1988), pp. 182–188. The pertinent works by Bianchi and Lie are: Luigi Bianchi, "Sulle Forme differenziali quadratiche indefinite," *Memorie della reale Accademia dei Lincei* 4 (5) (1888):539–603; Sophus Lie, "Neue Integrationsmethode der Monge-Ampèreschen Gleichung," *Archiv for Mathematik og Naturvidenskab* 2 (1877):1–9, or Sophus Lie, *Gesammelte Abhandlungen*, ed. Friedrich Engel and Poul Heegaard, 7 vols. (Leipzig: B. G. Teubner; Oslo: H. Aschehoug & Co., 1922–1960), 3:287–293; and Sophus Lie and Georg Scheffers, *Geometrie der Berührungstransformationen* (Leipzig: B. G. Teubner, 1896), pp. 376–384.

[86]Dirk J. Struik, "Frederick Shenstone Woods," unpublished manuscript of 1955, Wesleyan University Archives.

[87]Frederick S. Woods, "Spaces of Constant Curvature," *Annals of Mathematics* 2 (1901–1902):71–112.

[88]Frederick S. Woods and Frederick H. Bailey, *Course in Mathematics for Students of Engineering and Applied Science*, 2 vols. (Boston: Ginn and Co., 1907–1909).

[89]James E. Oliver to Felix Klein, 26 September 1892, Klein Nachlass XI, NSUB, Göttingen.

DURING THE ERA OF KLEIN AND HILBERT, MATHEMATICAL ACTIVITY IN GÖTTINGEN CENTERED AROUND THE AUDITORIUM, BUILT IN 1865.

topic closely related to the ideas Klein and Lie had avidly explored around 1870).[90] Thereafter, he returned to Cornell for what would be a career lasting more than forty years.

In his early mathematical work, Snyder focused on metrical (as opposed to projective) structures that arise in line and sphere geometry. As the well-known geometer Arthur Coble later noted, this initial interest colored much of Snyder's subsequent work, in particular his research on Cremona transformations and the birational geometry of curves.[91] Up until the 1920s, Snyder's prolific output and his talents as a teacher made him, together with Frank Morley of the Johns Hopkins, one of the most influential algebraic geometers in the nation. Together with Henry White, in fact, Snyder emerged as a principal heir to Klein's geometric legacy.

Of course, not all of the Americans who studied under Klein in Göttingen went on to lead careers as distinguished as those of Snyder, Osgood, or Bôcher. If thus far in our account of the emergence of the American mathematical research community, we have, understandably enough, focused on the success stories, we pause here to reflect briefly on the failures. A most

---

[90]See Virgil Snyder, "Ueber die lineare Complexe der Lie'schen Kugelgeometrie," (Doctoral Dissertation, University of Göttingen, 1895).

[91]Arthur B. Coble, "Virgil Snyder, 1869-1950," *Bulletin of the American Mathematical Society* 56 (1950):468–471 on p. 468. See, also, Archibald, pp. 218–223.

THE GÖTTINGEN MATHEMATISCHE GESELLSCHAFT WAS FOUNDED BY KLEIN AND HEINRICH WEBER IN 1892. PICTURED IN THIS PHOTOGRAPH, DATING PROBABLY FROM THE 1901–1902 ACADEMIC YEAR, ARE (LEFT TO RIGHT, FRONT ROW) MAX ABRAHAM, GEORG SCHILLING, DAVID HILBERT, FELIX KLEIN, KARL SCHWARZCHILD, GRACE CHISHOLM YOUNG, DIESTEL, ERNST ZERMELO; (LEFT TO RIGHT, SECOND ROW) FANLA, HANSEN, CONRAD MÜLLER, JOHN DOWNEY, ERHARD SCHMIDT, YOSHIYE, SAUL EPSTEEN, FLEISHER, FELIX BERNSTEIN; (LEFT TO RIGHT, THIRD ROW) OTTO BLUMENTHAL, GEORG HAMEL, HEINRICH MÜLLER.

striking case in point involved James Harrington Boyd. Arriving in Göttingen for the Winter Semester of 1890–1891, Boyd enrolled in Klein's course on analytical mechanics (also attended by Van Vleck; T. M. Blakslee, who held a professorship of mathematics at Des Moines College in Iowa; and later World's Chess Champion, Emanuel Lasker) as well as the second installment of a three-semester course on linear differential equations (attended by Van Vleck and Bôcher). The following semester, the summer of 1891, Boyd gave his first talk in Klein's seminar—speaking on the theory of Schwarz *s*-functions—and Klein suggested that he investigate a particular aspect of this subject in his dissertation research.[92] Apparently, though, Klein had already developed some grave doubts as to Boyd's ability to bring this research to a successful conclusion, and he may well have heaved a sigh of relief when the young man returned to the United States after the end of the term. If this was indeed the case, Klein must have felt little pleasure on receiving a letter from Boyd a full year later in which he explained that

> I shall sail on July 2d on the Maasdam for Rotterdam and expect to arrive in Göttingen about the 15th of July.
>
> I have finished the first part of my work in all its details (for integral λ, μ, ν) with many elegant figures which cost me much time and pains to make.
>
> I have done something on the Markoff work but not as much as I would that I had but as much as circumstances would allow.
>
> I hope to remain long enough to bring my work in Göttingen to a satisfactory end both to *you* and myself.[93]

Boyd did return to Göttingen (having earned an Sc.D. in Princeton) in the interim) and, during the winter of 1892–1893, he attended Klein's lectures on higher geometry as well as the parallel seminar. In the latter, he gave another presentation dealing with *s*-functions, and around Christmas, he submitted a draft of his dissertation work, only to have it rejected by Klein as unsatisfactory. Boyd persisted and, in April of 1893, handed Klein a new, revised version. Rather than meeting with the student face-to-face as he normally did, however, Klein opted to answer Boyd in writing. "In response to the present provisional portion of your work that now lies before me,"

[92]Schwarz *s*-functions map the upper half plane conformally onto circular arc triangles. Klein had employed such functions in a crucial way in his theory of the icosahedral equation to forge a link between geometric Galois theory and function theory. See Hermann Amandus Schwarz, "Ueber diejenigen Fälle, in welchen die Gaussische hypergeometrische Reihe eine algebraische Function ihres vierten Elementes darstellt," *Journal für die reine und angewandte Mathematik* 75 (1872):292–335; Klein, "Weitere Untersuchungen über das Ikosaeder," *Mathematische Annalen* 12 (1877):503–560, or *Klein*: *GMA*, 2:321–380 on pp. 330–335 and 350; and Jeremy Gray, *Linear Differential Equations and Group Theory from Riemann to Poincaré* (Boston: Birkhäuser, 1986), pp. 97–108.

[93]James H. Boyd to Felix Klein, 28 June 1892, Klein Nachlass, NSUB, Göttingen (Boyd's emphasis).

Klein explained, "I regret very much that I can give you no better account than I did at Christmas time."[94] This ominous opening preceded a devastating critique in which Klein stated that he could only interpret the introductory paragraphs of the submitted work as revelatory of the author's total lack of understanding of the concept of a group. He went on to say that "the only sentence in paragraph 3 that makes any sense is the one I finally got through to you after constant repetition. All the rest is either unclear, irrelevant, or false. But even more, everywhere a result is stated its derivation is missing, not to mention the imprecisions that appear in the statements themselves. Finally, those passages the details of which we did not discuss together are full of the most obvious mistakes."[95]

Having expressed his opinion as to the quality of Boyd's work, Klein went on to spell out the consequences in the plainest possible language. Klein facetiously suggested that as a first option he himself could sit down and write up the dissertation in about two week's time. Or he could spend ten times as long carefully explaining everything point-by-point to Boyd. Klein concluded coldly that "[y]ou yourself would not entertain the first possibility, whereas you might perhaps be very pleased to see the second come about. To this I have to say that I reject the second possibility just as definitely as the first."[96]

As this letter reveals, Klein did not suffer fools gladly. His message to the young man was clear and unequivocal—Boyd did not have the wherewithal to earn the degree—and the unstated implication was just as easily inferred—he should abandon his plan for a Göttingen degree and return home gracefully. Apparently unwilling to take "no" for an answer, however, Boyd labored on. He contacted Klein's closest collaborator, Robert Fricke (then a *Privatdozent* in Göttingen), to inquire about the possibility of attending the second semester of his advanced course on the theory of automorphic functions. He also enrolled in the second half of Klein's two-semester course on higher geometry during the Summer Semester of 1893[97] and submitted yet another version of his thesis work to Klein that June. Once again, the impetuous American's submission brought an uncharacteristic written response from Klein:

> Following my letter from mid-April, I really thought you would have left this matter aside in order to pursue what you now lack, namely, an understanding of the elements.
>
> Instead of doing that, you have contacted Dr. Fricke. We *Dozenten* have no right, in accordance with the principles of our German universities, to hinder you from pur-

[94] Draft of Felix Klein to James H. Boyd, mid-April, 1893, Klein Nachlass, NSUB, Göttingen (our translation).

[95] *Ibid.*

[96] *Ibid.*

[97] Snyder and Woods also attended this course.

> suing your studies in any way you wish, even when we consider this completely pointless. And, in fact, in your case I do not know what could be less purposeful than for you to enroll in the second half of the *Automorphen* without having attended the first. That is the same unfortunate tendency that I complain about with regard to you and many of your compatriots: that you want to attend highly specialized lecture courses before you have a secure grasp of the elementary things.[98]

Klein would repeat the last complaint three months later before an American audience at his final Evanston Colloquium lecture. (See Chapter 8.)

In fact, Klein mentioned his imminent departure for the Chicago Congress in his letter to Boyd, and he noted further that the newly constituted Mathematics Department at the University of Chicago would be a logical place for Boyd to continue his research. He even pointed out that he planned to lecture on s-functions in Chicago (see the discussion of the fifth Evanston lecture in Chapter 8), so that the Chicago professors—Moore, Bolza, and Maschke—would be directly informed about the particular problems Boyd had been working on. In closing, however, Klein warned Boyd not to dawdle, for "when I recommend a topic to a seminar participant, this is always done for a limited time only; it is not the case that someone may regard such a theme as, in some sense, their private property, hindering me or someone else from finishing off the subject."[99]

Just how much of this advice Boyd took to heart is unclear, but he did go on to Chicago for further graduate studies later that fall. Predictably enough, he fared no better there than he had in Göttingen. After spending two years at Chicago as a Tutor and then seven years in rank as an Instructor, Boyd left mathematics for Harvard's Law School and became a practicing attorney beginning in 1904.[100]

As Boyd's unhappy plight shows, it took more than ambition to obtain a doctorate under Felix Klein, for, while he gave generously of his knowledge of and enthusiasm for mathematics, he also expected his students to possess the mathematical maturity necessary to do creative mathematics on their own. He certainly showed no willingness as a *Doktorvater* to "hold his students' hands" when it came time for them to write their dissertations. Even if he had been so inclined, he was far too busy and much too impatient to take the time for this. Thus, the fact that so many Americans were able to bring their doctoral theses to a successful conclusion under Klein's supervision may be taken as a significant sign of the improved standards in mathematics

[98] Draft of Felix Klein to James H. Boyd, 19 June 1893, Klein Nachlass, NSUB, Göttingen (our translation).

[99] *Ibid.*

[100] For further information on Boyd, see James McKeen Cattell, *American Men of Science: A Biographical Directory*, 1st ed. (New York: Science Press, 1906).

instruction that had been established at several leading universities in the United States during the course of the 1880s. Indeed, scarcely ten years earlier such an impressive crop of native grown mathematical talent stemming from the New World would have been nearly unimaginable.

The foregoing discussion—in its positives as well as its negatives—has aimed to chronicle the broader mathematical activities that engaged Klein's American students during roughly the period from 1884 to 1892. In keeping with this objective, we have primarily viewed events from afar. Our purpose has been to provide an overview, and we have largely done so by consciously neglecting the specific mathematical ideas that percolated through the work of these individuals as well as the experiential side of the students' sojourns abroad. We next turn to address these facets of the foreign study tours of the Americans, beginning with a detailed discussion of two lectures Henry White delivered in Felix Klein's seminar on hyperelliptic functions during the Summer Semester of 1888. Our objective in the section which follows is twofold. First, these lectures dealt with Clebsch's approach to algebraic curve theory, a topic central to Klein's teaching legacy and certainly one familiar to most of his American students. Second, their style of presentation clearly reflected the characteristic pedagogical perspectives that Klein sought to impart to his students and which he himself had elevated to a virtual art form. We then turn, in the present chapter's closing three sections, to an examination of some of the more personal elements in our story. In these sections, we hear the students speak of their foreign educational experiences as they describe their reactions to student life in Germany, relate their difficulties with "the awful German language,"[101] indicate what they liked and did not like about the teaching styles of Klein and other professors, and share their joys and anxieties through the various stages of their doctoral work.

Stepping to the Podium: Two Seminar Lectures by Henry White

As indicated earlier, the seminars run by Felix Klein constituted one of the most important proving grounds for young American mathematical talent. The presentations his students made in this venue tended to cover a wide range of topics, and they were delivered with a variety of goals in mind. (See Table 5.1 for a complete list.) For example, Mellen Haskell twice gave lectures that dealt directly with his dissertation research. Similarly, the talk given by Henry Thompson in the same seminar in which White spoke addressed a topic closely related to the theme he eventually developed in his *Doktorarbeit.* Of quite a different nature, Henry Tyler's lecture that semester essentially reported on a piece of mathematical literature—the first chapter

---

[101] A phrase made famous in Mark Twain's hilarious essay, "The Awful German Language." For a modern-day account of the same theme, see Gordan Craig, *The Germans* (New York: New American Library, 1983), pp. 310–332.

of the book by Clebsch and Gordan on the theory of Abelian functions.[102] White's presentations also shared something of this quality, but, in his case, both *Vorträge* aimed additionally to illuminate a particular mathematical method, namely, Clebsch's application of Abel's Theorem to problems in algebraic geometry.[103]

White began his first lecture with a statement of Clebsch's formulation of Abel's famous result for curve-theoretic purposes.[104] Let $C_n \subset \mathbf{P}^2(\mathbb{C})$ be a curve of order n as determined by a homogeneous polynomial equation $f = 0$, which has only cusps and double points as singularities. This *ground curve* C_n (which remains fixed throughout) has genus p. To C_n Clebsch associated p linearly independent, everywhere finite integrals expressed in the form:

$$(u_k)_y^x = \int_y^x \phi_k d\tilde{\omega}, \quad k = 1, 2, \ldots, p.$$

Here, the ϕ_k (or *adjoint* forms with respect to f) are homogeneous polynomials of degree $n-3$ that vanish at the d double points and r cusps of C_n; the integrals are evaluated along a fixed (homotopically trivial) path in C_n joining $x = (x_1, x_2, x_3)$ to $y = (y_1, y_2, y_3)$; and $d\tilde{\omega} = \frac{x_k dx_l - x_l dx_k}{\frac{\partial f}{\partial x_m}}$ (which is independent of permutations of $k, l, m = 1, 2, 3$). Clebsch's invariant-theoretic set-up served as the starting point for much of Klein's work on hyperelliptic and Abelian functions as well as for White's own dissertation.[105]

---

[102] Alfred Clebsch and Paul Gordan, *Theorie der Abel'schen Functionen* (Leipzig: B. G. Teubner, 1866).

[103] Clebsch had introduced this technique in his classic paper, "Ueber die Anwendung der Abel'schen Functionen in der Geometrie," *Journal für die reine und angewandte Mathematik* 63 (1864):189–243. In their detailed account of research in the theory of algebraic functions, Alexander Brill and Max Noether placed this paper in the mainstream of a tradition that began with Riemann (and his young follower, Gustav Roch) and ran through the authors' own work, which they characterized as lying in a geometric-algebraic direction. See Alexander Brill and Max Noether, "Die Entwicklung der Theorie der algebraischen Functionen in älterer und neuerer Zeit," *Jahresbericht der deutschen Mathematiker-Vereinigung* 3 (1892–1893):111–565, on p. 287. Interestingly enough, they placed Klein's research in a different category, one dominated by invariant theory. In what was certainly a curious collection of names, they identified Weber, Christoffel, Frobenius, Schottky, and Noether himself as this approach's other leading representatives.

[104] White's lectures are recorded in the Protokollbuch für das Seminar über hyperelliptische Funktionen, Sommer Semester 1888, Mathematisches Institut der Universität Göttingen. For a discussion of the original formulation by Abel and its historical background, see Morris Kline, *Mathematical Thought from Ancient to Modern Times* (New York: Oxford University Press, 1972), pp. 644–651; and Roger Cooke, "Abel's Theorem," in *The History of Modern Mathematics*, 1:389–421.

[105] White's dissertation (see footnote 59 above for the reference) dealt with a special case arising in Klein's invariant-theoretic approach to Abelian functions. In "Zur Theorie der Abelschen Funktionen," *Mathematische Annalen* 36 (1890):1–83, or *Klein*: *GMA*, 3:388–473, Klein set forth a general form for integrals of the third kind in n-space: $P_{\xi\eta}^{xy} = \int_y^x \int_\eta^\xi d\omega_z d\omega_\zeta \frac{\psi(z, \zeta)}{(u_z v_\zeta - u_\zeta v_z)^2}$. Here z and ζ are points in n-space and the *canonical curve* ψ is an algebraic form that de-

Suppose the mn pairs of points $y_i = (y_{i1}, y_{i2}, y_{i3})$, $x_i = (x_{i1}, x_{i2}, x_{i3})$, $i = 1, 2, \ldots, mn$, are determined by the intersection of C_n with two curves, C_m and C'_m of order m. As formulated by Clebsch, Abel's Theorem states that $\sum_{i=1}^{mn}(u_1)_{y_i}^{x_i} = \sum_{i=1}^{mn}(u_2)_{y_i}^{x_i} = \cdots = \sum_{i=1}^{mn}(u_p)_{y_i}^{x_i} = 0$. Clebsch also established in his paper the converse—that the vanishing of these p sums implies that the two sets of endpoints lie on curves of order m—a result of decisive importance for applications to algebraic geometry as White demonstrated in his lectures.

Before he turned to these applications, however, White first noted that the converse of Abel's theorem is normally used by fixing some of the x_i (or identifying them with the corresponding y_i) in which case one refers to the remaining x_i's (or y_i's) as *residual* with respect to the fixed set.[106] (It is often convenient to let $y_1 = y_2 = \cdots = y_{mn}$, in which case the lower endpoints trivially lie on a curve of order m. In this situation, reference to the endpoints y_i is normally omitted, a convention adopted below.)

White continued with an elementary example illustrating how this theorem can be applied in the case of a nonsingular cubic curve C_3 (for which the genus $p = 1$, and where we write $u = u_1$). Taking three collinear points $p_1, p_2, p_3 \in C_3$, he fixed three arbitrary lines ℓ_1, ℓ_2, ℓ_3 passing through p_1, p_2, p_3, respectively. These three lines then determine six further points of intersection on C_3—ℓ_1 with p_4, p_5; ℓ_2 with p_6, p_7; and ℓ_3 with p_8, p_9—and, White asserted, these six points lie on a conic. The proof is immediate, since the $m = 1$ case of Abel's theorem yields

$$u^{p_1} + u^{p_2} + u^{p_3} = u^{p_1} + u^{p_4} + u^{p_5} = u^{p_2} + u^{p_6} + u^{p_7} = u^{p_3} + u^{p_8} + u^{p_9} = 0.$$

Adding the last three sums and subtracting the first yields $\sum_{i=4}^{9} u^{p_i} = 0$, precisely the condition for $p_4, p_5, \ldots, p_9$ to lie on a conic (that is, on a curve C_2) by Clebsch's result.[107] (See Figure 5.1.)

pends on the ground curve under investigation. White considered the case where ψ is given by the total, nonsingular intersection of $n-2$ hypersurfaces in $n-1$ space: $f_1(x_1, \ldots, x_n) = 0, \ldots, f_{n-2}(x_1, \ldots, x_n) = 0$, and he exhibited ψ as a simultaneous covariant of the f_i. For further details on Klein's general theory and its connections with the work of Weierstrass and others, see Brill and Noether, "Die Entwicklung der Theorie der algebraischen Functionen," pp. 462–470.

[106]If two sets of points on C_n are residual with respect to a third, then they are called *coresidual.* A necessary and sufficient condition for two sets of points on C_n, say $x_1, x_2, \ldots, x_j$ and $y_1, y_2, \ldots, y_j$, to be corresidual is that $\sum_{i=1}^{j}(u_k)_{y_i}^{x_i} = 0$, $k = 1, 2, \ldots, p$. Sylvester had already developed this basic idea in what he called a theory of residuation, published in George Salmon, *Treatise on the Higher Plane Curves*, 2nd ed. (Dublin: Hodges, Foster, and Co., 1873). Of course, this theory employed none of the analytic tools so central to Clebsch's approach.

[107]One can also express this by saying that there is a dependence relation between the six points, since, in general, five points determine a conic.

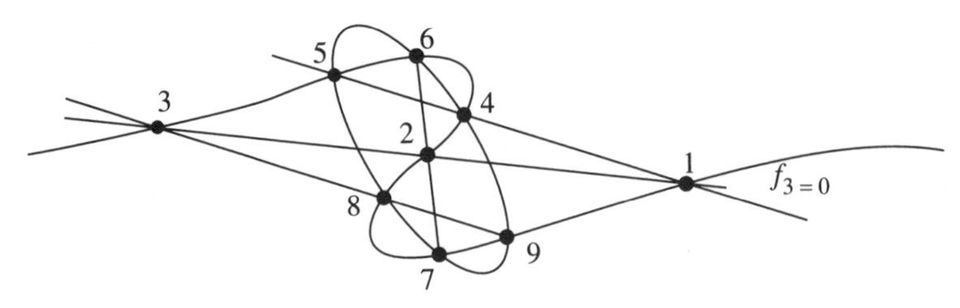

FIGURE 5.1

This simple example served as the point of departure for a whole series of geometrical deductions. For instance, White considered the special case when the six points lying on a conic collapse into three—say, $p_4 = p_5$, $p_6 = p_7$, and $p_8 = p_9$—yielding a conic tangent to the ground curve C_3. For the subsequent discussion, he fixed the endpoints of the integral u and allowed its path to vary, thereby treating the integrals u_k as infinitely-many-valued functions of their endpoints. In particular, for $p = 1$ this means that u must now be computed in terms of its fundamental periods a, b, given by $\int_A du = a$, $\int_B du = b$. (Here, A and B are two canonical closed curves—meridan and longitudinal curves—on the associated Riemann surface, a torus.) Thus, letting $\Omega = \{ma + nb \mid m, n \in \mathbb{Z}\}$, the condition for six points of a nonsingular cubic to lie on a conic now reads: $\sum_{i=1}^{6} u^{p_i} \equiv 0 \pmod{\Omega}$.[108]

In the case of a tangential conic, when $p_4 = p_5$, $p_6 = p_7$, and $p_8 = p_9$, the above condition reduces to $2(u^{p_4} + u^{p_6} + u^{p_8}) \in \Omega$, or equivalently $u^{p_4} + u^{p_6} + u^{p_8} \equiv \frac{ia+jb}{2} \pmod{\Omega}$ for $i, j \in \{0, 1\}$. When $(i; j) = (0; 0)$, p_4, p_6, and p_8 are collinear, and the conic degenerates to a double line. When $(i; j) = (0; 1)$, $(1; 0)$, and $(1; 1)$, three independent systems of tangential conics result. Following Riemann, the pair $(i; j)$ for $(i; j) \neq (0; 0)$ is called the *characteristic* of the system.

White proceeded to consider the systems of tangential conics for a nonsingular quartic curve C_4. These have characteristics expressible as $(i_1, i_2, i_3; j_1, j_2, j_3)$, where $i_s, j_t \in \{0, 1\}$. (Note that the genus p of C_4 is 3.) Neglecting the case in which the conic degenerates into a double line, there are thus 63 independent systems. Exactly one conic from each of these systems passes through a fixed point $p_0 \in C_4$.

In his second lecture as transcribed by Osgood, White discussed the system of 28 bitangents associated with a nonsingular quartic curve C_4. The three conditions for four points $p_1, p_2, p_3, p_4 \in C_4$ to lie on a line are $\sum_{i=1}^{4} u_k^{p_i} \equiv 0 \pmod{\Omega}$ for $k = 1, 2, 3$. If, however, $p_1 = p_2$ and $p_3 = p_4$, then the line is a bitangent, and these three conditions reduce to $2(u_k^{p_1} + u_k^{p_3}) \in \Omega$. As before, this leads to a consideration of characteristics—here, 63 pos-

[108]In the case of a curve C_n of genus p, where $\Omega = \{m_1a_1 + \cdots + m_pa_p + n_1b_1 + \cdots + n_pb_p \mid m_i, n_i \in \mathbb{Z}\}$, the corresponding condition for the coresiduality of the two sets of points $y_1, y_2, \ldots, y_j$ and $x_1, x_2, \ldots, x_j$ appears as $\sum_{i=1}^{j} (u_k)_{y_i}^{x_i} \equiv 0 \pmod{\Omega}$, $k = 1, 2, \ldots, p$.

Wir wollen jetzt einige Sätze über die Berührungspunkte einer C_3 mit einer C_4 beweisen. Wir bemerken zunächst, dass es 64 Systeme von C_3 giebt, welche die C_4 in 6 Punkten berühren. Diese entsprechen den 64 möglichen $\frac{A}{2}$ in den Gleichungen

$$u_i^{(1)} + u_i^{(2)} + \cdots + u_i^{(6)} = \frac{A_i}{2}.$$

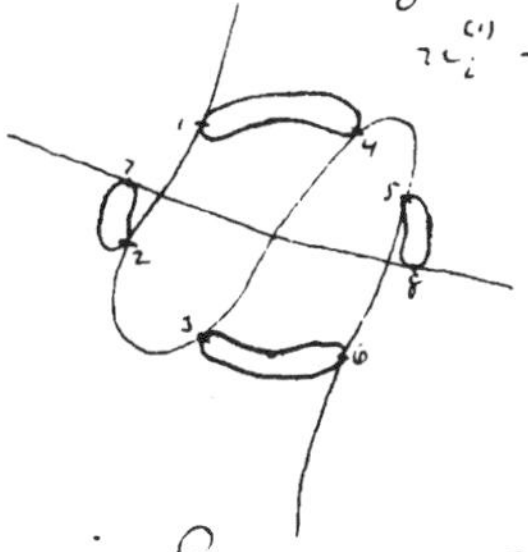

Eine Doppeltangente wird durch die Gleichungen, $u_i^{(7)} + u_i^{(8)} = \frac{A_i'}{2}$

Für jede solche giebt es ein und nur ein A_i so beschaffen, dass $\frac{A_i + A_i'}{2} \equiv 0$, also im Ganzen 28 solche A_i. Bei jedem von diesen haben wir

$$u_i^{(1)} + u_i^{(2)} + \cdots\cdots u_i^{(8)} \equiv 0,$$

FIGURE 5.2

sibilities. This problem is merely a special case of the more general one of determining the number of tangential curves C_{n-3} with $\frac{n(n-3)}{2}$ points of contact determined by these pairs of bitangents with a C_n of genus p. Again, following the lead of Riemann, White briefly touched on the problem of inverting the p integrals, the role played by the Θ-functions in this inversion process, and the decisive importance of characteristics. In particular, Riemann distinguished the characteristics $(i_1, \ldots, i_p ; j_1, \ldots, j_p)$, where $i_s, j_s \in \{0, 1\}$ by parity—even or odd—depending on whether or not the quantity $\sum_{s=1}^{p} i_s j_s \equiv 0 \pmod 2$, respectively. He had further shown that the odd characteristics correspond to actual solutions and that there are $2^{p-1}(2^p - 1)$ such cases. Thus, since $p = 3$ for a nonsingular quartic C_4, there are 28 solvable characteristics, the number of bitangents given by Plücker's formulas.

After presenting these results, White examined the connection between the 28 bitangents and the 63 systems of tangential conics associated with the quartic curve he had discussed in his first lecture. He took as an example the system with characteristic $(1, 1, 1 ; 1, 1, 1)$ and indicated that six pairs of bitangents correspond to this system. Equivalently, $(1, 1, 1 ; 1, 1, 1)$ can be represented as a sum $(i_1, \ldots, i_p ; j_1, \ldots, j_p) + (i_1', \ldots, i_p' ; j_1', \ldots, j_p')$ of odd characteristics in exactly six ways. Since this combinatorial result holds for any of the 63 possible characteristics, each of the 63 systems of conics contains six pairs of bitangents. Thus, in each such case the eight points of contact determined by these pairs of bitangents with C_4 must lie on a conic section. He then gave a simple counting argument to show that by fixing any such set of eight contact points, there will be 315 conics that pass through them.

These excerpts from White's two lectures shed considerable light on the mathematical and pedagogical ideals that animated Klein's seminars. In particular, White's sources—Riemann and Clebsch—not only defined a central chapter in nineteenth-century mathematics, namely, the geometric approach to algebraic function theory, but they also evoked Klein's almost mystical reverence for the Göttingen mathematical tradition. As noted in Chapter 4, Klein felt that he had penetrated to the innermost depths of Riemann's thought in the exposition of his Leipzig lectures of 1881.[109] In Klein's considered opinion, Clebsch's application of transcendental methods in algebraic geometry represented a giant stride forward, but he also thought that Clebsch's disciples, Alexander Brill and Max Noether, had inadvertently undermined their mentor's promising approach by prematurely abandoning it in favor of a more rigorous theory.[110]

White's lectures reflected both Klein's interest in exposing his students to a particular mathematical legacy and his preoccupation with a methodological orientation and style of mathematics that had fallen distinctly out of favor during this period. Klein always tried to minimize the problematic aspects of Riemann's theory. What interested him were its architectural features—a quality he also greatly admired about Weierstrass' approach to function theory—and the uses that could be made of them in algebraic geometry. To be sure, White's lectures presented but a fleeting glimpse of these latter possibilities, but this, too, was a trademark of Klein's teaching style. Indeed, White's whole performance was vintage Klein: it turned on a few carefully chosen examples that illustrated a central theoretical result (Abel's Theorem); avoided all but the most essential details; paid careful attention to the concrete, *anschauliche* features arising from the simplest representative situations; passed from these to interesting singular cases (like tangential conics) via the principle of continuity; moved from details readily *seen* in concrete cases to make a sudden inductive leap (no proofs please!) to a general theory encompassing the cases; and, of course, evoked a sense of the vast vistas open for exploration through the open-ended character of the whole presentation. Arthur Cayley sensed this last aspect of Klein's mathematical vision when he vividly described the *Ikosaeder* as that "tract of beautiful country, seen at first glance in the distance, but which will bear to be rambled through and studied in every detail of hillside and valley, stream, rock, wood, and flower."[111] Contrary to what the Americans found in a conventional lecture

[109]Felix Klein, *Über Riemanns Theorie der algebraischen Funktionen und ihrer Integrale. Eine Ergänzung der gewöhnlichen Darstellungen* (Leipzig: Verlag B. G. Teubner, 1882), or *Klein: GMA*, 3:499–573.

[110]See Klein's remarks on this in *Riemannsche Flächen*, pp. 191–192 as well as the critique of Clebsch's work and that of the book by Clebsch and Gordan, *Theorie der Abel'schen Functionen*, in Alexander Brill and Max Noether, "Die Entwicklung der Theorie der algebraischen Functionen," pp. 318–347, followed by the discussion of the work done after 1870 in the tradition of Brill and Noether, pp. 347–366.

[111]Quoted in the translator's preface to Felix Klein, *Lectures on the Icosahedron and the*

course or seminar, the whole thrust of Klein's teaching style was directed toward one overriding, pedagogical objective: awakening interest in and understanding of the broader significance of a mathematical theory. To gain a firsthand impression of how Klein's students responded to this approach, we turn now to consider the experiences of two Americans who certainly admired Klein and who were, to varying degrees, inspired by him, but who left Göttingen in search of a different kind of environment in which to complete their mathematical studies.

Studying Outside Göttingen

It goes without saying that not all the Americans who went abroad during this period gravitated to Göttingen, but even among those who did, some found Klein's idiosyncratic approach less than ideal. One of these was Harry Tyler, who went to Göttingen along with William Osgood, but decided to leave after just one year in order to test the waters in Erlangen. He did not arrive there empty-handed in the fall of 1888, however; Klein had given him a thesis topic concerning the Abelian integrals associated with a curve with only double point singularities.[112] After having sampled the mathematical life in both Göttingen and Erlangen, Tyler passed on some interesting reflections on events and personages in the two German university towns in a letter to Osgood written on 1 March 1889. He also mentioned a mutual friend's disappointment with the mathematical atmosphere in Berlin.[113] "I'm not much surprised at this," admitted Tyler,

> and am very glad—mathematically—that I gave up [on going to] Berlin. I am even glad that I came here. I'm very glad I went to Göttingen first—a first semester here would have been a mournful experience for me—and I hope I don't underestimate any of the attractions of G[öttingen] mathematically. I think I appreciate them much better in the light of different experience. I understand in a measure the superiority which Klein and Gordan each credited to himself or to the other when I met the latter in Göttingen, and although I have at present more admiration for K[lein] I feel better satisfied not to have worked with him alone.[114]

Solution of Equations of the Fifth Degree, trans. George Gavin Morrice, 2nd ed. (London: Kegan Paul, Trench, Trübner & Co., 1913), p. v.

[112]Tyler reported on the progress he had made on this topic in Harry W. Tyler to Felix Klein, 8 December 1888, Klein Nachlass XII, NSUB, Göttingen.

[113]This letter from Harry W. Tyler to William F. Osgood was first dated 20 February 1889. He interrupted it, however, on 1 March to write: "I have postponed this in order to write to Demplin [the mutual friend], who is about leaving Berlin to see the rest of Europe. He writes that Weierstrass has really been lecturing since Christmas, but that 'the work at the University was hardly what I hoped for.' " Letter held in Osgood Nachlass, NSUB, Göttingen.

[114]*Ibid.*

Of course, Klein and Gordan knew one another very well, having collaborated together on several papers dealing with the Galois theory of modular equations during the 1870s. Klein acknowledged his debt to Gordan in the preface to his *Lectures on the Icosahedron.* "If now a far-reaching theory has grown," he ventured, "I attribute this result primarily to Professor Gordan. ... In this place I must report ... that Professor Gordan has spurred me on when I flagged in my labours, and that he has helped me ... over many difficulties which I should never have overcome alone."[115] Gordan, on the other hand, was one of Felix Klein's greatest admirers, despite (or perhaps precisely because of) the latter's radically different orientation and talents. When, for instance, the Berlin faculty passed over Klein and appointed H. A. Schwarz and Georg Frobenius to succeed Weierstrass and Kronecker in 1892, Gordan wrote to Klein: "I am sorry to hear that you were not called to Berlin, as your all-embracing spirit would have instilled order in the mathematical relationships in Germany."[116]

Regardless of the mutual admiration of Klein and Gordan, for his part, Tyler felt that invariant theory, Gordan's forte, suited his interests better than did Klein's peculiar blend of potential theory, Abelian functions, and projective Riemann surfaces. He thus took up certain aspects of the theory of resultants under Gordan's direction, although, initially, he neither planned to study long in Erlangen nor to work on a dissertation there. Only a month or so after taking up this topic, however, Gordan informed him that it would be acceptable as a doctoral thesis. This came as a big surprise to Tyler, who knew perfectly well that "[t]he subject has been rather exhaustively discussed for ages, so whatever novelty there is in the substance, it will not be so profound as to impose upon anybody."[117] He reported to Osgood on 20 February 1889 that the *Arbeit* was "approximately in order," but fretted about the oral examination, which required proficiency in two subjects other than mathematics.[118]

Back in Göttingen, Osgood was busy mulling over his future. He surely appreciated the advantages of spending at least a semester or two at another university, and he probably felt that he had little to gain by staying in Göttingen. As noted, the spring of 1889 found Klein gearing up to do mathematical physics, a subject in which the young American had no strong interest. Osgood had already spent two pivotal years in Göttingen, and he

[115] Felix Klein, *Lectures on the Icosahedron*, pp. viii–ix.

[116] Paul Gordan to Felix Klein, 16 April 1892, Klein Nachlass IX, NSUB, Göttingen (our translation).

[117] Harry W. Tyler to William F. Osgood, 20 February 1889, Osgood Nachlass, NSUB, Göttingen.

[118] *Ibid.* Tyler stated: "Of course it's not altogether easy or safe for one to expect to prepare for examination in Mathematics and 2 *Nebenfächern* at one Semester's notice, so although I am not unlikely to make the attempt, I shan't work for the degree alone and shan't allow failure against such odds—if it comes to that—to break me up altogether. I shall probably try Physics and Chemistry as *Nebenfächer*."

still sensed a need for something beyond Klein's lofty perspectives, namely, precision and rigor. Whether or not he had considered going to Erlangen earlier, Tyler's letter clearly piqued his interest. About two months later, he wrote back asking the latter's advice as to whether he should consider taking his degree there.

Tyler responded to Osgood's queries very frankly on 28 April 1889: "I think in the first place that it is much better for you or anyone else who has 3 years abroad not to spend the whole time in Göttingen unless for special reasons of great importance. I do not think the incidental loss comparable with the gain in broader knowledge of mathematics and mathematicians. So I wouldn't stay in Göttingen even if it were somewhat better than any other university unless my special work there were too interesting to leave."[119] Tyler's observations about Göttingen largely reflected his current opinion of Klein. On the one hand, he expressed admiration for his external qualities: "I know of nobody who can approach him as a lecturer .... He's certainly acute, fertile in resource, not only understands other people, but makes them understand him, and seems to have a very broad firm grasp of the philosophical relations and bearings of different subjects, as well as great versatility and acquaintance with literature."[120] On the other hand, Tyler felt that these very qualities tended to "make one overrate him."[121] Moreover, he noted some serious drawbacks for those who chose Klein as their thesis adviser: "[s]o busy a man cannot and will not give a student a very large share of his time and attention; so too he will not study out or interest himself especially in the painstaking elaboration of details, preferring to scatter all sorts of seed continually and let other people follow after to do the hoeing."[122]

Tyler followed this assessment of Klein as a personality and as a potential adviser with some pragmatic and insightful remarks about Klein's potential-theoretic approach to function theory. While admitting that he "like[d] Klein's lectures and *Hefts* [sic] because they weave so many threads in together," he felt that "that very quality robs them of other merits which I especially prize."[123] To Tyler's way of thinking, Klein

> attains what he wishes—philosophical exposition of the subject *im Grossen und Ganzen* [as a whole]—but it would seem ridiculous to claim—what he certainly would not claim for himself—that he does not sacrifice completeness of detail, and that this is not a real sacrifice. I think this broad view of things is very attractive and something that any student may well go to Göttingen for, but there seems to me danger that in attempting to follow in Klein's direc-

[119]Harry W. Tyler to William F. Osgood, 28 April 1889, Osgood Nachlass, NSUB, Göttingen.

[120]*Ibid.*

[121]*Ibid.*

[122]*Ibid.*

[123]*Ibid.*

> tion, he will produce only rubbish, whereas he might really have been of some use if he had stooped to the drudgery of "hoeing."[124]

As indicated earlier, Klein's introduction of potential functions on Riemann surfaces played a key role in his whole approach to the theory of functions of a single complex variable. Tyler's reaction, however, might well have reflected the general skepticism then surrounding Riemann's theory as well as one of its key tools, the Dirichlet Principle.[125]

Tyler's unequivocal opinions regarding Klein and the situation in Göttingen reveal how strongly the mathematical atmosphere in Erlangen had affected him during his brief time there. He certainly suffered from no delusions of grandeur about Erlangen's importance as a mathematical center, but, to his mind, precisely therein lay its decisive advantage over Göttingen. "[A]nyone coming here from Klein," he wrote Osgood, "would be sure to look at mathematical things from a new standpoint and as matters are now would be practically certain of a degree of interest and attention almost out of the question in Göttingen, and especially valuable when one is beginning original work. I have been and am still embarrassed by the opportunities."[126] Tyler made it clear that the mathematics courses at Erlangen could not compare with the offerings at most major universities in Germany, but he felt more than compensated by the almost daily contact he enjoyed with Gordan and Noether. (The latter, who had been stricken with polio, had to be brought to class in a wheel-chair, a task that Tyler and another student performed on alternating days.) Tyler had become fairly well acquainted with the personality quirks of both men, and he described these to Osgood in some detail. "Both men are so peculiar and so irreconcilable," he wrote, "that ... [personal relations] must be cultivated with some tact especially if one tries to divide his attentions equally. So far as I know N[oether] like G[ordan] confines himself to pure mathematics—though he studied Physics with Kirchhoff—and both I think run to *Tiefe* [depth] rather than *Breite* [breadth], as compared with Klein. If they have so much in common, that's about all. G[ordan] is outspoken, irascible, exasperating, violent; N[oether] is taciturn, serious, equable, patient."[127]

Having painted such a vivid and detailed portrait of the mathematical environs of Erlangen, Tyler closed his long letter by offering his friend the following summary advice:

> come here if you want ... detailed work in pure mathematics. If you want to work especially with Gordan I

[124] *Ibid.*

[125] See note 87 in Chapter 4 regarding the Dirichlet Principle and its subsequent vindication at the hands of David Hilbert.

[126] Harry W. Tyler to William F. Osgood, 28 April 1889, Osgood Nachlass, NSUB, Göttingen.

[127] *Ibid.*

> wouldn't suggest any preparation unless the first volume of his book. If you had anything underway very likely it wouldn't interest him. For Noether, on the other hand, I think it would be worthwhile to have something yourself to propose—in Abelian Functions if you like or any of his subjects that you know from the *Annalen* as well as I could tell you. I wouldn't advise you to come unless you feel sure your tastes will lie in these directions. I do not see the least reason to doubt your being able to make the Ph.D. in two Semesters here, or even in one if necessary.[128]

Soon after writing this, Tyler passed his examinations in Erlangen and took his degree.[129] He then returned to the United States and joined the faculty at MIT, eventually becoming Chair of the Mathematics Department. As Dirk Struik recalled, he played an instrumental role in transforming it into one of the major institutions for mathematics in the country.[130]

Following Tyler's advice, Osgood did go to Erlangen in the fall of 1889, and with a research topic that fit in well with Max Noether's interests. In an important paper of 1880, Noether had treated the differentials of p independent, everywhere finite integrals as homogeneous coordinates in the space $\mathbf{P}^{n-1}(\mathbb{C})$ in order to handle the corresponding algebraic structures by means of the linear invariant theory for p variables.[131] This theory likewise served as the basis for Klein's work on Abelian functions during the late 1880s. Klein already had Henry White hard at work on his thesis research in this area, and he turned a deaf ear when Noether suggested that White might come to Erlangen to work under him.[132] Apparently, Klein had no such misgivings about Osgood's leaving, but before the latter's departure he gave him another special topic in Abelian functions to investigate, and this served as the basis for the thesis he completed the following year with Noether's help.[133]

Osgood seemingly felt that this work represented little more than a rite of passage into higher mathematics.[134] On the successful completion of his

---

[128] Harry W. Tyler to William F. Osgood, 28 April 1889, Osgood Nachlass, NSUB, Göttingen. The book Tyler referred to was Paul Gordan, *Vorlesungen über Invariantentheorie*, ed. Georg Kerschensteiner, 2 vols. (Leipzig: B. G. Teubner, 1885–1887).

[129] Tyler's Erlangen thesis was entitled "Beziehungen zwischen der Sylvester'schen und der Bézout'schen Determinante."

[130] Dirk J. Struik, "The MIT Department of Mathematics during its First Seventy-Five Years: Some Recollections," in *A Century of Mathematics in America—Part* II, pp. 163–178 on p. 166.

[131] Max Noether, "Ueber die invariante Darstellung algebraischer Functionen," *Mathematische Annalen* 17 (1880):263–284.

[132] Max Noether to Felix Klein, 5 April 1889, Klein Nachlass X, NSUB, Göttingen. Klein answered on 28 April 1889 that he could not afford to let White go in view of the scarcity of students in Göttingen at the time. See Felix Klein to Max Noether, 28 April 1889, Klein Nachlass XII, NSUB, Göttingen.

[133] See note 59 above for the full reference to Osgood's thesis.

[134] As his Harvard colleague, J. L. Walsh, recalled, the only time Osgood ever mentioned his dissertation in his presence, he dismissed it by saying that Klein and Noether had essentially written it for him. See Walsh "William Fogg Osgood, p. 81.

doctoral examination, he did write his fiancée in Göttingen, however, to congratulate her on her future title of "Frau Dr."[135] They married soon afterward and moved to Boston following Osgood's appointment to an instructorship at Harvard. He remained there, a major figure in the Department, until his retirement in 1933.

Of course, Tyler and Osgood both knew that Erlangen was in no way comparable to Göttingen or other leading centers for mathematics in Germany. Its greatest virtue seemed to lie in its very isolation, which created an environment conducive for advanced students, especially those undertaking dissertation research. Moreover, Klein's close personal relations with Noether and Gordan, two senior members of the Clebsch school, made Erlangen a natural move for his advanced students. Although both Erlangen professors enjoyed international reputations for their contributions to algebraic geometry and invariant theory, they nevertheless tended to attract few students on their own, in part because their style of mathematics had fallen out of fashion by the 1890s.

Studying with Sophus Lie in Leipzig

The fact that Tyler and Osgood both went from Göttingen to Erlangen only highlights the delicately spun web of connections that Klein had developed thus far in his career. Paul Gordan and Max Noether had, after all, been close friends and allies for some twenty years, and Klein was one who placed considerable value on such relationships. Much of his success within the German mathematical community stemmed, in fact, from his ability to maintain older, tried-and-true alliances while cultivating new ones. In one notable instance, however, Klein found himself unable to salvage a relationship that may have meant more to him than any other—his friendship with the Norwegian mathematician Sophus Lie.

Against all odds, Klein had managed to secure for Lie the professorship in geometry that he had relinquished when he accepted the offer from Göttingen. At that time, Lie had real misgivings about exchanging the quiet environs of his native Norway for the drab city life of Leipzig, but he also knew that this move represented a once-in-a-lifetime chance to enhance his reputation in the world of mathematics. Although Lie did earn considerable acclaim during his tenure in Leipzig from 1886 to 1898, these years turned out to be a most unhappy chapter in his life. Lie became increasingly convinced that a number of mathematicians—but especially Wilhelm Killing—had been using his ideas without due acknowledgement. In the early 1890s, Klein, too, unwittingly crossed the line that made him just one more enemy in Lie's mind, when he published both a paper on non-Euclidean geometry and a set

[135] William Fogg Osgood to Theresa Ruprecht, 1 July, 1890, Osgood Nachlass, NSUB, Göttingen.

of lecture notes in which he inadvertently failed to mention some of Lie's earlier results.[136] Lie accused Klein of all manner of treachery in this affront in private correspondence, while in the preface to the third volume of the *Theorie der Transformationsgruppen*, he hinted at his animosity in a more public forum. There, the Norwegian wrote that "I am not a student of Klein, nor is the opposite the case, even if it perhaps comes closer to the truth. ... I have a high regard for Klein's talent and I will never forget the interest he took in my scientific efforts from the very beginning, but I do believe that he, for example, does not sufficiently distinguish between ... the introduction of a concept and its utilization."[137]

While an analysis of the various factors—personal, scientific, psychological, and medical—that prompted Lie's break with Klein falls well beyond the scope of the present study,[138] this rift and Lie's public pronouncement made a marked impression in mathematical circles. Even as late as 1922, Virgil Snyder referred to Lie's attack in a review of the first volume of Klein's collected works. There, Snyder noted that the third volume of Lie's *Transformationsgruppen* "contains several statements regarding the contributions of Klein to the development of non-Euclidean geometry and to the foundations of geometry that are sadly out of keeping with the worth and dignity of that monumental work."[139] It is thus ironic, in view of these circumstances, that during the mid-1890s, as Klein stepped back from his role as the leading mentor in Germany for America's fledgling mathematicians (compare the discussion at the close of Chapter 8), his former friend, Sophus Lie, assumed this position. In fact, between 1892 and 1898, Lie attracted nearly twenty American auditors, including Leonard Eugene Dickson, G. A. Miller, and H. F. Blichfeldt, three prominent names in American group theory.[140] In all, six students from the United States actually completed doctoral dissertations under Lie's direction during his twelve-year tenure at Leipzig.[141]

One of the first of these, James M. Page, arrived sometime after 1885. A

---

[136]See Felix Klein, "Zur Nicht-Euklidischen Geometrie," *Mathematische Annalen* 37 (1890): 544–572, or *Klein*: *GMA*, 1:353–383; and the lithographed notes of his lectures on higher geometry delivered in 1893.

[137]Sophus Lie, *Theorie der Transformationsgruppen. Unter Mitwirkung von Friedrich Engel*, 3 vols. (Leipzig: B. G. Teubner, 1888–1893), 3:xvii (our translation).

[138]For more details, see David E. Rowe, "Der Briefwechsel Sophus Lie—Felix Klein, eine Einsicht in ihre persönlichen und wissenschaftlichen Beziehungen," *NTM. Schriftenreihe für Geschichte der Naturwissenschaften, Technik und Medizin* 25 (1988):37–47. The medical aspects are addressed in Bernd Fritzsche, "Einige Anmerkungen zu Sophus Lies Krankheit," *Historia Mathematica* 18 (1991):247–252.

[139]Virgil Snyder, "Review of F. Klein, *Gesammelte Mathematische Abhandlungen*, vol. 1," *Bulletin of the American Mathematical Society* 28 (1922):125–129 on p.127.

[140]These three collaborated in writing G. A. Miller, H. F. Blichfeldt, and L. E. Dickson, *Theory and Applications of Finite Groups* (New York: John Wiley, 1916).

[141]Records from the *Procancellariatsbuch der Universität Leipzig*, Leipzig University Archives. The authors wish to thank Walter Purkert for making this information available to them.

native Virginian, Page had studied mathematics at Randolph-Macon College in Ashland before going abroad. He apparently had received solid preparation since, just two years later, he had completed work on a dissertation which dealt with the theory of transformation groups in four-space. In his first years at Leipzig, Lie spent a good deal of time assisting his doctoral students, and Page clearly benefited from this consideration. Lie must also have felt that his investment had paid off, for in his report on Page's dissertation he explained that

> [t]his work deals with and resolves an important and rather difficult problem in the theory of continuous transformation groups. It is true that the author was largely shown a way that offered good prospects for success. Nevertheless, the task of carrying the entire investigation through required not only considerable energy but also a significant amount of finesse in dealing with the computational techniques. The work must therefore be regarded as a true scientific accomplishment. The presentation is as clear as can be expected taking into account the novelty and difficulty of the subject.[142]

After earning his degree, Page returned to Virginia, serving as a private school master from 1888 to 1895 before breaking into the ranks at the University of Virginia in 1896. By 1901, he had not only risen to a full professorship but also published a number of papers and a textbook dealing with transformation groups and their connections with the theory of ordinary differential equations.[143]

Another student who worked under Lie, Edgar O. Lovett, had earned a doctorate from the University of Virginia before going to Leipzig for the Winter Semester of 1895–96.[144] Lovett's mathematical interests centered on celestial mechanics, and between 1892 and 1895 he had even worked as an astronomer at Virginia's Leander McCormick Observatory. Coming from this

---

[142] *Ibid.*

[143] See, for example, James M. Page, "On the Primitive Groups of Transformations in Space of Four Dimensions," *American Journal of Mathematics* 10 (1888):293–346; and *Ordinary Differential Equations: An Elementary Text Book with an Introduction to Lie's Theory of the Group of One Parameter* (London: Macmillan, 1897). For further information on Page, see James McKeen Cattell, *American Men of Science: A Biographical Directory*, 2nd ed. (New York: Science Press, 1910).

[144] At Virginia, Lovett most likely worked with the mathematician, Charles Scott Venable, and the astronomer, Ormond Stone. In Leipzig, Lovett, like other students, also profited from Friedrich Engel's presence. One student who actually earned his degree under Engel's direction was Arthur Graham Hall. A Michigan native, Hall studied mathematics at the University of Michigan and was appointed to an instructorship there in 1894. He went to Leipzig in 1900 and published his dissertation two years later. See Arthur Graham Hall, *Bestimmung der Differentialgleichungen aller endlichen continuirlichen Gruppen von Punkttransformationen in der Ebene* (Leipzig: Breitkopf und Härtel, 1902). This biographical information may be found in the "Lebenslauf" which accompanied Hall's dissertation.

background, Lovett was naturally drawn to the potential applications of Lie's theory to this field. In 1898, he earned a second doctorate, this time from Leipzig, with a dissertation treating Lie's theory of contact transformations and its ties to perturbation theory.[145] After giving a Summer Quarter course at the University of Chicago in 1898, Lovett joined the Princeton faculty in the fall. He accepted a position ten years later as Professor of Mathematics and Celestial Mechanics at Rice Institute.[146]

Charles Leonard Bouton came to Leipzig one year after Lovett, having just taken an A.M. from Harvard. He studied for three semesters under Lie, earning his degree in 1898 for a dissertation on differential invariants.[147] Afterward, he returned to Harvard, where he introduced two advanced classes: one on transformations in geometry, and the other dealing with applications of transformation groups to differential equations. According to his Harvard colleagues, "[a]ll of Bouton's subsequent scientific work bore the clear impress of Lie's genius."[148] Two of Lie's other American doctoral students, John Van Etten Westfall and David Andrew Rothrock, arrived in Leipzig only about a year before Lie's departure. Although both completed dissertations in 1899, neither went on to accomplish much of significance in mathematical research.[149]

Among Lie's doctoral students during the late 1890s, one young foreigner did go on to play a major role in elevating mathematics on America's West Coast. Hans Blichfeldt, a native of Denmark, had passed the preliminary examination at Copenhagen University with high honors in 1888 when he was only fifteen.[150] In that same year, his family emigrated to the United States, and young Blichfeldt spent the next six years working at various jobs in the lumber industry in Wyoming, Oregon, and Washington. He entered Stanford University as a special student in the fall of 1894, studying a challenging curriculum that covered calculus, quaternions, higher plane curves, differential equations, analysis, solid geometry, and invariant theory in addition to courses in English, German, and physics. After earning the A.B. in 1896, he remained at Stanford, taking an A.M. the following year. His graduate work covered projective geometry, curve tracing, vector analysis, and the theories of functions and substitutions, a course of study that reflected

[145] Edgar O. Lovett, "The Theory of Perturbation and Lie's Theory of Contact Transformations," *Quarterly Journal of Pure and Applied Mathematics* 30 (1898):47–149.

[146] Lovett served continuously on the faculty at Rice until his retirement in 1941. He died in 1957. On Lovett, see Cattell, *American Men of Science*.

[147] This appeared as Charles L. Bouton, "Invariants of the General Linear Differential Equation and Their Relation to the Theory of Continuous Groups," *American Journal of Mathematics* 21 (1899):25–84.

[148] William F. Osgood, Julian L. Coolidge, and George H. Chase, "Charles Leonard Bouton," *Bulletin of the American Mathematical Society* 28 (1922):123–124.

[149] Westfall taught for six years as an Instructor at the University of Iowa before accepting a position as Superintendent of the Bureau of Statistics for Equitable Life. Rothrock continued to teach at the University of Indiana (see Cattell *American Men of Science*).

[150] On Blichfeldt, see Halsey Royden, "A History of Mathematics at Stanford," in *A Century of Mathematics in America—Part* II, pp. 237–277 on pp. 237–248.

contemporary standards in applied mathematics based on the tradition of the *École polytechnique.*[151]

Sometime during his student days at Stanford, Blichfeldt learned about Lie's theory of transformation groups and struck on the idea of studying under the guidance of the master himself. Encouraged by his friends and by Professor Rufus Green, he spent the academic year of 1897–1898 attending Lie's lectures, alongside Miller and Gerhard Kowalewski. Blichfeldt quickly established himself as one of Lie's stars, taking his doctoral degree *summa cum laude* in 1899 for a dissertation "On a Certain Class of Groups of Transformations in Space of Three Dimensions."[152] Thereafter, he returned to Stanford as an Instructor, becoming Executive Head of the Department from 1927 until his retirement in 1938. A prolific writer, Blichfeldt followed in the footsteps of Hermann Minkowski by advancing the new geometrical approach to number theory,[153] and he also made notable contributions to group theory.[154] Thanks to his work and energy during the so-called Blichfeldt era, Stanford emerged as one of the leading mathematical centers on the West Coast of the United States.

Page, Lovett, Bouton, Blichfeldt, these and other students who journeyed to Leipzig to hear Lie's lectures, encountered a man whose teaching style and personality contrasted sharply with Klein's punctilious manner.[155] Lie lectured almost exclusively on his majestic theory of transformation groups and its applications to differential equations. That he did so with a certain panache graphically emerges from this portrait by Kowalewski, one of his finest Leipzig students:

> Lie never wore a tie. ... And even the finest tie would never have found its place, [because] right at the outset of his lecture he would remove his collar with a deft movement, often remarking in the process: "I love freedom [Ich

---

[151]*Ibid.*, p. 239. At Stanford, the program in mathematics was run by Robert Allardice, an A.M. from Edinburgh University, and Rufus Green, an M.S. from Indiana University.

[152]Hans F. Blichfeldt, "On a Certain Class of Groups of Transformations in Space of Three Dimensions," *American Journal of Mathematics* 22 (1900):113–120. Kowalewski remembered Blichfeldt as one of the "élite troops" in Lie's classes. See Gerhard Kowalewski, *Bestand und Wandel. Meine Lebenserinnerungen zugleich ein Beitrag zur neueren Geschichte der Mathematik* (Munich: Verlag R. Oldenbourg, 1950), p. 55.

[153]Hans F. Blichfeldt, "A New Principle in the Geometry of Numbers, with Some Applications," *Transactions of the American Mathematical Society* 15 (1914):227–235; and "The Minimum Value of Quadratic Forms and the Closest Packing of Spheres," *Mathematische Annalen* 101 (1929):605–608. For further details, see C. G. Lekkerkerker, *Geometry of Numbers*, Bibliotheka Mathematica, vol. 8 (Groningen/Amsterdam: Wolters-Noordhoff/North-Holland, 1969).

[154]Hans F. Blichfeldt, "The Finite, Discontinuous Primitive Groups of Collineations in Four Variables," *Mathematische Annalen* 60 (1905):204–231.

[155]G. A. Miller, citing the opinion of an unnamed Frenchman, characterized their differences in a way many would have agreed with: "Klein is a gentleman while Lie is just a great fellow." See George A. Miller, "Some Reminiscences in Regard to Sophus Lie," *American Mathematical Monthly* 6 (1899):191–193 on p. 193.

> liebe die Freiheit]." Then he would start his lecture with the words, "Please, gentlemen, show me your notes, so that I can remember what I did last time." Then someone in the front row would jump up and show him their notebook, while Lie nodded in satisfaction saying: "So, now I remember."[156]

Lie clearly cultivated an extemporaneous approach. He directly queried his students in class, posing basic questions related to his theory. If the individual questioned failed to give an adequate answer, he took this in stride, knowing that he could turn to the star members of the audience. Likewise, he depended on these stars to bail him out on those not so rare occasions when he found himself entangled in a calculation. According to G. A. Miller, in fact, Lie "[s]ometimes ... had to pay quite heavily for his lack of preparation, being unable ... to prove simple things in his own theory .... He kept in good spirits on such occasions but he generally could not maintain enough self-possession to work his way out of the difficulties during the rest of the hour."[157] To all those participating in his courses, including himself apparently, Lie applied the saying "You must exercise with the fundamental ideas of my theory like a soldier with his rifle."[158] In short, Lie liked to teach, and, perhaps because he was a foreigner who spoke a rather broken German, he seemed to appeal particularly to students from outside Germany.[159] Besides the Americans, he attracted a number of Frenchmen, Russians, and representatives from other countries as well. These students formed a small, but devoted cadre which thrived on the informal atmosphere in his classes.[160] Lie's success as a teacher clearly stemmed as much from his personal warmth as from his legendary mathematical talents.

The Women Make Their Mark

Although none of the foreign students drawn to Lie in the 1890s happened to be women, this decade nevertheless marked an important turning point in educational opportunities for foreign women on German soil. Prior to 1894, the status of female students within the German universities fell outside the bounds of academic discourse since, from an institutional standpoint, the very concept of a "female student" was generally regarded as an

[156] Kowalewski, *Bestand und Wandel*, p. 54 (our translation).

[157] Miller, p. 191.

[158] Kowalewski, *Bestand und Wandel*, p. 95. (our translation).

[159] He even claimed that they were better prepared than his German students, although one must wonder whether this assessment also applied to the Americans. On at least two occasions Lie offered special instruction to American students during summer vacations designed to familiarize them with "mathematical phrases rather than to teach [them] much mathematics." See Miller, p. 192

[160] Lie "would talk [to them] for hours ... asking questions in regard to their work as well as in regard to the mathematicians in their country." See *ibid.*, pp. 191–192.

oxymoron. This linguistic difficulty aside, a number of foreign women had, in fact, pursued science and mathematics in Germany even before 1894. The most famous of these by far was Sonya Kovalevskaya, who studied with Leo Königsberger in Heidelberg and privately with Karl Weierstrass in Berlin before receiving her doctorate *in absentia* from Göttingen in 1874.[161] Such cases, however, were entirely exceptional and were sanctioned neither by policies of the respective educational ministries nor by those of particular faculties. Not until 1908 did the German universities concede to women the general right to matriculate. Interestingly, the initial sequence of events which ultimately led to the official enrollment of women as students, owed largely to the persistent efforts of a handful of women, many of them Americans.[162]

Christine Ladd-Franklin, who, as we saw in Chapter 3, had completed all of the requirements for a doctorate at Hopkins only to be denied her degree, helped to organize the assault on the academic barricades through her work with the Association of Collegiate Alumnae (ACA). In 1888, she suggested the idea of establishing an ACA European Fellowship to help support the efforts of young women wishing to study overseas. Accompanying her husband, Fabian Franklin, to Germany during his sabbatical leave from the Johns Hopkins in 1891, she hoped to participate with him in university courses there. At Göttingen, however, she was forced to watch from the sidelines, while he attended Klein's classes on algebraic equations and Riemann surfaces and lectured on Hilbert's famous paper of 1890, "Über die Theorie der algebraischen Formen."[163] Taking interest in her case, Klein entered a motion at a faculty meeting on behalf of Ladd-Franklin in an effort to gain her admission as a regularly matriculated student, but to no avail: she could attend as an auditor (*Gastzuhörerin*) or not at all. (She opted instead to do some experimental work on vision in the laboratory of Georg Elias Müller.)[164]

During this same year, Ruth Gentry traveled to Germany to study mathe-

[161]See Ann Hibner Koblitz, *A Convergence of Lives. Sofia Kovalevskaia: Scientist, Writer, Revolutionary* (Boston: Birkhäuser, 1983); and Roger Cooke, *The Mathematics of Sonya Kovalevskaya* (New York: Springer-Verlag, 1984).

[162]Margaret W. Rossiter, *Women Scientists in America* (Baltimore: Johns Hopkins University Press, 1982), pp. 37–43. The policies regarding women auditors were somewhat more liberal in Leipzig, where Klein had at least one female auditor, a young woman from Cambridge, Massachusetts named Alice Hayes. See Alice Hayes to Felix Klein, 7 November 1896, Klein Nachlass IX, NSUB Göttingen. For an overview of participation by women in research-level mathematics in the United States, consult Della Dumbaugh Fenster and Karen Hunger Parshall, "Women in the American Mathematical Research Community, 1891–1906," in *The History of Modern Mathematics*, vol. 3, ed. Eberhard Knobloch and David E. Rowe (Boston: Academic Press, Inc., 1994), forthcoming.

[163]David Hilbert, "Ueber die Theorie der algebraischen Formen," *Mathematische Annalen* 36 (1890):473–534, or David Hilbert, *Gesammelte Abhandlungen*, 3 vols. (Berlin: Julius Springer, 1932-1935), 2:199–257. Franklin spent part of his stay studying with Hilbert in Königsberg (see David Hilbert to Felix Klein, 13 April 1892 in *Der Briefwechsel David Hilbert–Felix Klein* (1886–1918), ed. Günther Frei, Arbeiten aus der Niedersächsischen Staats- und Universitätsbibliothek Göttingen, vol. 19 (Göttingen: Vandenhoeck & Ruprecht, 1985), pp. 79–80.

[164]Rossiter, p. 40.

matics on an ACA Fellowship. Gentry had earned her undergraduate degree in 1890 from the University of Michigan (presumably taking classes with either Cole or Haskell) before going on to Bryn Mawr, where James Harkness and Charlotte Angas Scott served as her mentors.[165] After a good deal of persistent effort, Gentry received permission to attend courses as an auditor at the University of Berlin. From there, she wrote to Klein on 21 July 1891 asking for permission to attend his lectures. Klein may have had considerable sympathy for the plight of aspiring women who wished to study mathematics, but her letter apparently found him in no mood to buck the system at that point. After learning that he could not honor her request, she wrote back, thanking him for his reply and expressing her regret that she would not be admitted officially. She then asked "whether you will give me private lessons. I do not know whether you ever give private lessons or not, but if there is any hope for me please do not refuse. For a long time I have hoped that some day I might study with Professor Klein, and now that I am here it is hard for me to give up the idea."[166] Klein apparently declined this request as well, and Gentry returned to Bryn Mawr where she completed her dissertation in 1896.[167]

As women from the United States continued to clamor for the right to enter the German universities, an opportunity to champion their cause soon presented itself to Klein. On 8 April 1893, his former student, Heinrich Maschke, wrote to inform him of a promising young woman from the University of Chicago:

> Miss Mary F. Winston, is applying for a fellowship that she would use to go to Germany next year. She has ... talent, thinks independently, and, at any rate, is above average. ... Bolza and I hope to convince her ... to go to Göttingen , and at the same time we are urging her not to go to Berlin in order to preserve her from becoming rigid. Now there is first of all the question, whether female students or graduates (*Habilitanten*) will be admitted at Göttingen, or whether, if this is not the case, you believe that through your influence you could succeed in having an exception made in this case.[168]

[165]On Charlotte Scott, consult Patricia C. Kenschaft, "Charlotte Angas Scott, 1858–1931," *College Mathematics Journal* 18 (1987):98–110, and "Charlotte Angas Scott (1858–1931)," in *Women of Mathematics: A Biobibliographic Sourcebook*, ed. Louise S. Grinstein and Paul J. Campbell (New York: Greenwood Press, 1987), pp. 193–203. She had come to the United States after having done her mathematical studies at Girton College, Cambridge. (She actually earned her B.S. (1882) and D. Sc. (1885) from the University of London, since Cambridge did not grant degrees to women at this time.) Taking a position at Bryn Mawr in 1885, she had a forty-year career there, supervising the doctoral dissertations of seven women.

[166]Ruth Gentry to Felix Klein, 21 July 1891, Klein Nachlass IX, NSUB, Göttingen.

[167]See Judy Green and Jeanne LaDuke, "Women in the American Mathematical Community: The Pre-1940 Ph.D.'s," *The Mathematical Intelligencer* 9 (1) (1987):11–23 on pp. 13–15.

[168]Heinrich Maschke to Felix Klein, 8 April 1893, Klein Nachlass X, NSUB, Göttingen (our translation).

Klein often enjoyed testing the limits of what he could accomplish, and Maschke's query clearly posed just such a challenge. He surely knew from the earlier episode involving Ladd-Franklin that the University's *Kurator*[169] would fight him every step of the way. So, before he left to attend the World's Fair in Chicago in 1893 (see Chapter 7 below), Klein sounded out officials in the Ministry of Culture and obtained assurances that they, at least, would do nothing to stand in the way of his plans. With this information in hand, Klein put the Ministry on notice that, the *Kurator's* position notwithstanding, he and his colleagues expected to bring this issue up at the beginning of the next semester. In a letter of 1 August 1893 to Friedrich Schmidt(-Ott), an official in the Prussian Ministry of Culture, Klein wrote that "[o]ur *Kurator* has declared with respect to the admission of women to our lectures that he must for his part hold to the applicable stipulations until such time as the Ministry decides something different. Thus, as soon as an actual case arises, probably in the coming semester, my colleagues and I intend to submit a formal petition."[170]

During his visit to Chicago, Klein had the opportunity to meet Mary Frances "May" Winston in person, as she attended both the Mathematics Congress held in conjunction with the World's Fair and Klein's Evanston Colloquium lectures that followed it.[171] He learned that she had studied classical languages and mathematics at the University of Wisconsin, graduating in 1889 at the age of twenty. After teaching for a year at Downers College in Fox Lake, Wisconsin, she had applied for a fellowship to study mathematics at Bryn Mawr. Although she failed to win it, Winston did come in a close second and received encouragement to reapply the following year from Charlotte Scott. The second time around brought her success, but she remained only one year at Bryn Mawr, lured back to the Midwest by the opening of the University of Chicago near her hometown of Forreston, Illinois. Despite the opportunity of participating in the Chicago experiment, Winston's real ambition remained to study in Germany. Klein told her he would do what he could to help her should she decide to come to Göttingen, but he was pointedly cautious and avoided making any promises.[172]

Unfortunately, the situation grew even more complicated when the ACA

---

[169] The *Kurator* was the official representative of the Ministry of Culture at each of the German universities and was responsible for carrying out its policies.

[170] Felix Klein to Friedrich Schmidt, Klein Nachlass XI, Nr. 726., NSUB, Göttingen (our translation).

[171] On Winston, see, for example, Betsey S. Whitman, "Mary Frances Winston Newson," *Mathematics Teacher* 76 (1983):576–577.

[172] Klein mentioned to Winston that if the Prussian Ministry turned down her application to study at Göttingen, she should consider going to Zurich or Leipzig. See Winston's paper delivered at a meeting of the Women's Alliance of the Unitarian-Universalist Church, St. Petersburg, Florida, around 1952, in the Mary F. Winston Papers, Sophia Smith Collection, Smith College, Northhampton, Massachusetts (hereinafter *SSC*).

turned down her application for a European fellowship. After learning of this, Christine Ladd-Franklin sent Winston $500 toward the cost of her trip.[173] With the steadfast support of her family and friends, a meager subsistence budget, and no guarantee that she would ever set foot in a German lecture hall, Winston boarded a ship bound for Antwerp late in September of 1893. She was bursting with excitement and anticipation when she arrived in Göttingen and took a room at a boarding house in the Obere Karspüle. Her first day there, she met four other American women, two of whom hailed from Chicago and looked familiar from the ACA meetings she had attended. She also spoke with her landlady, whom she described as "a very bright woman," and made arrangements to take German lessons with her.[174] Although Klein himself had not yet returned from his trip to the United States, the following evening Winston called on Frau Prof. Klein, whom she found "*very* quiet and modest and as kind as she can be."[175] After conversing for a while in a mixture of English and German, Anna Klein introduced her to the family's two youngest daughters (then aged six and eight) and invited her to return the following morning so that she could give her a tour of the University. Winston eagerly accepted the invitation, especially since Mrs. Klein explained to her that if she went alone, she would be received as a student rather than as a lady. Having played the part of the Frau Prof. since she was sixteen, Anna Klein (née Hegel, a granddaughter of the famous philosopher) certainly knew her way around German academic circles.

The following day, she accompanied Winston on a stroll through the botanical gardens that separated the Klein home from the Auditorium building. Once inside, she showed her the Mathematics Library and Reading Room and introduced her to Ernst Ritter, Klein's assistant. Ritter then continued the tour with some brief remarks on the collection of models and calculating machines housed in the department. Winston was delighted to find that she "understood every word he said," even if what he had to say seemed relatively elementary.[176] Later, Anna Klein also arranged invitations for the young *Amerikanerin* to the homes of Professors Heinrich Weber and Woldemar Voigt.

In the meantime, Winston remained on pins and needles waiting to learn of her fate. With Klein away, the whole matter had to be placed on hold until his return. The situation looked gloomy, and she began to plot contingencies in the event of a debacle. She learned from a friend that Ruth Gentry had left Berlin for Paris and that she now regretted not having gone there from the very first. Winston still hoped that everything would work out for her in Göttingen, but consoled herself with the thought that Paris still had much to

---

[173] Rossiter, p. 41.

[174] Mary. F. Winston to family, 10 October 1893, *SSC*.

[175] Mary F. Winston to family, 15 October 1893, *SSC* (Winston's emphasis).

[176] *Ibid.*

offer: "So it may not be such a misfortune if I fail to be admitted here."[177] Several days passed, and her dreams faded with them since she still had no news from Klein. As doubts crowded in on her, she suddenly received a message that Klein had just arrived the night before and was on his way over to call on her. Winston could scarcely contain her excitement:

> [H]e said that I was to write out an application in my own name and that it was to be sent to Berlin together with that of two other ladies, one in Math. and one Physics, who were also asking admission, and *in the meantime* we are to be admitted to the lectures. That, of course, means a great deal but "there's many a slip" and everything is to be done very quietly. We are to go to Prof. Klein's private office before the regular meeting time for changing classes so as not to meet the students in the halls and from there we are to go into the class. The gentleman in the Physics department ... came also and Prof. K. sat down and with an occasional suggestion from the other gentleman, proceeded practically to write my application for me.[178]

Two days later, Winston had just finished breakfast when one of the other two female applicants, Grace Chisholm, an Englishwoman recently graduated from Girton College, Cambridge, came to pay her a visit. A year older and probably a bit better trained than the young American, Chisholm had already attended mathematics courses at Cambridge and thus had "grown quite accustomed to braving the awful men."[179] The following day they got together with the third young intruder, Margaret Maltby, who came to Göttingen from Wellesley to study physics.

The petition granting admission to the three foreigners still had to pass through the hands of von Maier, the *Kurator*, and, predictably enough, he recommended that the Ministry deny the request. Appealing to a report that had been filed in December of 1891 and which set forth arguments against Christine Ladd-Franklin's request for recognition as a student, the *Kurator* affirmed his position in favor of the *status quo*. In private, the University's highest administrative official expressed his opinion in rather less diplomatic language, accusing Klein of propagating a notion "worse than social democracy, which only seeks to abolish the difference in possessions. You want to abolish the difference between the sexes!"[180] But Klein and his allies had apparently played their cards right, and only five days after the petition's submission, the Ministry gave its approval. Von Maier, who had grown in-

[177] *Ibid.*

[178] Mary F. Winston to family, dated 17 October 1893, *SSC* (Winston's emphasis).

[179] *Ibid.*

[180] Klein Nachlass XXII L (Personalia), S. 6, NSUB, Göttingen (our translation).

FROM LEFT TO RIGHT, MISS JOHNSON, MARY FRANCES WINSTON, AND ANNIE MACKINNON.

creasingly weary of being repeatedly outmaneuvered by Klein and others, resigned his post four months later.

Winston learned of the news on a Sunday. Home alone, her housemates all at church, she received

> a portentous-looking document about a foot and a half long, from Prof. Klein. Of course I knew it was the answer to my application, but ... I was unable to read it, for it was written in German script. So I had to go down and have someone decipher it for me. I soon learned that it was favorable and my joy was great but everyone was gone, off to church. ... I felt the want of someone to sympathize with me. ... Just at that moment Miss Chisholm appeared. I do not know why it had not suggested itself to *me* to go and find *her* but somehow it hadn't. We fairly fell into each other's arms with rejoicing.[181]

The Ministry's proviso to the effect that permission had been granted as an exception to the normal rule could hardly dampen their spirits, nor were they troubled by the fact that the issue of matriculation remained unresolved. Winston's optimism soared, and she even pondered the possibility that they might take their doctoral degrees in Göttingen. She and Chisholm spent the afternoon together "talking all the time as fast as we could" about the possibilities which the future might hold.[182] During their conversation, Chisholm confided that she had felt rather miffed at having been turned down for a fellowship at Chicago the previous spring. Later that day, around dinner time, Anna Klein called to congratulate the new student and to wish her luck with her studies. This gesture and her "sweet and unassuming" manner made a strong impression on Winston, particularly since "she was the wife of the most influential man in the University."[183]

After this round of torment and exhilaration, the young women readied themselves for their classes, which began two days later. Winston enrolled in three of the offerings in mathematics—a four-hour course on hypergeometric functions with Klein and two two-hour courses on function theory and algebraic number theory given by Heinrich Weber—but planned to drop one of Weber's courses in order to have plenty of time for independent reading. Grace Chisholm, who had never heard Klein lecture before, left this account of her experience:

> Miss Winston and I made for the Sanctum and found Klein there working till lecture time. Klein, instead of beginning with his usual "Gentlemen!" began "Listeners!" [*meine Zuhörer*] with a quaint smile; he forgot once or

---

[181] Mary F. Winston to her family, 29 October, 1893, *SSC* (Winston's emphasis).

[182] *Ibid.*

[183] *Ibid.*

> twice and dropped into "Gentlemen!" again, but afterwards he corrected himself with another smile. He has the frankest, pleasantest smile and his whole face lights up with it. He spoke very slowly and distinctly and used the blackboard very judiciously. Mr. Woods said he never heard anyone lecture so well and neither have I. I found my notes afterwards perfectly clear though queerly spelt; but I understood as well as at an English lecture.[184]

Winston was also delighted to find that she could follow Klein's lectures without difficulty, though she confessed that she had problems taking notes. This proved to be a rather insignificant stumbling block, however, since she had plenty of time available for studying the official *Ausarbeitung* prepared by Ernst Ritter. She found Weber's lectures more difficult to follow in detail, but took satisfaction in the fact that they were not altogether unintelligible.[185]

Of course, being able to understand a mathematical lecture delivered in German was one thing; the ability to present such a lecture oneself, quite another. For Winston, the first woman to speak in Klein's seminar, there was more than a little at stake.[186] (See Table 5.1.) Though understandably nervous for several days beforehand, she managed to acquit herself well on the occasion. In a letter to her family, she reported that "[i]t went off reasonably well. I cannot say that I cast much honor on the feminine sex thereby but on the other hand I do not think that anyone will draw the conclusion from it that women cannot learn Mathematics."[187]

In fact, Klein was impressed by all three of the women who came to Göttingen for this historic experiment. In a report written on 10 December 1893 to Friedrich Althoff, the man largely in charge of university affairs in Prussia, Klein described both their progress and his impressions. "You will not be surprised when I tell you," he wrote, "that we have had the most positive experiences: our three women are not only exceptionally diligent and conscientious, but their accomplishments are in no way inferior to those of our best students; indeed, to some extent they serve as a model for them. Thus, my friends and I are altogether inclined to continue the course we have set upon."[188] While clearly more progressive than most of his colleagues when it came to extending opportunities to women and others outside the mainstream of German academic life, Klein certainly did not advocate a radical break with the past. He believed rather that Germany should adopt

---

[184] Ivor Grattan-Guinness, "A Mathematical Union: William Henry and Grace Chisholm Young," *Annals of Science* 29 (1972):105–183 on p. 110.

[185] Mary F. Winston to her family, 29 October 1893, *SSC.*

[186] In Mary F. Winston to her family, 21 December 1893, *SSC.*, Winston mentioned her distinction of being the "first."

[187] *Ibid.*

[188] Felix Klein to Friedrich Althoff, 10 December 1893, Althoff Nachlass, B Nr. 92, Bl. 92, Zentrales Staatsarchiv Merseburg (our translation).

FROM LEFT TO RIGHT, (FIRST ROW) MARY FRANCES WINSTON AND GRACE CHISHOLM; (SECOND ROW) CHARLES JACCOTTET, POUL HEEGAARD, AND GINO FANO CA. 1895.

an English-style system of colleges for women and that, concurrently, they should be allowed to take certain specific courses only available at the universities, assuming that it was appropriate for them to do so. In his opinion, courses offered by the faculties of medicine, law, and theology would *not* fall into this category.[189]

Winston, Chisholm, and Maltby knew how much hinged on their respective performances, and they took much satisfaction from the realization that their success might pave the way for generations of women to follow. Filled with intellectual adventure, their mission abroad created a natural bond of solidarity between them. Over the next few years, the friendship between the two mathematicians, Winston and Chisholm, especially deepened. They saw one another often, both during and after class sessions, and they took several trips together. On one occasion, Winston even accompanied her English friend on a trip home to London.

In 1895, their second year together in Göttingen, Winston and Chisholm were joined by another young woman who had come from the United States to study under Felix Klein, Annie MacKinnon. MacKinnon had just taken her doctorate from Cornell and had won an ACA fellowship for European study. A year earlier, Klein had made her acquaintance when he stopped in Ithaca during his whirlwind tour of the United States. (See Chapter 8 below.) MacKinnon gently reminded him of that meeting in a letter dated 9 July 1894 and posted shortly before her departure for Europe. She wrote that "[d]oubtless you remember a reception given in your honor last summer at the home of Prof. Oliver of Ithaca, and possibly you may recall the fact that at this reception a young woman asked you concerning the admission of women to the University of Göttingen. The young woman is now about to sail for Germany for the purpose of spending the coming year there in the study of mathematics. I am the young woman, and with your approval I shall be pleased to study at Göttingen."[190] She reported further on having read about Göttingen's decision to admit women in the *Nation*, and expressed her thanks to Klein for his efforts. "With many other Americans," she added, "I greatly appreciate your attitude toward the admission of women to your University."[191]

Only a few weeks after MacKinnon penned these words, Klein received a letter from Bryn Mawr's Charlotte Scott on behalf of her student, Isabel Maddison. Scott knew of Klein's concern "that applicants for admission [be] qualified not only to follow the work, but also to put it to good use,"[192] and she was confident that Maddison met these requirements. The following semester, the winter of 1894–1895, her young protégée was one of thirteen

---

[189]Klein expressed this opinion in *ibid.*

[190]Annie MacKinnon to Felix Klein, 9 July 1894, Klein Nachlass X, NSUB, Göttingen.

[191]*Ibid.*

[192]Charlotte Angas Scott to Felix Klein, 30 July 1894, Klein Nachlass XI, NSUB, Göttingen.

students enrolled in Klein's course on number theory: nine men (including Percey Smith of Yale and the Canadian J.C. Fields, after whom the Fields Medal was named) and four women. From this point onward, women comprised a notable contingent not only in nearly all of Klein's classes but also in those of his colleagues.[193] Thus, after only a year, the women had made their presence felt within Göttingen's mathematical community.

A highpoint for the first group of women who came to Göttingen was reached on 26 April 1895, when Grace Chisholm passed her doctoral examination, the first woman to accomplish the feat at a Prussian University. Although a singular event in more ways than one, Chisholm's final ordeal and her reaction to it certainly reflected elements that would have struck a familiar chord to any of the Americans who took their degrees in Germany. For this reason, her vibrant and detailed description of the examination procedure and the emotional highs and lows she went through that day serves as a kind of impressionistic portrait of this rite of passage within the German academic world. Not surprisingly, Chisholm initially felt that the appointed hour would never arrive:

> I went round to May Winston, who has been my good angel all through a very trying time, and between us we managed to pass two of the slowest hours I have ever known. We looked up things in Klein's lectures, and expounded to one another our ideas on a variety of subjects which at that time seemed likely to occur in the examination, and which as a matter of fact never came at all. Indeed, I must tell you that Klein did not ask me one single question which I had thought likely or prepared for, and indeed the whole examination was as unlike anything I had expected as you can imagine.[194]

As the minutes slowly ticked by, Chisholm and Winston sat on a windowsill awaiting the carriage that would take them to the *Aula*,[195] where the examination was scheduled to begin at 5:30. Earlier that day, Winston had ordered the carriage for five o'clock. Around quarter past five, however, Chisholm noticed a coach driving away. When Winston went to see whether it might have been theirs, the maid informed her that the coachman had simply asked for the "*Gentleman* who had ordered the carriage."[196] Winston suddenly realized that "in ordering the coachman in livery, [she] had mentioned it was to drive to the Aula, and the man had said: 'I suppose it is a doctor examination,' and she had said yes, and then thinking he might imagine it was hers, which

[193] By 1900, women were applying by the dozens for special permission to matriculate at Göttingen.

[194] Grattan-Guinness, p. 127.

[195] The *Aula* is an auditorium used for official functions of the University.

[196] Grattan-Guinness, p. 128.

in fact would have never entered his head, she added: 'It is not for me but for another person.' Now you see how the mistake came about, and how it is not so easy to do a thing till now supposed to be an exclusively male performance!"[197] Unfortunately, Chisholm had little time to enjoy this humorous mishap, for as she later recalled, " [i]t [was] so late I had to go on my own legs as fast as I could, and of course I lost my way, but after wandering round several triangles and squares I got to the Aula very hot and five minutes late. The pedell comforted me however by telling me the professors were not yet ready, and I had five minutes to cool down and recover my equanimity."[198]

After this nerve-wracking experience, she was led into the examination room to face several other members of the faculty as well as her inquisitors: Klein, the physicist Woldemar Voigt, and the astronomer Wilhelm Schur. Klein began by asking her to tell him about her studies in Cambridge, and as she spoke, he interjected a series of questions. He inquired, for example, about Pascal's hexagon and quizzed her on the properties of cubic curves and surfaces. This part of the oral went very well, since she was, as she put it "most at home" with geometry.[199] Things proceeded less smoothly with differential equations: "I was stupid, and if he had been inclined to badger me I should have come off badly although I really knew the things, but he led me on so gently that I recovered myself and got on better."[200]

Throughout the test, those present talked among themselves and paid only scant attention to the proceedings, until Klein asked a question that captured the audience's attention. "You have in your dissertation had a good deal to do with space of many dimensions," Chisholm recalled him saying. "Now of late we have heard a good deal of curious talk about the fourth dimension, and its connexion with the spirits and ghosts and so on. What is your point of view when you consider our space merely as a part of space of more dimensions?"[201] As this question proved, these oral examinations could range widely in subject matter—from Pascal's hexagon to ghosts—and the candidate needed to be open-minded and alert. Apparently, Chisholm acquitted herself well, for her response to this query effectively ended Klein's portion of the examination.

The ordeal was still far from over; Chisholm had yet to deal with the questions posed by Klein's two colleagues. She managed to get through the physics portion fairly well, although she lost control of her German answering

---

[197] *Ibid.*

[198] *Ibid.*

[199] *Ibid.*

[200] *Ibid.*

[201] *Ibid.* Klein had plenty of experience with this theme. Indeed, he may have even felt a little guilty about the fate of his former Leipzig colleague, the astrophysicist Zöllner, who turned to spiritualism around the time that Klein told him about how space curves become "unknotted" in spaces of higher dimension. See, Felix Klein, *Vorlesungen über die Entwicklung de Mathemak im* 19. *Jahrhundert*, 2 vols. (Berlin: Springer Verlag, 1926–1927), 1:169–170.

a query about weights in a vacuum. The jump to astronomy, however, proved a bit too much in her emotionally spent state: "The first few questions I answered in a very dazed sort of way, fortunately recovering myself presently and answering fairly on planets and comets. I do not think it was more than another quarter of an hour when Schur declared himself satisfied."[202]

The excruciating test over, the rest transpired very rapidly. The *Dekan*[203] asked her to wait a moment in an adjacent room. A bell rang after a brief pause, and she was escorted back in to hear the *Dekan* announce: "Miss Chisholm, the philosophical Faculty is *very satisfied* with your examination: I give you back your dissertation and certificates and inform you that you have made the examination *Magna cum laude.*"[204] Her moment of triumph had arrived, but after such an ordeal she felt as if in a trance:

> I was almost stupified: Klein and Voigt shook hands with me and Schur held the door open as I went out. When I got into the pedell's room I had to tell the pedell, just so as to believe it myself. He was smiling and holding in his hands a card, which he then put into mine, and taking me to the table showed me the most exquisite bouquet of roses and carnations sent me with the card of congratulations by Professor and Mrs. Keilhorn. The next moment Miss Winston arrived and the pedell left us alone together; we used the occasion to execute a war dance of triumph.[205]

Winston stayed on in Göttingen until the following summer, when she, too, satisfied the requirements for the doctorate *magna cum laude.*[206] While in Göttingen, she also wrote a short note on the Riemann *P*-function, which Klein thought sufficiently important to warrant publication in the *Mathematische Annalen.*[207] In a letter to her family back home, Winston expressed a sentiment that might well have come from any number of her compatriots in foreign study. "If I had time," she said, "I should like to write an article on

---

[202] *Ibid.*

[203] The position of *Dekan* of the Philosophical Faculty was filled each year by election within the Faculty. The *Dekan* presided over Faculty meetings and represented the Faculty in all official matters.

[204] Grattan-Guinness, p. 129. Chisholm's emphasis.

[205] *Ibid.*

[206] Her dissertation was printed as Mary F. Winston, *Ueber den Hermiteschen Fall der Laméschen Differentialgleichung* (Göttingen: W. Fr. Kästner, 1897). Following Klein's advice, Winston tried at first to get her dissertation published in the United States. When her efforts met with no success, she sent her study back to Göttingen to have it printed there. See Mary F. Winston to Felix Klein, 28 November 1896, Klein Nachlass XII, NSUB, Göttingen. This letter also indicated that Oskar Bolza had proofread her thesis.

[207] Mary F. Winston, 'Eine Bemerkung zur Theorie der hypergeometrischen Function,' *Mathematische Annalen* 46 (1895):159–160. Winston felt very pleased about this, although she had no illusions about the originality of the paper: "Klein told me himself that he put it in [*Mathematische Annalen*] on account of the importance of the thing itself and not on account of the brilliance of the performance." See Mary F. Winston to her family, 18 October 1894, *SSC*.

'Why do we go to Europe?' to prove that there is no reason to go except to learn to value and *to improve* our own country."[208]

The Kleinian Legacy

Klein founded no school of American mathematics, and yet his influence on the emergent mathematical research community in the United States was far more pervasive than that of Sylvester. First of all, his German students Oskar Bolza and Heinrich Maschke emigrated to the United States and established themselves ultimately at what would become America's premier training ground in research-level mathematics, the University of Chicago. Moreover, no fewer than six of his students went on to become President of the American Mathematical Society, and thirteen served as Vice President (see Table 5.3). None of these Americans, however, entered deeply into the distinctly Kleinian geometric approach to complex analysis that characterized the work of his closest disciples, most notably Robert Fricke, the real author of the four large Klein-Fricke volumes on elliptic modular functions and automorphic function theory.[209] On the contrary, Klein's American students imported only selected portions of their mentor's legacy: for Cole, these were substitution groups and algebraic equations; for Snyder and White, algebraic and projective geometry; for Bôcher and Van Vleck, certain problems arising from potential theory; and for Osgood, Riemannian function theory. Klein expressed some bitterness in later years that the research he and his contemporaries had cultivated, and which had captivated the attention of his American students in Göttingen, had fallen out of fashion. "When I was a student," he reminisced in his lectures on the development of nineteenth-century mathematics, "the Abelian functions were considered—as a consequence of the Jacobian tradition—the indisputable summit of mathematics, and each of us had the understandable ambition to contribute something further here. And now [ca. 1916]? The younger generation hardly knows the Abelian functions anymore."[210] In Klein's view, this had come about because the subject had simply grown too complex. Younger mathematicians, increasingly disenchanted with classical pursuits, had turned to newer fields like axiomatics and set theory, which demanded much less prior study. In a similar vein, Klein also criticized the fascination with group theory that had taken hold in the United States, remarking that "[f]or many sensibilities [this field] is especially attractive, since here also one can work without having to know much other mathematics or without having to combine various ideas with

---

[208] Mary F. Winston to family, 2 December 1894, *SSC* (Winston's emphasis).

[209] Felix Klein and Robert Fricke, *Vorlesungen über die Theorie der elliptischen Modulfunktionen*, 2 vols. (Leipzig: B. G. Teubner, 1890–1892; reprint ed., New York: Johnson Reprint Corporation, 1966); and *Vorlesungen über die Theorie der automorphen Funktionen*, 2 vols. (Leipzig: B. G. Teubner, 1897–1912; reprint ed., New York: Johnson Reprint Corporation, 1965).

[210] Klein, *Vorlesungen über die Entwicklung der Mathematik im 19. Jahrhundert*, 2:312.

one another."[211] As for Klein's own pet projects, Eric Temple Bell actually put his finger on the main source of difficulty when he wrote that "few of Klein's ... contemporaries were willing to assimilate his singularly personal methods. ... Klein's mathematics demanded too much knowledge of too many things for mastery in a reasonable time, and in addition it frequently presupposed a facility in spacial [sic] linguistics beyond the capacities of most mathematicians."[212]

In sum, Klein's influence on his American students had less to do with his specific research programs than it did his general ability to inspire them and to train them to do mathematical research. E. H. Moore may have sensed this when he wrote to Klein in 1904 on the occasion of the University of Chicago's "Festival of Recognition of the Indebtedness of American Universities to the Ideals of German Scholarship." After offering his regrets that Klein had been unable to attend this "successful" and "historic" event,[213] Moore, by then the most prominent American mathematician and the most influential teacher of his generation, offered a sincere expression of his gratitude. "Certainly in the domain of mathematics," he wrote, "German scholars in general and yourself in particular have played, by way of example and counsel and direct and indirect inspiration, quite the leading role in the development of creative mathematics in this country, and on behalf of my colleagues here I wish to express our most grateful recognition and appreciation of our profound debt."[214] To many Americans—not just Moore—Klein represented an emissary of mathematical culture at large, something they very much wanted to transplant to the United States.

---

[211] *Ibid.*, p. 338. Eric Temple Bell was a good deal more blunt in his criticism of this trend in American research in *The Development of Mathematics* (New York: McGraw-Hill Book Company, 1945), pp. 445–446. On this score, however, it should be noted that, with the exception of Cole, none of Klein's American students strayed very far into the camp of the "modernists."

[212] Bell, *The Development of Mathematics* , pp. 511–512.

[213] E. H. Moore to Felix Klein, 23 March 1904, Klein Nachlass X, NSUB, Göttingen.

[214] *Ibid.*

TABLE 5.1
LECTURES DELIVERED IN KLEIN'S SEMINARS, 1881–1896

Winter Semester (WS) 1880–1881: Seminar on Geometry and Function Theory[215]

Irving Stringham, "On Regular Bodies in 4-Dimensional Space"
____, "Groups of Motions of 4-Dimensional Bodies"

Summer Semester (SS) 1886: Seminar on Regular Bodies and Triangle-Functions

Mellen W. Haskell, "Resolvents of the Fourth Degree"
____, "Linear Differential Equations"

WS 1886–1887: Seminar on Group Theory and Algebraic Equations

Mellen W. Haskell, "On the Real Elements of the Curve $\lambda^3\mu+\mu^3\nu+\nu^3\lambda=0$ and its Associated Class Curve"
Heinrich Maschke, "The Group of the 28th-Degree Equation for the Straight Lines on F^3" (two lectures)
____, "Finite Groups of Binary and Ternary Linear Substitutions"
Oskar Bolza, "The Group of the 28th-Degree Equation for the Bitangents of a Plane C_4"

SS 1887: Colloquium on Group Theory and Algebraic Equations

Mellen W. Haskell, "Symmetry Lines on the Projective Riemann Surface of the Curve $\lambda^3\mu + \mu^3\nu + \nu^3\lambda = 0$"
Oskar Bolza, "Representation of the Rational Invariants of 6th-Degree Binary Forms by Theta-Modules"
____, "On the Theta-Modules of Binary Forms that Admit Linear Transformations"
Heinrich Maschke, "On the Group Problem for Burkhardt Modules"
____, "On Ternary and Quaternary Forms and ... Problems Arising from the Trisection of Periodic Functions"

WS 1887–1888: Seminar on Hyperelliptic Functions

Henry Dallas Thompson, "On the Zeros of the Θ-Functions"

SS 1888: Seminar on Hyperelliptic Functions

Henry Seely White, "Applications of Abel's Theorem Following Clebsch"

[215] All of the lectures listed in this table were delivered in German. We have translated the original German titles for the English-speaking reader.

Henry Dallas Thompson, "The Connection Between the Hyperelliptic θ and σ (for $p = 3$) in Various Cross-Cut Systems"
Henry Seely White, "Bitangents of the C_4, etc. Following Clebsch"
Harry W. Tyler, "On the First Chapter of the Book by Clebsch and Gordan"

WS 1888–1889: Seminar on Abelian Functions

Henry Seely White, "Norming Third-Order Integrals of Space Curves"
William Fogg Osgood, "The New Type of Riemann Surfaces"

WS 1889–1890: Seminar on Partial Differential Equations of Physics, on Cyclides and Lamé Functions

Maxime Bôcher, "On Families of Cyclides"

SS 1890: Seminar on Bessel Functions, Spherical Functions, and Hypergeometric Functions

Maxime Bôcher, "Bessel Functions in Astronomy"
——, "Hypergeometric Series in Potential Theory and Perturbation Theory"
Haskell Curry, "Reduction of B to Σ Following Binet"

SS 1891: Seminar on Hypergeometric and Automorphic Functions

James Boyd, "Straight-Lined Triangles given by the Hypergeometric Function"
Edward Burr Van Vleck, "On Rational Transformations of the Hypergeometric Function"

WS 1891–1892: Seminar on Hypergeometric and Lamé Functions

Fabian Franklin, "On Hilbert's Paper on the Theory of Algebraic Forms (*Mathematische Annalen* 36 (1890))"
Edward Burr Van Vleck, "Generalization of the Lamé Polygon Mapping by Hypergeometric η with Complex λ, μ, ν"

WS 1892–1893: Seminar on Various Subjects

James Boyd, "On the Theory of the Schwarz s-Function"
Frederick Shenstone Woods, "Pseudominimal Surfaces, I"
——, "Pseudominimal Surfaces, II"

WS 1893–1894: Seminar on Linear Differential Equations and the P-Function

Frederick Shenstone Woods, "Minimal Surfaces"
Virgil Snyder, "Sphere Geometry"
Mary Frances Winston, "The Connectivity Formulas of the Principal Branches of the P-Function"

SS 1894: Seminar on Linear Differential Equations and Spherical Functions

Mary Frances Winston, "Spherical Functions as Special Cases of the Hypergeometric Functions"
Virgil Snyder, "Vibrations on a Circular-Formed Membrane"

WS 1894–1895: Seminar on the Foundations of Analysis for Functions of a Single Variable

Mary Frances Winston, "On the Concept of a Function"
Isabel Maddison, "On the Concept of a Curve"
Charles A. Noble, "On English Textbooks"
Annie MacKinnon, "Further Examples of Differentiability of Functions"

SS 1895: Seminar on the Foundations of Analysis for Functions of Several Variables

Mary Frances Winston, "Tschebyscheff's General Theory"
Charles A. Noble, "Differential Equations"
Annie MacKinnon, "Calculus of Variations"
Isabel Maddison, "Continuity and Many-Valuedness of $f(x, y)$"

WS 1895-1896: Seminar on Number Theory

Annie MacKinnon, "The Rational Points Associated with the Conic $ax^2 - by^2 = c$"
——, "Classifying Types in the Field $\sqrt{-m}$"

SS 1896: Seminar on Number Theory

Annie MacKinnon, "On Smith's Curve"

TABLE 5.2
DOCTORAL DISERTATIONS WRITTEN BY AMERICANS UNDER KLEIN'S SUPERVISION

1. Henry Burchard Fine (1886) — "On the Singularities of Curves of Double Curvature"

2. Mellen Woodman Haskell (1890) — "Über die zu der Kurve $\lambda^3\mu + \mu^3\nu + \nu^3\lambda = 0$ im projektiven Sinne gehörende mehrfache Überdeckung der Ebene"

3. Maxime Bôcher (1891) — "Ueber die Reihenentwicklungen der Potentialtheorie"

4. Henry Seely White (1891) — "Abelsche Integrale auf singularitätenfreien einfach überdeckten, vollständigen Schnittkurven eines beliebig ausgedehnten Raumes"

5. Henry Dallas Thompson (1892) — "Hyperelliptische Schnittsysteme und Zusammenordnung der algebraischen und transzendenten Thetacharakteristiken"

6. Edward Burr Van Vleck (1893) — "Zur Kettenbruchentwicklung Laméscher und ähnlicher Integrale"

7. Frederick Shenstone Woods (1895) — "Ueber Pseudominimalflächen"

8. Virgil Snyder (1895) — "Ueber die linearen Komplexe der Lieschen Kugelgeometrie"

9. Mary Frances Winston (1897) — "Ueber den Hermiteschen Fall der Laméschen Differentialgleichung"

Table 5.3
Students of Klein at the American Mathematical Society

Presidents

1. William Fogg Osgood (1905–1906)
2. Henry Seely White (1907–1908)
3. Maxime Bôcher (1909–1910)
4. Henry Burchard Fine (1911–1912)
5. Edward Burr Van Vleck (1913–1914)
6. Virgil Snyder (1927–1928)

Vice Presidents

1. Henry Burchard Fine (1892–1893)
2. Henry Seely White (1901)
3. Maxime Bôcher (1902)
4. William Fogg Osgood (1903)
5. Alexander Ziwet (1903)
6. Oskar Bolza (1904)
7. Washington Irving Stringham (1906)
8. Heinrich Maschke (1907)
9. Edward Burr Van Vleck (1909)
10. Mellen Woodman Haskell (1913)
11. Virgil Snyder (1916)
12. Frank Nelson Cole (1921)
13. Harry Walter Tyler (1923)

Chapter 6
Changes on the Horizon

Essentially forced abroad to pursue advanced studies, Klein's American students returned from Germany to find a rapidly changing situation in American higher education. By the early 1890s, a number of schools had come to embrace the German notion that research and teaching represented complementary facets of the university professor's calling. As a consequence, these American institutions sought to engage academics who had earned doctoral degrees for pieces of original research, and they made research productivity a key criterion in promoting and otherwise rewarding the members of their faculties. Thus, unlike Sylvester's students, who had emerged from the heady atmosphere of the Johns Hopkins of the late 1870s and early 1880s only to find themselves in an environment largely unsupportive of their newly acquired research goals, Klein's students came back into an academic climate considerably more conducive to such aims.

Just as significant for the future course of American mathematics, however, this *Wanderlust* generation seized the fresh opportunities afforded them at home. They created brand-new graduate programs and strengthened old ones; they published their own work and promoted that of others; and they established new venues for conveying mathematical information locally and regionally. Through these and other activities, this group effectively reformed the structure and defined the function of what would become a mathematical *profession*. At the same time, they laid the groundwork for the creation of a national organization which would represent the interests of the newly emergent community of American research mathematicians. In the vanguard of these changes, universities like Cornell and Clark, but especially the University of Chicago, set new standards for higher education in the United States.

Changes on the American Scene after 1876

When the Johns Hopkins University opened in the fall of 1876, it stood, as we have seen, as the first American university expressly dedicated to the advancement, rather than the mere diffusion, of knowledge. As such, it took as a primary mission the training of future researchers through graduate education (while not neglecting education at the lower, undergraduate level). The production of new and original contributions to research thus represented an institutional goal set for both the faculty and the advanced students. This emphasis on graduate studies and research marked a significant departure from the more traditional, collegiate notion of higher education, namely, the disciplining of the mental and moral faculties of young minds through a liberal education.[1]

Despite the novelty of the Hopkins approach, some older institutions had already introduced modest graduate programs into their curricula as early as mid-century. As we saw in Chapter 1, Yale had established its Department of Philosophy and the Arts in 1847 for nominally graduate studies, spurred by "(1) the demand by graduates and others for instruction beyond that given in the college; (2) the existence of endowed scholarships for graduate instruction and the fact that they would be improved by providing regular instruction for their holders; (3) the desire to avoid crowding the undergraduate course and interfering with the tutoring of the mind; [and] (4) the good example graduate students would provide for the undergraduates."[2] In 1861, this department had produced the first three Ph.D.'s earned at an American institution of higher education: one in philosophy and psychology, one in classical languages and literature, and one in physics.[3] Ten years later, the University of Pennsylvania became only the second American school to award the doctoral degree, and Harvard followed in 1872 when it simultaneously launched its graduate school officially.[4]

Thus, graduate education hardly thrived in the United States prior to the founding of the Johns Hopkins. Only forty-four students nationwide pursued advanced studies in 1870, and these at only a handful of schools including Yale, Harvard, Michigan, and Princeton.[5] By 1900, however, as many as 5,668 students across the country were engaged in graduate work, a phenom-

[1]The relationship between American colleges of the earlier half of the nineteenth century and American universities as they developed over the century's second half is complicated. As Colin B. Burke has pointed out in his book, *American Collegiate Population: A Test of the Traditional View* (New York and London: New York University Press, 1982), more traditional scholarship has tended to denigrate the former while exalting the latter. The present study, which concerns the development of mathematics at the research level, necessarily focuses on the graduate setting in the United States and should not be seen as part of this debate.

[2]Brooks Mather Kelley, *Yale: A History* (New Haven: Yale University Press, 1974), p. 182. At this point, the graduate instruction in mathematics was given by Hubert Newton's predecessor, Anthony D. Stanley, and consisted of calculus or analytical mechanics. This certainly fell far short of the course of study Benjamin Peirce offered at Harvard at essentially the same time. Recall the discussion in Chapter 1.

[3]*Ibid*, p. 185. Yale only adopted the designation "university" in 1887. Many other older institutions also changed their names to "university" during the latter half of the nineteenth century.

[4]*Ibid.*, pp. 257–258; and Samuel Eliot Morrison, ed., *The Development of Harvard University Since the Inauguration of President Eliot* 1869-1929 (Cambridge: Harvard University Press, 1930), p. 452.

[5]Kelley, p. 257.

enal increase in a span of only thirty years.[6] According to Charles Eliot, Harvard's President from 1869 to 1909, this rapid development resulted in large part from the example set by the Hopkins Trustees, administration, faculty, and students. Speaking at the celebration honoring the Baltimore institution on its twenty-fifth anniversary in 1901, Eliot offered this assessment of its role in the history of graduate education:

> your first achievement here, with the help of your colleagues, your students, and your trustees, has been ... the creation of a graduate school, which has not only been itself a strong and potent school, but which has lifted every other university in the country in its departments of arts and sciences. I want to testify that the graduate school of Harvard University ... did not thrive, until the example of Johns Hopkins forced our faculty to put their strength into the development of our instruction for graduates. And what was true of Harvard was true for every other university in the land which aspired to create an advanced school of arts and sciences.[7]

While Eliot may have been justified in characterizing Hopkins as initial the spark which ignited the late nineteenth-century American movement in higher education, it took more than one spark to light the fire which had spread across the United States by the turn of the century.

Among the East Coast schools, Eliot's own institution, Harvard, assumed the leading role in these developments during his forty-year presidency there. As noted in Chapter 1, a mid-century policy change at Harvard had allowed Benjamin Peirce to focus his teaching activities more tightly on courses aimed at advanced students. His pedagogical efforts, sustained over several decades, had resulted in a small but solid mathematics program, perhaps the most successful one at the undergraduate level in the country by 1880. His son, James Mills Peirce, a distant relative, Benjamin Osgood Peirce, and William Byerly all elected to pursue the course of studies Peirce had fashioned and all went on to assume positions on the Harvard faculty. Of these students, B. O. Peirce became an active researcher specializing in mathematical astronomy, while J. M. Peirce and Byerly devoted themselves less to the production of original research and more to the preparation of advanced textbooks aimed at the continued improvement and up-grading of the Harvard curriculum.[8]

---

[6]Laurence R. Veysey, *The Emergence of the American University* (Chicago: University of Chicago Press, 1965), p. 269.

[7]*Johns Hopkins University Celebration of the Twenty-Fifth Anniversary of the Founding of the University and Inauguration of Ira Remsen, LL.D. As President of the University* (Baltimore: Johns Hopkins University Press, 1902), p. 105.

[8]For more on the history of mathematics at Harvard, consult Garrett Birkhoff, "Mathematics at Harvard, 1836–1944," in *A Century of Mathematics in America—Part* II, ed. Peter Duren *et al.* (Providence: American Mathematical Society, 1989), pp. 3–58.

The Harvard mathematical program and the initiatives taken by Benjamin Peirce and his protégés clearly met with real, if limited, success. Many figures important in the development of mathematics in America took undergraduate (and, in some cases, graduate) degrees there: James Oliver, William Story, Irving Stringham, Mellen Haskell, Frank Cole, William Osgood, and Maxime Bôcher, among others. When Harvard acquired two of Klein's strongest American students, Osgood and Bôcher, in the early 1890s, it also secured the talent necessary to launch a serious graduate program in mathematics. During the next decade, Osgood and Bôcher steadily guided Harvard's development into a major center for research in both real and complex analysis. The university's preeminent position in analysis was fully assured in 1912 with the appointment of George D. Birkhoff, himself a product of Harvard's undergraduate program.

If Osgood and Bôcher returned from Germany to an atmosphere in Cambridge congenial to their newly acquired research goals, two other Klein students, Henry Fine and H. D. Thompson, settled into a very different environment in Princeton. There, the research ethic they had acquired abroad would only slowly infiltrate the institution's educational philosophy. Impetus in this direction came in 1905 when University President Woodrow Wilson set up a preceptorial system ostensibly to infuse the faculty with a group of enthusiastic and vital young scholars and thereby spark the *under*graduate program. The Princeton preceptor, reminiscent of the Oxbridge tutor, was charged with meeting regularly with a small group of students to discuss material of mutual interest. As Chair of the Mathematics Department from 1904 to 1928, Fine capitalized on Wilson's new directives to augment the faculty with fresh Ph.D.'s of real research promise. As we shall sketch in Chapter 10, his strategy built the foundation for what would become, by the 1920s, Princeton's strong research tradition especially in the fields of topology and differential geometry.[9]

While such changes animated East Coast institutions like Harvard and Princeton, similar transformations were taking place at a number of state-funded schools in the Midwest and elsewhere. In 1862, the Morrill Act had made federal support—in the form of so-called land grants—available to states and allowed for either the creation of new *land-grant* universities or

[9]Wilson's two predecessors, James McCosh and Francis Patton, fostered an atmosphere of discipline and religious conservatism at Princeton which retarded its transformation into a true university. Nevertheless, the first tentative steps toward "university status" were made during McCosh's regime. See, for example, Veysey, pp. 21–56; and Frederick Rudolph, *The American College and University*: *A History* (New York: Alfred A. Knopf, 1962), pp. 297–300. On subsequent developments in the Princeton Mathematics Department, see William Aspray, "The Emergence of Princeton as a World Center for Mathematical Research, 1896–1939," in *History and Philosophy of Modern Mathematics*, ed. William Aspray and Philip Kitcher, Minnesota Studies in the Philosophy of Science, vol. 11 (Minneapolis: University of Minnesota Press, 1988), pp. 346–366 and reprinted in *A Century of Mathematics in America—Part* II, pp. 195–215. On the preceptorial system, consult Woodrow Wilson, "The Preceptorial System at Princeton," *Educational Review* 39 (1910):385–390.

the enhancement of preexistent ones, on the condition that these institutions incorporate agricultural and mechanical training into their curricula.[10] This infusion of funds helped to strengthen certain state universities from coast to coast, and by the 1880s these revitalized institutions had begun to make their presence felt in American higher education in general and in research-level mathematics in particular through judicious hiring and program-building. In Wisconsin, for instance, the state legislated the creation of the University of Wisconsin as early as 1848 and subsequently devoted all of its land-grant money to the institution's further development. Wisconsin melded strong programs in agriculture and engineering to its preexistent collegiate curriculum, in keeping with the spirit of utility and public good inherent in this federal support. Assuming the reins of leadership in 1903, the geologist Charles R. Van Hise "wished to fuse scientific research with the earlier aim of practical training" and succeeded in making his university "a center of revivified utilitarian faith and practice."[11] Klein's student E. B. Van Vleck contributed to this revivification, at least relative to mathematics, when he brought his German training and commitment to research to Wisconsin first in 1893–1895 and then permanently in 1906 during Van Hise's regime. Under Van Vleck, Wisconsin's Mathematics Department eventually emerged as an important center of activity within the American mathematical research community.

Unlike Wisconsin, other states chose to devote their land-grant funds to the establishment of essentially new institutions. The Massachusetts legislature, for example, chartered its school for engineering and applied science in 1861, and by the time the Massachusetts Institute of Technology (MIT) opened in 1865, it had also benefited from federal land-grant funds. Under the leadership of physicist and geologist William Barton Rogers, from 1865 to 1872 and then again from 1878 to 1882, MIT embraced not only the utilitarian tenets of a school of applied science but also the notion of the advancement of scientific knowledge. It was into this atmosphere that F. S. Woods and H. W. Tyler brought their German doctorates in the 1890s.[12]

On the West Coast, the California state government used its Morrill Act

[10]For more on the Morrill Act and the federal government's support of science in general, see A. Hunter Dupree, *Science in the Federal Government: A History of Policies and Activities* (Baltimore: Johns Hopkins University Press, 1986). The notion of governmental support for higher education was hardly new. In colonial times and into the nineteenth century, the now common distinction between public and private educational institutions had yet to be drawn. So, for example, the General Court of Massachusetts came to the financial aid of Harvard College over one hundred times prior to 1789 and continued to allocate money to it consistently until 1823. See, among many other possible sources, Rudolph, pp. 185–188.

[11]Veysey, pp. 104–105.

[12]During the Tyler years, C. L. E. Moore, H. B. Phillips, F. L. Hitchcock, Norbert Wiener, and Dirk Struik all joined the MIT mathematical staff. See Dirk Struik, "The MIT Department of Mathematics During Its First Seventy-Five Years: Some Recollections," in *A Century of Mathematics in America—Part* III, ed. Peter Duren *et al.* (Providence: American Mathematical Society, 1989), pp. 163–178.

money to create the University of California at Berkeley in 1868. After surviving a rather tumultuous infancy during which Daniel Coit Gilman's research-oriented ideals clashed with the utilitarian and populist objectives of the politically powerful California Grange, this university, too, grew into a vocation- yet research-oriented school. Its educational mission had developed sufficiently by 1882 to accommodate the mathematical ideals Irving Stringham brought there fresh from his training under Sylvester in Baltimore and Klein at Leipzig. The young mathematician nevertheless realized that he faced an uphill battle. In a letter to Klein written in 1888, he wistfully observed that "[t]he plants of intellectual culture grow but slowly, and on new, raw ground like that of California they can hardly flourish without very great efforts. I impatiently await the time when it will be possible to bring some important researchers in the field of mathematics to the California coast."[13] Eight years later, Stringham did succeed in bringing to Berkeley, if not a major figure at least another of Klein's students, Mellen Haskell.

Whereas Wisconsin, MIT, and Berkeley partially utilized their federal monies to promote and encourage original research, other state universities made their marks without an infusion of land-grant funds. As a case in point, the University of Michigan had emphasized research on and off since the presidency of Henry P. Tappan began in 1852. Although Tappan's persistent attempts in the 1850s to model his university on its German counterparts resulted in his ouster, the academic climate had moderated somewhat by the time James B. Angell took over the position in 1871, and research and studies of more immediately practical concern were allowed to coexist. This made it possible not only for F. N. Cole to bring the "new mathematics" there from 1888 to 1895 but also for Alexander Ziwet to find a congenial climate for work in the applied mathematics and engineering tradition he had absorbed at the polytechnical institute in Karlsruhe and at the United States Coast and Geodetic Survey.[14] At Michigan and other state-supported universities at the end of the nineteenth and into the twentieth century, mathematical research ultimately found a fairly comfortable home.[15]

The mathematical cause also got an important boost thanks to the efforts of a handful of mathematical enthusiasts associated with Columbia College in New York City. Like Princeton, Columbia was somewhat slower to move into the ranks of the leading research-oriented institutions in the United States. Its President, Frederick A. P. Barnard, inspired by his own obser-

---

[13]Irving Stringham to Felix Klein, 19 August 1888, Klein Nachlass XI, Niedersächsische Staats- und Universitätsbibliothek, Göttingen (hereinafter abbreviated NSUB, Göttingen).

[14]For further information on the institutional development of the University of Michigan in the context of the emergence of the American university, see Veysey, pp. 100–102. Ziwet would also study later under Felix Klein.

[15]For a glimpse at the subsequent history of research mathematics at Berkeley, see Robin Rider, "An Opportune Time: Griffin C. Evans and Mathematics at Berkeley," in *A Century of Mathematics in America—Part* II, pp. 283–302.

vations in Europe and by Eliot's actions at Harvard, tried to create a true university there during his regime from 1864 to 1889, but a hostile Board of Trustees continually blocked his efforts. Columbia did not succeed in making this transition until after the official founding of its graduate school in the 1890s.[16] The imminent shift in institutional direction, however, was strongly foreshadowed by events which took place in its Department of Mathematics in 1888. That year, the New York (later American) Mathematical Society (NYMS and AMS, respectively) officially came into existence.

In 1887 at the urging of his professor John H. Van Amringe, the twenty-two-year-old Columbia College student Thomas Fiske had embarked on a six-month mathematical pilgrimage to Cambridge University.[17] Armed with letters of introduction to Arthur Cayley, his disciple, Andrew R. Forsyth, and their Cambridge colleague, James W. L. Glaisher, Fiske had entered not only the Cambridge lecture hall but also the broader sphere of British mathematics. In particular, he had attended the regular meetings of the London Mathematical Society (founded in 1865) as Glaisher's guest and had come away with lasting impressions of the importance of the shared mathematical experience. As Fiske himself explained, "[o]n my return to New York I was filled with the thought that there should be a stronger feeling of comradeship among those interested in mathematics, and I proposed to my classmates and friendly rivals, [Harold] Jacoby and [Edward] Stabler, that we should try to organize a local mathematical Society."[18]

By November of 1888, the three graduate students had drafted and distributed a proposition which would come to represent the birth certificate of the American Mathematical Society. It read:

> It is proposed by some recent students of the graduate school of Columbia College to establish a mathematical society for the purpose of preserving, supplementing, and utilizing the results of their mathematical studies. It is believed that the meetings of the society may be rendered interesting by the discussion of mathematical subjects, the criticism of current mathematical literature, and the

[16]Veysey, pp. 99–100.

[17]For a detailed account of the founding and history of the first fifty years of the American Mathematical Society, see Raymond C. Archibald, *A Semicentennial History of the American Mathematical Society* 1888–1938 (New York: American Mathematical Society, 1938; reprint ed., New York: Arno Press, 1980). We do not attempt to recount the full story of the NYMS and its development into the AMS here, since Archibald documented that process so fully in his book.

[18]Archibald, *Semicentennial History*, p. 4. Jacoby earned his Columbia Ph.D. in 1896 in astronomy and served on the Columbia faculty in various capacities beginning in 1894. Stabler finished his doctorate in 1888 and held an actuarial position at the Manhattan Life Insurance Company from 1894 through 1898. From 1902 through 1931 when he retired, he was employed by the Century Leather Company and the U. S. Leather Company. See James McKeen Cattell, *American Men of Science: A Biographical Dictionary*, 1st ed. (New York: Science, Press, 1906) and subsequent editions.

> solution of problems proposed by its members and correspondents. It is also intended that original investigations to which members may be led shall be brought before the society at its meetings. It is hoped that this society may elicit your interest and be favored with your advice. It is earnestly desired that you will assist in its organization by being present at its first meeting hereby called for Thanksgiving Day at 10:00 A.M.[19]

Only three people other than the three organizers expressed interest in the undertaking through their attendance at this first meeting: the Columbia mathematics professor John Van Amringe, his colleague in astronomy and geodesy John K. Rees, and one of their graduate students James Maclay. A year later, the resultant New York Mathematical Society's (NYMS) membership had increased to a mere sixteen.

Hardly a large-scale operation, the Society decided to expand its sphere of influence in 1891 through the publication of a bulletin modeled on the British journal *Messenger of Mathematics.* In order to help insure the success of such a venture, Fiske, as Society Secretary, worked up a mailing list of college and university professors as well as others interested in mathematics and sent them invitations both to join the Society and to subscribe to its bulletin. By the end of 1891, the membership had burgeoned to over two hundred, and the first three numbers of the *Bulletin of the New York Mathematical Society* had appeared in print, largely as a result of Fiske's efforts. The organization had indeed begun to reach America's mathematical public.[20] Four years later, Frank Nelson Cole was appointed to a position on the Columbia faculty and soon assumed the key offices of Secretary and *Bulletin* editor of what had become in 1894 the *American* Mathematical Society. By 1902, David Eugene Smith, a historian of mathematics and professor at Columbia's associated Teachers' College, had joined forces with Cole in the AMS to serve as the Society's first Librarian. Smith and Cole remained in these offices until 1920, thereby insuring the stability of the Society's operations and Columbia's role as the focal point of the AMS well into the twentieth century. The organizational activities of Fiske and others at an evolving Columbia University thus resulted in a full-fledged national society, which represented the interests of a fast-growing community of research-oriented mathematicians. (See Chapter 9 for further details on its growth as a national organization.)

---

[19] *Ibid.*

[20] *Ibid.*, p. 5. One student who would thrive in this atmosphere at the end of the decade was Edward Kasner. An undergraduate at the City College of New York, Kasner took a Master's degree and doctorate at Columbia in 1897 and 1899, respectively. From 1900 to 1910, he served on the faculty at Barnard and, in 1910, became a professor at Columbia. His principal area of mathematical research was in differential geometry. He became a member of the National Academy of Sciences in 1917. On his life and work, consult Jesse Douglas, "Edward Kasner," Biographical Memoirs: National Academy of Sciences 31 (New York: Columbia University Press, 1958), pp. 180–209.

Mathematics at Cornell and Clark Universities

In spite of the sometimes dramatic changes taking place at the state and Ivy League universities during the last quarter of the nineteenth century, it was the handful of schools created from the private fortunes of several American entrepreneurs which most enhanced research-level mathematics in the United States. According to Laurence Veysey in his history of higher education in the United States, one of those, Cornell University, "was the first major university in America, discounting a few tentative experiments, to be created on a reformed basis from the ground up. Its founding inaugurated an era in private educational philanthropy."[21] Chartered by the New York State legislature in 1865, Cornell had come into existence thanks both to Morrill Act funds and to $500,000 of the money Ezra Cornell had made on his way to becoming the controlling stockholder in Western Union. With historian and educational reformer Andrew D. White at its helm from 1867 to 1885, "Cornell contained in its outlook not only the practical vocationalism of the land-grant idea but also the science, technology, and spirit of scholarship of the new university movement."[22] Although not convinced that the United States of the 1870s was ready to support education at the graduate level, White nevertheless moved Cornell in that direction during his regime, and his successor, Charles Kendall Adams, maintained this orientation throughout his seven-year tenure in office.

In mathematics, White hired James Edward Oliver as an Assistant Professor in 1871. Oliver had pursued Benjamin Peirce's Lawrence School course of study prior to assuming a post at the Nautical Almanac Office in Cambridge and so had received essentially the best applied mathematical training available on American shores. A quiet and unpretentious man, Oliver personally tended to do mathematical research for his own edification and enjoyment rather than for the reputation gained through publication, but he nevertheless valued and appreciated the research ethic which had begun to infiltrate the nation's institutions of higher education by the 1880s. As a result, he worked steadily to build up not only a strong faculty but also a respectable graduate program at Cornell during his tenure as Departmental Chair there from 1873 to his death in 1895.

In his annual report to the President of the University for 1886–1887, for example, Oliver stressed the importance of research and publication, while outlining some of the difficulties he and his colleagues—Lucien Augustus Wait, George William Jones, James McMahon, and Arthur Safford Hathaway—experienced in trying to accomplish these nobler academic ends. He assured Adams that in the Department of Mathematics, "[w]e are not unmindful of the fact that by publishing more, we could help to strengthen the university, and that we ought to do so if it were possible. Indeed, every one

[21] Veysey, pp. 99–100.

[22] Rudolph, p. 266

of us five is now preparing work for publication or expects to be doing so this summer."[23] Still, he reminded the President somewhat chidingly that "such work progresses very slowly because the more immediate duties of each day leave us so little of that freshness without which good theoretical work can not [sic] be done."[24]

On average, the mathematicians at Cornell taught from seventeen to twenty hours each week during the 1886–1887 academic year, and although their numbers increased by two—Duane Studley and George Egbert Fisher—the following year, Oliver stated in no uncertain terms that his "department's whole teaching force, composed of only about one-eleventh of all active resident professors, has to do about one-ninth of all the teaching in the university."[25] The academic environment at Cornell, unlike that at the Johns Hopkins, did not facilitate the realization of the dual goals of teaching *and* research, which increasingly came to define the emergent scientific and mathematical professions in the last decades of the nineteenth century. Cornell's mathematicians were expected to teach long hours first and to do as much research as their remaining time allowed.

In spite of their already heavy teaching load at the undergraduate level, Oliver and his colleagues also aspired to develop and maintain a viable graduate program. The Chair reported proudly on his department's achievements toward the accomplishment of its self-imposed mission in 1887–1888. As he told Adams,

> eleven graduate students have taken more or less of their work with us. Allowing for such as were partly in other departments or remained but part of the year, we find that the mathematical department has had about one-seventh of all the graduate work in the university. This would seem to be our full share of this *desirable kind of teaching*, when it is considered that the higher mathematics is difficult, abstract, and hard to popularize; [and] that of course we can not [sic] attract students to it by laboratories and large collections (except of books), nor by the prospect of lucrative industrial applications.[26]

The course of study which these graduate students followed consciously exceeded the level of sophistication of, for example, the successful series of

---

[23]Annual Report to the President of Cornell University for 1886–1887, as quoted in Florian Cajori, *The Teaching and History of Mathematics in the United States* (Washington: Government Printing Office, 1890), p. 180. Karen Hunger Parshall discussed Cajori's book and the telling glimpse it provides of American mathematics in 1890 in "A Century-Old Snapshot of American Mathematics," *The Mathematical Intelligencer* 12 (3) (1990):7–11.

[24]Annual Report to the President of Cornell University for 1886–1887, as quoted in Cajori, p. 180.

[25]Annual Report to the President of Cornell University for 1887–1888, as quoted in Cajori, p. 186.

[26]*Ibid.* Our emphasis.

college-level mathematics textbooks Oliver co-authored with Wait and Jones,[27] but it could not rival contemporaneous programs abroad. Cornell's advanced students sampled from both pure and applied mathematics, ranging, on paper at least, from the general theory of forms (given by McMahon) and number theory (taught by Hathaway) to mathematical optics and rational dynamics (offered by Oliver and Wait). As their course texts, they read principally from the best British books of the day but occasionally also from some of the modern-day Continental classics.[28] It was in his continuing efforts both to supplement further his program's offerings and to increase his own awareness of advances in mathematics that Oliver spent the 1889–1890 academic year studying in Göttingen with Klein. In 1894, Oliver even tried to import some of this German scholarship to Ithaca in the person of Klein's student Ernst Ritter. Unfortunately, for the Mathematics Department, however, Ritter developed a severe fever during his Atlantic crossing and died on Ellis Island a few days after landing in the United States.[29] The next year, 1895, brought more tragedy as well as triumph to the school. Its guiding mathematical spirit, Oliver, died, but Klein's student Virgil Snyder returned to his *alma mater* and, in so doing, solidified the graduate program Oliver had worked to establish there.

As this sequence of events suggests, a viable graduate program in mathematics developed only slowly and hesitantly at Cornell, a university which struggled early on to define its respective commitments to teaching and research. At Clark University, another institution born of private endowment, a very different situation obtained, however. This school opened in Worcester, Massachusetts in 1889 and initially supported *only* programs at the graduate and post-graduate levels. It thus sounded its commitment to advanced teaching and research from the start and immediately set up strong programs in mathematics as well as in other disciplines.

Jonas Gilman Clark, originally a New England manufacturer of chairs and tinware, had struck it rich first as an importer to and then as a furniture manufacturer in the California of the gold rush years. Following a nervous breakdown in 1860, however, Clark liquidated his lucrative West Coast businesses, reinvested partially in San Francisco real estate, and eventually moved

---

[27] James E. Oliver, Lucien Wait, and George Jones, *A Treatise on Trigonometry*, 4th ed. (Ithaca: G. W. Jones, 1890); and *A Treatise on Algebra*, 2nd ed. (New York: Dudley F. Finch, 1887).

[28] For example, Oliver listed the following as among the course texts adopted since the 1870s: George Salmon, *Lessons Introductory to the Modern Higher Algebra*, 4th ed. (Dublin: Hodges, Figgis, & Co., 1885); Thomas Muir, *The Theory of Determinants with Graduated Sets of Exercises for Use in Colleges and Schools* (London: Macmillan & Co., 1885); Georges Halphen, *Traité des Fonctions élliptiques et de leurs Applications*, 3 vols. (Paris: Gauthier-Villars, 1886–1891); and Peter Lejeune-Dirichlet, *Vorlesungen über Zahlentheorie*, ed. Richard Dedekind (Braunschweig: F. Vieweg und Sohn, 1880). See Cajori, pp. 184–185.

[29] See *Bulletin of the American Mathematical Society* 2 (1895):22; and Felix Klein, "Ernst Ritter," *Jahresbericht der deutschen Mathematiker-Vereinigung* 4 (1894–1895):52–54.

back East to New York City. He gradually built up real estate holdings there before branching out into the promising but less costly Worcester market.

Although the management of his substantial fortune occupied a large part of his time after his return from the West Coast, Clark nevertheless began to cultivate seriously an interest in higher education both at home and abroad. He gathered information on the domestic scene through books, shorter publications, and even interviews with the graduates of various American schools. Furthermore, he visited and studied the organization and goals of some of the foremost foreign institutions while on European tours with his wife. Motivated perhaps by the example of his California friend Leland Stanford, Clark ultimately decided to make higher education more than a hobby and to found his own university. By 1887, the Massachusetts legislature had signed the act of incorporation for "an institution for the promotion of education and investigation in science, literature and art, to be called Clark University" and to be located in that state in the city of Worcester.[30]

From the beginning, Clark conceived of his new university as something "higher than Harvard" with a mission shaped by teaching as well as by original research.[31] In his first meeting with the Board of Trustees, he put forth the French and Prussian universities as well as Cornell and Hopkins as examples of institutions worthy of emulation, while effectively dismissing the British universities as anemic manifestations of the union of Church and State. The university he envisioned would start out as a four-year program for undergraduates built from his $1,000,000 bequest, together with matching funds from the broader community. After four years, with the graduation of the first class and the accumulation of the necessary additional capital, it would grow into a full-fledged university with a complete graduate component.

Clark clearly had in mind a new institution which would, in time, rival the Johns Hopkins, but his largely Harvard-educated Board feared that such a strong school in Worcester would detract from their *alma mater* nearby. They thus counseled him to limit the new undertaking in scope by concentrating on just one area of study, and that at the graduate level. In this way, they believed, Clark University would complement—rather than weaken—Harvard, although they did not use this rationale in presenting their ideas to the institution's benefactor. By agreeing with this plan in February of 1888, Clark unwittingly fell prey to the Trustees' hidden agenda and undermined not only his own conception of the school but also his authority over the Board he had assembled. This internal tension only heightened after April

[30]William A. Koelsch, *Clark University* 1887–1987: *A Narrative History* (Worcester: Clark University Press, 1987), pp. 8–9. The present account of the early history of Clark University follows Koelsch's treatment as found on pp. 1–40.

[31]*Ibid.*, p. 9. Many of the standard sources on the history of higher education claim that Clark initially planned to start a college which would provide a relatively inexpensive education for boys from less privileged households. Koelsch's research, based largely on archival sources, uncovered no basis for this conclusion.

of 1888 when the Hopkins psychologist G. Stanley Hall accepted the Clark presidency.

Clark and Hall saw eye to eye on the University's overarching philosophy. Both wanted to create an important educational force in the United States which would challenge the best of the European schools. Both also set original research as one of the highest priorities for the faculty and students. The two men differed, however, in their conception of the optimal way to realize these goals. Hall agreed with the Trustees, holding that Clark University, unlike the Johns Hopkins, should begin solely as a graduate school.[32] He convinced Clark that the undergraduate component should only come into being after the first Ph.D.'s graduated, thereby providing a ready instructional force. Unfortunately, Hall really had little commitment to this plan, favoring and consistently working for a purely graduate-level program. Worse still, the citizenry of Worcester failed totally to meet Clark's expectations of financial support for their city's new university. While Hall and the Trustees believed that Clark would completely bankroll the venture, the businessman awaited a monetary response from the community. These differing expectations remained unreconciled when Clark University opened in October of 1889 with its graduate departments in mathematics, physics, chemistry, biology, and psychology.

Irrespective of the internal tensions between the benefactor, the Trustees, and the President, Hall had succeeded in attracting to the University some three dozen students taught by a staff of nineteen spread over five departments.[33] In mathematics, Hall, like Gilman following Sylvester's resignation in 1883, had initially tried to secure Felix Klein as his first Professor of Mathematics. He interrupted his European faculty-finding tour to report to Jonas Clark on 14 November, 1888 that he

> [had] learned on all sides that Professor Klein, of whom we have often spoken as about the very best mathematician in Europe, is widely so considered here by those experts most competent to judge. I lately spent several hours with him

[32]For the story of Hall's life, including a discussion of his educational philosophy, see G. Stanley Hall, *Life and Confessions of a Psychologist* (New York: Appleton-Century-Crofts, Inc., 1923); and Dorothy Ross, *G. Stanley Hall: The Psychologist as Prophet* (Chicago: University of Chicago Press, 1972), pp. 186–230. As Veysey pointed out, Clark was not the only institution of higher education on American shores devoted solely to graduate training. The Catholic University of America, founded in Washington, D. C. in 1887, shared this distinction (p. 166). For more detail on the role of Clark University in the history of graduate education in America, see W. Carson Ryan, *Studies in Early Graduate Education: The Johns Hopkins, Clark University, The University of Chicago* (New York: Carnegie Foundation for the Advancement of Teaching, 1939), pp. 47–90; Veysey, pp. 165–171; and Edmund C. Sanford, *A Sketch of the History of Clark University* (Worcester: Clark University Press, 1923).

[33]These numbers were culled from Louis N. Wilson, *Clark University Directory of Alumni, Faculty and Students* (Worcester: Clark University Press, 1915). Thirty-two students in Wilson's directory are listed as having begun at Clark in 1889. The faculty figure includes those in both permanent and nonpermanent ranks. Nine held the rank of Assistant Professor or above.

> talking about the possibility of his joining us at Worcester. He is inclined to come if he could have $5000 per year, which was offered him at Baltimore ... .
>
> ... He told me he was chiefly attracted by the opportunity of doing only very advanced work for a very few men, with whom he could carry on his researches.[34]

As before with the Johns Hopkins, however, negotiations between Klein and the new university in Worcester came to naught in the end. Whereas in 1883 Klein had yet to recover fully from the strains of his mathematical encounter with Poincaré and seemed genuinely tempted by the Hopkins offer, in 1888, he had recently moved to Göttingen and was intent on galvanizing his position there. Thus, rather than jeopardizing his political position in Germany by seriously entertaining an outside offer, Klein hoped to serve the Prussian Ministry as a sort of unofficial traveling cultural emissary. It soon became clear in negotiations over the latter possibility with ministerial official Friedrich Althoff that the Ministry actually favored having foreign students come to Klein rather than having him journey abroad, so after another couple of months of discussion, Klein decided to remain in Göttingen.[35] Forced to look elsewhere for his mathematician, Hall settled on his former Hopkins colleague,William E. Story, appointing him Professor and Acting Head of the Department in 1889.[36]

By the time the University opened later that fall, Story had been joined by two other Hopkins mathematicians: his own graduate student and recent Ph.D., Henry Taber, and the visiting German mathematician and student of Felix Klein, Oskar Bolza. The mathematics faculty expanded further during the next two years, picking up Henry White and the Berlin- and Paris-trained Frenchman, Joseph de Perott. Together this team covered many of the areas of then current interest in mathematical research, with Story lecturing on geometrical topics, Bolza treating the calculus of variations and elliptic function theory, Taber dealing particularly with the theory of hypercomplex number systems (now called algebras), White handling geometry and invariant theory, and de Perott concentrating on number theory. While these five men comprised undoubtedly the strongest Department of Mathematics yet organized in the United States, their potential as a collective entity ultimately

[34]G. Stanley Hall to Jonas G. Clark, 14 November 1888, in N. Orwin Rush, ed., *Letters of G. Stanley Hall to Jonas Gilman Clark* (Worcester: Clark University Library, 1948), p. 21. A portion of this quotation appeared in Roger Cooke and V. Frederick Rickey, "William E. Story of Hopkins and Clark," in *A Century of Mathematics in America—Part* III, pp. 29–76 on pp. 47–48. See, also, Constance Reid, "The Road Not Taken: A Footnote in the History of Mathematics," *The Mathematical Intelligencer* 1 (1) (1978):21–23.

[35]For the substance of the discussions between Friedrich Althoff and Felix Klein, see 22L Personalia, p. 3, Klein Nachlass, Niedersächsische Staats- und Universitätsbibliothek (hereinafter NSUB, Göttingen).

[36]Roger Cooke and Frederick Rickey discussed the setting up of the first Clark Mathematics Department at some length in *A Century of Mathematics in America—Part* III, pp. 47–57.

went unrealized. As we shall soon see, internal political tensions and the organization of the University of Chicago coincided to thwart any chances that Clark University had of dominating research in its various areas of specialization. After just three years, Clark's real potential had faded into a lost opportunity. In mathematics, as well as in physics, anthropology, and other areas, the mantle of leadership passed to the upstart university on the shores of Lake Michigan.

The Founding of the University of Chicago

Like the Johns Hopkins University and Clark University before it, the University of Chicago owed its existence largely to the private fortune of one man, Standard Oil Company magnate John D. Rockefeller. As an educational philanthropist, however, Rockefeller differed markedly from his two notable predecessors. First of all, Rockefeller, unlike Hopkins, was very much alive during the first four and a half decades of the new university's existence. He repeatedly heard and responded to appeals for substantial donations to the university in Chicago and ultimately gave upwards of $35,000,000 to the cause over the years, whereas Johns Hopkins bequeathed a generous but finally inadequate $3,500,000 to found a university in Baltimore.[37] Furthermore, Rockefeller, unlike Clark, was still very much involved in his own business empire when plans for Chicago were being drawn up and later implemented. He sought out and chose educational advisers in whom he could trust—men like Frederick T. Gates, Thomas W. Goodspeed, and William Rainey Harper—and left the job of building the university totally in their hands. In stark contrast to Clark's involvement in Worcester, Rockefeller was so detached from the Chicago venture that he refused to consider the suggestion that the University bear his name, and he visited the campus only twice—on its fifth and tenth anniversaries—during the first ten years of its history.[38] No particular educational vision, no special need for a pet project motivated Rockefeller to back the Chicago institution. He was a devout Baptist desirous of supporting his denomination in the most fruitful way possible, but he was initially unsure as to what form that support might take.

After much discussion and debate in 1888 and 1889, the American Baptist Education Society resolved to establish a major university in Chicago, and the main proponents of this plan, Gates and Goodspeed, eventually convinced

[37]Thomas Wakefield Goodspeed, *A History of the University of Chicago Founded by John D. Rockefeller: The First Quarter-Century* (Chicago: University of Chicago Press, 1916), p. 293. The much abbreviated history of Chicago's founding which follows was drawn from this source as well as from Richard J. Storr, *A History of the University of Chicago: Harper's University, The Beginnings* (Chicago: University of Chicago Press, 1966). The archival material upon which these two books were largely based may be found in the University of Chicago Archives, William Rainey Harper Papers and University Presidents' Papers 1889–1925 (hereinafter denoted UC Archives, Harper Papers and UPP 1889–1925, respectively).

[38]Rockefeller was, however, a member of the University's Board of Trustees.

Rockefeller that theirs was the project he had been looking for. By May of 1889, Rockefeller had made his first pledge of $600,000 "for the endowment fund for a college in Chicago on condition that an additional $400,000 be raised elsewhere for land and buildings."[39] By June of 1890, the citizenry of the second largest and fastest growing city in the United States, together with the Baptist denomination nationwide, had met Rockefeller's challenge. Several weeks later, Rockefeller committed an additional $1,000,000 to the school with $800,000 specifically earmarked for graduate studies.[40] Finally, by September of 1890, the thirty-four-year-old Yale Divinity School Professor of Semitic Languages William Rainey Harper had been offered the University's first presidency, a post he accepted early in 1891. With the financial and administrative machinery in place, Harper was free to concentrate on building a major university according to his own carefully thought out educational plan.

No newcomer to educational circles, Harper had entered Muskingum College in his hometown of New Concord, Ohio at the tender age of ten and had graduated three years later in 1870.[41] He worked in his father's dry goods store for several years following graduation and taught an occasional class in Hebrew at Muskingum. At seventeen, he went eastward to pursue graduate studies at Yale, where he received his Ph.D. two years later for a thesis on the comparative philology of prepositions of Latin, Greek, Sanskrit, and Gothic. The next several years found him teaching Latin and Greek in preparatory schools, but in 1879 the Baptist Union Theological Seminary in the Chicago suburb of Morgan Park presented him with the opportunity to teach his first love, Hebrew. In Morgan Park, Harper had not only his first contact with the city of Chicago but also his introduction into an active circle of educators which included Col. Francis W. Parker, among others.[42] An indefatigable teacher, Harper had soon single-handedly organized a summer session in Hebrew at the seminary and had set up auxiliary summer schools in Worcester, New Haven, and Philadelphia. As if this did not occupy enough of his time, he signed on in 1883 to teach Hebrew at Chautauqua in upstate New York.

Originally established in 1874 by Reverend John Heyl Vincent as a two-week-long summer school for the training of Methodist Sunday School teachers, Chautauqua had come to represent a major movement in popular education by the 1880s.[43] It gradually transcended its denominational origins

[39] Storr, p. 31.

[40] *Ibid.*, p. 47.

[41] Details on Harper's life may be found in Goodspeed, pp. 98–105; Storr, pp. 18–19; and particularly Joseph E. Gould, *The Chautauqua Movement*: *An Episode in the Continuing American Revolution* (n.p.: State University of New York, 1961), pp. 13–38.

[42] Parker became Principal of Chicago's Normal School in 1883 where he instituted many progressive educational reforms. See, for example, Lawrence A. Cremin, *The Transformation of the School*: *Progressivism in American Education* 1876–1957 (New York: Alfred A. Knopf, 1961), pp. 128–135.

[43] On Chautaugua and its history, see Gould, pp. 3–12 and 72–100.

and expanded its offerings into a Liberal Arts College with over one hundred teachers who satisfied the educational needs of over two thousand students in subjects varying from music and art to mathematics and from the languages and their literatures to the practical arts.[44] Furthermore, local groups, the Chautauqua Literary and Scientific Circles, sprang up across the United States and inspired interested townspeople to work their way through a prescribed, four-year course of study. What had, by 1883, nominally become Chautauqua University also eventually incorporated into its operations a full-scale correspondence school and a press for the dissemination of readings and other relevant information. Harper contributed to all facets of this innovative, grass roots, educational movement, first as a Lecturer and then as Principal of the Liberal Arts College continuously from 1887 to 1898.

Dividing his time between Morgan Park and Chautauqua until 1886, Harper made a major career move in that year when he accepted a professorship at Yale. Given all of the teaching and textbook writing he had done prior to his appointment at his *alma mater* (seven textbooks by 1885), he had found little opportunity to engage in research. The return to New Haven, however, rekindled his purely scholarly interests, and he embarked upon what would become a lifelong research project, a linguistic study of the Bible's Old Testament in the finest German, philological tradition. The educational philosophy absorbed through his association with Chautauqua and his own growing sense of the importance of scholarly research deeply influenced the ideals which Harper would set down for the projected university in Chicago as early as 1890 in a document entitled *Official Bulletin No.* 1.

In this plan directed to the Board of Trustees, Harper described a university consisting of three parts: "The University Proper, The University Extension, and The University Publication Work."[45] As we have seen, Gilman had already incorporated the first and third of these components into his overall plan at the Johns Hopkins, but the second, the University Extension, represented a new departure in American higher education. Whether as a result of his experiences at Chautauqua or his understanding of the extension system in place at England's Cambridge University, Harper committed himself and his new institution to providing advanced education to those unable to attend classes formally.[46] This tripartite structure became quadripartite with the adjunction of "The University Libraries, Laboratories, and Museums" by the time *Official Bulletin No.* 2 came out in April of 1891. In outlining this new division, Harper forshadowed the institutionalization

[44] *Ibid.*, p. 24.

[45] Goodspeed, p. 134.

[46] Scholars differ in opinion on the effect of Chautauqua on Harper's educational philosophy in general and on his insistence on an extension program in particular. Gould went so far as to characterize the University of Chicago as a Chautauqua transferred to the shores of Lake Michigan. See Gould, pp. 55–71. Goodspeed, on the other hand, noted that Harper was well aware of the British extension system. See Goodspeed, p. 136.

of the departmental library, a feature which scientists and mathematicians in particular would come to hold sacred.[47] Of the several branches of his university, though, Harper's "University Proper" had the greatest impact on research-level mathematics.

The University of Chicago, like the Johns Hopkins, had an undergraduate college as well as a graduate school from the very beginning; and like Hopkins, Chicago placed special emphasis on the training of future researchers. Rockefeller had made specific financial provisions for graduate studies as early as 1890 in his second pledge, so Harper carefully crafted plans for this aspect of the University's curriculum in order to fulfill both the benefactor's and his own wishes. Writing in 1892 in a report intended for the Board, Harper crystallized his developing vision of graduate education. He opened with the fundamental assertion "that the *university* idea is to be emphasized. It is proposed to establish, not a college, but a university."[48] Toward the realization of this goal, he explained that "[a] large number of professors have been selected with the understanding that their work is to be exclusively in the Graduate Schools."[49] These professors would seek primarily "not to stock the student's mind with knowledge of what has already been accomplished in a given field, but rather so to train him that he himself may be able to push out along new lines of investigation."[50] Since, "[i]t is only the man who has made investigation who may teach others to investigate," Harper argued, the Graduate School faculty would consist of research—as opposed to mere college—professors, for "[w]ithout this [research] spirit in the instructor and without his example students will never be led to undertake the work."[51] Furthermore, as a researcher himself, Harper recognized the impossibility of this mission in the absence of a conducive working environment. Thus, he warned the trustees that "if the instructor is loaded down with lectures, he will have neither time nor strength to pursue his investigations. Freedom from care, time for work, and liberty of thought are prime requisites in all such work. To this end, ... [i]t is expected that Professors and other instructors will, at intervals, be excused entirely for a period from lecture work, in order that they may thus be able to give their entire time to the work of investigation."[52] Finally, given the preeminent role that the production of original research played in this scheme, the "[p]romotion of younger men in the departments will depend more largely upon the results of their work as investigators than upon the efficiency of their teaching, although the latter will by no means be overlooked. In other words, it is proposed in this institution

[47] Goodspeed, p. 135.

[48] *Ibid.*, p. 145 (Harper's emphasis).

[49] *Ibid.*

[50] *Ibid.*

[51] *Ibid.*

[52] *Ibid.*, pp. 145–146.

to make the work of investigation primary, the work of giving instruction secondary."[53]

The basic philosophy which Harper articulated in this report was not altogether new. At Hopkins, Clark, Yale, and Harvard, similarly inspired principles had emphasized the primacy of research in the graduate school setting. What was new was the recognition that different sets of rules would have to apply to the graduate and undergraduate faculties. For the graduate faculty, teaching loads would be lighter, leave time would be provided, and promotion would be contingent upon research productivity. At Chicago, financed generously by Rockefeller and the city's well-to-do, Harper had the resources to implement such policies within his graduate school. As we shall see in Chapter 9, for mathematics as for the other academic disciplines, this system profoundly and positively affected research activities within the University.

Building a Department of Mathematics at Chicago

In order to realize his educational and institutional goals, Harper needed to secure the right faculty. He had to assemble a group of people who shared not only his vision for the new university but also his standards of excellence in research. In particular, the success of his venture hinged on locating and securing a professor for each of the academic departments to act as his authority, guide, and counsel in making further decisions and appointments. Like Gilman, Harper sought experts who could pass judgment on technical matters relating to their fields of specialization, but unlike Gilman, he expected his faculty to work within an already highly detailed institutional structure rather than to define that level of detail through their day-to-day activities.[54] In choosing as his mathematical adviser the recent Yale Ph.D. Eliakim Hastings Moore, Harper took a chance on a fresh and untried talent. His gamble, however, paid off.

Harper had met Moore in New Haven as early as the summer of 1889. At that time, Harper held a Chair in the Divinity School as well as a professorship in the undergraduate college, while Moore, only five-and-a-half years his junior, was at the end of a two-year stint as Tutor in Mathematics. As Principal of Chautauqua's College of Liberal Arts, Harper was always on the lookout for able instructors, so naturally, a Yale Ph.D. with teaching experience such as E. H. Moore would have represented a fine catch. Unfortunately for Chautauqua, though, this young mathematician had developed altogether different ideas about how to spend his summers.

Born on 26 January 1862, E. H. Moore led an uncommonly privileged life

[53] *Ibid.*, p. 146.

[54] Daniel Coit Gilman, *The Launching of a University and Other Papers: A Sheaf of Remembrances* (New York: Dodd, Mead & Co., 1906), pp. 48–49; and Veysey, pp. 159–165.

ELIAKIN HASTINGS MOORE (1862–1932)

for a boy growing up in rural Ohio.[55] When the Civil War ended in 1865, his father, who had risen to the rank of Lieutenant Colonel in the Union forces, returned to the Methodist ministry which took him and his young family to many of the small towns dotting the Ohio countryside. One fixed point in this somewhat itinerant life, however, was Athens, Ohio, the home of Ohio University as well as Moore's grandfather, Eliakim Hastings Moore, Sr. As a banker, University Treasurer, and county official, the elder E. H. Moore, and so his entire family, occupied an important place in the local society. In fact, Moore's district accorded him one of its highest honors by sending him to Washington in 1869 as its Republican Representative to the Forty-First Congress.[56]

The young Moore attended Woodward High School in Cincinnati and there came under the eye of Ormond Stone, who, as Director of the Cincinnati Observatory, secured his services as a summer assistant. Stone, an observational astronomer with a mathematical bent, "had keen insight in the choosing of able students, and, once they came under his influence, the ability of turning their interests permanently to scientific careers."[57] Whether or not Stone's example and guidance played a significant role in the subsequent career of his summer helper, E. H. Moore did proceed to Yale in the fall of 1879 where he studied, among other things, mathematics and astronomy. The professor in charge of both of these subjects was Hubert Anson Newton.

As noted in Chapter 1, Newton had recognized and profited from the advantages of a European mathematical education as early as the 1850s and had brought to Yale's chair of mathematics not only a deep appreciation of research-level mathematics but also an undiminished sense of the value of European training for gifted American students. As an undergraduate from 1879 to 1883, E. H. Moore, nicknamed "Plus Moore" because of his consistently high academic achievements, must have struck Newton as just such a budding talent. In fact, Newton took Moore on as a doctoral student in 1883 and supervised his 1885 thesis on "Extensions of Certain Theorems of Clifford and Cayley in the Geometry of n Dimensions."[58] With Newton's

---

[55]On E. H. Moore's life, see Gilbert A. Bliss, "Eliakim Hastings Moore," *Bulletin of the American Mathematical Society* 39 (1933):831–838; Raymond C. Archibald, *A Semicentennial History of the American Mathematical Society* (New York: American Mathematical Society, 1938), pp. 144–150; Leonard E. Dickson, "Eliakim Hastings Moore," *Science* 77 (1933):79–80; and Charles C. Gillispie, ed., *Dictionary of Scientific Biography*, 16 vols., 2 supps. (New York: Charles Scribners' Sons, 1970–1990), s.v. "Moore, Eliakim Hastings," by Ronald S. Calinger (hereinafter abbreviated *DSB*). See, also, Karen Hunger Parshall, "Eliakim Hastings Moore and the Founding of a Mathematical Community in America, 1892–1902," *Annals of Science* 41 (1984):313–333; reprinted in *A Century of Mathematics in America—Part* II, pp. 155–175.

[56]*Biographical Dictionary of the American Congress:* 1774–1927 (Washington, D.C.: United States Government Printing Office, 1928), p. 1327.

[57]Charles P. Olivier, "Ormond Stone," *Popular Astronomy* 41 (1933):294–298 on p. 296.

[58]E. H. Moore, "Extensions of Certain Theorems of Clifford and Cayley in n Dimensions," *Transactions of the Connecticut Academy of Arts and Sciences* 7 (1885):9–26.

further encouragement Moore continued his studies in Germany in 1885–1886.

After a summer of intensive language study at Göttingen, he proceeded to the University of Berlin. There, seventy-year-old Karl Weierstrass continued to attract large audiences, while Leopold Kronecker also drew significant numbers of auditors. Although it remains unclear just how much these or other German mathematicians influenced him directly, "[t]here is no doubt," according to his student and later colleague Gilbert Ames Bliss, "that the year abroad affected greatly ... Moore's career as a scholar. It established his confidence in his ability to take an honorable place in the ... circle of mathematicians, acquainted him at first hand with the activities of European scientists,"[59] and, perhaps most important of all, imbued him with the research ethic that so thoroughly pervaded German mathematics.

Upon his return to the United States, however, Moore accepted a teaching-intensive position at the academy for secondary instruction associated with Northwestern University in Evanston, Illinois. Not surprisingly, he found the job far from conducive to the pursuit of mathematical research and left it after just one year to take the two-year tutorship at Yale. In spite of the fact that this was no less time-consuming, it was a post at an institution with some real mathematical resources. Moore used these means to his advantage during the next two years in New Haven. He succeeded in extending some of the work he had done in his thesis and presented his new research in two papers published in the *American Journal of Mathematics.*[60] Although this research probably played little rôle in securing him an assistant professorship of mathematics at Northwestern University in 1889, Moore went to Evanston committed to the further pursuit of new mathematical results. He firmly believed that while teaching defined one of the duties of the mathematics professor, the production of original research represented a function of equal, if not greater, consequence.

This attitude toward research, which was symptomatic of the changes taking place in American mathematics in the nineteenth century's final decade, came through clearly in Moore's belated response to William Harper's offer of a teaching position at Chautauqua for the summer of 1890. On 20 September 1890, Moore admitted somewhat sheepishly from Illinois that

> Your note of last April should have received, though it did not demand, an earlier answer. You remember our conversation at the end of the summer of 1889 in which I

[59]Bliss, "Eliakim Hastings Moore," p. 833.

[60]See E. H. Moore, "Algebraic Surfaces of Which Every Plane Section Is Unicursal in the Light of n-Dimensional Geometry," *American Journal of Mathematics* 10 (1888):17–28; and "A Problem Suggested in the Geometry of Nets of Curves and Applied to the Theory of Six Points Having Multiply Perspective Relations," *American Journal of Mathematics* 10 (1888):243–257. For more on Moore's work in these and other papers, consult Gilbert A. Bliss, "The Scientific Work of Eliakim Hastings Moore," *Bulletin of the American Mathematical Society* 40 (1934):501–514.

> expressed the feeling that now that I am at Evanston the summer *time* is more valuable to me than it was while I was at New Haven, owing to the fact that during the year at Evanston I have to rely so entirely on my own library for the book-needs of my mathematical study. Hence I could not spare the summer for (relatively) elementary teaching, though I did feel that if the teaching were of the more advanced college courses the gain to me in experience would offset the loss of time, and with respect to that I feel still the same.[61]

The Chautauqua of the late 1880s and early 1890s may have embraced, under Harper's direction, relatively high standards of research in theology and philology in its advanced courses on the Bible, but its instruction in mathematics remained at an elementary level.[62] Thus, it offered little challenge for an aspiring mathematical researcher like Moore. As Moore's letter of 20 September went on to reveal, though, word of a very real and intriguing challenge had already begun circulating in Chicago. He closed "by heartily congratulating you [Harper] and the Chicago University and the generous men who have made the combination so promising."[63]

While Moore apparently entertained no expectations at this time relative to the new university to be founded in Chicago, just five months later on 16 February 1891, he found himself under consideration for a position there and responded favorably to Harper's overtures. As an Assistant Professor just turned twenty-nine, he wrote in all sincerity that

> it goes without saying that it is a pleasure and is appreciated as a high compliment to me to be thought of by you with reference to the University of Chicago. I am not so bound to the Northwestern as to be unable to consider a proposition to go to Chicago, my engagement being for no definite term of years. The University of Chicago will certainly have from the very outset a very high standing, if the financial strength necessary to realize the high ideals is present, and probably that is fully assured. ... It is possible that you may have in mind work of a nature more attractive to me than anything Northwestern may have.[64]

Harper very quickly assured Moore that he did indeed have something more attractive in mind, since the University of Chicago planned to stress original

[61]E. H. Moore to W. R. Harper, 20 September, 1890, UC Archives, UPP 1889–1925, Box 46, Folder 26 (Moore's emphasis). This appears to be the earliest extant piece of correspondence between Moore and Harper.

[62]Gould, p. 34.

[63]Moore to Harper, 20 September 1890.

[64]E. H. Moore to W. R. Harper, 16 February 1891, UC Archives, UPP 1889–1925, Box 17, Folder 2.

research and graduate instruction. Anyone considered seriously for the faculty at Harper's university had to share unequivocally his sense of its mission. Questioning Moore on this point, Harper elicited this emphatic response on 23 March 1891: "In reply to your letter of Feb. 26 and in continuation of my brief note of March 3, ... let me repeat that the two points suggested in your letter: the recognition of the desirability of investigation, and graduate instruction, if not at once, very soon: are of decided weight in my eyes."[65]

This reply must have provided Harper with further evidence of Moore's potential, for their negotiations concerning the mathematics professorship continued. They met in Chicago during the third week of April, and Moore recorded the substance of their conversation in a letter the following week:

> My pleasant hour with you a week ago remains an enjoyable spot in memory. After due consideration I am prepared to state now that a definite offer to me on the conditions informally discussed by us that evening but at a salary of $4000 for a scholastic year would be accepted by me.
>
> I appreciate fully your unwillingness to make me an offer that will not be accepted, and I write this merely as a statement of what I would accept if offered, and do not wish to be understood as in any sense urging or beseeching you to make such an offer. Whether it will seem to you and the trustees for the interest of Chicago University to make such an offer, is for you to judge. The status would be full professor (with a salary above the ordinary, just as certain head professors will receive a salary above the ordinary), though I should hope to have a voice in the organization of the department of mathematics in its higher work.[66]

Although Harper apparently had no other real candidate in the wings, he remained unwilling to appoint Moore immediately to the headship of the department, probably because of the latter's age, his basically untried mathematical talent, and his administrative inexperience. Furthermore, Harper drove a hard bargain financially and chose to let Moore reconsider his salary demand for a while before resuming negotiations. After this usual sort of jockeying, however, the President and the mathematician came amicably to terms. On 29 February 1892, Moore signed his letter of intent with Chicago and became Full Professor and Acting Head of its Department of Mathematics.[67]

---

[65]E. H. Moore to W. R. Harper, 23 March 1891, UC Archives, UPP 1889–1925, Box 14, Folder 15.

[66]E. H. Moore to W. R. Harper, 29 April 1891, UC Archives, UPP 1889–1925, Box 14, Folder 15.

[67]E. H. Moore to W. R. Harper, 29 February 1892, UC Archives, UPP 1889–1925, Box 14, Folder 15. Moore's initial salary was apparently that originally offered him, namely $3500. Relative to his mathematics faculty, at least, Harper would prove almost immovable on salary

Moore set to work immediately formulating his plan for the new department. Two days later on 2 March, he sent Harper a fairly detailed blueprint complete with library and apparatus needs. "It seems to me," Moore wrote, "we shall need in all five men to attend to 400–500 students in the undergraduate and graduate work. (The advanced courses must be offered; the classes will be small.) To start with fewer would probably require us to make hasty choice of some one the last day. As to type of men: strong teachers, with devotion to the subject and with ambition, ability and determination to do original work."[68] Moore contended that of these five two should be hired as Tutors or at some comparable, nonpermanent rank, and, in addition to himself, two should be hired in the professorial ranks. As he explained to Harper, "the advanced work of the department can hardly be allowed to rest on what I can do alone, supplemented by the excess work of tutors."[69] Therefore, "the other two professors, say, one associate—, and one assistant— ... should be men of *tried* ability as teachers along the higher lines; they would divide their work (chiefly) between sophomores and advanced courses. With three of us then to furnish the body of higher courses, and with two tutors to do (excess) advanced work, the department would start out fairly well equipped."[70] With a permanent faculty of three, having distinct and diverse interests, Moore's Chicago—unlike Sylvester's Hopkins—would have breadth as well as depth at the graduate level.

As for the department's needs in areas other than personnel, Moore lobbied hard for a $2000 appropriation for books and equipment, instead of Harper's proposed $1500. He argued that the department required a full $1500 for books and journals and an additional $500 for mathematical models and other pedagogical aids.[71] Concerned with the overall finances of the University, however, Harper firmly limited the size of the permanent faculty to two and the amount of the operational budget to $1500. At least initially, the President's tight fiscal grasp of the situation constrained the ambitious Acting Head of the department, and Moore focused on the problem of finding one, not two, colleagues.

Already in his acceptance letter of 29 February, Moore had proposed a viable professorial candidate, his friend Henry Seely White. Moore had met White as early as the spring of 1890 in Evanston. At that time White, fresh from Göttingen, had taken an interim position in Illinois as a preparatory

---

negotiations. For more on the formation of the first Chicago Department of Mathematics, see Karen V. H. Parshall, "The One-Hundredth Anniversary of Mathematics at the University of Chicago," *The Mathematical Intelligencer* 14 (2) (1992):39–44.

[68]E. H. Moore to W. R. Harper, 2 March 1892, UC Archives, UPP 1889–1925, Box 14, Folder 15.

[69]*Ibid.*

[70]*Ibid* (Moore's emphasis).

[71]Undoubtedly, the models Moore had in mind were those manufactured by the Darmstadt firm of L. Brill. Recall the discussion of these in Chapter 4.

school instructor. He moved on to Clark as an Associate in the fall and there demonstrated his ability to teach research-level mathematics through his courses on higher plane curves, invariant theory, Abelian integrals, and theta functions, all courses related to his own thesis research.[72] Furthermore, he showed promise as an original researcher not only in his thesis (published in 1891) but also in the paper based on it which had appeared in the *Mathematische Annalen* in 1890.[73] In recommending White to Harper, Moore argued more generally that "[o]ne of Klein's men would be apt to have the broadest mathematical horizon."[74]

Whether the recommendations from the Clark faculty proved uninspiring or whether White himself discouraged his consideration, his name had dropped out of contention by the beginning of April, and Moore had presented several new candidates, most notable among them, William Fogg Osgood and Maxime Bôcher. (Although White did not end up on the Chicago faculty, he did leave Clark for the Chicago area, succeeding Moore at Northwestern.) Like White, both of these young men satisfied the "Klein criterion," although as noted, Osgood had actually taken his degree under Max Noether at Erlangen. At the time of Chicago's search for a faculty, they were both serving as instructors at Harvard, and, according to Moore, both looked "likely soon to secure permanent positions there."[75] While they seemed of equal potential, Moore evidently decided to go after Bôcher, for in a letter on 10 May 1892, he told Harper that "Bôcher's answer ought to be here in two days now."[76] Moore found himself back at square one when that reply did come, apparently in the negative. The question then became, who, among "Klein's men," was available and at a sufficiently advanced level to assume an associate or assistant professorship in a new university with such high expectations? At Clark, another mathematician besides White met this standard.

As noted, Oskar Bolza had earned his degree under Klein in June of 1886, but he had left Germany one year later hoping for brighter professional prospects in the United States. Arriving in 1888 to face what he described as "an unknown future in the New World,"[77] Bolza had spent the Winter Term

---

[72]Moore to Harper, 29 February 1892.

[73]See Henry Seely White, "Abel'sche Integrale auf singularitätenfreien, einfach überdeckten, vollständigen Schnittcurven eines beliebig ausgedehnten Raumes," *Nova Acta der Kaiserlichen Leopoldinisch-Carolinischen deutschen Academie der Naturforscher* 57 (1891):41–128; and "Ueber zwei covariante Formen aus der Theorie der abel'schen Integrale auf vollständigen singularitätenfreien Schnittcurven zweier Flächen," *Mathematische Annalen* 36 (1890):597–601.

[74]Moore to Harper, 2 March 1892.

[75]E. H. Moore to W. H. Harper, 25 March 1892, UC Archives, UPP 1889–1925, Box 14, Folder 15.

[76]E. H. Moore to W. R. Harper, 10 May, 1892, UC Archives, UPP 1889–1925, Box 14, Folder 15.

[77]On Bolza's life and work, see Gilbert A. Bliss, "Oskar Bolza–In Memoriam," *Bulletin of the American Mathematical Society* 50 (1944):478–489; and Gillispie, ed., *DSB*, s.v. "Bolza,

of 1889 at the Johns Hopkins before taking a position on the first mathematics faculty at Clark later that year. With a three-year contract at the rank of Associate, his future, while still far from secure, must have at least looked brighter than before. Unfortunately, the ever-worsening political climate at Clark had permanently altered this initial optimism by his last year in Worcester.

Clark University had found itself woefully underendowed from the outset in light of Jonas Clark's unfulfilled expectation of outside matching funds and Hall's lavish promises of money for both faculty and equipment. In May of 1891, the benefactor unwittingly compounded these fiscal problems by conceding to Hall's request to be given greater control over the university's finances. Despite the President's published statements to the contrary, his own mismanagement of funds—coupled with the absence of frank and open communication between himself, Clark, and the Trustees—led quickly to difficulties.[78] The situation finally came to a head during the third year of operation. By January of 1892, President Hall fell victim not only to his fiscal excesses but also to his double-dealing and general lack of diplomacy. Among other things, Hall broke promises of laboratory equipment to several of his professors, forbade the faculty to convene in order to discuss university matters, and allegedly disciplined faculty members without prior consultation with their departmental heads.[79] Not surprisingly, he plummeted from favor among the faculty as a result. Bolza detailed the events which had led up to what he viewed as the inevitable crisis in a letter to Klein written somewhat after the fact. As he explained:

> The new academic year began with a fairly considerable cutback in appropriations; in the library [there were] not only no new purchases but also reductions, for example, half of our mathematics journals were cut. It was, however, not so much the financial difficulties as the politics, which the President pursued in the face of these difficulties—in good German, his endless lies [*endlosen Lügereien*]—which heightened the general embitterment.
>
> The discontent had gradually become so widespread that most of us earnestly considered leaving at the first opportunity; a few already had outside offers. In order

Oskar," by Ronald S. Calinger. The most comprehensive source on his life, however, is his autobiography, Oskar Bolza, *Aus meinem Leben* (Munich: Verlag Ernst Reinhardt, 1936). For this quotation, see p. 19. Most of the biographical information on Bolza comes from this source.

[78] Koelsch, p. 40. Not surprisingly a rich folklore has grown up around this crisis at Clark. Standard works on the history of higher education, such as Laurence Veysey's *The Emergence of the American University*, as well as Edmund Sanford's history of Clark University, place the blame squarely on Jonas Clark's shoulders, citing his withholding of funds and overall meddlesomeness. Koelsch disagrees on the basis of his thorough scouring of the archival evidence.

[79] Typescript of the history of the crisis of confidence at Clark University written by Professors Warren Lombard (physiology), Franklin Mall (anatomy), and Henry Donaldson (neurology), p. 17, UC Archives, UPP 1889–1925, Box 28, Folder 3. See, also, Koelsch, pp. 35–39.

Oskar Bolza (1857–1942)

> to forestall the impending crisis, several of us decided to try to improve the situation by seeking an extension of the faculty's rights. ... Next followed several very stormy faculty meetings, the upshot of which was a collective resignation of the faculty (21st January).[80]

Nine (of eleven) permanent faculty members, including Bolza, resigned *en masse* "owing to lack of confidence in the President of Clark University."[81] After a month of intense negotiations between the Trustees, the faculty, and Hall, the resignations were withdrawn, but ill-feelings remained. Harper visited the Clark campus in the midst of this intense bickering. His arrival with its attendant call to Chicago (known as "Harper's raid") thus proved extraordinarily well-timed. As Hall later recounted bitterly in his autobiography:

> Dr. Harper, learning of the dissatisfaction here, had ... met and engaged one morning the majority of our staff, his intentions and even his presence being unknown to me. Those to whom we paid $4000, he gave $7000; those we paid $2000, he offered $4000, etc., taking even instructors, docents, and fellows. This proved really to be the nucleus and, I think, the turning point in the early critical stage of the development of the Chicago institution. ... Thus Clark had served as a nursery, for most of our faculty were simply transplanted to a richer financial soil.[82]

While he greatly exaggerated Harper's role in the Clark catastrophe and overestimated the salaries offered at Chicago, Hall quite correctly held that this combination of circumstances had cost Clark dearly. The University saw its promising potential evaporate with the loss of so many talented and carefully selected faculty members. In a cruel twist, the University of Chicago—and not Clark—moved into position to assume the leadership role in research-level education in the United States. As for mathematics, the transplanted faculty member, Oskar Bolza, proved instrumental in this shift, although his commitment to Chicago hardly came immediately.

Harper first had E. H. Moore sound out Bolza during the troubled winter

---

[80] Oskar Bolza to Felix Klein, 15 May 1892, Klein Nachlass, Box 8, NSUB, Göttingen (our translation).

[81] Typescript of the crisis at Clark, p. 17. Besides Bolza and the authors of this document, the other signatories of the resignation letter were: Albert A. Michelson in physics, C. O. Whitman in zoology, John U. Nef in chemistry, George Baur in osteology and palentology, and Franz Boas in anthropology. Of those at the rank of Instructor or above, only William Story and Edmund Sanford declined to sign this document.

[82] Hall, pp. 295–297, as reprinted in *American Higher Education: A Documentary History*, ed. Richard Hofstadter and Wilson Smith, 2 vols. (Chicago: University of Chicago Press, 1961), 2:759–761. In all, fifteen Clark faculty members accepted Harper's call to Chicago, most notably Michelson, Whitman, Nef, and Bolza. The average salary offered to department heads was $4000. See Goodspeed, pp. 211–213. For more specifics on the impact of "Harper's raid" on mathematics at Clark, see Cooke and Rickey, pp. 58–59.

of 1892. At that time, Bolza withdrew his name from further consideration, having resolved to return to Germany. He did suggest the name of an alternative candidate, however, that of his long-time friend Heinrich Maschke.[83] Like Bolza, Maschke had left behind the poor prospects for congenial employment within the German academic setting and had taken a position as an electrician at the Weston Electric Company in Newark, New Jersey in 1891.[84] Given that Maschke had never held a university-level position, his candidacy must not have appeared as strong to Harper and Moore as that of, say, Bôcher. Even so, the Chicago team did focus on Maschke after attempts to secure Bôcher failed in mid-May, and an offer went out to him on 1 June 1892. Maschke's response of 6 June, while positive, was not unqualified. He wrote: "Your esteemed favor of the 1st inst. recd. and in reply I beg to say that I would be very happy to devote myself again to pure mathematics. The general conditions you mention would suit me very well; you will, however, allow me the question whether it would not be possible for you to make the salary $2500 instead of $2000. Would you be kind enough as to write me about that point again?"[85]

Whether or not Harper and Moore felt that this point of negotiation would ultimately thwart their efforts to snare Maschke, they left him dangling and once again focused on Bolza. This time, they approached him with the proposition of coming and trying Chicago out for a year before making a permanent commitment to it.[86] Bolza's friend and Clark colleague Henry Taber heard of this renewed bid and urged him not to pass up such a singular opportunity. On the force of Taber's argument, Bolza arranged to travel to Chicago to meet with Harper and Moore personally on 15 June. He rendezvoused first with Maschke in New York to inquire as to the status of his friend's negotiations with Chicago and, on learning that they were still open, convinced himself that both had a very good chance of landing permanent academic posts at the same institution. Unfortunately, Harper insisted that Chicago could afford to hire either Bolza or Maschke, but not both, despite Moore's continued insistence on the need for three department members in the professorial ranks. Bolza forced Harper's hand when he refused to accept his offer of an associate professorship unless Maschke were given an assistant professorship. His gamble paid off. Thanks to a new gift, Harper had the money needed to fund the additional salary, and an offer went out to Maschke.[87] Thus, in one

[83]Bolza, *Aus meinem Leben*, p. 24.

[84]On Maschke's life, see Oskar Bolza, "Heinrich Maschke: His Life and Work," *Bulletin of the American Mathematical Society* 15 (1908):85–95.

[85]Heinrich Maschke to W. R. Harper, 6 June 1892, UC Archives, Harper Papers, Box 14, Folder 10.

[86]Bolza, *Aus meinem Leben*, pp. 24–25.

[87]*Ibid.* Maschke accepted the "assistant professorship under conditions agreed with Dr. Bolza" in a telegram to Harper dated 17 June 1892 and held in UC Archives, Harper Papers, Box 14, Folder 10.

HEINRICH MASCHKE (1853–1908)

quick maneuver, the two friends became colleagues, and Moore got the three professorial positions he had wanted all along.

In fact, in its final form, the Mathematics Department conformed remarkably closely to the first blueprint Moore had laid before Harper early in March. In addition to the three professors, who would concentrate their efforts on the department's advanced offerings, the President had already appointed two younger men in the tutorial ranks to take on the more elementary instruction. The first of these, Harris Hancock, was a young man from a prominent Virginia family, who had begun his study of mathematics at the University of Virginia. He next moved to the Johns Hopkins, where he earned a B.A. in 1888 and completed much of the work preliminary to the Ph.D. Desirous of a European degree, though, he traveled first to Cambridge and then to Berlin, where he planned to take a doctorate in June of 1892 under the direction of Lazarus Fuchs.[88] Moore backed Hancock's candidacy, describing him as "an ideal beginning of a line of mathematical tutors in the University, unless our expectations are disappointed, which is unlikely."[89] An offer soon followed which Hancock accepted on 5 April to become the second official member of the department. (All would not remain harmonious, however. See Chapter 9 for an account of subsequent conflicts between Hancock and Moore.)

The third mathematician to join the faculty, Jacob W. A. Young, was also hired to teach primarily lower division courses. After graduating from Bucknell University in 1887, Young went to Berlin for the 1888–1889 academic year with the intention of working toward a doctorate. As Clark's William Story explained in his letter recommending Young for a Chicago position, however, "finding himself not well prepared for such work, he returned to America and came to us. I think this is the experience of most Americans who go to Germany to study mathematics, and it seems to me that the difficulty they meet is to be found not altogether in an insufficient preparation—although the mathematical training given in most American colleges is ridiculously poor,—but quite as much in a fundamental difference of method of instruction and habits of thought here and there."[90] Young had performed well at Clark in spite of his less than successful sojourn abroad and had shown, in Story's view, "good mathematical ability, a lively interest in his studies, and great industry,"[91] particularly in his special areas of expertise, number theory and the theory of groups. On the basis of Story's endorsement as well as on recommendations from Henry White and George

---

[88]Hancock outlined his academic career in a letter of application to W. R. Harper dated 10 September 1891 and housed in the UC Archives, UPP 1889–1925, Box 37, Folder 12.

[89]E. H. Moore to W. R. Harper, 22 March 1892, UC Archives, Harper Papers, Box 14, Folder 15.

[90]W. E. Story to W. R. Harper, 6 April 1892, UC Archives, Harper Papers, Box 15, Folder 23. Felix Klein would come to much this same conclusion in 1893. See Chapter 8 below.

[91]*Ibid.*

W. Hill, Young was offered and accepted a tutorship early in the summer of 1892.[92] The Chicago Mathematics Department was finally complete one month later with the Bolza-Maschke coup.[93]

Moore had found and secured a staff of mathematical promise, but this by no means assured that the team, once assembled, would meld together to form the first-rate department envisioned by Harper and Moore. Would the junior members, Hancock and Young, learn to teach elementary courses effectively but without neglecting their own development into original researchers? Would the senior members, Moore, Bolza, and Maschke, succeed both in bringing students up to the research level in mathematics and in maintaining the highest standards of productivity in their own work? Would the individual goals and ideals of the department members, but particularly the professors, coalesce to define a harmonious working unit or ultimately work at cross purposes? The answer to this last question seemed particularly uncertain to the department's newly appointed Associate Professor, Bolza. He worried that the expectations he and Maschke held based on their common experiences within the German educational system would run counter to those Moore had absorbed from the very different American educational milieu. Mulling over his concerns in a letter to Felix Klein on 15 January 1893, Bolza wrote:

> The spirit of the university is, to be sure, our German ideal of university freedom, yet in many ways [it is] diametrically opposed to ours. Everything is reminiscent more of Cambridge and Oxford than of a German university: outwardly, the style of the buildings; the daily chapel services; on all festive occasions professors and students must wear caps and gowns (in exactly the Oxford style); faculty meetings are opened with a prayer. And then the

[92] E. H. Moore to W. R. Harper, 11 May 1892, UC Archives, Harper Papers, Box 14, Folder 15. Young remained at Chicago for his entire career, serving successively as Instructor (1894–1897), Assistant Professor (1897–1908), and Associate Professor (1908–1926). Shifting his interest from the theory of groups to pedagogy early on, Young was made Emeritus Professor of Mathematical Pedagogy in 1926. He died in 1948. See "Mathematics at Chicago: 1892–1968," UC Archives, General Archival Files, Mathematics Department. Dated 17 October 1968, this account, although anonymous, was most probably written by A. Adrian Albert.

[93] It is important to note here the contrast in size between the initial department at Chicago and that at Hopkins. With five instructors as opposed to three (or more accurately, perhaps, two-and-a-half given Craig's official status as a graduate student), Chicago was able to give its students greater breadth of mathematical training. Furthermore, during Chicago's second year of operation, the mathematics faculty increased again when James Harrington Boyd came to serve as Tutor or Associate. (Recall the discussion of Boyd in Chapter 5.) See *The University of Chicago Annual Register July*, 1893-*July*, 1894 *with Announcements for* 1894–1895 (Chicago: University of Chicago Press, 1894), p. 106; and H. E. Slaught and G. A. Bliss, "The Department of Mathematics," *The University Record*, n.s., 15 (Chicago: University of Chicago Press, 1929):97–104 on p. 97.

> rules and regulations and examinations! Reading the catalog, one concludes that the sole purpose of learning is to pass exams.[94]

As he worried about the seeming contradiction between the outward Englishness of the new university and the inner Germanness of its avowed educational philosophy, Bolza nevertheless professed his faith in the assembled faculty and its commitment to research:

> Fortunately all of this is not meant as literally as it would appear on paper, but is, as the President himself says, "flexible," and the great majority of men filled with a true scientific spirit who are already at the university assume that as far as postgraduate work is concerned, all of these limitations on freedom will shortly be lifted. ...
>
> At any rate, the developments of the first few years should determine whether our university will develop into a real university or into an English training ground for passing exams.[95]

Bolza's conclusion here would soon ring prophetic. The sense of common mission shared by Harper and his faculty very quickly gave substance to the ideals of "a real university" rather than those of an "English training ground for passing exams." The developments, particularly of the first few years, proved stunningly decisive for Chicago's Department of Mathematics . By the end of the summer of 1893, the Chicago mathematicians had staged the Mathematical Congress associated with the World's Columbian Exposition and had arranged for Felix Klein's subsequent Evanston Colloquium Lectures. These two events, totally unique and unprecedented in the American context, served not only to focus the mathematical spotlight immediately on Chicago but also to galvanize Moore and his colleagues into a research-oriented department with national vision and ambition.

[94] Oskar Bolza to Felix Klein, 15 January 1893, Klein Nachlass, Box 8, NSUB, Göttingen (our translation.)

[95] *Ibid.*

Chapter 7
The World's Columbian Exposition of 1893 and the Chicago Mathematical Congress

Studies, both ancient and modern, in the history of mathematics have often tended to focus exclusively on the contributions of various individuals to a vast body of mathematical knowledge, while giving little or no consideration to the cultural context in which the mathematical developments took place.[1] Whereas such works clearly play an indispensable role in documenting the growth of mathematical ideas, they also serve to reinforce or perpetuate a widespread myth that the history of mathematics involves nothing more than recording the "discovery" of completely disembodied ideas. The present study clearly approaches its subject matter from a very different angle. It aims to underscore how certain mathematical ideas and activities issued from a larger matrix of scientific, educational, economic, social, and cultural factors, all of which entered crucially into the emergence of the American mathematical research community.

The relevance of these broader, external factors ought never be overlooked, but in the heady atmosphere of Chicago during the dynamic era of the early 1890s, their effects became literally palpable. The notion that mathematics was an integral and sublime part of human culture found concrete expression in the form of the International Mathematical Congress staged in conjunction with the World's Columbian Exposition of 1893.[2] In the autumn

[1]As an extreme example of this so-called internalistic, historiographical approach, see Nicolas Bourbaki, *Éléments d'Histoire des Mathématiques* (Paris: Masson, 1984). Dirk Struik's *A Concise History of Mathematics*, 4th rev. ed. (New York: Dover Publications, Inc., 1987), on the other hand, represents the more externalistic approach to the subject, by taking into account the broader cultural context. In the nineteenth century, Moritz Cantor and H. G. Zeuthen were the two leading representatives of the externalist and internalist approaches to the history of mathematics, respectively. Jesper Lützen and Walter Purkert discuss their methodological squabbles in "Conflicting Tendencies in the Historiography of Mathematics: Moritz Cantor and H. G. Zeuthen," in *The History of Modern Mathematics*, vol. 3, ed. Eberhard Knobloch and David E. Rowe (Boston: Academic Press, Inc., 1994).

[2]We discuss the cultural implications of mathematics at the Fair in "Embedded in the Culture: Mathematics at the World's Columbian Exposition of 1893," *The Mathematical Intelligencer* 15 (2) (1993):40–45.

of that year, the mathematicians of the University of Chicago stood poised and ready to advance research-level mathematics in the United States from a state of passive receptivity to one of active participation in the quest for new knowledge. Into their midst stepped Felix Klein, who embraced their cause (while propagandizing for his own) with all of his considerable energy and enthusiasm. In the next two chapters, we offer a detailed account of the atmosphere in Chicago prior to, during, and immediately after the dramatic cultural events of 1893 captured the world's attention. Amid a wild cacophony of distinct voices, the mathematicians present in Chicago could barely be heard; nevertheless, their influence would make itself felt. Beating a quiet retreat from the Congress, they reconvened on the campus of Northwestern University, where they heard Klein deliver his two-week-long siren song in praise of mathematics, the Evanston Colloquium lectures. In a symbolic sense, the events surrounding Klein's visit to the United States represent the passing of the mathematical torch—at least insofar as the training of America's leading mathematicians was concerned—from Göttingen to Chicago and the hearkening of a new era in American mathematics.

The Chicago World's Columbian Exposition

When Chicago hosted the World's Columbian Exposition in 1893 and its accompanying Mathematical Congress, it was still animated to a large degree by the spirit that had rebuilt the city after the great fire of more than twenty years earlier. By 1875, it had already outgrown its western rivals, St. Louis and Cincinnati, so that only New York, Philadelphia, and Brooklyn remained larger. To the urbane Easterner, Chicago epitomized crass materialism—a dirty, uncouth, and uncultured metropolis with nothing more to offer than its lumber, grain, and cattle. Although a clearly exaggerated view, there was no denying that Chicago had its problems. Not only did bossism and political corruption plague the city but also rampant slum housing dominated its landscape. Almost fifty percent of its population lived in apartments that housed ten or more people, a circumstance that prompted Jane Addams to found Hull House in 1889. On top of this, the Midwestern center had a reputation for political radicalism and anarchy going back at least to the Haymarket incident of 1886. None of these circumstances made Chicago a very likely site for the Columbian Exposition, arguably the greatest World's Fair ever held.

Although a number of cities vied for the honor of hosting the 1893 World's Fair, by 1889 only New York and Chicago remained in the running.[3] These

[3]Much has been written on the Chicago Fair from a cultural, from an historical, and from an architectural point of view. Compare, for example, Reid Badger, *The Great American Fair: The World's Columbian Exposition and American Culture* (Chicago: Nelson Hall, 1979); Robert W. Rydell, *All the World's a Fair: Visions of Empire at American International Expositions*, 1876–1916 (Chicago: University of Chicago Press, 1984), pp. 38–71; J. Seymour Currey, *A Century of Marvelous Growth*, 5 vols. (Chicago: Clarke Publishing Co., 1912), especially volume 3 *Chicago and Its History and Builders*; and Stanley Appelbaum, *The Chicago World's Fair of* 1893: *A*

two traditional rivals went on to wage an infamous propaganda war with hopes of inducing Congress to settle the issue in their respective favors. In this contest, New Yorkers claimed to have experience on their side; they had, after all, staged the first World's Fair on American soil in 1853, just two years after London's great Crystal Palace extravaganza. Obviously an attempt to cash in on the London success, the New York Fair, replete with its own none too spectacular crystal palace, attracted relatively few spectators and hardly any foreigners, despite the promotional skills of P. T. Barnum and Horace Greeley. By 1889, however, this failed effort had apparently long since been forgotten, and the New Yorkers warned that if the Columbian Exposition were held in Chicago, it would amount to nothing more than a cattle show. During the heat of this campaign, in fact, a New York journalist dubbed Chicago the "Windy City," not because of the stiff breeze that often comes off of Lake Michigan but rather for the hot air expended by Chicagoans in their efforts to bring the Fair to their hometown.[4]

Yet, however immodestly its citizens may have sometimes behaved, only a biased or uninformed observer could have dismissed Chicago's manifest civic pride as nothing more than idle boasting. The founding of the University of Chicago had marked the culmination of a whole series of cultural achievements that had altered the face of the city during the 1880s and early 1890s. In 1889, Dankmar Adler and Louis Sullivan completed work on their impressive Civic Auditorium building, and two years later the Chicago Symphony gave its inaugural concert under the baton of Theodore Thomas. In the meantime, a handful of architects, associated with the indigenous American school founded by William LeBaron Jenney, were busy transforming downtown Chicago into a modern-day metropolis.[5]

Unlike his Eastern counterparts Henry Hobson Richardson and Richard Morris Hunt, both of whom had studied at the *École des Beaux-Arts*, Jenney had attended the *École polytechnique*. There, he had acquired a strong engineering background that enabled him to develop the no-nonsense, functionalist style which would come to dominate the burgeoning Loop area of downtown Chicago. In Jenney's studio, young men like Sullivan, Martin Roche, William Hollabird, and Daniel Burnham got their start. Today, this Chicago school is generally regarded as the first expression of a truly modernist approach to architecture. It was, therefore, one of the great ironies of the Chicago Fair that the Eastern architects of the *Beaux-Arts* tradition celebrated one of their last triumphs on its grounds—the famed Court of

---

Photographic Record (New York: Dover Publications, Inc., 1980). Many of the particulars about the Fair which follow may be found in several of these—as well as other—sources.

[4]Badger, p. 51.

[5]On the Chicago school and its significance, see Carl W. Condit, *The Chicago School of Architecture: A History of Commercial and Public Building in the Chicago Area,* 1875–1925 (Chicago: University of Chicago Press, 1964); and Sigfried Giedion, *Space, Time, and Architecture*, 3rd ed. (Cambridge, MA: Harvard University Press, 1954).

Honor, later dubbed the "White City." Many years afterward, Louis Sullivan bitterly recalled (although with more nostalgia than historical accuracy) that the Columbian Exposition had marked the end of the Chicago architectural tradition.[6]

After the city of Chicago won the right to host the Fair, its organizers were immediately faced with even greater obstacles that threatened to turn the whole venture into a giant fiasco. To begin with, the problem of obtaining a suitable site hounded Frederick Law Olmsted, the renowned landscape architect of New York's Central Park. Olmsted favored a location on the northern lakeshore, but the Illinois Central Railroad, which was lobbying for a southern site, simply refused to run its lines to the north. This left the Fair Company with only two choices, Washington or Jackson Park, both on the south side of town. Of the two, Washington Park seemed preferable, but the Chicago Parks Board refused to allow any landscape alterations there. The Fair Company was thus forced to reconcile itself to what it deemed the less desirable Jackson Park site. In fact, the area was little more than a vast marshland, and the landscaping that began in the spring of 1891 proved nothing less than a Herculean effort that required the removal and relocation of over a million cubic yards of earth.

Out of this emerged a beautiful wooded island surrounded by a lagoon, one of the few remnants of the Fair that has managed to survive to the present day. The lagoon was connected by a system of waterways to a central basin which served as the focal point for the buildings of the Court of Honor. In addition to the sixty-one acres of waterways, the Fair's engineers also had to construct drainage and sewage systems, provide a water supply, erect some fifty-odd buildings, make landscape refinements, and finish the decorating and painting. The responsibility of the Chicago Fair Company, all of this had to be completed by October of 1892 in order to accommodate the steadily incoming exhibits prior to the opening-day ceremonies on 1 May 1893. Under the supervision of Daniel Burnham, who later became famous both for his leadership of the City Beautiful Movement and for his urban planning designs for Chicago and Washington, D.C., twelve thousand workers toiled around the clock under electric lights, sometimes accompanied by the sweet strains of the Chicago Symphony, in their efforts to get in under the wire. Burnham exhorted his men on from a small, shack-like office on the fairgrounds, invoking a spirit of cooperation and self-sacrifice, and declaring that the Columbian Exposition would be the third great event in American history (after the Revolutionary and Civil Wars).[7]

[6]Sullivan wrote about his impressions of the impact of the Fair on Chicago architecture in *The Autobiography of an Idea* (New York: Press of the American Institute of Architects, Inc., 1926), pp. 317–325. In particular, he remarked that the "damage wrought by the World's Fair will last for half a century from its date, if not longer [p. 325]." This was also quoted in Appelbaum, p. 7. This "prophecy" was made, however, a good thirty years after the Fair had closed.

[7]Badger, p. 70. On Burnham, see Thomas S. Hines, *Burnham of Chicago, Architect and Planner* (Chicago: University of Chicago Press, 1979).

The Court of Honor with the statue of the Republic in the foreground.
World's Columbian Exposition, Chicago, Illinois, 1893.

This sense of *esprit de corps* aside, all kinds of financial problems plagued the undertaking. Even before the Fair had opened, the federal government had withheld $750,000 pending the posting of security to pay for judges and the cost of prize awards.[8] The Fair Company, unable to come up with the necessary funds, was only bailed out when the Illinois Central Railroad decided to put up $1,000,000 in bonds. By opening day, the economic depression of 1893 was well underway, and on 9 May, the Chemical National Bank in Chicago declared bankruptcy. Two days later, Columbian National, located directly on the fairgrounds, also went under. In the meantime, constant rain contributed to light attendance throughout the month of May, which averaged only about 30,000 paying customers daily. In an effort to overcome this slow start and to attract more visitors, the Fair Company added special events like parades, fireworks, and concerts with bandmaster John Philip Sousa and pianist Ignance Padarewski. One might have thought that with 65,000 exhibits, "Little Egypt" belly dancing on the Midway, and Buffalo Bill's Wild West Show just outside the fairgrounds, extra attractions were the last thing this Fair needed.

What ultimately saved the day was the extraordinary financial success enjoyed by the concessions along the Midway Plaisance, a $\frac{7}{8}$-mile-long thoroughfare connecting Jackson and Washington Parks. (Today, it forms part of the University of Chicago campus.) The Chicago Midway was the prototype for what has since become a standard feature at American fairs and carnivals. Its success also helps explain why, as cultural events, modern-day World's Fairs are not nearly as interesting as their nineteenth-century forerunners. Certainly the organizers of the Chicago Fair would have been appalled to learn that the true legacy of the Columbian Exposition would be found on its Midway rather than in the *Beaux-Arts* ideals of its Court of Honor. Yet under the shrewd management of twenty-one-year-old Sol Bloom, the Midway enabled the Fair Company not only to survive a potential financial disaster but also to squeeze out a slight profit in the end. Once word of its attractions got out, attendance figures skyrocketed: the June figures tripled those in May; the total figure in August reached 3.5 million; September drew 4.5 million; and during the closing month of October, 6.8 million visitors went through the turnstiles, a figure thirty percent higher than the best month at the Paris Fair of 1889 with its Eiffel Tower.[9]

That the Chicago effort surpassed the Parisian show must have gratified its American organizers for they had drawn heavily on the 1889 exposition, mirroring such details as the decorative material used in the construction of

[8]The details here and on the following page may be found in various of the sources cited in note 3 above. We have, therefore, chosen not to reference each fact separately.

[9]Badger, p. 131.

The Giant Ferris Wheel on the Midway at the World's Columbian Expositon, Chicago, Illinois, 1893. One of the buildings of the University of Chicago may be seen in the center left of the photograph.

the major buildings and the sculptural motifs that adorned the water basin of the Court of Honor. If the Parisians had ultimately succeeded in staging a more harmonious and graceful affair, the Chicagoans made up for it with the stunning variety, extent, and scope of their extravaganza. Spread over 633 acres, an area four times the size of any previous fair, the Columbian Exposition still had difficulty making room for everything.[10] Chicago had the giant 250-foot Ferris wheel on the Midway with its thirty-six wood-veneered cars, each of which held sixty people. It also boasted the world's largest telescope housed in the world's largest building, Post's gigantic Hall of Manufactures and Liberal Arts. For those whose tastes ran along strictly epicurean lines, there were such oddities as the eleven-ton cheese from Ontario and the 1500-pound chocolate Venus de Milo from New York.

The exhibits themselves were among the most exotic features of the Fair, however. At the Palace of Fine Arts (which today houses the only remaining architectural vestige of the 1893 Fair, Chicago's Museum of Science and Industry), the seventy-four galleries displaying over 9,000 works of art went up in only slightly more than a month, and with predictable results. Often many of the more significant exhibits were overlooked amidst all the hoopla. Thus, while thousands thronged around General Electric's seventy-foot tower of light bulbs, Edison's kinetograph, tucked away in a corner of Electricity Hall, went virtually unnoticed. Not that it would have been easy to sort out the really significant and interesting from this vast conglomeration of gimmicks and stunning inventions. A conservative estimate had it that to see everything just once and *quickly* would have required about three weeks and over 150 miles of walking![11] Henry Adams, who visited for a fortnight, gave this lucid and inimitable description of what he encountered:

> [there was] matter of study to fill a hundred years ... at Chicago, educational game started like rabbits from every building, and ran out of sight among thousands of its kind before one could mark its burrow. The Exposition itself defied philosophy. One might find fault till the last gate closed, one could still explain nothing that needed explanation. As a scenic display, Paris had never approached it, but the inconceivable scenic display consisted in its being there at all—more surprising, as it was, than anything else on the continent, Niagara Falls, the Yellowstone Geysers, and the whole railway system thrown in, since these are all natural products in their place; while since Noah's

---

[10]For a pictorial sense of the conglomeration of architectural styles and exhibits at the Fair, see Appelbaum, and James William Buel, *The Magic City: A Massive Portfolio of Original Photographic Views of the Great World's Fair and Its Treasures of Art, Including a Vivid Presentation of the Famous Midway Plaisance* (St. Louis: Historical Publishing Co., 1894; reprint ed., New York: Arno Press, 1974). The latter work, especially, contains photographs of the Krupp Building and the Women's Building, among many, many others. See below.

[11]Appelbaum, p. 5.

> Ark, no such Babel of loose and ill-joined, such vague and ill-defined and unrelated thoughts and half-thoughts and experimental outcries as the Exposition, had ever ruffled the surfaces of the Lakes.[12]

While Chicago may have outdone Paris in terms of sheer spectacle, it still had much in common with the 1889 Exposition. In particular, it borrowed the idea of concurrently staging "scientific" congresses. Apparently something of an afterthought in the case of the Columbian Exposition, its Congresses were conceived and promoted by Chicago lawyer Charles C. Bonney as an antidote to the "Windy City's" materialistic image.[13] Like the Fair proper, these meetings taken as a whole brought together myriad conflicting ideas, visions, and images.

Run by an independent organization, the World's Congress Auxiliary, the Congresses convened about seven miles north of the Jackson Park fairgrounds adjacent to the newly built structure located at Michigan and Adams that later became the Chicago Art Institute. In terms of attendance, the two most successful of the gatherings were the World's Parliament of Religions and the Congress of Representative Women, the latter drawing 150,000 spectators. In fact, women played an auspicious role at the Columbian Exposition, which featured a special Woman's Building designed by Sophie Hayden and prominently located on the lagoon just north of Jenney's Horticultural Building. It housed one of the Fair's finest works of art, a mural entitled "Modern Woman" by the American expatriate artist Mary Cassatt.[14] The Fair's most heavily attended Congress, however, the World's Parliament of Religions, attracted some 700,000 spectators and representative religious figures from all around the world. Practically all who took part in this six-day event judged it a great success; indeed, with such harmony abounding, many felt the millennium at hand.

Meanwhile, the Krupp armory exhibit evoked a rather different image back on the fairgrounds. Built at a cost of $1,500,000 and located along the lakeshore in a smaller replica of Villa Hügel, the family's palatial estate in Essen, it displayed the world's largest gun with a firing range of nearly thirteen miles. *Scientific American*, which reported on the Krupp Pavilion, began by noting the strong presence at the Fair of Kaiser Wilhelm II's revitalized country:

> Of all the foreign nations that are taking part in the World's Columbian Exposition at Chicago, Germany takes the lead,

---

[12]Henry Adams, *The Education of Henry Adams* (Boston: Houghton Mifflin Co., 1961), pp. 339–340. This is also quoted in part in Badger, p. 120.

[13]Badger, pp. 77–78.

[14]On the Woman's Building and its conception and exhibits, see Jeanne Madeline Weimann, *The Fair Women: The Story of The Woman's Building, World's Columbian Exposition, Chicago 1893* (Chicago: Academy, 1981).

> in extent, variety, cost, and superiority in almost every characteristic. Of the private exhibitions, Krupp, the great metal manufacturer of Germany, stands at the head. His exhibit is wonderful, and by its greatness almost dwarfs all other exhibits in the same line.[15]

It may seem incredible today that the Krupps would have lavished their wealth on such an ominous showcase of weaponry, but the firm certainly knew what it was doing. Alfred Krupp had staged one of his first great triumphs forty years earlier in London when he put his steel cannons on display in the Crystal Palace, and his son, Friedrich, realized that there was no better opportunity to gain trade contracts with foreign countries than by crushing the competition at a world's fair.[16] Apparently, no one protested the incongruity of such battlefield exhibits at an event intended to celebrate the promise of a new age of prosperity. In part, this was no doubt due to the belief that war represented a noble and chivalrous activity rather than a ghastly threat to civilized life. Beyond this, however, the expressions of international goodwill so often bandied about at the fairs only thinly camouflaged the competitive attitudes and subterranean hostilities that the leading nations harbored against one another. Henry Adams was right—the Chicago Fair was a murky confusion of voices on the brink of modernity. Its vitality was frenzied and directionless, a strange mixture of technical prowess coupled with a longing for the security of a passing age. Probably not a few of the 700,000 strong who saw themselves as ushering in a new era at the World's Parliament of Religions felt equally enthralled by the power of Fritz Krupp's vision of the future. Certainly only a few noticed that one of the widely varying aspects of late-nineteenth-century culture represented at the Fair was mathematics, but among those who did were several of the formative figures in the emergence of the discipline at the research-level in the United States.

Felix Klein at the Chicago World's Fair

Felix Klein had been waiting a good ten years for a suitable opportunity to lecture in the United States, and the Columbian Exposition finally gave him his chance. A year earlier, and two weeks before President Harper had even appointed Bolza and Maschke, Klein had received a letter from Harper offering to finance a twelve-week lecture course that would run concurrently with it. It had apparently occurred to Chicago's President that this might induce Klein to attend the Mathematics Congress, then in its early planning stages, and thereby boost his new school's mathematics program. "I am confident," he wrote Klein, "that you do not appreciate how highly you are estimated in

[15] "The Krupp Exhibit at the Great Fair," *Scientific American* 69 (July–Dec. 1893):33 and 40 on p. 33. This was also quoted in William Manchester, *The Arms of Krupp,* 1587–1968 (New York: Bantam Books, 1981), p. 247.

[16] See Manchester, pp. 76–79 and 247–248.

this country, nor can you estimate the inspiration your coming would furnish to scores of young men interested in this great department."[17] Although the exact terms remained vague—Harper merely assured him that the "compensation will be satisfactory"[18]—Klein eagerly awaited further details.

Writing to Bolza on 15 January 1893, Klein asked his former student to act as his negotiator. Bolza reported back that Harper was so busy that he only had one opportunity to speak with him, shortly before Christmas, and nothing further had been resolved. When Bolza finally got a chance to see him again, the energetic President no longer seemed quite as keen about an extended visit from Klein.[19] He informed Bolza that while he was still very interested in inviting Klein to Chicago, other departments stood in much greater need of development than mathematics. He offered to bring the matter before the Board of Trustees, but thought it unlikely that they would approve more than $2,000 for the venture. Bolza, who had suggested a figure of $3,000, pointed out that even an ordinary Full Professor made more than $2,000 a semester, adding that he doubted Klein would accept anything less than $2,500. Harper then reiterated that he would put the issue before the Trustees, but continued to express his doubts about a figure higher than $2,000. He also made it clear that he would rather drop the whole matter than risk a rejected offer.[20] E. H. Moore, who headed the local organizing committee, had, of course, been hoping all along that Klein would attend the Congress. On 29 April 1893, he wrote to the latter: "You know how *very* many admirers and friends you have among the promising younger mathematicians of this country and can see what a powerful stimulus to mathematics here it would be if you were to be present at the Congress and honor us with a paper."[21] Klein, however, had bigger ideas in mind than simply presenting a paper at a mathematics meeting.

This, apparently, was where matters stood in the spring of 1893 when Klein worked out an arrangement with the Prussian Ministry of Culture whereby he would attend the Mathematics Congress as an official representative of his government. Interestingly enough, after this, his negotiations with Harper quickly fell by the wayside; Klein obviously had more interest in enhancing his position and that of Göttingen than he did in earning a few thousand dollars lecturing in the United States. His appointment clearly signified that he now stood in the good graces of Friedrich Althoff, the head of higher education in Prussia and, in the years that lay ahead, Klein's most powerful and influential political benefactor.

[17]William R. Harper to Felix Klein, 3 June 1892, Klein Nachlass IX, Niedersächsische Staats- und Universitätsbibliothek, Göttingen (hereinafter abbreviated NSUB, Göttingen).

[18]*Ibid.*

[19]Oskar Bolza to Felix Klein, 15 January 1893, Klein Nachlass VIII, NSUB, Göttingen.

[20]*Ibid.*

[21]E. H. Moore to Felix Klein, 29 April 1893, Klein Nachlass X, NSUB, Göttingen (Moore's emphasis).

Relative to the planned lecture series, Klein realized that he would never have a better opportunity to sound his message than at the Fair, where he could engage the captive audience gathered for the Mathematics Congress. He therefore announced that he would offer a two-week cycle of lectures on higher mathematics as a sequel to the regular Congress sessions. The organizers could not have been more delighted. Northwestern's Henry White heartily endorsed Klein's plan, telling him "that your scheme for a comprehensive survey of recent mathematics is the most useful, though not the least onerous, that could be devised. If you have strength to carry it out, it will be in every way a worthy contribution to a noble object."[22] E. H. Moore was even more emphatic. After learning of Klein's decision, he replied that "[y]ou cannot easily understand, or rather imagine, how much pleasure your two letters brought to me personally and to all of us interested in the success of the Congress. We are deeply grateful to you and your government .... As to the September Colloquia: I esteem it a great privilege to be one of the circle of those to profit by the inspiration of your leadership through the domains of modern mathematics."[23]

Klein viewed the Mathematics Congress not only as a superb chance for the German *Reich* to demonstrate its growing dominance in his field but also as a golden opportunity for him to solidify his reputation as the leading representative of German mathematics.[24] Except for this latter possibility, he might not have given much thought to attending the Congress at all. On 26 June 1893, he wrote to Simon Newcomb, a kindred spirit whose ambition and drive were more or less on a par with his own, that "I understand that opinions differ as to the importance of the Chicago Congress. I myself would not have thought of going had not our government directly empowered me. In fact, they have developed a great enthusiasm for it. I just read in the paper that Helmholtz will be traveling to Chicago on August 21 so as to attend the Electrical Congress as President of the *Physikalisch-technische Reichsanstalt*."[25] In fact, the seventy-two-year-old Helmholtz had been even less eager to make the trip to Chicago than Klein. Earlier he had written a friend in New York that he would not be coming, not only because of his age but also because "[g]reat exhibitions ... have never attracted me, nor have I found that they taught one anything of importance that one did not know before, or that was worth the disturbance and excitement that it cost one."[26] After the German government approached him, however, a sense of duty

[22]Henry Seely White to Felix Klein, 5 June 1893, Klein Nachlass XII, NSUB, Göttingen.

[23]E. H. Moore to Felix Klein, 14 June 1893, Klein Nachlass X, NSUB, Göttingen.

[24]We highlight the various cultural agendas played out by the mathematicians at the Fair in "Embedded in the Culture."

[25]Felix Klein to Simon Newcomb, 26 June 1893, Newcomb Papers, Library of Congress, Washington, D. C. (our translation).

[26]Leo Königsberger, *Hermann von Helmholtz*, trans. Frances A. Welby (New York: Dover Publications, Inc., 1965), p. 410.

overrode his initial judgment, and the senior statesman of German science reluctantly decided to undertake the arduous journey. When he wrote back to inform his friend of his new plans, he expressed this startling opinion: "I am convinced that America represents the future of civilized Humanity, and that it contains a vast number of interesting men, while in Europe we have only chaos or the supremacy of Russia to look forward to."[27]

The Prussian *Kultusministerium* appeared intent to send off its biggest scientific names, and no one could have been happier about this than Klein. After agreeing to serve as a commissioner at the Fair, he began soliciting expository articles from Germany's leading mathematicians. He intended to present these at the Mathematics Congress and thereby to profile for those in attendance the recent progress of his countrymen in a variety of fields. He also engaged the services of his ever-faithful protégé, Walther von Dyck, whom he asked to prepare a collection of mathematical models and apparatus for inclusion in the German Universities Exhibit.[28] Dyck had just readied a similar display destined for the autumn meeting of the *Deutsche Mathematiker-Vereinigung* in Munich, so this meant mounting another exhibit with another catalogue describing another set of models and assorted paraphernalia.[29] By June, however, Klein appeared none too optimistic about Germany's chances of staging anything that would win high marks for its cultural mission abroad. In a letter to Bolza marked "Confidential and only to be discussed with Maschke," he wrote:

> I have the impression that the general German University Exhibit will only come together very imperfectly: the whole thing was taken up too late and lacks the right personalities to lead it. But I am sticking together with my friends [most notably Dyck, whom he described as the "soul behind the operation"] so that at least the mathematics [exhibit] will succeed .... My request of you and Maschke is a personal one, namely, that you not leave us in the lurch, that you deal with our undertaking as if it were your own, and that you doggedly carry it through even in the face of surrounding discouragement.[30]

This assessment of the overall situation was, as it turned out, quite accurate.

---

[27] *Ibid.*, p. 411.

[28] Several older American universities, like Harvard and Columbia, also mounted exhibits at the Fair, whereas officials at Cornell decided, against the wishes of the faculty, to confine their efforts to an exhibit in agricultural science. See James E. Oliver to Felix Klein, 25 January 1892, Klein Nachlass XI, NSUB, Göttingen.

[29] Walther von Dyck, ed., *Deutsche Unterrichtsausstellung in Chicago,* 1893: *Special-Katalog der mathematischen Ausstellung* (Berlin: n.p., 1893). Dyck had originally hoped to unveil this collection a year earlier at the Nuremberg meeting, but this had to be canceled due to a cholera epidemic.

[30] Draft of Felix Klein to Oskar Bolza, 6 June 1893, Klein Nachlass VIII, NSUB, Göttingen (our translation).

THE MATHEMATICS SECTION OF THE GERMAN UNIVERSITIES EXHIBIT

Friedrich Schmidt(-Ott), the government official charged with overseeing the German Universities Exhibit, later admitted that the entire project was both underfunded and disorganized as a result of time constraints.[31] Furthermore, even if Klein thought Schmidt(-Ott) not quite the "right personality" for the job, the truth is that with only about a year's notice and a paltry budget of 300,000 Reich Marks he had little chance to demonstrate his administrative prowess.

On the other side of the Atlantic, Maschke and Bolza awaited word from Klein, who admonished them to assure that the shipments reached their final destination on the fairgrounds. They found themselves in a ticklish situation when it came to satisfying their former mentor's request. Although more than willing to do what they could to help Klein, they also felt compelled to inform him that they could go only so far in supporting the "show" he was planning. In particular, they balked when it came to Klein's plans to have the survey articles he had gathered, which were written in German, read before the Congress. This was simply out of the question for Maschke and Bolza. "Our foreign status," wrote Bolza, "is already a touchy point, and we must avoid altogether anything that would call attention to it."[32]

Still, Maschke did his part in setting up the German mathematics exhibit, which was, if nothing else, a testimonial to Teutonic thoroughness.

[31] Friedrich Schmidt-Ott, *Erlebtes und Erstrebtes,* 1860–1950 (Wiesbaden: n.p., 1952), p. 28.

[32] Oskar Bolza to Felix Klein, 25 June 1893, Klein Nachlass VIII, NSUB, Göttingen (our translation).

It centered around seven glass cases filled with plaster and string models (many manufactured by the Darmstadt firm of L. Brill) as well as a variety of other mathematical instruments—harmonic analyzers, planimeters, calculating machines, to name a few. A special Gauss-Weber exhibit featured a colossal bust of Gauss and a replica of the telegraphic instrument designed by Gauss and Weber, which the exhibitors hoped would help counter the widespread belief in the United States that Samuel F. B. Morse had been the first to construct such a device. Portraits of Jacobi, Dirichlet, and Riemann adorned the walls which were filled with a whole library full of mathematics books: the collected works of Dirichlet, Gauss, Jacobi, Möbius, Riemann, and Weber; the entire series of mathematics and science publications of the Berlin, Göttingen, Leipzig, and Munich Academies of Science; complete sets of seven German journals devoted to one or another aspect of mathematics; around 500 textbooks (mostly the classics of the day); and to top it all off, copies of every doctoral dissertation or *Habilitationsschrift* written on a mathematical subject at a German university since 1850.[33] Such a grandiose display might well have made sense at a meeting like the International Congress of Mathematicians held in Heidelberg in 1904, the first to take place on German soil, and not just because it would have saved shipping all those volumes of erudite scholarship across the ocean and through the Great Lakes. On the Chicago fairgrounds, amid the sea of humanity that flowed past this monument to the German mathematical endeavor, who but the few mathematicians in attendance would have noticed anything more than some curious shapes and other geegaws in glass cases surrounded by ponderous-looking volumes with meaningless titles and authors? While it would be difficult to conceive of a more unlikely setting for a mathematical event that would mark a milestone in the history of American mathematics, the mathematicians present clearly felt otherwise. For them, the German mathematics exhibit indicated that their discipline occupied its own special niche in the pantheon of modern culture.

The Chicago Mathematics Congress

The Chicago Mathematics Congress defies measurement by the conventional yardsticks used to determine the relative success of a mathematics meeting. Although only nominally an international Congress with only a handful of mathematicians from outside the Midwest in attendance, it succeeded in showing that significant mathematical activity was taking place in the United States, and not just at the traditional institutions of higher learning on the East Coast. For E. H. Moore and his Chicago colleagues, the Congress and subsequent Colloquium represented unique opportunities

[33]See Dyck, ed., *Deutsche Unterrichtsausstellung in Chicago,* 1893, pp. iii–ix, and *Amtlicher Bericht über die Weltausstellungen in Chicago* 1893, *erstattet von Reichskommisar,* 2 vols. (Berlin: Reichsdruckerei, 1894), p. 988.

to put Midwestern mathematics on the map, and as such they were rife with symbolic significance.

As it turned out, only four foreigners attended—Klein, Eduard Study, the Austrian Norbert Herz, and Bernard Palidini of Pisa—and of these only Klein and Study actually participated.[34] Of the forty-five mathematicians present on opening day, at most a dozen could be considered serious researchers, whereas many of the others scarcely had any idea of what modern mathematics entailed.[35] That, however, was precisely what Klein had come to tell them.

The festivities commenced on the morning of Monday, 21 August, in a joint session of the scientific Congresses in mathematics, astronomy, and astrophysics. During these opening ceremonies, Charles C. Bonney, President of the World's Congress Auxiliary, joined other Fair officials in greeting the scientists. The organizers of the three Congresses then responded, and Klein gave a brief speech in which he communicated the goodwill of the German government. Helmholtz's introduction followed and elicited a great cheer from those in attendance. Taking slightly less than an hour, these formalities finally gave way to the individual Congresses themselves. For their parts, the mathematicians and astronomers convened in a joint session chaired by the astronomer George W. Hough.[36]

During this brief gathering, Klein gave a somewhat longer speech in which he repeated his earlier invitation to visit the German Universities Exhibit. He also called attention to the survey articles he had brought with him. These, he asserted, "give collectively a fairly complete account of contemporaneous mathematical activity in Germany."[37] He then launched upon a sweeping survey of modern mathematics that shed considerable light on his own philosophical views:

> When we contemplate the development of mathematics in this nineteenth century, we find something similar to what has taken place in other sciences. The famous investigators of the preceding period, Lagrange, Laplace, Gauss, were each great enough to embrace all branches of mathematics and its applications. In particular, astronomy and mathematics were in their time regarded as inseparable.

---

[34] Study's presence owed to the fact that he happened to be job-hunting at this time; soon afterward he landed a temporary position at the Johns Hopkins.

[35] A list of the participants can be found in H. S. White, "A Brief Account of the Congress on Mathematics held at Chicago in August, 1893," in *Mathematical Papers Read at the International Mathematical Congress Held in Connection with the World's Columbian Exposition Chicago* 1893, ed. E. H. Moore *et al.* (New York: Macmillan & Co., 1896), pp. vii–xii on pp. ix–xii (hereinafter denoted *Congress Papers*).

[36] *Ibid.*, pp. vii–xii.

[37] Felix Klein, "The Present State of Mathematics," *Congress Papers*, pp. 133–135, or Felix Klein, *Gesammelte Mathematische Abhandlungen*, 3 vols. (Berlin: Springer-Verlag, 1921–1923), 2:613–615 on p. 613 (hereinafter abbreviated *Klein*: *GMA*).

> With the succeeding generation, however, the tendency to specialization manifests itself. . . . [T]he developing science departs at the same time more and more from its original scope and purpose and threatens to sacrifice its earlier unity and to split into diverse branches.[38]

Concomitant with this fragmentation within mathematics, Klein noted that "the attention bestowed upon it by the general scientific public diminishes. It became almost the custom to regard modern mathematical speculation as something having no general interest or importance, and the proposal has often been made that, at least for purpose of instruction, all results be formulated from the same standpoints as in the earlier period. Such conditions were unquestionably to be regretted."[39]

Always sensitive to this problem and the dangers posed by the rapid proliferation of research in pure mathematics, Klein adamantly argued for the unity of all knowledge, a notion he believed was embodied in the very ideals of German *Wissenschaft*. Yet, even in the face of these tendencies, Klein saw reason for optimism, and he unhesitatingly claimed that

> [t]his is a picture of the past. I wish on the present occasion to state and to emphasize that in the last two decades a marked improvement from within has asserted itself in our science, with constantly increasing success . . . . This unifying tendency, originally purely theoretical, comes inevitably to extend to the applications of mathematics in other sciences, and on the other hand is sustained and reinforced in the development and extension of these latter.[40]

For Klein, the basis for internal unity stemmed largely from the group concept and the notion of an analytic function of a complex variable. As for the utility of pure mathematics in promoting progress in the sciences, he gave two examples to illustrate this point. The first concerned the application of group theory to the classification of crystallographic structures. Although much had been done along these lines in earlier decades, not until the 1880s did Evgenii Federov and Arthur Schoenflies spell out this approach explicitly.[41] By 1891, Schoenflies had already written a textbook on the subject,

[38] *Ibid.*, p. 613–614.

[39] *Ibid.*, p. 614.

[40] *Ibid.*

[41] Klein's interest in and involvement with the crystallographic researches of Federov and Schoenflies is documented in J. J. Burckhardt, "Der Briefwechsel von E. S. von Federow und F. Klein, 1893," *Archive for History of Exact Sciences* 9 (1972):85–93; and J. J. Burckhardt, "Der Briefwechsel von E. S. von Federow und A. Schoenflies," *Archive for History of Exact Sciences* 7 (1971):91–141. Erhard Scholz gives the historical background to these group-theoretic studies in "Crystallographic Symmetry Concepts and Group Theory (1850–1880)," in *The History of Modern Mathematics*, ed. David E. Rowe and John McCleary, 2 vols. (Boston: Academic

and Klein asked him to write up some of the historical background for a paper on "Krystallographie und Gruppentheorie" presented to the Chicago Congress.[42] The other example to which Klein alluded involved connections between various astronomical problems and the theory of linear differential equations, a theme elaborated upon in an article by Heinrich Burkhardt.[43] In his address, Klein further underscored the importance of George William Hill's methods in establishing this linkage. (See Chapter 1.) Hill had used infinite determinants to approximate solutions of a linear differential equation in the neighborhood of an essential singularity, work that had important consequences for lunar theory.

In closing, however, Klein could not resist propagandizing for Göttingen mathematics:

> Speaking, as I do, under the influence of our Göttingen traditions, and dominated somewhat, perhaps, by the great name of *Gauss*, I may be pardoned if I characterize the tendency that has been outlined in these remarks as a *return to the general Gaussian programme.* A distinction between the present and the earlier period lies evidently in this: that what was formerly begun by a single mastermind, we now must seek to accomplish by united efforts and cooperation.[44]

After alluding to Poincaré's efforts to foster this kind of alliance in France, Klein went on to note that "[f]or similar purposes we three years ago founded in Germany a mathematical society, and I greet the young society in New York and its Bulletin as being in harmony with our aspirations. But our mathematicians must go further still. They must form international unions, and I trust that this present World's Congress at Chicago will be a step in that direction."[45]

Having set the tone and laid out the broader dimensions of his mission abroad, Klein undoubtedly felt that the meeting had gotten off to a satisfactory start. In fact, this joint session lasted only about a half an hour before the mathematicians and astronomers voted to meet separately. After posing for group photographs, the mathematicians convened in an adjacent room and

---

Press, Inc., 1989), 2:3–27; and Erhard Scholz, *Symmetrie - Gruppe - Dualität: Zur Beziehung zwischen theoretischer Mathematik und Anwendungen in Kristallographie und Baustatik des* 19. *Jahrhunderts*, Science Networks - Historical Studies, vol. 1 (Basel: Birkhäuser, 1989).

[42] Arthur Schoenflies, "Gruppentheorie und Krystallographie," *Congress Papers*, pp. 341–349.

[43] Heinrich Burkhardt, "Ueber einige mathematische Resultaten astronomischer Untersuchungen, insbesondere über irreguläre Integrale linearer Differentialgleichungen," *Congress Papers*, pp. 13–34.

[44] Klein, *The Present State of Mathematics*, p. 615 (Klein's emphasis).

[45] *Ibid.*

spent the remainder of the morning on organizational matters. E. H. Moore presided over this most auspicious planning session in his capacity as chair of the local committee. A nominating committee consisting of John Monroe Van Vleck, Henry T. Eddy, and Oskar Bolza then put forth the following candidates for the Executive Committee, all of whom were unanimously elected by the Congress participants: William Story of Clark University, President; E. H. Moore, Vice-President; Harry W. Tyler of MIT, Secretary; and Klein and Henry Seely White. The main item of business on this new committee's agenda was the establishment of a program for the next five days. It resolved to open each morning session at 9:30 with summary accounts of the numerous papers communicated by mathematicians not in attendance and to follow this with anywhere from one to four lectures. This would leave the afternoons of Tuesday, Wednesday, and Friday open so that participants could visit the German Universities Exhibit and hear Klein's lecture-demonstrations of the models and other apparatus on display there. With these details out of the way, properly mathematical matters could quickly come to the fore.

E. H. Moore opened the Tuesday morning session by moving that the Congress elect Klein as its "Honorary President," a motion it passed by acclamation. Then, a printed program was distributed which listed the various speakers as well as the papers to be presented *in absentia*. (See Table 7.1.) Strikingly modern and readable, the papers and lectures presented in Chicago seemed crafted to communicate with the widest possible audience. Most of them were prepared as reports rather than as original research work, and their subject matter mirrored fairly closely the era's mainstream fields of research: elliptic and Abelian integrals and functions, invariant theory, hypercomplex number systems, finite groups, and Lie theory. Instead of plunging into various possible thickets of technicalities, the speakers and authors consciously motivated their presentations with historical remarks and emphasized the broader contexts that informed developments in these fields. If many of these mathematicians were specialists, they certainly knew how to avoid the standard pitfalls that often plague the expert when addressing a more general audience.

Klein began the session on Tuesday morning by trotting out some of his biggest names: David Hilbert on invariant theory, Heinrich Weber on Diophantine equations, and Adolf Hurwitz on the reduction of binary quadratic forms. Hilbert's paper gave a summary of his monumental work in algebraic invariant theory during the preceding five years. In it, he pointed out that this subject had quickly evolved into a major research field due to the discovery of a variety of formal methods for constructing invariant expressions, but he hastened to add that his own approach followed an entirely different line of reasoning. This enabled him to prove such important results as his celebrated finite basis theorem. Typically enough, Hilbert said very little about the work of the leading invariant theorists who had preceded him, although he ended

THE PARTICIPANTS AT THE OPENING PLENARY SESSION OF ASTRONOMERS AND MATHEMATICIANS.

his essay with a rather illuminating historical observation:

> One can normally distinguish with ease three clear developmental stages in the history of a mathematical theory: the naive, the formal, and the critical. Thus in the theory of algebraic invariants, the founders, Cayley and Sylvester, are also to be regarded as representatives of the naive period: they had the immediate joy of first discovering how to exhibit the simplest invariant expressions and to apply them elegantly to the solution of equations of the first four degrees. The inventors and perfecters of the symbolic calculus, Clebsch and Gordan, are the representatives of the second stage, whereas the critical period finds its expression in the theorems 6-13 listed above.[46]

These theorems were, of course, Hilbert's own.

Eduard Study gave the most interesting of the four lectures actually delivered on Tuesday, but since he spoke in German it probably failed to make much of an impression on many in the audience.[47] In his remarks, Study described a good deal of nineteenth-century work on hypercomplex number systems: William Clifford's biquaternions, Sylvester's nonions (recall Chapter 3), and the relevant work on Lie groups and Lie algebras undertaken by some of Sophus Lie's leading followers—Friedrich Schur, Theodor Molien, Friedrich Engel, Georg Scheffers, and Study himself. He also mentioned the fundamental result of Frobenius: that any associative system of hypercomplex numbers with division is isomorphic to either the real numbers, the complex numbers, or the quaternions.[48]

The Wednesday morning session opened with a paper on number theory and geometry written by Minkowski that anticipated by two years certain results he presented in his well-known *Geometrie der Zahlen.*[49] For example, the Americans were treated, in the case of $\mathbb{R}^3$, to his famous theorem on closed, convex subsets V that are symmetric about the origin, which states that if the volume of $V \geq 2^3$, then V must contain at least one point with integral coordinates other than $(0, 0, 0)$. Perhaps immodestly, Minkowski

[46] David Hilbert, "Ueber die Theorie der algebraischen Invarianten," *Congress Papers.*, pp. 116–124 on p. 124, or David Hilbert, *Gesammelte Abhandlungen*, 3 vols. (Berlin: Julius Springer, 1933–1935), 2:376–383. For a discussion of pre-Hilbertian invariant theory, see Karen Hunger Parshall, "Toward a History of Nineteenth-Century Invariant Theory," in *The History of Modern Mathematics*, 1:157–206. The impact of Hilbert's ideas on the area of invariant theory is analyzed in Karen Hunger Parshall, "The One-Hundredth Anniversary of the Death of Invariant Theory?," *The Mathematical Intelligencer* 12 (4) (1990):10–16.

[47] Eduard Study, "Ältere und neuere Untersuchungen über Systeme complexer Zahlen," *Congress Papers*, pp. 367–381.

[48] Georg Frobenius, "Ueber lineare Substitutionen und bilineare Formen," *Journal für die reine und angewandte Mathematik* 84 (1878):1–63, or Jean-Pierre Serre, ed. *Ferdinand Georg Frobenius: Gesammelte Abhandlungen*, 3 vols. (New York: Springer-Verlag, 1968), 1:343–405. As noted in Chapter 2, this result was proved independently by Charles Sanders Peirce, but appeared three years after the publication of Frobenius' paper.

[49] Hermann Minkowski, *Geometrie der Zahlen* (Leipzig: B. G. Teubner, 1896).

The Congress of Mathematicians at the World's Columbian Exposition, Chicago, Illinois, 1893. From left to right (bottom row) James E. Oliver and William E. Story; (second row) William B. Smith, Henry S. White, Felix Klein, Harry W. Tyler, and Thomas F. Holgate; (third row) Arthur G. Webster, Clarence A. Waldo, Eduard Study, John M. Van Vleck, Henry T. Eddy, James B. Shaw, James MacMahon, and John Kleinheksel; (top row) Edwin M. Blake, Herbert G. Keppel, Frank Loud, Henry Taber, Oskar Bolza, E. H. Moore, and Heinrich Maschke.

offered the opinion that "this theorem on nowhere concave bodies with middle point appears to me to be among the most fruitful in all of number theory."[50] After indicating several of its applications, he concluded by mentioning that the theorem could be generalized, an allusion perhaps to the fact that it holds in spaces of arbitrary dimension.

Felix Klein found Minkowski's fusion of geometry and number theory highly congenial, not least because it reconfirmed his lifelong faith in the fertility of such approaches. Yet, while in Chicago, Klein was especially on the lookout for new developments in applied mathematics, and he learned about some of these from the then President of Indiana's Rose Polytechnic Institute, Henry Turner Eddy, who spoke about recent work in modern graphical methods.[51] Eddy represented the growing strength of engineering science in the United States. A graduate of Yale's Sheffield Scientific School, he went on to take his doctorate from Cornell and to emerge as one of the first of a new generation of American scientific technologists.[52]

Eddy's lecture evinced an impressive knowledge of descriptive geometry in general and of graphical statics in particular. He framed his remarks by indicating how work in this field had traditionally relied on two principal techniques: the method of the reciprocal frame with its associated force diagrams and the notion of the equilibrium polygon or catenary. The first method utilized Newtonian force parallelograms in order to display the reciprocal relationship between the structure of a framework and the forces acting at its joints. This technique had first been employed by James Clerk Maxwell and had later been greatly elaborated by Luigi Cremona and others. The second approach grew from work of the seventeenth-century mathematician Pierre Varignon, but its real importance for statics became apparent only through later work of Carl Culmann. Culmann had studied Poncelet's projective geometry in Metz and had brought this tradition into the engineering schools in Germany. Along with Franz Reuleaux, Ferdinand Redtenbacher, and Franz Grashof, he developed sophisticated mathematical analyses of the efficiency of machines, although by the 1890s this tradition at the German *Technische Hochschulen* had increasingly come under attack by a younger generation of engineers who found it far removed from the needs of modern technology.[53] Regarding the second edition of Culmann's textbook

[50] Hermann Minkowski, "Ueber Eigenschaften von ganzen Zahlen, die durch räumliche Anschauung erschlossen sind," *Congress Papers*, pp. 201–207 on p. 205 (our translation).

[51] Henry T. Eddy, "Modern Graphical Developments," *Congress Papers*, pp. 58–71.

[52] His influential study of 1878, *Researches in Graphical Statics*, appeared in German translation just two years later. Henry Turner Eddy, *Researches in Graphical Statics* (New York: D. Van Nostrand, 1878); *Neue Constructionen aus der graphischen Statik* (Leipzig: B. G. Teubner, 1880). On Eddy, see Edwin Layton, "Mirror-Image Twins: The Communities of Science and Technology in 19th-Century America," *Technology and Culture* 12 (1971):562–580 on p.575.

[53] The Chicago World's Fair played an important role in stimulating this debate. Alois Riedler, one of the chief protagonists for the praxis-oriented engineers in Germany, traveled to Chicago as an official emissary of the German government, and afterward filed a report in

Graphische Statik (1875), Eddy indicated in his lecture that "its publication in English is not a matter of great importance now. It is too learned for practical use by busy men."[54]

With his prejudices clearly enunciated, Eddy proceeded to devote the major part of his lecture to a synopsis of the second edition of Maurice Lévy's four-volume study, *La Statique graphique et ses Applications aux Constructions.*[55] Eddy regarded this lengthy study as nothing less than a masterpiece. "The subject is treated with a detail, precision and elegance such as especially distinguish the best French scientific treatises," he wrote. "It is not too much to say that it is a work of such magnitude and acumen as to make it a monument of intellectual and mathematical power."[56] Besides the fact that Lévy had given Eddy and his work rather prominent attention, the American was particularly pleased that the Frenchman's exposition of the theory never lost sight of the practical applications for engineering.

Eddy not only appreciated applied mathematics fully, but he also had a solid command of the mathematical methods needed by engineers. Thus, while he may have been known for his expertise in graphical statics, he was certainly very familiar with devices like planimeters, various kinds of mechanical integrators, slide rules, indicator diagrams for steam engines, and graphical methods in electrodynamics. His lecture also revealed his talent for promoting the cause of engineering mathematics in general. He described a recent report of the British Association for the Advancement of Science which detailed the use of graphical methods in engineering and science, and which contained a ninety-five-page list of published articles dealing with such methods. Eddy asserted that this list ran to over 2,000 papers, "a fact which shows how widespread is the professional use of graphics and how it has come into use as a common medium of expression in England, where conservatism in methods is more persistent than in any other country where great constructions are common, except perhaps Germany."[57] Eddy's dynamism and his scientific sophistication left an impression on Klein, who returned to Germany with new ideas about how to forge stronger ties between scientists and engineers in his own country.[58]

which he pointed to American engineering schools as the model Germany should strive to emulate. Klein offered a very different opinion of American science and mathematics. Riedler and Klein clashed on several occasions after their return to Germany. For a detailed study of this conflict, see Susann Hensel, "Die Auseinandersetzungen um die mathematische Ausbildung der Ingenieure an den Technischen Hochschulen in Deutschland Ende des 19. Jahrhunderts," in S. Hensel, K.-N. Ihmig, and M. Otte, *Mathematik und Technik im* 19. *Jahrhundert in Deutschland*, Studien zur Wissenschafts-, Sozial-, und Bildungsgeschichte der Mathematik, vol. 6 (Göttingen: Vandenhoeck & Ruprecht, 1989), pp. 1–111, esp. pp. 55–73.

[54]Eddy, p. 61. The first edition was Carl Culmann, *Die graphische Statik* (Zürich: n.p., 1866).

[55]Maurice Lévy, *La Statique graphique et ses Applications aux Constructions*, 2nd ed., 4 vols. (Paris: Gauthier-Villars, 1886–1888).

[56]*Ibid.*, p. 63.

[57]*Ibid.*, p. 70.

[58]See Karl-Heinz Manegold, *Universität, Technische Hochschule und Industrie: Ein Beitrag zur Emanzipation der Technik im* 19. *Jahrhundert unter besonderer Berücksichtigung der Bestrebungen Felix Kleins* (Berlin: Duncker & Humblot, 1970), pp. 116–120.

Thursday's session, devoted to function theory, began with a paper by Charles Hermite and proceeded to a contribution from Sylvester's fourth and Klein's first American student, Irving Stringham. Stringham's work, "A Formulary for an Introduction to Elliptic Functions,"[59] opened by noting how the classical approach to elliptic functions inaugurated by Abel and Jacobi had been thrust into the shadows of late by Weierstrass' introduction of the functions σ, $\wp$, $\wp'$ in place of the Jacobian functions sn, cn, and dn.[60] As he pointed out, one significant advantage of Weierstrass' $\wp$-function lay in certain connections with invariant theory that had been hinted at by Arthur Cayley as early as 1846.[61]

Prompted by pedagogical considerations, Stringham questioned whether, despite the obvious virtues of the $\wp$-function, "it has been demonstrated that its introduction renders the Jacobian functions henceforth useless for the purposes of study and application."[62] To counter this view, he showed how to situate these Jacobian functions within the larger context of the transformation theory of elliptic integrals, that is, integrals $\int \frac{dz}{\sqrt{V(z)}}$, where $V(z)$ is a real polynomial of degree 3 or 4 having distinct roots. This theory focuses on transformations of z that produce a certain normal form of the elliptic differential $\frac{dz}{\sqrt{V(z)}}$. By appropriate choices of these transformations, five different types of normal forms emerge. The classical case, originally treated by Legendre but with restrictions on the variables, produces a normal form of the type $\frac{dz}{\sqrt{(1-z^2)(1-k^2z^2)}}$, where $k \in \mathbb{C}$ is the modulus of the integral. The transformation associated with Weierstrass' theory, first discovered by Charles Hermite and published by Cayley, reduces the differential to the normal form $\frac{dz}{\sqrt{4z^3-g_2z-g_3}}$, where g_2 and g_3 are certain invariants of $V(z)$. Stringham also mentioned a third normal form, the one preferred by Klein and exploited in one of his most important papers, that is, $\frac{dz}{\sqrt{z(1-z)(1-\sigma z)}}$, where σ is a root of the original $V(z)$.[63]

Following this presentation, Stringham next indicated how to introduce the elliptic functions in each of these cases. In particular, he pointed out

[59]Irving Stringham, "A Formulary for an Introduction to Elliptic Functions," *Congress Papers*, pp. 350–366.

[60]For the definitions of the Weierstrassian and Jacobian functions and a discussion of the theory surrounding them, see E. T. Whittaker and G. N. Watson, *A Course in Modern Analysis*, 4th ed. (Cambridge: University Press, 1927; reprint ed., Cambridge: University Press, 1963), pp. 429–535. We define sn and $\wp$ in note 73 in Chapter 4.

[61]Arthur Cayley, "On the Reduction of $du/\sqrt{U}$, when U is a Function of the Fourth Order," *Cambridge and Dublin Mathematical Journal*, 1 (1846):70–73, or Arthur Cayley and A.R. Forsyth, ed., *The Collected Mathematical Papers of Arthur Cayley*, 14 vols. (Cambridge: University Press, 1889–1898), 1:224–227 (hereinafter cited as *Math. Papers AC*).

[62]Stringham, p. 351.

[63]See Felix Klein, "Ueber die Transformation der elliptischen Funktionen und die Auflösung der Gleichungen fünften Grades," *Mathematische Annalen* 14 (1878–1879):111–172, or *Klein: GMA*, 3:13–75. Stringham also gives two further normal forms which we omit here.

that in four of the five cases a series of simple variable changes allows for the passage from one normal form to another. After this, he showed how to recover the basic Jacobian elliptic functions sn, cn, and dn within the framework of any of the other systems and actually illustrated the procedure from the point of view of Klein's approach. Of course, none of these musings was likely to lead to any new breakthroughs in research, as the author seemed well aware. Nevertheless, they do suggest that Stringham saw the forest as well as the trees, and his paper, written with intelligence and taste, makes interesting reading even today.

As Stringham's paper exemplified, by the early 1890s a number of Americans had reached the level of comprehending, assimilating, and appreciating many of the most advanced realms of pure mathematics. Still, very few had reached the stage where they could significantly extend the frontiers of research. As Heinrich Burkhardt underscored in his paper on linear differential equations in astronomy, however, at least one individual, George William Hill, had succeeded spectacularly even at this more rarefied level. Burkhardt, a *Privatdozent* in Göttingen from 1889 to 1897, had served as Klein's principal assistant in carrying out the Kleinian research program for hyperelliptic functions.[64] In his survey article, Burkhardt discussed the theory-driven work on linear differential equations with complex solutions undertaken by Lazarus Fuchs, Gösta Mittag-Leffler, and others.[65] He rightly remarked that Hill's approach had grown not so much out of these researches as out of observationally inspired investigations on the mutual gravitational effects that the sun and earth exert on the moon.

Shifting gears from mathematical astronomy to function theory proper by afternoon, Oskar Bolza stepped to the podium and spoke to his assembled colleagues about Weierstrass' theory of hyperelliptic integrals of the first and second kinds.[66] In spirit, his lecture resembled the paper submitted by Stringham, as both strove to give the audience an overview of important aspects within the related fields of elliptic and hyperelliptic function theory. In actuality, there was an unmistakable and critical difference between their two presentations. Its expository qualities notwithstanding, Stringham's paper had virtually no connections with recent mathematical research, whereas Bolza's was steeped in ongoing developments in a field (to which he himself had made noteworthy contributions) at the very heart of mathematical activity during this period.

---

[64]Klein expressed his appreciation of Burkhardt's gifts and accomplishments in the "Vorbemerkungen zu den Arbeiten über hyperelliptische und Abelsche Functionen" which he wrote to introduce his papers on this subject. See *Klein*: *GMA*, 3:317–322 on pp. 321–322.

[65]For a detailed discussion of Fuchs's work and its subsequent implications for nineteenth-century mathematics, see Jeremy Gray, *Linear Differential Equations and Group Theory from Riemann to Poincaré* (Boston: Birkhäuser, 1986), pp. 56–120.

[66]Oskar Bolza, "On Weierstrass' Systems of Hyperelliptic Integrals of the First and Second Kind," *Congress Papers*, pp. 1–12.

In keeping with his training and general orientation, Bolza surveyed Weierstrass' theory by drawing heavily on the Riemannian approach as championed by Klein. He began with a general hyperelliptic curve of genus p and showed how to construct a so-called canonical system of $2p$ integrals of the second kind satisfying the necessary bilinear relations. He then indicated how to transform any given canonical system into another and demonstrated the effect such a transformation has on the Weierstrass Θ-function. Finally, he discussed three special canonical systems: the more traditional ones of Riemann-Clebsch and of Weierstrass, and a new one developed by Klein which drew upon projective geometry and invariant theory.

From function theory, the Congress moved on to Friday's group-theoretic session, which opened with a paper by Clark University's Joseph de Perott on Listing's construction of the Galois group of order 660 associated with the modular equation of degree $p = 11$. This contribution was followed by a paper from Frank Nelson Cole that, in some sense, marked the first clear statement on American shores of the problem of determining and classifying all finite simple groups.[67] Cole began by assessing the then current state of group theory:

> Despite the great advances of the past fifty years, the Theory of Groups remains to-day in many respects in a very unfinished state. It is true that we possess an accurate system of general classification on the one hand and an elaborate knowledge of special cases on the other. But between these two extremes lies a vast middle ground, the exploration of which is extremely slow and difficult.[68]

In this connection Cole mentioned the difficulties researchers had encountered in their attempts to determine all groups of a given order, all simple groups, etc. He proceeded to chart a possible course for the theory's continued development:

> In an abstract and intricate theory like that of groups, too much must not be expected in the way of general development from the accumulation and study of individual examples. No amount of such experimentation could have led to our modern knowledge. Progress is from abstract to abstract. Nevertheless, in the absence of a general method, something may be accomplished by the tentative, step-by-step process, especially within moderate limits where the labor involved is not incommensurate with the value of the result. Thus, it is of some scientific interest to obtain all the groups of lower degrees and orders, and the simple groups below any convenient order.[69]

[67] As we saw in Chapter 2, however, Cayley had stated the more general problem of finding all finite groups of a given order n in his paper "The Theory of Groups," *American Journal of Mathematics* 1 (1878):50–52 on p. 51, or *Math. Papers AC*, 10:401–403.

[68] Frank Nelson Cole, "On a Certain Simple Group," *Congress Papers*, pp. 40–41.

[69] *Ibid.*

Following these remarks, Cole went on to enumerate all simple groups of order 660 or less, a result he had recently obtained independently of Otto Hölder.[70] This enumeration depended on Cole's discovery of a certain simple subgroup G of order 504 realized as a subgroup of the symmetric group S_9. (In modern notation, $G \cong PSL_2(8)$.) Cole's simplicity proof employed a now-standard argument from Klein's *Ikosaeder*,[71] namely, that any normal subgroup of a group G must be a union of complete conjugacy classes of G. Since Cole's group had five conjugacy classes consisting of 168 elements of order 9, 56 of order 3, 63 of order 2, and 216 of order 7, plus the identity, it follows that the only possible orders for a normal subgroup are

$$168\alpha + 56\beta + 63\gamma + 216\delta + 1 \quad \text{where } \alpha, \beta, \gamma, \delta \in \{0, 1\}.$$

Since this integer must also divide 504, its only possible solutions are $\alpha = \beta = \gamma = \delta = 0$ or $\alpha = \beta = \gamma = \delta = 1$, proving G is simple.

Another contribution to this session, the paper on crystallographic groups alluded to by Klein in his opening remarks, displayed Arthur Schoenflies's notable abilities as an expositor. Schoenflies began with an overview of nineteenth-century investigations by mineralogists such as Johann Friedrich Christian Hessel and August Bravais, who had first deduced the thirty-two known crystal classes on the basis of assumptions regarding the symmetry properties of their constituent molecules.[72] Because speculations about the molecular structure of matter generally met with disfavor, at least in mid-century Germany, it took several decades before Bravais's theory found many adherents there. By the 1880s, this situation had begun to change, and a variety of competing theories had arisen regarding the symmetry properties of molecular systems in crystals. These systems, as Schoenflies indicated, could be distinguished from one another by virtue of the corresponding symmetry groups admitted. For example, Bravais's theory, which assumed both congruence and parallel orientation of the individual molecules, led to a group of translations in three-space. If, as in the rival theory of Leonhard Sohncke, however, the molecules are assumed congruent but no longer in parallel orientation, the group consists of certain orientation-preserving isometries of

[70] Frank Nelson Cole, "Simple Groups as far as Order 660," *American Journal of Mathematics* 15 (1893):303–315; and Otto Hölder, "Ueber einfache Gruppen," *Mathematische Annalen* 40 (1892):55–88 as well as "Die Gruppen der Ordnungen p^3, pq^2, pqr, p^4," *Mathematische Annalen* 43 (1893):301–412. For a discussion of Hölder's work, consult B. L. van der Waerden, *A History of Algebra from al-Khwārizmīto Emmy Noether* (Berlin: Springer-Verlag, 1985), pp. 155–157, or more particularly, Julia Nicholson, "Otto Hölder and the Development of Group Theory and Galois Theory," (unpublished Dissertation, Oxford University, 1993).

[71] Felix Klein, *Vorlesungen über das Ikosaeder und die Auflösung der Gleichungen vom fünften Grade* (Leipzig: B. G. Teubner, 1884), pp. 17–18.

[72] Schoenflies, "Gruppentheorie und Krystallographie," *Congress Papers*, pp. 341–349. See the discussion in Scholz, *Symmetrie - Gruppe - Dualität*, pp. 110–154.

$\mathbb{R}^3$. Finally, in the most general system where the molecules may be mirror images of one another, the resultant group is the full group of isometries. For this most general case, Federov and Schoenflies proved that exactly 230 crystallographic groups exist, each of which corresponds to exactly one crystal type.

Schoenflies ended his survey by conceding that contemporaneous crystallographers had shown little willingness to drop Bravais's theory. This, he felt, was a mistake, and he registered a strong protest against their conservative attitude.[73] Felix Klein, who always argued that the potential applications of pure mathematics were practically limitless, later recalled with great satisfaction how Schoenflies's general theory quickly came back into vogue after 1912 with Max von Laue's discovery of the phenomenon of X-ray diffraction in crystals.[74]

Moving from finite to continuous groups, Franz Meyer presented, in tabular form, a summary classification of Lie groups of transformations of $\mathbb{R}^2$ that would soon appear in the third and final volume of the classic treatise of Lie and Engel.[75] Whereas Lie and Engel would exhibit the twenty-seven types of associated infinitesimal transformations (their Lie algebras), however, Meyer actually exhibited the transformation groups themselves. His first table, containing all possible projective groups in the plane, was followed by a list of corresponding geometric structures that remain invariant for each of the thirty types of projective transformations. Thus, Meyer presented a detailed and, in a sense, comprehensive classification scheme that contained, as a very special case, the sorts of structures studied by Klein and Lie in their early papers on the theory of W-curves.[76] Following in Lie's footsteps, Meyer displayed thirty types of characteristic differential equations left invariant by the corresponding projective groups in his third table. This thereby rounded out the list of those given in Lie's lectures on differ-

[73]In his words, "[o]ne can comprehend why crystallographers would remain attached to a theory if in all cases it possesses the advantage of greater simplicity and graspability; it is unjustified, however, to reject the more general structures in which the molecules have screw-shaped configurations simply because one regards them as 'unnatural' .... The decision as to whether the general structures can claim a physical validity ... depends on one thing only: whether the molecular effects of the general structures are in agreement with the remaining properties of matter, or whether it is possible to regard these as necessary physical properties of crystals, that is, as mathematical consequences of the molecular structure." See Schoenflies, pp. 348–349 (our translation).

[74]Felix Klein, *Vorlesungen über die Entwicklung der Mathematik im* 19. *Jahrhundert*, 2 vols. (Berlin: Springer-Verlag, 1926–1927), 1:344–345.

[75]Franz Meyer, "Tabellen von endlichen continuierlichen Transformationsgruppen," *Congress Papers*, pp. 187–200. The Lie-Engel work was published as Sophus Lie, *Theorie der Transformationsgruppen*, 3 vols. (Leipzig: B. G. Teubner, 1888, 1890, 1893).

[76]See Felix Klein and Sophus Lie, "Ueber diejenigen ebenen Kurven, welche durch ein geschlossenes System von einfach unendlich vielen vertauschbaren linearen Transformationen in sich übergehen," *Mathematische Annalen* 4 (1871):50–84, or *Klein*: *GMA*, 1:424–458. On W-curves, see Thomas Hawkins, "Line Geometry, Differential Equations, and the Birth of Lie's Theory of Groups," in *The History of Modern Mathematics*, 1: 275–329 on pp. 277–289.

ential equations as prepared by his student, Georg Scheffers.[77] Meyer also enumerated eighty-one types of homogeneous transformation groups in three variables and indicated at each stage where these belonged in his tabulation of projective transformations in the plane.

With these papers behind them, the conferees next heard from the remaining Chicago mathematicians. Maschke's lecture highlighted his recent work on a certain quaternary group (that is, subgroup G of $GL_4(\mathbb{C})$) of order $336 = 2 \cdot 168$ arising as the Galois group of the eighth-degree Jacobian modular equation. He mentioned straightaway Klein's contributions, noting his description of G as a subgroup of a quaternary group arising from line geometry as well as his realization that G contains a subgroup of order 168.[78] Using work of Francesco Brioschi, Maschke worked out the (absolute) invariants of G, proving that these are generated by seven invariants satisfying three algebraic relations. (Bolza later reported that Maschke had obtained this result as early as 1888, but he apparently delayed announcing it in hopes of obtaining the full invariants for Klein's larger "line-geometric" group.[79])

Coming last but certainly not least on Friday, E. H. Moore lectured on a doubly-infinite system of simple groups, unquestionably the most important contribution to mathematical research to issue from the Chicago Congress.[80] Interestingly enough, Moore picked up right where Frank Nelson Cole had left off, and true to Cole's prediction that future progress in finite group theory would depend on fertile new ideas of an abstract nature, Moore came armed with just this type of machinery. He began, as Cole had done, by enumerating the various simple groups of order 600 or less. In so doing, however, he called attention to a new class, a generalization of the groups associated to the modular equation for the transformation of elliptic integrals at the prime q.[81] While this known class of simple groups has order $\frac{q(q^2-1)}{2}$, Moore's new type of simple group has order

[77] Sophus Lie, *Vorlesungen über Differentialgleichungen mit bekannten infinitesimalen Transformation* (Leipzig: B. G. Teubner, 1891).

[78] Heinrich Maschke, "The Invariants of a Group of $2 \cdot 168$ Linear Quaternary Substitutions," *Congress Papers*, pp. 175–186. Klein studied the Jacobian equation of degree 8 both in Felix Klein, "Ueber die Auflösung gewisser Gleichungen vom siebenten und achten Grade," *Mathematische Annalen* 15 (1879):251–282, or *Klein*: *GMA*, 2:390–425; and "Zur Theorie der allgemeinen Gleichungen sechsten und siebenten Grades," *Mathematische Annalen* 28 (1886–1887):499–532, or *Klein*: *GMA*, 2:439–472.

[79] Oskar Bolza, "Heinrich Maschke: His Life and Work," *Bulletin of the American Mathematical Society* 14 (1908): 85–95 on pp. 89–90.

[80] E. H. Moore, "A Doubly-Infinite System of Simple Groups," *Congress Papers*, pp. 208–242.

[81] Moore's proof followed arguments roughly analogous to those already given by Klein-Fricke for the special case $n = 1$, $q > 2$. See Felix Klein and Robert Fricke, *Vorlesungen über die Theorie der elliptischen Modulfunctionen*, 2 vols. (Leipzig: B. G. Teubner, 1890, 1892), 1:419–450.

$$\frac{q^n(q^{2n}-1)}{2}, \qquad \text{for} \quad q > 2, \quad (q, n) \neq (3, 1),$$

and

$$q^n(q^{2n}-1), \qquad \text{for} \quad q = 2, \; n > 1.$$

Now denoted $PSL_2(q^n)$, this new class contains Cole's group as the special case $(q, n) = (2, 3)$. (See Chapter 9 for more discussion of Moore's group-theoretic work.) Moore thus succeeded in finding a two-parameter family of simple groups, only a few of which had been known prior to his lecture. Unbeknownst to him, however, Émile Mathieu had defined these groups back in 1860, but without showing that they were, in fact, simple.[82] This oversight aside, Moore's Congress contribution marks an early milestone in what would later become an American speciality—the classification of finite simple groups.[83]

Compared with Friday's rich program, the Congress's closing session proved rather anticlimactic. After five morning papers, Klein gave the afternoon's only scheduled talk, speaking ostensibly on the development of the theory of groups during the preceding twenty years. Although the subject sounded promising enough, judging from the abstract Klein published in the *Congress Papers*, he quickly wandered from his chosen topic. He apparently devoted some of his remarks instead to a report on the "higher geometry" course he had offered in Göttingen the previous summer.[84] In that course, he had focused primarily on Lie's theory of continuous groups and its applications to the Riemann-Helmholtz-Lie space problem. He would soon take up the latter subject in the eleventh of his Evanston Colloquium lectures.

Leaving group theory behind, Klein turned to some of the larger educational issues that had by now begun to capture his attention. In particular, he discussed certain reforms in mathematics instruction recently implemented in Göttingen as well as related efforts to expose future teachers to a wide range of mathematical applications in astronomy, physics, and technology.

---

[82]See Moore's note of 29 October 1895 appended to E. H. Moore, "A Doubly Infinite System of Simple Groups," p. 242, in which he points out that he had only recently become aware of Mathieu's paper, "Nombre de Valeurs que peut acquérir une Fonction quand On y permute ses Variables de toutes les Manières possibles," *Journal de Mathématiques*, 2nd ser. 5 (1860):9–42. These groups are, of course, not to be confused with the five sporadic groups that today bear Mathieu's name.

[83]On the classification problem, see Daniel Gorenstein, "The Classification of the Finite Simple Groups. A Personal Journey: The Early Years," in *A Century of Mathematics in America—Part* I, ed. Peter Duren *et al.* (Providence: American Mathematical Society, 1988), pp. 447–476.

[84]Felix Klein, *Einleitung in die höhere Geometrie, Vorlesungen gehalten im Wintersemester* 1892–93 *und Sommersemester* 1893, 2nd ed. (Leipzig: B. G. Teubner, 1907). The portion on Lie's theory was omitted in the edition prepared by Wilhelm Blaschke, namely, Felix Klein, *Vorlesungen über Höhere Geometrie*, 3rd ed. (Berlin: Springer Verlag, 1926).

Although he thought a merger of the institutes of technology with the universities would serve ideally to bridge the gap between theory and praxis, and although he had actually proposed this plan to the Ministry of Culture in a lengthy memorandum written five years earlier, Klein now recognized the impracticality of such a gigantic institutional reform.[85] He therefore favored the establishment of special institutes at the universities which would enable highly trained engineers, mathematicians, and physicists to pool their talents toward the solution of problems demanding sophisticated methods. "There can be no question," Klein concluded, "that the progressive development of our culture requires more and more men who have full control of scientific fundamentals both from the technical and the mathematical-physical sides."[86] It was with these thoughts in mind that the Mathematical Congress came to a close.

The Congress Adjourns

Whatever doubts Klein and others may have harbored initially about the Congress, by its end, all agreed that E. H. Moore and his colleagues had staged a rewarding and highly significant mathematical meeting. Several leading European mathematicians—including Hermite, Hilbert, Hurwitz, Minkowski, Max Noether, Pincherle, and Weber—had supported the Congress by submitting papers *in absentia*. Klein's efforts, in particular, had produced an impressive collection of articles, most of them written by the younger generation of German mathematicians. In the face of such stiff competition, the contributions of the American contingent inevitably appeared more modest, but those in the Midwest like Cole, Eddy, Moore, and the German émigrés, Bolza and Maschke, had nevertheless managed to make a respectable showing for the United States. The Chicago mathematicians—in sharp contrast with the members of Chicago's dynamic young architectural school—had also not been compelled to battle a band of leading members of the "Eastern establishment" for a place in the limelight. Since the Easterners had tended to stay home (with the notable exception of Clark University's faculty, Harry Tyler, and a few others), the triumph was the Midwesterners' to savor alone. Thus, with the Chicago Congress, the mathematicians of the Midwest sent a loud and clear signal to their compatriots back East, as well as to those abroad, that higher mathematics in the New World could flourish even in the "hinterlands" of the United States. Having sown the seeds for this fertile event, the Chicago mathematicians could now contemplate reaping their harvest: a

[85] On Klein's ideas and efforts in this area, see Manegold, pp. 103-109.

[86] Felix Klein, "Ueber die Entwicklung der Gruppentheorie während der letzten zwanzig Jahre," *Congress Papers*, p. 136 (our translation). This paper was not reproduced in *Klein: GMA*.

Congress proceedings filled with an unusual array of expository articles on research-level mathematics.[87]

More than anything, the success of the Chicago Congress had to do with its timing. Coming on the heels of the founding of the University of Chicago and a half-step ahead of the formation of a national mathematics organization in the United States, the Congress embodied the groundswell of interest in mathematics at the research level that would surge through the country in the years immediately ahead. Moreover, as Klein's opening remarks clearly indicated, the Chicago Congress acted as a harbinger of a new era of international cooperation in mathematics: shortly afterward European national organizations began laying plans for the First International Congress of Mathematicians, to be held in Zurich in 1896.[88] Yet, with the close of the precedent-setting Chicago Congress, the mathematical festivities in the greater Chicago area were far from over. Many looked forward with excitement and anticipation to the next two weeks when Felix Klein would finally deliver a series of lectures in the United States.

---

[87]On the part this played in fostering the emergence of the American Mathematical Society, see Raymond C. Archibald, *A Semicentennial History of the American Mathematical Society* (New York: American Mathematical Society, 1938), p. 7. Unfortunately, financial difficulties prevented the Congress proceedings from appearing before 1896. See the discussion in Chapter 9 below.

[88]Donald J. Albers, G. L. Alexanderson, and Constance Reid, *International Mathematical Congresses: An Illustrated History* 1893–1986 (New York: Springer-Verlag, 1987).

TABLE 7.1
PROGRAM OF THE CHICAGO MATHEMATICAL CONGRESS

TUESDAY, AUGUST 22
PAPERS

David Hilbert (Königsberg University): "Invariantentheorie"

Heinrich Weber (Göttingen University): "Ganzzahlige Gleichungen"

Eugen Netto (Giessen University): "Kronecker's arithmetisch-algebraische Tendenzen"

Adolf Hurwitz (ETH, Zurich): "Ueber die Reduction der binären quadratischen Formen"

Artemas Martin (U. S. Coast and Geodetic Survey): "On Fifth-Power Numbers whose Sum is a Fifth-Power"

LECTURES

Alexander Macfarlane (University of Texas): "On the Definitions of the Trigonometric Functions"

Eduard Study (Marburg University): "Complexe Zahlen"

Henry Taber (Clark University): "Concerning Matrices and Multiple Algebra"

Albert M. Sawin (Evansville, Wisconsin): "On the Algebraic Solution of Equations"

WEDNESDAY, AUGUST 23
PAPERS

Hermann Minkowski (Bonn University): "Zahlentheorie und Geometrie"

Alfred Pringsheim (Munich University): "Geltungsbereich der Taylorschen Reihe"

William H. Echols (University of Virginia): "On Interpolation Formulae and their Relation to Infinite Series"

Salvatore Pincherle (Bologna University): "Résumé de quelques Résultats relatifs à la Théorie des Systèmes récurrentes de Fonctions"

Max Noether (Erlangen University): "Singuläre Puncte einer algebraischen Curve"

Matyáš Lerch (Prague): "Sur une Intégrale définie qui représente la Fonction $\zeta(s)$ de Riemann"

LECTURES

Henry T. Eddy (Rose Polytechnic Institute): "Modern Graphical Methods"

George Bruce Halsted (University of Texas): "Some Salient Points in the History of Non-Euclidean and Hyper-Spaces"

Alexander Macfarlane (University of Texas): "The Principles of the Elliptic and Hyperbolic Analysis"

Thursday, August 24
Papers

Charles Hermite (Sorbonne, Paris): "Sur quelques Propositions fondamentales de la Théorie des Fonctions elliptiques"

Irving Stringham (University of California): "A Formulary for an Introduction to Elliptic Functions"

Martin Krause (Dresden Polytechnic): "Ueber die Transformation der hyperelliptischen Functionen"

Lothar Heffter (Giessen University): "Allgemeine Theorie der linearen Differentialgleichungen"

Heinrich Burkhardt (Göttingen University): "Lineare Differential gleichungen in der Astronomie"

Robert Fricke (Göttingen University): "Automorphe Functionen und Zahlentheorie"

Lectures

Oskar Bolza (University of Chicago): "On Weierstrass' Systems of Abelian Integrals of the First and Second Kinds"

Eduard Study (Marburg University): "Spherical Trigonometry"

Friday, August 25
Papers

Joseph de Perott (Clark University): "A Construction of Galois' Group of 660 Elements"

Frank Nelson Cole (University of Michigan): "Concerning the Formation of Groups"

Arthur Schoenflies (Göttingen University): "Krystallographie und Gruppentheorie"

Franz Meyer (Clausthal Mining Academy): "Die continuirlichen Gruppen der Ebene"

Edouard Weyr (Prague Polytechnic Institute): "Sur l'Équation des Lignes géodésiques"

Victor Schlegel (Hagen, Germany): "Einige Sätze vom Schwerpunkt"

Lectures

Heinrich Maschke (University of Chicago): "On a Quaternary Group of 2520 Linear Substitutions"

Eliakim Hastings Moore (University of Chicago): "A Doubly-Infinite System of Simple Groups"

SATURDAY, AUGUST 26
PAPERS

Victor Schlegel (Hagen, Germany): "Der pythagorische Lehrsatz in mehrdimensionalen Räumen"

Émile Lemoine (Paris): "La Géométrographie ou l'Art des Constructions géométriques"

Émile Lemoine (Paris): "Règle des Analogies dans le Triangle et Transformation continue"

Maurice d'Ocagne (Paris): "Nomographie: Sur les Équations représentables par trois Systèmes rectilignes de Points isoplèthes"

T. M. Pervouchine (Kasan): "Note Concerning Arithmetical Operations Involving Large Numbers" (communicated by Prof. A. Wassilieff of Kasan University)

LECTURES

Felix Klein (Göttingen University): "Concerning the Development of the Theory of Groups during the Last Twenty Years"

Chapter 8
Surveying Mathematical Landscapes: The Evanston Colloquium Lectures

Just as the Chicago Congress transcended the bounds of an ordinary mathematics meeting, so the Colloquium lectures Klein delivered in Evanston from 28 August to 9 September 1893 represented far more than a conventional series of mathematics lectures. Charged with strong opinions and novel insights, they reflected the highly personal views of a man whose life and career made him especially qualified to speak about the developments of the preceding twenty-five years as well as about the principal figures who contributed to them. They encapsulated, like no other work of Klein's, save his lectures on the development of nineteenth-century mathematics,[1] a sense of his all-encompassing vision of the mathematical landscape. Unique in style and scope for their day, these colloquia have retained their distinctiveness right up to the present.

Many of the mathematicians who attended the Chicago Congress extended their visit to take in this second meeting hosted at Northwestern University in Evanston by Henry Seely White. Besides the Chicago mathematicians, the audience included Story and Taber from Clark, Fabian Franklin of the Johns Hopkins, Cornell's James E. Oliver, Tyler of MIT, the elder Van Vleck from Wesleyan, his son E. B. (called "Ned" by his father and friends), fresh from Göttingen with a doctorate in hand, and the young Chicago Fellow, Mary Winston. No fewer than nine of those present had studied with Klein in Germany.[2] Yet even those in the audience who had attended his lectures

[1]Felix Klein, *Vorlesungen über die Entwicklung der Mathematik im* 19. *Jahrhundert*, 2 vols. (Berlin: Springer-Verlag, 1926–1927).

[2]Bolza, Franklin, Maschke, Oliver, Study, Tyler, E. B. Van Vleck, Clarence Waldo, and White had studied with Klein, and Winston and Ziwet would work with him in the years ahead. For a complete list of colloquium participants, see Felix Klein, *The Evanston Colloquium Lectures on Mathematics* (New York: Macmillan & Co., 1893; reprint ed., New York: American Mathematical Society, 1911), pp. vii–viii. (This, along with several addtional works by Klein, will appear in a modern edition as *The Erlangen Program, Evanston Colloquium Lectures, and Other Selected Works*, ed. Jeremy Gray and David E. Rowe (New York: Springer-Verlag), forthcoming.) During most of Klein's stay in the Chicago metropolitan area, White entertained his former teacher at his home on Foster Street, just minutes away from the lecture hall where he delivered his Evanston Colloquium lectures.

Henry Seely White (1861–1943)

in Göttingen would very likely not have anticipated the vast panorama of mathematical ideas they were about to view.

As we saw in Chapter 5, the Americans who studied under Klein were exposed, first and foremost, to his work on hyperelliptic and Abelian functions, potential theory, and linear differential equations. Most of them learned relatively little about other fields—ranging from algebraic geometry to Galois theory and the theories of elliptic modular and automorphic functions—to which he had made fundamental contributions. Moreover, Klein rarely presented even slightly complicated proofs in his lectures and nearly always focused on the broader dimensions of a theory.[3] In the latter respect, he went even further in his Evanston colloquia, spreading his wings as never before in an effort to take in sweeping mathematical vistas. Although much of the information he conveyed was tailored to an American audience and was certainly familiar to the experts who were present on this occasion, even they must have been astonished to see this terrain from such dizzying heights. Klein's lectures, which marked his first serious attempt to sketch an overview of certain major trends in mathematical research, were a novel experience for both audience and speaker alike. They represented a largely untried form of scientific popularization to which Klein had clearly given much thought, and it was no accident that he chose to deliver them abroad. In Germany, lectures like these would not only have appeared presumptuous to members of the mathematical community, but they would also have clashed with élitist ideals of *Wissenschaft* which rejected anything smacking of popularization. In America, however, the land of "unlimited possibilities,"[4] and in the quintessentially American atmosphere of Chicago amidst the celebrations of the Great Fair, Klein had an ideal time and place to explore this new form. He also enjoyed a receptive audience eager to accompany him on this two-week intellectual journey.

The Evanston Colloquium Lectures

After a day off following the close of the Congress, Klein began his Evanston lectures on Monday, 28 August, delivering one each day for two weeks (Sundays excepted). He spoke at a leisurely rate and in English, posing questions to his audience as he went along; lengthy informal discussions followed each talk. Alexander Ziwet of the University of Michigan recorded the lectures and, together with Klein, subsequently revised them for publication. Ziwet's text, however, provides only a rather condensed version of what

[3]To a lesser extent, this was characteristic of David Hilbert and Hermann Minkowski as well, both of whom stressed conceptual clarity in contrast to Minkowski's successor at Göttingen, Edmund Landau, an infamous fanatic when it came to pedantic precision.

[4]This phrase, "unbegrenzte Möglichkeiten," has been one of the most commonly used German descriptions of life in the United States.

Klein actually said during these two weeks.[5] As Klein himself admitted, the book by itself gives no sense of the atmosphere on this historic occasion. He later described the mood as one of "scientific enthusiasm"; by the end of the second week, he and his American colleagues were meeting together in five-hour-long sessions.[6]

As Klein made clear from the outset, he intended in his lectures "not so much to give a complete account of any subject, as to *supplement* the mathematical views that I find prevalent in this country."[7] With this in mind, he began his opening lecture by dividing mathematicians into three categories: (1) logicians, (2) formalists, and (3) intuitionists. By employing this scheme, he hoped to characterize and highlight important stylistic differences observable between mathematicians independent of their particular subject matter, and he defined his terms by means of carefully chosen examples. Klein considered Weierstrass as a leading exemplar of category (1), despite the fact that his rigorous approach had almost no connection with mathematical logic as practiced by pioneering figures like George Boole and Charles S. Peirce. He indicated that the tendency toward formalism was already quite familiar in the United States: it characterized not only the approach Sylvester followed and taught at Hopkins but also the work of Arthur Cayley and Paul Gordan. As representative intuitionists (Klein's terminology having obviously nothing in common with intuitionism in the sense of L. E. J. Brouwer), he chose Lord Kelvin and Christian von Staudt, surely one of the few instances when these two men have been mentioned in the same breath. Clebsch, on the other hand, was more eclectic, combining categories (2) and (3). Curiously enough, Klein considered his own work a manifestation of categories (1) and (3), whereas, from today's perspective, "logician" is probably the last label anyone would pin on Felix Klein.

Predictably enough, Klein's first six lectures stressed the role of *Anschauung* in modern geometry and function theory. In view of his opening remarks, it is clear that in Klein's opinion American mathematicians had relatively little exposure to this characteristically Kleinian style, despite his own considerable influence on them. He therefore hoped to inculcate a certain appreciation for the intuitive approach to mathematical research and, perhaps, even win some converts along the way. Turning first to his sources, he paid homage to his erstwhile mentor, Alfred Clebsch, in a tribute by no means devoid of critical remarks:

---

[5]Klein approved the text prepared by Ziwet before it went to press. In fact, he had page proofs in hand when he set sail for Germany in early October. The volume appeared in January of 1894. See Henry White to Felix Klein, 5 October 1893, and 14 January 1894, Klein Nachlass XII, Niedersächsische Staats- und Universitätsbibliothek, Göttingen (hereinafter abbreviated NSUB, Göttingen).

[6]Felix Klein, "Bericht über die Reise nach Chicago zwecks Teilnahme am mathematischen Congresse," Klein Nachlass I C, NSUB, Göttingen.

[7]Klein, *Evanston Colloquium*, p. 2 (Klein's emphasis).

> However great the achievement of Clebsch's in making the work of Riemann more easy of access to his contemporaries, it is my opinion that at the present time the book of Clebsch [actually *Theorie der Abelschen Funktionen* by Clebsch and Gordan] is no longer to be considered the standard work for an introduction to the study of Abelian functions. The chief objections to Clebsch's presentation are twofold: they can be briefly characterized as a lack of mathematical rigour on the one hand, and a loss of intuitiveness, of geometrical perspicuity, on the other.[8]

Part of what motivated this criticism was almost certainly Klein's desire to distance himself from the Clebsch school, whose members—Gordan, Alexander Brill, and Max Noether, among others—were tied to a mathematical tradition that, to his mind at least, was on its way out. Sensing major new changes on the horizon within both the older German mathematical community and the younger one just developing in the United States, Klein dropped any number of hints that suggested a promising future for both countries. As a sign of the times, he had already put his stock in a young upstart named Hilbert, although in trying to promote the young mathematician's career he had encountered much resistance from his older colleagues.[9] Not wishing to break with them completely, he preferred postponing Hilbert's appointment as Associate Editor of the *Mathematische Annalen*. However, he made it clear to Hilbert that his alliance with the older members of the Clebsch school had become nothing more than a marriage of convenience.[10]

Behind these internal politics lurked a substantive issue of considerable import to Klein. While he admired Clebsch and his followers for their achievements in showing how to apply Riemann's theory of Abelian integrals and their inverse functions to the study of algebraic curves, deep down he still felt that their approach obscured the key physico-geometrical ideas central to the Riemannian legacy and its furtherance.[11] Clebsch's closest disciples sought to circumvent Riemann's approach and thereby avoid the stumbling block represented by the Dirichlet principle. Coming from this point of view, Brill and Noether sought to establish the central results of algebraic geometry without recourse to transcendental methods. Their efforts in the name of rigor amounted to supplanting Riemann's intuitive and, to Klein's mind at

---

[8] *Ibid.*, p. 4.

[9] See David E. Rowe, "Klein, Hilbert, and the Göttingen Mathematical Tradition," in *Science in Germany: The Intersection of Institutional and Intellectual Issues*, ed. Kathryn M. Olesko, *Osiris*, 2nd series, 5 (1989):189–213 on pp. 195–197.

[10] Felix Klein to David Hilbert, 4 October 1894, in *Der Briefwechsel David Hilbert–Felix Klein* (1886–1918), ed. Günther Frei, Arbeiten aus der Niedersächsischen Staats- und Universitätsbibliothek Göttingen, 19 (Göttingen: Vandenhoeck & Ruprecht, 1985), pp. 110–111.

[11] As noted in Chapter 4, Klein believed he had rediscovered these ideas through prolonged pondering of the master's work.

least, intrinsically superior ideas.[12] "For these reasons," Klein continued, "it seems to me best to begin the theory of Abelian functions with Riemann's ideas, without, however, neglecting to give later the purely algebraical developments."[13]

From Riemann and Clebsch, Klein turned to another giant of his youth, Sophus Lie. In his second and third lectures, Klein focused on Lie's early researches on geometry and the theory of partial differential equations, work he considered a paradigmatic and instructive example of how great mathematics can emerge through intuition and unconscious inspiration rather than cold calculation. He cautioned his listeners that

> [t]o fully understand the mathematical genius of Sophus Lie, one must not turn to the books recently published by him in collaboration with Dr. Engel [*Theorie der Transformationsgruppen*, the third volume of which was about to appear], but to his earlier memoirs, written during the first years of his scientific career. There Lie shows himself the true geometer that he is, while in his later publications, finding that he was but imperfectly understood by the mathematicians accustomed to the analytical point of view, he adopted a very general analytical form of treatment that is not always easy to follow. Fortunately, I had the advantage of becoming intimately acquainted with Lie's ideas at a very early period, when they were still, as the chemists say, in the "nascent state," and thus most effective in producing a strong reaction.[14]

After alluding to Lie's debt both to Plücker and Monge, Klein went on to discuss the importance of Lie's new sphere geometry (as opposed to that of Darboux and Moutard), its relationship to his theory of contact transformations, and its connection with Plücker's line geometry via the line-to-sphere transformation.[15] Klein observed relative to this latter contact transformation that the two systems of lines on a hyperboloid of one sheet have the property that each line of the one system intersects all the lines of the others. It follows that the image of this quadric surface under the line-to-sphere transformation will be a surface enveloped by two systems of spheres with the property that each sphere of the one system is tangent to all spheres of

---

[12] See, for example, Klein's remarks in Felix Klein, *Riemannsche Flächen*: *Vorlesungen, gehalten in Göttingen* 1891/92, ed. Günther Eisenreich and Walter Purkert, Teubner-Archiv zur Mathematik, vol. 5 (Leipzig: B. G. Teubner, 1986), pp. 51–52 and 191–192.

[13] Klein, *Evanston Colloquium*, p. 5.

[14] *Ibid.*, p. 9.

[15] These mathematical ideas are discussed in Felix Klein, *Vorlesungen über höhere Geometrie*, ed. Wilhelm Blaschke, 3rd ed. (Berlin: Springer-Verlag, 1926), pp. 105–117; and David E. Rowe, "The Early Geometrical Works of Sophus Lie and Felix Klein," in *The History of Modern Mathematics*, ed. David E. Rowe and John McCleary, 2 vols. (Boston: Academic Press, Inc., 1989), 1:209–273 on pp. 239–255. Recall also the brief discussion in Chapter 4 above.

the other. These surfaces are the familiar cyclides of Dupin. Klein further mentioned the generalized cyclides introduced by Darboux and Moutard, pointing out that Maxime Bôcher had studied these fourth-order surfaces in his prize-winning dissertation on potential theory. They had originally come to Lie's attention during his Paris sojourn of 1870, and soon thereafter he discovered that they arise as images of Kummer surfaces under the line-to-sphere transformation. This realization had enabled Lie to determine the asymptotic curves of the Kummer surface, as noted in Chapter 4, and Klein had proceeded to work out the mathematical ramifications in considerable detail.[16]

Moving next to a discussion of Lie's general theory of contact transformations and its connection with Lagrange's theory of partial differential equations, Klein became necessarily more sketchy in his remarks. He first presented a number of examples, including that of a gearing mechanism he had demonstrated earlier at the German Universities Exhibit, and then showed how Lie's geometric concepts, and in particular his notion of a surface element, had led naturally to one of the key properties of Lie's line-to-sphere transformation: the correspondence between the asymptotic curves of one surface and the lines of curvature of another. Klein tied up his two lectures on Lie by emphasizing the thematic coherence of the Norwegian's work. He acknowledged that "[a]t the present time Lie is best known through his theory of continuous groups of transformations, and at first glance it might appear as if there were little connection between this theory and the geometrical considerations that engaged our attention in the last two lectures."[17] Klein wished to stress to his audience, however, that " [*i*]*t has been the final aim of Lie from the beginning to make progress in the theory of differential equations*; and as subsidiary to this end may be regarded both the geometrical developments considered in these lectures and the theory of continuous groups."[18] As we saw in Chapter 5, many young Americans had already found or would soon find an ideal vantage point from which to form their own judgments about the motivation behind Lie's mathematics, namely, as students in his classes at Leipzig. In fact, during the five-year period after Klein delivered these lectures, Lie attracted ever greater numbers of talented Americans up until his return to Norway in 1898.

Klein followed this tribute to his Norwegian friend with a discussion of some of his own work in the fourth lecture, entitled "On the Real Shape of Algebraic Curves and Surfaces." Setting the historical stage, he noted that

---

[16]Felix Klein and Sophus Lie, "Ueber die Haupttangentenkurven der Kummerschen Fläche vierten Grades mit 16 Knotenpunkten,"originally published in the *Monatsberichte der Berliner Akademie der Wissenschaften* (1870):891–899, or Felix Klein, *Gesammelte Mathematische Abhandlungen*, 3 vols. (Berlin: Springer-Verlag, 1921–1923), 1:90–97 (hereinafter abbreviated *Klein: GMA*).

[17]Klein, *Evanston Colloquium*, p. 24.

[18]*Ibid.* (Klein's emphasis.)

Newton "had a very clear conception of projective geometry" as evidenced by the classification scheme in which he derived all third-order plane curves from five fundamental types by means of central projection.[19] In Klein's opinion, though, Möbius' 1852 paper "Ueber die Grundformen der Linien der dritten Ordnung" marked the most important study of cubic curves in modern times.[20] As he remarked, this article and the work of Clebsch had led him to the problem of classifying all real third-order surfaces in $\mathbf{P}^3(\mathbb{C})$. Klein succeeded in deriving these surfaces in 1873 using a general method which hinged on applying the principle of continuity to a certain surface with four real conical points.[21]

At this juncture in his lecture, as in several of his other talks, Klein drew on his earlier demonstrations of the models on display at the Fair in an effort to capture as vividly as possible the geometrical flavor of his method. He called attention specifically to the model of Clebsch's diagonal surface. This was an example of a cubic surface (contained in $\mathbf{P}^4(\mathbb{C})$) defined by the equations $x_1^3+x_2^3+x_3^3+x_4^3+x_5^3=0$ and $x_1+x_2+x_3+x_4+x_5=0$.[22] (See Figure 8.1.) Since 1849, when Cayley and George Salmon showed that the most general cubic surface has 27 (not necessarily real) lines, the incidence configurations for the lines of the various types of cubic surfaces had captured the interest of geometers.[23] At the Fair, Klein had shown his audience an intriguing *ad hoc* way to obtain the diagonal surface from the cubic surface with four conical points by using a suitable deformation process. In this Evanston lecture, however, he stressed that such isolated results had little intrinsic value if they could not be situated in a larger theoretical context. This prompted him to call attention to the fact that the method he had introduced in 1873 recovered *all* possible types of real third-order surfaces. Quite clearly, he thought that American mathematicians would benefit more by setting their

---

[19] *Ibid.*, p. 25. Newton's work, "Enumeratio linearum tertii ordinis," appears in *The Mathematical Papers of Isaac Newton*, ed. D. T. Whiteside and M. A. Hoskins, 8 vols. (Cambridge: University Press, 1976–1981), 2:135–161.

[20] August Ferdinand Möbius, "Ueber die Grundformen der Linien der dritten Ordnung," *Abhandlungen der königlichen sächsischen Gesellschaft der Wissenschaften* 1 (1852):1–82, or August Ferdinand Möbius, *Gesammelte Werke*, 4 vols. (Leipzig: B. G. Teubner, 1886), 2:89–176.

[21] Felix Klein, "Ueber Flächen dritter Ordnung," *Mathematische Annalen* 6 (1873):551–581, or *Klein:GMA*, 2:11–43.

[22] See *ibid.*, pp. 29–32, and Felix Klein, *Vorlesungen über die Entwicklung der Mathematik im* 19. *Jahrhundert*, 1:166–167; and Wolf Barth and Horst Knörrer, "Algebraische Flächen," in *Mathematische Modelle*, ed. Gerd Fischer, 2 vols. (Berlin: Akademie Verlag, 1986), 2:7–24 on pp. 10–11.

[23] See Arthur Cayley, "On the Triple Tangent Planes of Surfaces of the Third Order," *Cambridge and Dublin Mathematical Journal* 4 (1849):118–132, or Arthur Cayley and A. R. Forsyth, ed., *The Collected Mathematical Papers of Arthur Cayley*, 14 vols. (Cambridge: University Press, 1889–1898), 1:445–456 (hereinafter cited as *Math. Papers AC*). For a modern discussion of this subject, consult Yu. I. Manin, *Cubic Forms: Algebra, Geometry, Arithmetic* (Amsterdam: North-Holland Publishing Company, 1974).

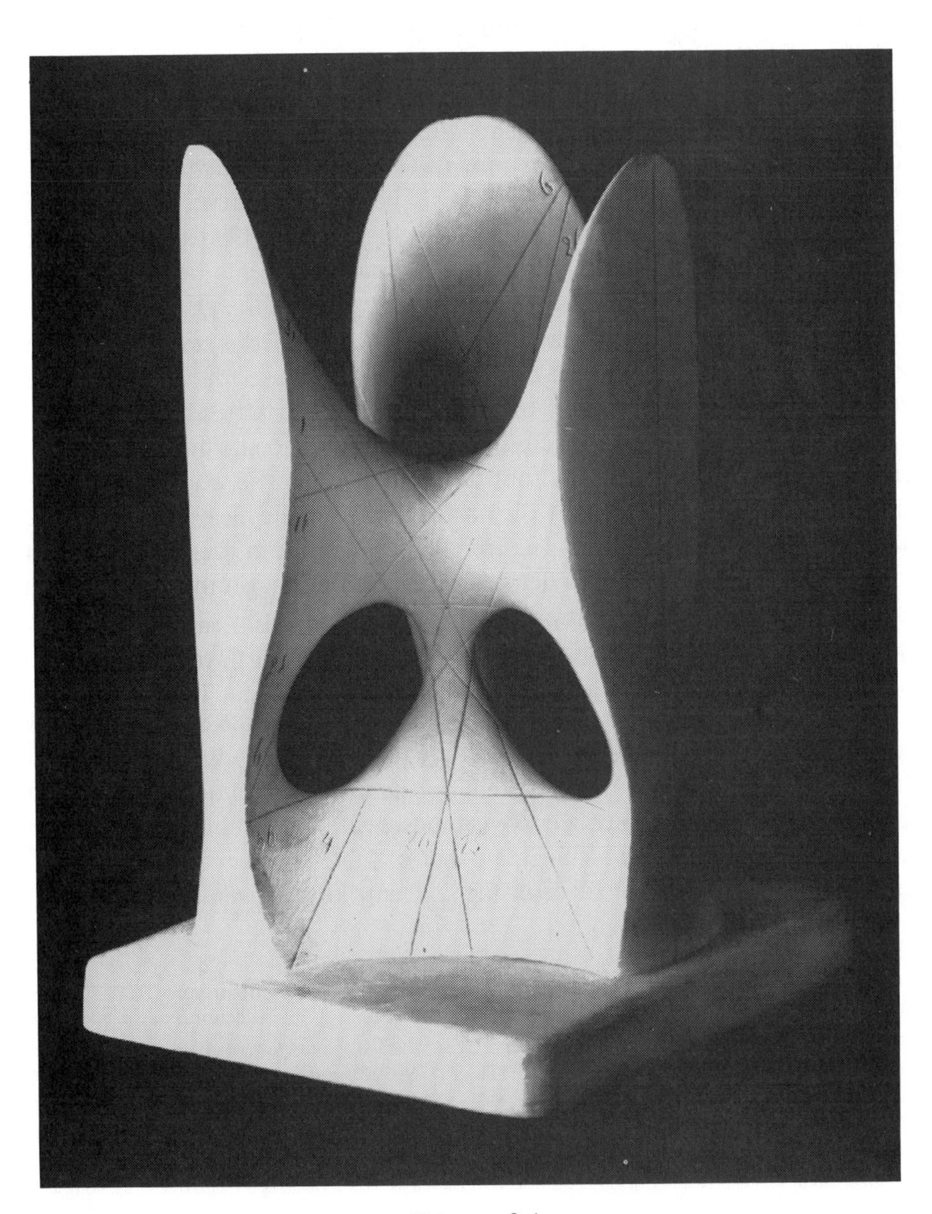

FIGURE 8.1

research sights on general goals such as this than by focusing their energies on potentially narrower theories resulting from experimentation with specific examples.

Klein next discussed an important paper written by H. G. Zeuthen in 1874, which set forth a related approach for studying plane quartic curves.[24] Five years earlier, Carl Friedrich Geiser had noted the following connection between quartic curves and cubic surfaces. Given a cubic surface S and a point $p \notin S$, the family of tangents from p to S form a cone of degree six. In the case where $p \in S$, this cone degenerates into the tangent plane E and a quartic surface Q, so that $E \cap Q$ will be a quartic curve. Furthermore, every quartic curve C can be obtained in this fashion. Under a projection, the 27 lines of S map to 27 of the 28 double tangents belonging to C (the missing double tangent being determined by one of the principal tangent lines to the surface S through p).[25] In particular, if the surface has four conical points, then the quartic derived from it will have four double points determined by the intersection of two conics, as illustrated in Figure 8.2. By employing the principle of continuity, Zeuthen had shown that the four outer regions pass over into the four branches of a quartic whose 28 double tangents have the property that they are all real. While clearly appreciative of this particular result, in his lecture Klein underscored the importance of the *method* from which it was derived. He called attention to the surprising fact that Zeuthen's argument was perfectly analogous to the one he had given earlier for obtaining Clebsch's diagonal surface from a cubic surface with four conical points.

Leaving this observation behind, Klein moved on to make some remarks about the topology of higher-order curves in the plane, including a description of Axel Harnack's well-known theorem on the maximum number of distinct topological components such a curve can possess.[26] Hilbert had made certain refinements of this classic result in a paper of 1891, and he later highlighted this whole constellation of ideas in the sixteenth of his twenty-three "Mathematical Problems."[27]

---

[24]H. G. Zeuthen, "Sur les différentes Formes des Courbes planes du quatrième Ordre," *Mathematische Annalen* 7 (1874):410–432.

[25]This result first appeared in C. F. Geiser, "Ueber die Doppeltangenten einer ebenen Curven vierten Grades," *Mathematische Annalen* 1 (1869):129–138, and is discussed in Jeremy Gray, *Linear Differential Equations and Group Theory from Riemann to Poincaré* (Basel: Birkhäuser, 1986), pp. 215–216.

[26]Harnack's theorem states that a real curve of genus p has at most $p+1$ real components, and that such a curve having $p+1$ components always exists. See Axel Harnack, "Ueber die Vieltheiligkeit der ebenen algebraischen Curven," *Mathematische Annalen* 10 (1876):189–198. Harnack had studied under Klein in Erlangen, where he wrote a dissertation on the application of elliptic functions to third-order curves.

[27]David Hilbert, "Ueber die reellen Züge algebraischer Curven," *Mathematische Annalen* 38 (1891):115–138, or David Hilbert, *Gesammelte Abhandlungen*, 3 vols. (Berlin: Julius Springer, 1932–1935), 2:415–436. On subsequent developments, see O. A. Olejnik, "Zum sechzehnten Hilbertschen Problem," in *Die Hilbertschen Probleme*, ed. P. S. Alexandrov, Ostwalds Klassiker

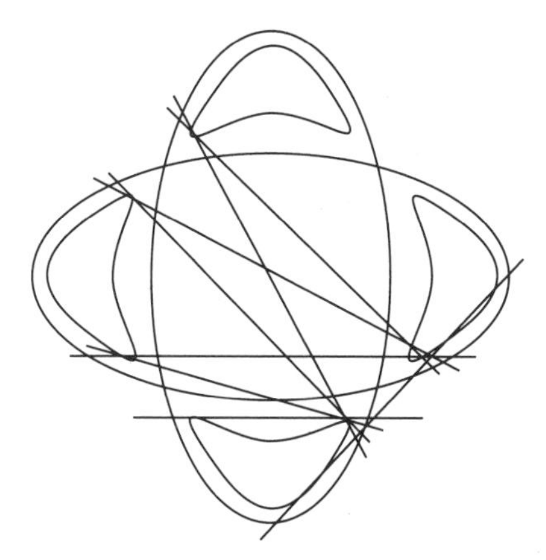

FIGURE 8.2

Klein continued his tour through contemporary algebraic geometry, turning to consider various models for the Kummer surface. He observed that several models of algebraic surfaces, including those that Plücker had constructed in connection with certain second-order line complexes, can be viewed as special cases of the Kummer surface. He also advertised the work of his former student Karl Rohn as the most important on this particular subject.[28]

Not surprisingly, Klein did not conclude this lecture without mentioning his *Lieblingsobjekt*, the projective Riemann surfaces he had introduced using dualized class curves.[29] He illustrated his main ideas on the topic by looking at the examples of an ellipse (genus $p = 0$) and a nonsingular cubic with two real components ($p = 1$). In the latter case, the dual curve C^* has class 3, so that exactly three tangents can be drawn to it from an arbitrary point in the plane. Clearly, all three are real for points outside the oval or inside the triangular branch, whereas only one of the three tangents is real when a point is chosen in the region bounded by its two components. Thus, the two imaginary tangents correspond to two sheets over this annular region that

---

der exakten Wissenschaften, vol. 252 (Leipzig: Akademische Verlagsgesellschaft Geest & Portig, 1979), pp. 233–249. Work directed toward the solution of the sixteenth problem spurred research on the topological properties of real algebraic manifolds in the twentieth century.

[28]One of Rohn's models for the Kummer surface utilizes the generating lines on a hyperboloid of one sheet. He took four lines from each of the two sets of generators, shaded the alternate regions, and then glued a copy of the shaded regions to the original along the boundary. In so doing, he produced a closed surface without boundary having 16 real nodal points. See Karl Rohn, "Die verschiedenen Gestalten der Kummer'schen Fläche," *Mathematische Annalen* 18 (1881):99–159; and "Die Flächen vierter Ordnung hinsichtlich ihrer Knotenpunkte und ihrer Gestaltung," *Mathematische Annalen* 29 (1887):81–96.

[29]Recall the definition of class curves given in note 63 of Chapter 4. Egbert Brieskorn and Horst Knörrer give various examples of dual curves in their book, *Plane Algebraic Curves*, trans. John Stillwell (Boston: Birkhäuser Verlag, 1986), pp. 583–586.

meet together on the boundary curve, and the resulting surface is a torus. (See Figure 8.3.) In the early 1870s, Klein had employed this picture to give an *anschauliche* representation of all nine inflection points of the general cubic: the three real ones corresponding to the cusps; the other six given by the small circles distributed symmetrically over the front and back of the torus.[30] In his fourth Evanston lecture, Klein also alluded to his notion of symmetric Riemann surfaces, that is, surfaces which can be mapped onto themselves by means of an orientation-reversing conformal transformation which leaves the real points of the the surface fixed. These surfaces arise as the Riemann surfaces for real algebraic curves, and since he could enumerate the number of possible types of symmetric surfaces constructible for any given genus p, this provided a general classification scheme for all real plane curves.[31]

In bringing this *tour de force* to a close, Klein suggested that "it would be interesting to investigate *all* algebraic configurations [that is, structures with complex dimension greater than one] so as to arrive at a truly geometrical intuition of these objects,"[32] but he added that this program would only bear fruit if undertaken in the proper *anschauliche* spirit: "I wish to insist in particular on what I regard as the principal characteristic of the geometrical methods that I have discussed to-day: these methods give us an actual mental image of the configuration under discussion, and this I consider as most essential in all true geometry."[33] In Klein's considered view, "the so-called synthetic methods, as usually developed, do not appear ... very satisfactory. While giving elaborate constructions for special cases and details they fail entirely to afford a general view of the configurations as a whole."[34] With regard to this last point, the Americans in the audience evidently had no reason to question Klein's verdict. The influence of leading synthetic geometers—such as Theodor Reye of Strassburg or Breslau's Rudolf Sturm—on American mathematicians had been and would remain negligible, particularly when measured against Klein's own record.

Still, influence can take many forms and operate on a variety of levels. Klein's Evanston lectures, and particularly this fourth lecture with its swift series of fleeting allusions, probably served more to dazzle than to inform or guide the audience, apart from those who happened to be specialists in the field. This, again, speaks meaningfully to the nature of Klein's influence on American mathematics, both within the context of these lectures and beyond. Obviously, several of his more specific ideas, like those involving his projective Riemann surfaces, failed to attract any serious attention

---

[30] Klein, "Ueber eine neue Art der Riemannschen Flächen (Erste Mitteilung)," *Mathematische Annalen* 7 (1874):558–566, or *Klein*: *GMA*, 2:89–98 on p. 92.

[31] Klein first presented these results in *Ueber Riemanns Theorie der algebraische Funktionen und ihrer Integrale* (Leipzig: B. G. Teubner, 1882), or *Klein*: *GMA*, 3:499–573 on pp. 561–569.

[32] Klein, *Evanston Colloquium*, p. 32 (Klein's emphasis).

[33] *Ibid.*

[34] *Ibid.*

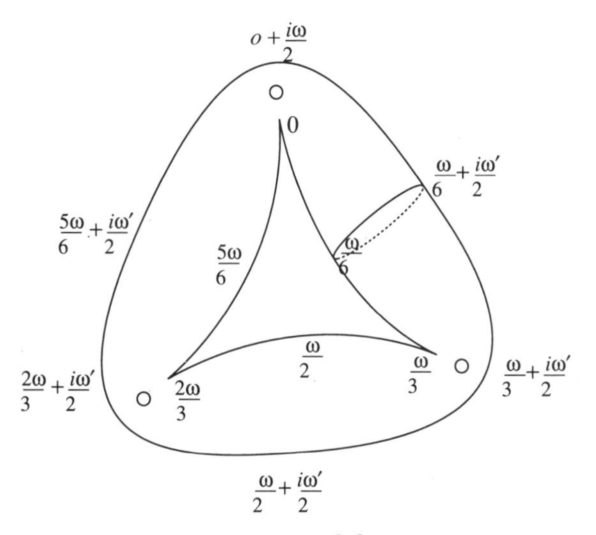

FIGURE 8.3

within the American mathematical community (or outside it either, for that matter). Nevertheless, his broad-based approach to geometry—more familiar than ever now that Haskell had published his translation of the *Erlanger Programm*—clearly appealed to many American mathematicians, even those who had little or no idea about fields like line or sphere geometry.[35] Even if few Americans could fully appreciate the rich historical sources that motivated Klein's mathematics, and if none (except possibly Osgood) sought to emulate him as a mathematician, they still regarded him as one of the greatest living mathematicians, an altogether eminent authority on nearly every aspect of almost every field, whose mathematical judgment they trusted implicitly. Although Klein knew his own limitations, he appreciated the adulation of his students and peers, and, in the final analysis, he gave the Americans what really mattered to them most—inspiration.

After this excursion into algebraic geometry, Klein turned in his fifth lecture to one of the central themes connecting the Göttingen tradition of Gauss and Riemann with that of Klein and his followers, namely, hypergeometric functions. A hypergeometric function is a solution of a *hypergeometric* (or *Gaussian*) differential equation, that is, one of the form

$$z(1-z)\frac{d^2w}{dz^2}+(\gamma-(\alpha+\beta+1)z)\frac{dw}{dz}-\alpha\beta w=0, \quad \text{where } \alpha,\ \beta,\ \gamma\in\mathbb{R}.$$

These solutions can be represented as analytic continuations of the hypergeometric series or by means of Riemann P-functions.[36] Although the classical theory generally treated elliptic functions as second in importance only to

[35]Felix Klein, "A Comparative Review of Recent Researches in Geometry," trans. M. W. Haskell, *Bulletin of the New York Mathematical Society* 2 (1893):215–249.

[36]For a modern treatment of these matters, see E. T. Whittaker and G. N. Watson, *A Course of Modern Analysis* (Cambridge: University Press, 1927), Chapter 16.

the algebraic and elementary transcendental functions, Klein impressed upon his audience the vital importance of the hypergeometric functions in applications to astronomy and physics.[37] In so doing, he underscored what he viewed as the necessary and healthy interplay between mathematics and the exact sciences. Ironically enough, however, as a teacher, Klein had probably contributed as much as anyone to reinforcing the purist orientation of American mathematics, a trend that would continue to characterize mathematical research in the United States for decades to come.

If Klein's fifth effort remained sketchy, his sixth, perhaps the best known of all twelve lectures, gave a crystal clear impression of his thoughts "On the Mathematical Character of Space-Intuition, and the Relation of Pure Mathematics to the Applied Sciences"—two loosely related topics that went to the heart of his philosophical vision.[38] Although Klein constantly stressed the underlying unity of all mathematics, he also argued that a profound duality—naive *vs.* refined intuition—shaped most mathematical ideas. He had already mentioned his views on this at the Chicago Congress, but he proceeded to spell out the distinction more carefully in Evanston. Not unexpectedly, he drew on historical examples to illustrate his point. Klein viewed Euclid's *Elements* as a prototype of refined intuition, while he considered the work of Newton and other pioneers of the differential and integral calculus as a byproduct of naive intuition. He further argued that the latter approach generally antedated the former when one traced the evolution of a mathematical theory. (The parallel with the views expressed by Hilbert in his paper on invariant theory was by no means accidental.[39]) With regard to calculus, he characterized the contemporary emphasis on rigor as akin to the *critical* spirit that animated Euclid. Since mathematical ideas naturally evolved from a naive to a refined state, Klein was convinced that the *Elements* represented the culmination of an earlier phase of development in ancient Greek math-

[37]After alluding to the geometric significance of the Riemann mapping theorem, he discussed the conformal representation of the upper half-plane as a circular-arc triangle [*Kreisbogendreieck*] and pointed out its connection with the hypergeometric functions. Oddly enough, he referred to Hurwitz's recent researches in this area but failed to mention the contributions of his former Göttingen colleague H. A. Schwarz (other than to say that Cayley had named the ratio of two solutions to the hypergeometric differential equation the Schwarzian derivative). One of the most fundamental papers on hypergeometric functions from this period is Hermann Amandus Schwarz, "Ueber diejenigen Fälle, in welchen die Gaussische hypergeometrische Reihe eine algebraische Function ihres vierten Elementes darstellt," *Journal für die reine und angewandte Mathematik* 75 (1872):292–335, or *Gesammelte Mathematische Abhandlungen*, 2 vols. (Berlin: Akademie der Wissenschaften, 1890), 2:211–259. This paper is discussed in detail in Gray, pp. 97–108.

[38]That Klein himself attached particular importance to this lecture may be deduced from the fact that he had it reprinted in his collected works. See Felix Klein, "On the Mathematical Character of Space-Intuition and the Relation of Pure Mathematics to the Applied Sciences," *Evanston Colloquium*, pp. 41–50, or *Klein*: *GMA*, 2:225–231.

[39]Hilbert later elaborated further on some his own philosophical ideas in a lecture course given in the Winter Semester of 1919–1920. See David Hilbert, *Natur und mathematisches Erkennen*, ed. David E. Rowe (Basel: Birkhäuser, 1992).

ematics, a conviction that appears to have been widely accepted by scholars at this time.[40]

Klein located the cause of the tension between these two forms of intuition "in the fact that *the naive intuition is not exact, while the refined intuition is not properly intuition at all, but arises through the logical development from axioms considered as perfectly exact.*"[41] The importance of the latter was, of course, unequivocal, and Klein illustrated the point by noting how helpless intuition became if confronted with the task of finding the limit points when a family of geometrical figures is continuously inverted with respect to one another as in the theory of automorphic functions. In certain prescribed cases—corresponding to functions of the Fuchsian class—a family of circular-arc polygons will produce a limiting circle or *Grenzkreis.* (Recall the discussion in Chapter 4.) Yet, as Klein pointed out early in his correspondence with Poincaré, examples with very complicated boundary figures exist, and, indeed, their limiting curves may not even be twice differentiable.[42]

While nearly all mathematicians agreed about the importance of refined intuition, only a few of Klein's contemporaries subscribed entirely to his views regarding the role of the naive variety. Moritz Pasch, for example, took the position that while the axioms for modern geometry gradually evolved as a distillation of our knowledge of spatial relationships, the axiomatic system ultimately took on a life of its own, independent of all recourse to the imagination. Hilbert later promoted this viewpoint in even more radical terms in his *Grundlagen der Geometrie.*[43] Klein certainly recognized the significance of Hilbert's subsequent achievement, but he flatly rejected the formalist assumptions underlying it, namely, the notion that it made no difference in geometry whether one spoke of points, lines, and planes, or tables, chairs, and beer mugs.[44] Already in his sixth Evanston lecture, he emphatically

---

[40]To support his position, Klein cited H. G. Zeuthen, *Die Lehre von den Kegelschnitten im Altertum*, trans. R. v. Fischer-Benzon (Copenhagen: Höst, 1885); and George Allman, *Greek Geometry from Thales to Euclid* (Dublin: Hodges, 1889). This view is now standard. See, for example, Wilbur R. Knorr, *The Evolution of the Euclidean Elements* (Dordrecht: Reidel, 1975).

[41]Klein, *Evanston Colloquium*, p. 42 (Klein's emphasis).

[42]Felix Klein to Henri Poincaré, 2 July 1881, in *Klein*: *GMA*, 3:596–599. Klein noted that such examples appeared in Friedrich Schottky, "Ueber die konforme Abbildung mehrfach zusammenhängenderebener Flächen," *Journal für die reine und angewandte Mathematik* 83 (1877):300
-351.

[43]The relevant works are Moritz Pasch, *Vorlesungen über neuere Geometrie* (Leipzig: B. G. Teubner, 1882); and David Hilbert, *Grundlagen der Geometrie*, 1st ed. (Leipzig: B. G. Teubner, 1899).

[44]Regarding the foundations of geometry, Klein stated that "*the axioms of geometry are—according to my way of thinking—not arbitrary, but sensible statements, which are, in general, induced by space perception and are determined as to their precise content by expediency* [his italics]." See Felix Klein, *Elementary Mathematics from an Advanced Standpoint. Geometry*, trans. E. R. Hedrick and C. A. Noble (New York: Dover Publications, Inc., 1939), p. 187. Hilbert's views on the foundations of geometry differed from Klein's more in emphasis than in substance, however. For a comparison of the two, see David E. Rowe, "The Philosophical

asserted that when doing research "it is always necessary to combine the intuition with the axioms,"[45] and he reinforced this by referring to examples he had discussed in his earlier lectures. "I do not believe," he remarked, "that it would have been possible to derive the results discussed in my former lectures, the splendid researches of Lie, the continuity of the shape of algebraic curves and surfaces, or the most general forms of triangles, without the constant use of geometrical intuition."[46]

In the remainder of this lecture, Klein took up the age-old question of the relationship between pure mathematics and its applications in the sciences. Here, his positivism came strongly to the fore:

> I believe that the more or less close relation of any applied science to mathematics might be characterized by the degree of exactness attained, or attainable, in its numerical results. Indeed, a rough classification of these sciences could be based simply on the number of significant figures averaged in each. Astronomy (and some branches of physics) would here take the first rank; the number of significant figures attained may here be placed as high as seven, and functions higher than the elementary transcendental functions can be used to advantage. Chemistry would probably be found at the other end of the scale, since in this science rarely more than two or three significant figures can be relied upon. Geometrical drawing, with perhaps 3 to 4 figures, would rank between these two extremes; and so we might go on.[47]

As this passage suggests, Klein argued that only those parts of pure mathe-

---

Views of Klein and Hilbert," in *Proceedings of the* 1990 *Tokyo Symposium on the History of Mathematics*, ed. Chikara Sasaki (Basel: Birkhäuser), to appear. Also consult Michael-Markus Toepell, *Über die Entstehung von David Hilberts "Grundlagen der Geometrie"* (Göttingen: Vandenhoeck & Ruprecht, 1986).

[45] Klein, *Evanston Colloquium*, p. 45.

[46] *Ibid.* Klein's final remarks on the role of intuition in mathematical research came in an oft-cited passage that probably drew relatively little notice at the time, but which carried sinister overtones that reverberated into another era: "Finally, it must be said that the degree of exactness of the intuition of space may be different in different individuals, perhaps even in different races. It would seem as if a strong naive space-intuition were an attribute pre-eminently of the Teutonic race, while the critical, purely logical sense is more fully developed in the Latin and Hebrew races. A full investigation of this subject, somewhat on the lines suggested by Francis Galton in his researches on heredity, might be interesting." *Ibid.*, p. 46. For a discussion of Klein's further views on this subject, see David E. Rowe, "'Jewish Mathematics' at Göttingen in the Era of Felix Klein," *Isis* 77 (1986):422–449. Ludwig Bieberbach, the principal architect of *Deutsche Mathematik,* tried to legitimize his stand by citing the views of his illustrious teacher, the passage quoted above serving as the prosecution's exhibit A. On Bieberbach, see Herbert Mehrtens, "Ludwig Bieberbach and 'Deutsche Mathematik,' " in *Studies in the History of Mathematics*, ed. Esther R. Phillips, MAA Studies in Mathematics, vol. 26 (Washington: Mathematical Association of America, 1987), pp. 195–241. Consult, also, Herbert Mehrtens, *Moderne—Sprache—Mathematik* (Frankfurt am Main: Suhrkamp, 1990), pp. 308–317.

[47] Klein, *Evanston Colloquium*, pp. 46–47.

matics with results lying within the experimental range of a given scientific discipline had any utility for that particular field. Moreover, much of pure mathematics clearly lay well outside the bounds of any possible application. In this connection, Klein mentioned the distinction between commensurable and incommensurable magnitudes, indicating that such notions could not be applied meaningfully to the periods of planets, which can only be determined within a certain range of accuracy. Thus, one can only inquire whether the ratio of the periods of two planets can be approximated by two *small* integers.[48] He further criticized Gustav Kirchhoff's contention that in spectroscopy only those wavelengths of light are absorbed which correspond *exactly* to the emitted wavelengths of some chemical element. Klein cited George Stokes's opinion with approval, that the absorption takes place *in the vicinity* of the corresponding wavelengths.[49]

Considerations such as these led Klein to suggest the possibility of constructing an abridged system of mathematics—built on the work of figures like Gauss and Tchebycheff—suitable for applications in the sciences. Although such an approach apparently made no great impression on American mathematicians, in later years Klein himself took up this whole panoply of ideas in earnest. In 1901, he devoted a series of lectures to what came to be called *Präzisions- und Approximationsmathematik*, that is, pure mathematics, on the one hand, and an *exact* theory of approximative relationships, on the other.[50] Carl Runge, who taught in Göttingen from 1905 to 1927, was strongly influenced by this philosophy and became the foremost exponent of *Approximationsmathematik* in Germany.[51]

Leaving these more philosophical reflections behind, Klein shifted his attention in his seventh lecture to algebra and number theory, addressing the ancient problem of squaring the circle. Klein's former student, Ferdinand Lindemann, had finally laid this issue to rest in 1882 when he proved the transcendence of π, a breakthrough made possible by certain techniques developed nine years earlier by Charles Hermite in his proof of the transcendence of e.[52] Although, in one sense, these results brought an important

[48] Here, Klein was reacting against a tradition that extended from Nicole Oresme (1320–1382) to Poincaré, and beyond to modern-day mathematical algorithms that generate chaotic structures. On Oresme's contributions, consult Edward Grant, "Nicole Oresme and the Commensurability or Incommensurability of Celestial Motions," *Archive for History of Exact Sciences* 1 (1961):420–458.

[49] Klein, *Evanston Colloquium*, p. 47.

[50] Klaus T. Volkert discusses Klein's views within a larger historical and philosophical context in *Die Krise der Anschauung*, Studien zur Wissenschafts-, Sozial-, und Bildungsgeschichte der Mathematik, vol. 3 (Göttingen: Vandenhoeck & Ruprecht, 1986), pp. 226–242.

[51] Klein's 1901 lectures on "Prinzipien der Anwendung der Differential- und Integralrechnung auf Geometrie" were first published in book form as the third volume of *Elementarmathematik vom höheren Standpunkte aus* (Berlin: Springer-Verlag, 1928). Regarding Klein's influence on Runge, see Gottfried Richenhagen, *Carl Runge* (1856–1927): *Von der reinen Mathematik zur Numerik* (Göttingen: Vandenhoeck & Ruprecht, 1985), pp. 146–147.

[52] Ferdinand Lindemann, "Über die Zahl π," *Mathematische Annalen* 20 (1882):213–225; and Charles Hermite, "Sur la Fonction exponentielle," *Comptes Rendus* 77 (1873):18–24, 74–79, 226–233, and 285–293.

chapter in the history of number theory to a close, Klein intimated that the arguments supporting them lacked any quality of self-evident finality. He was therefore pleased to report on the very latest developments in this direction, namely, the recent progress made by Hilbert, Hurwitz, and Gordan in simplifying the proofs originally given.[53] In his opinion: "[t]he problem has ... been reduced to such simple terms that the proofs for the transcendency of e and π should henceforth be introduced into university teaching everywhere."[54] Not that he had any confidence that this would diminish the number of would-be circle-squarers. He cautioned that "this class of people has always shown an absolute distrust of mathematicians and a contempt for mathematics that cannot be overcome by any amount of demonstration."[55]

Klein continued his tour of number theory, exploring two of its central topics in his eighth lecture: the geometric treatment of the composition of binary quadratic forms and Kummer's theory of ideal numbers. For the former, he drew on material from a series of fifteen lectures on number theory he had delivered the previous spring in Göttingen. Although the theory of binary quadratic forms had had its inception in Gauss's *Disquisitiones arithmeticæ*, it was chiefly through Dirichlet's lectures that it had entered into the mainstream of modern mathematics. Two forms are considered equivalent in the classical theory if one is carried into the other by a unimodular linear change of variables. While Gauss had already made use of a so-called lattice of lines to represent a form with a given discriminant D,[56] Klein suggested the idea of identifying a given equivalence class of forms (which must necessarily have the same D) with a lattice of points in the complex plane.[57] In setting forth these ideas and their ramifications to his American audience, Klein emphasized how the geometry of these point lattices and their multiplicative properties provided Kummer's theory of ideal numbers with a new sense of tangibility and self-evidence. Conceptual clarity, one of the hallmarks of the Göttingen mathematical tradition, had here, to Klein's mind at least, scored another resounding triumph. Klein reiterated his point that mathematics is best served when its practitioners strive to place their work in its natural theoretical framework, by concluding that "[t]he whole difficulty encountered by every one when first attacking the study of Kummer's ideal

---

[53] David Hilbert, "Ueber die Transcendenz der Zahlen e," *Mathematische Annalen* 43 (1894): 216–219; Adolf Hurwitz, "Beweis der Transcendenz der Zahl e," *Mathematische Annalen* 43 (1894):220–221; and Paul Gordan, "Transcendenz von e und π," *Mathematische Annalen* 43 (1894):222–224.

[54] Klein, *Evanston Colloquium*, p. 53.

[55] *Ibid.*, pp. 52–53.

[56] Carl Friedrich Gauss, *Disquisitiones Arithmeticæ*, trans. Arthur A. Clarke, rev. William C. Waterhouse *et al.* (New York: Springer-Verlag, 1966), Section 5.

[57] Klein first presented this idea in "Ueber die Komposition der binären quadratischen Formen," *Nachrichten der Gesellschaft der Wissenschaften zu Göttingen, Mathematisch- Physicalische Abteilung* (1893):106–109, or *Klein*: *GMA*, 3:283–286.

numbers is therefore merely a result of his mode of presentation."[58]

Just as Klein felt that he had succeeded in casting Kummer's theory in the proper light, so, too, did he believe that he had made significant conceptual progress with regard to Galois theory, and he sketched these ideas in the ninth of his lectures. As he saw it, "[f]ormerly the 'solution of an algebraic equation' used to mean its solution by radicals. All equations whose solutions cannot be expressed by radicals were classed simply as *insoluble*, although it is well known that the Galois groups belonging to such equations may be very different in character."[59] Noting that such ideas could still be found in 1893, Klein opined that "ever since the year 1858, a very different point of view should have been adopted. This is the year in which Hermite and Kronecker, together with Brioschi, found the solution of the equation of the fifth degree, at least in its fundamental ideas."[60] And then came Klein, who gave this subject a novel geometric and invariant-theoretic twist by introducing the icosahedral equation, an equation of degree sixty with Galois group isomorphic to the rotation group of the icosahedron.

Klein referred his listeners to his classic *Vorlesungen über das Ikosaeder* for details, but he did pause to point out some of the basic properties of the icosahedral equation. In particular, he noted that it played the same role in the theory of quintic equations that the equations of the form $\eta^n = z$, where z is a known quantity, played in the theory of equations solvable by radicals. Whereas the latter case engendered a cyclic group of linear substitutions in one variable of order n, the general quintic led to a group of sixty substitutions in two variables. Klein pursued this analogy to some depths before going on to mention further work on higher-order equations undertaken by Gordan, Burkhardt, Maschke, and himself.[61] In the course of this survey, he set forth the following general program: given an equation with Galois group G, determine the group G' of linear substitutions in the least number

[58]Klein, *Evanston Colloquium*, p. 66. For a discussion of Kummer's work on the theory of ideal factors, see Harold M. Edwards, *Fermat's Last Theorem: A Genetic Approach to Algebraic Number Theory* (New York: Springer-Verlag, 1977), pp. 76–151.

[59]*Ibid.*, p. 67 (Klein's emphasis).

[60]*Ibid.*

[61]Paul Gordan, "Ueber Gleichungen siebenten Grades mit einer Gruppe von 168 Substitutionen," *Mathematische Annalen* 20 (1882):515–530; and 25 (1885):459–521; Heinrich Burkhardt, "Untersuchungen aus dem Gebiete der hyperelliptischen Modulfunctionen. Dritter Theil," *Mathematische Annalen* 41 (1893):313–343; Heinrich Maschke, "Ueber die quaternäre, endliche, lineare Substitutionsgruppe der Borchardtschen Moduln," *Mathematische Annalen* 30 (1887):496–515; "Aufstellung des vollen Formensystems einer quaternären Gruppe von 51840 linearen Substitutionen," *Mathematische Annalen* 33 (1889):317–344; and "Ueber eine merkwürdige Configuration gerader Linien im Raume," *Mathematische Annalen* 36 (1890):190–215; and Felix Klein, "Ueber die Auflösung gewisser Gleichungen vom siebenten und achten Grade," *Mathematische Annalen* 15 (1879):251–282, or *Klein: GMA*, 2:390–425; and "Zur Theorie der allgemeinen Gleichungen sechsten und siebenten Grades," *Mathematische Annalen* 28 (1887):499–532, or *Klein: GMA*, 2:439–472.

of variables that is isomorphic to G.[62] This again signaled Klein's broader aim in these lectures, namely, to show how past mathematical achievements tie in with present research interests. By illuminating these connections in as many fields as possible, Klein hoped to impart to his American audience a deeper appreciation of some of the many threads that made up the tapestry of mathematics as it then appeared and thereby to stimulate and guide their work.

In the lecture that followed, "On Some Recent Advances in Hyperelliptic and Abelian Functions," Klein had the opportunity to push and expound upon another of his pet programs: the extension of his *Stufentheorie* to arbitrary elliptic functions as well as hyperelliptic and Abelian functions. (As we saw in Chapter 4, this theory had originally arisen in Klein's work on elliptic modular functions.) Although Fricke had already laid out much of the groundwork for this approach in the Klein-Fricke volumes on elliptic modular functions, Klein underscored the recent progress made in this field by his students, Bolza and Ernst Ritter, as well as by Heinrich Burkhardt and Wilhelm Wirtinger. He also called attention to the work done by his American students, White and Osgood, on algebraic curves and their corresponding Abelian functions.[63] In connection with his own research on the Θ-functions associated with a quartic curve of genus 3, Klein made this characteristic observation: "Here, as elsewhere, there seems to reign a certain pre-established harmony in the development of mathematics, what is required in one line of research being supplied by another line, so that there appears to be a logical necessity in this, independent of our individual disposition."[64]

Whereas, as this pronouncement suggests, this tenth lecture carried certain philosophical overtones, Klein's penultimate presentation sounded a conciliatory note in the direction of Sophus Lie. Klein opened his survey of "The Most Recent Researches in Non-Euclidean Geometry" by distinguishing three possible approaches to the subject. The first of these, exemplified in the work of Lobachevsky and Bolyai, simply replaced Euclid's parallel postulate by another axiom and proceeded systematically to construct a space with these requisite properties. The second approach, one that Klein himself had

---

[62] Klein, *Evanston Colloquium*, pp. 72–73.

[63] Oskar Bolza, "Darstellung der rationalen ganzen Invarianten der Binärform sechsten Grades durch die Nullwerthe der zugehörigen θ-Functionen," *Mathematische Annalen* 30 (1887): 478–495; Heinrich Burkhardt, "Grundzüge einer allgemeinen Systematik der hyperelliptischen Funktionen erster Ordnung: Nach Vorlesungen von F. Klein," *Mathematische Annalen* 35 (1890): 198–296; and "Beiträge zur Theorie der hyperelliptischen Sigmafunktionen," *Mathematische Annalen* 32 (1888):351–442; Wilhelm Wirtinger, "Untersuchungen über Abel'sche Functionen vom Geschlechte 3," *Mathematische Annalen* 40 (1892):261–312; Henry Seely White, "Abel'sche Integrale auf singularitätenfreie, einfach überdeckten, vollständigen Schnittcurve eines beliebig ausgedehnten Raumes," *Nova Acta der kaiserlichen Leopoldinisch-Carolinischen deutschen Academie der Naturforscher* (1891):43–128; and William Fogg Osgood, "Zur Theorie der zum algebraischen Gebilde $y^m = R(x)$ gehörigen Abel'schen Functionen," (Dissertation, Erlangen University, 1890).

[64] Klein, *Evanston Colloquium*, p. 83.

adopted in the early 1870s, began with the complex projective plane and an embedded conic. The generalized Cayley metric then led to the now familiar classification scheme of elliptic, parabolic, and hyperbolic geometries.

The third possible approach Klein mentioned stemmed from Riemann's famous *Habilitationsvortrag*, "On the Hypotheses Which Form the Basis for Geometry."[65] Helmholtz had taken Riemann's ideas a step further by pointing out that Riemann's quadratic differential form followed necessarily from the assumption that rigid bodies could be freely transported throughout the ambient space. Klein contrasted this point of view with his own by suggesting that whereas he had grounded his own ideas in optical properties, Helmholtz had taken a basic principle of mechanics as his starting point. He then indicated how Lie had come to take up the so-called Riemann-Helmholtz (or, more appropriately, the Riemann-Helmholtz-Lie) space problem.

Since Lie considered this one of the most important and direct applications of his theory of continuous groups, Klein took pains to point out and explain its significance.[66] In part, he emphasized this aspect of Lie's work in order to make amends for his failure to do so in his lithographed lectures on non-Euclidean geometry of 1889–1890, an omission that had greatly upset Lie.[67] While this may have helped ease Klein's conscience, the olive branch he proffered to Lie on this occasion came too late. Back in Leipzig, the third volume of Lie's *Theorie der Transformationsgruppen* was about to come off the press with its stinging preface. (Recall Chapter 5.) In private, Lie could be even more brutal. Writing to Adolf Mayer probably one or two years later, Lie compared Klein to "an actress, who in her youth dazzled the public with glamorous beauty but who gradually relied on ever more dubious means [*verwerflichere Mittel*] to attain success on third-rate stages."[68] Presumably,

[65]Bernhard Riemann, "Ueber die Hypothesen welche der Geometrie zu Grunde liegen," *Abhandlungen der königlichen Gesellschaft der Wissenschaften zu Göttingen* 13 (1868):132–152, or *Bernhard Riemann's Gesammelte Mathematische Werke*, ed. Heinrich Weber, 2nd ed. (Leipzig: B. G. Teubner, 1892), pp. 272–287.

[66]See Sophus Lie, "Ueber die Grundlagen der Geometrie," *Berichte über die Verhandlungen der königlichen sächsischen Gesellschaft der Wissenschaften zu Leipzig*, 3 (1890):355–418, or *Gesammelte Abhandlungen*, ed. Friedrich Engel and Poul Heegaard, 7 vols. (Oslo: H. Aschehoug & Co. and Leipzig: B. G. Teubner, 1922–1960), 2:414–476.

[67]This circumstance prompted Klein to remark: "I have the more pleasure in placing before you the results of Lie's investigations as they are not taken into due account in my paper on the foundations of projective geometry ... nor in my lectures on non-Euclidean geometry ... [*Evanston Colloquium*, p. 88]." He explained this by noting that Lie's lengthy articles in the *Berichte* of the Saxon Academy of Sciences had only appeared after his own lectures, and Lie's shorter article of 1886 "had somehow escaped my memory." Klein referred here to "Zur Nicht-Euklidischen Geometrie," *Mathematische Annalen* 37 (1890):544–572, or *Klein*: *GMA*, 1:353–383, and to his lithographed lectures on "Höhere Geometrie" delivered in the Summer Semester of 1893.

[68]Sophus Lie to Adolf Mayer, undated, probably 1894–1895, Mayer Nachlass, Universität Leipzig (our translation). The complex circumstances surrounding this falling out between Lie and Klein are discussed in David E. Rowe, "Der Briefwechsel Sophus Lie-Felix Klein, eine Einsicht in ihre persönlichen und wissenschafthen Beziehungen," *NTM Schriftenreihe für Geschichte*

Lie had in mind the American stage—with its Chicago Congress and Evanston Colloquium lectures—when he wrote these words, although his judgment was clearly tinged by a mixture of envy, jealousy, and contempt for Klein's enormous popularity.

Setting personal feelings aside, Klein tried to set the record straight in his eleventh lecture by laying out the historical development as he saw it. He mentioned Helmholtz's introduction of the monodromy axiom, which stipulated closed orbits for points that move about a fixed axis. The example of the logarithmic spiral—studied by Klein in connection with the theory of W-curves—pointed to the necessity of invoking such an axiom when considering motions in the plane.[69] Lie showed, however, that in three-space Helmholtz's other axioms already implied the monodromy condition. Unfortunately for Lie, his arguments made little impression on the physicist Helmholtz, who saw them as the typical hair-splitting of the mathematician.[70]

Having touched upon this perennial bone of contention between mathematicians and physicists, Klein moved on to discuss both geometry in the large and the problem of classifying various possible space forms, highlighting the novel work of William Kingdon Clifford. He asserted first that one could not deduce the infinity of space by showing that its curvature is zero and concluded that

> our geometrical demonstrations have no absolute objective truth, but are true only for the present state of our knowledge. These demonstrations are always confined within the range of the space-conceptions that are familiar to us; and we can never tell whether an enlarged conception may not lead to further possibilities that would have to be taken into account. From that point of view we are led in geometry to a certain modesty, such as is always in place in the physical sciences.[71]

These musings would appear all the more prescient some two decades later when Albert Einstein began to develop the geometrical foundations of his general theory of relativity. Einstein's new geometry of space-time generalized the flat, space-time geometry that Klein's younger colleague Minkowski had conjured up as a mathematical model to describe the phenomena in Einstein's special theory. Ironically enough, this famous space-time geometry of Minkowski is nothing more than a four-dimensional version of a pseudo-

---

der Naturwissenschaften, Technik und Medizin 25 (1988):37–47.

[69]Felix Klein and Sophus Lie, "Ueber diejenigen ebenen Kurven ... ," *Mathematische Annalen* 4 (1871):50–84, or *Klein*: *GMA*, 1:424–458.

[70]Klein discussed these matters with Helmholtz during the course of their voyage to the United States. See Leo Königsberger, *Hermann von Helmholtz*, trans. Frances A. Welby (Oxford: Clarendon Press, 1906), pp. 412–413.

[71]Klein, *Evanston Colloquium*, p. 93.

Euclidean geometry that Klein and (especially) Lie had employed extensively in their early research.[72]

In a more direct allusion to things to come, Klein gave his closing lecture on "The Study of Mathematics at Göttingen" largely for the benefit both of those planning to study there in the future and of those in a position to advise suitable candidates about study abroad. He opened with some general remarks about universities in Germany, the theoretical orientation generally pursued there, and his recent efforts to counterbalance this by strongly recommending that students of mathematics enroll in courses like astronomy, technical mechanics, and descriptive geometry. Advanced students, and particularly those who wished to pursue a doctorate, might occasionally take such courses, but they eventually focused their energy on purer, more specialized lectures and seminars.

Klein then turned to consider the special needs of foreign students. Most of the Americans and other foreigners who had come to Göttingen won his praise for their "great enthusiasm and energy."[73] He allowed, too, that for some years his advanced courses had mainly been sustained by American students. The primary message he wished to deliver, however, concerned prior, overall preparation. Klein recognized that Americans generally came to Göttingen and, indeed, to the various German universities in order to pursue advanced courses. Yet, he emphasized how, more often than not, "their preparation is entirely inadequate for such work."[74] He warned that "[a] student having nothing but an elementary knowledge of the differential and integral calculus, usually coupled with hardly a moderate familiarity with the German language, makes a decided mistake in attempting to attend my advanced lectures."[75] Given this state of affairs, Klein asked rhetorically whether the American student "[w]ould ... not do better to spend first a year or two in one of the larger American universities? Here he would find more readily the transition to specialized studies, and might, at the same time, arrive at a clearer judgement of his own mathematical ability; this would save him from the severe disappointment that might result from his going to Germany."[76]

As we saw in Chapter 5, Klein had been particularly desirous of students who could edit his lectures and help promote his various research programs, so he wanted to encourage only the best-prepared to come to Göttingen. Furthermore, echoing the views of Harry Tyler, he had to admit that the encyclopedic character of his higher lecture courses sometimes made these unsuitable for the American student whose first priority was to obtain the

[72]For a discussion of these developments, see Klein, *Vorlesungen über die Entwicklung der Mathematik im 19. Jahrhundert*, 1:143–145.

[73]Klein, *Evanston Colloquium*, pp. 96–97.

[74]*Ibid.*, p. 97.

[75]*Ibid.*

[76]*Ibid.*

doctorate. He went on to declare that

> I do not regard it as at all desirable that all students should confine their mathematical studies to my courses or even Göttingen. On the contrary, it seems to me far preferable that the majority of the students should attach themselves to other mathematicians for certain special lines of work. My lectures may then serve to form the wider background on which these special studies are projected. It is in this way, I believe, that my lectures will prove of the greatest benefit.[77]

Although not exactly a farewell to American mathematics, this statement betrayed a Klein groping for a new role. Still only dimly aware of what this might be, he was prepared to step back and let others shape the future course of higher mathematics in the United States by serving as mentor to the country's fledgling mathematicians. Before him sat the very men who would thenceforth assume much of that responsibility—Moore, Bolza, and Maschke—and there can be no doubt that, by the time Klein brought his Evanston Colloquium lectures to a close, all three had gained a new appreciation of the enormous opportunities and challenges that lay before them.

As for the other members of the audience, they went their separate ways after the colloquia. Henry Eddy soon accepted a position at the University of Minnesota; Henry White remained at Northwestern until 1905 when he moved to Vassar College; in 1895, Wisconsin's E. B. Van Vleck returned to his *alma mater*, Wesleyan; and Mary Frances Winston, who joined Klein in Göttingen the very next semester, became the first American woman to earn a doctorate at a Prussian university three years later. In 1903, Van Vleck, White, and F. S. Woods—the three undergraduates of Wesleyan's John Van Vleck who took doctorates under Klein in Göttingen—delivered the fourth series of AMS Colloquium lectures; it was published as the first volume in a new series, the *AMS Colloquium Publications*. Thus, in this particular instance, the Kleinian connection and the example of Klein's Evanston Colloquium lectures helped generate one of the significant new institutions that promoted the work of America's emergent mathematical community.

TOURING THE AMERICAN MATHEMATICAL SCENE

After having spent three intense weeks amid both the tumult of the Fair and the activity in Evanston, Felix Klein left Chicago for a whirlwind tour that took him to nearly all the leading universities and other institutions of higher learning on the East Coast. Like the Swiss-born naturalist Louis Agassiz some forty-seven years earlier, Klein wanted both to survey and to assess the American scientific scene.[78] Yet, unlike Agassiz, who spent the

---

[77] *Ibid.*, p. 98.

[78] See Robert V. Bruce, *The Launching of Modern American Science*, 1846–1876 (Ithaca: Cornell University Press, 1987), pp. 43–63.

rest of his career in Boston, Klein had less than a month at his disposal, and, in some sense, he had a good deal more to see. Whereas Agassiz had found a small but active community of scientists in Boston centered around Harvard and the American Academy of Arts and Sciences, he had noted the stark contrasts in the extent and intensity of support for science elsewhere in the United States. Philadelphia, as the former federal Capitol, boasted the American Philosophical Society, the University of Pennsylvania, botanical gardens, and museums, but in 1846 Boston had recently eclipsed it as America's premier scientific city. Running a poor third, New York impressed Agassiz more as the financial and trade center it was than as a major seat of scientific activity. Lesser still, Albany, New Haven, Princeton, Troy, Washington, D.C., and West Point encouraged scientific endeavors to some extent but paled in comparison even with New York.

In his survey of the state of American mathematics in 1893, Klein uncovered a somewhat different geographical profile. Whereas by mid-century several larger cities had amassed the requisite concentration of wealth and population to support general scientific inquiry, mathematics at century's end often thrived in newer institutions located outside of these traditional centers. Thus, Klein had witnessed firsthand the energy and enthusiasm of the mathematicians in upstart Chicago, and on leaving the Midwest, he moved on to the very different rural splendor and seclusion of Ithaca in upstate New York. While at Cornell, he was hosted by the senior mathematics professor there, James E. Oliver, who, as we have seen, had studied under Klein during the academic year 1889–1890 and who had also attended the Chicago festivities. Since he knew of Cornell's potential through his prior association with Oliver as well as through his then current contact with Oliver's protégé, Virgil Snyder, he was surely curious to see the school up close. Especially impressed by its engineering facilities, Klein envisioned the potentiality of the hybrid, land-grant/privately endowed university as a congenial academic environment for his future students on the successful completion of their doctorates.

After Cornell, Klein visited Clark University, Harvard, and the new facilities at the Massachusetts Institute of Technology. He also stopped at Yale to meet Willard Gibbs and at Wesleyan to see the elder Van Vleck on his way southward to New York City.[79] On his arrival there, the New York Mathematical Society fêted him at a special meeting held in his honor at Columbia College. The Society's President, Emory McClintock, introduced him as an "apostle, prophet, evangelist, and teacher of mathematics—excelling in each

[79]Van Vleck accompanied Klein on his trip to New Haven. See Felix Klein to J. Willard Gibbs, 17 September 1893, Yale University Library. While in Chicago, Van Vleck had apparently struck a deal with representatives of the German government that enabled him to purchase the mathematical models on display there for $750. Not that Wesleyan was the first to acquire such a collection, for when Bolza arrived at Clark in 1889 he wrote Klein that the University already had a complete set of the Brill firm's models.

office," and a local newspaper described him as "notable among great specialists in science ... [for his] wide and accurate knowledge on almost every high topic of public interest."[80] Klein acknowledged McClintock's adulatory remarks and proceeded to give "a most amazing exhibition of original research"[81] dealing with trigonometrical formulas in non-Euclidean geometry. This was followed by a period for open discussion and questions, during which "[h]e gave swift replies, covering wide fields, from the general trend of scientific thought in Germany at the present day to the most abstruse detail in some mathematical problem treated by some individual scientist."[82]

Klein's operational headquarters during his travels was Mayer's Hotel located in Hoboken, New Jersey, an establishment frequented by many German tourists, not least because of its convenient location near the piers for several transatlantic ships. It was from here that he wrote to the Naval Observatory's Simon Newcomb in the hopes of arranging a meeting.[83] Klein was eager to see Newcomb not just because of the latter's considerable international scientific reputation but, even more importantly, because of his practical experience as a leading mover in American scientific circles.[84] It was not a foregone conclusion that he and Newcomb would get along, however. Besides the fact that both men had more than healthy-sized egos, Newcomb and Klein's friend and former student, Fabian Franklin, had had a rather serious falling out which might well have affected the conviviality of the occasion. Nevertheless, things apparently did go well, since Newcomb even arranged for Klein to meet with Hopkins President Daniel Gilman, the man who probably could have secured the eminent German for his faculty ten years earlier. On Sunday, 1 October, Newcomb wrote to Gilman that "Professor Klein is here and intends to go to Baltimore Tuesday afternoon. He is so full of energy, enthusiasm, and well-considered ideas that it is a great pleasure to talk with him. If you could make it convenient to have him spend Tuesday evening with you I think you would feel well repaid. Probably the necessary arrangements could be made through Dr. Franklin."[85]

On the final leg of his journey, Klein visited Princeton, where he was received by Henry Fine and Henry Dallas Thompson. It must have been a weary Felix Klein who finally boarded the "Saale" headed for Bremen on 7 October. But Klein never seemed to rest unless his body absolutely demanded

[80]Newspaper clipping entitled "Welcome to a German Scientist," Daniel Coit Gilman Papers, Ms. 1, The Milton S. Eisenhower Library, The Johns Hopkins University (hereinafter referred to as Gilman Papers).

[81]*Ibid.*

[82]*Ibid.*

[83]Felix Klein to Simon Newcomb, 24 September 1893, Newcomb Papers, Library of Congress, Washington, D. C.

[84]Albert E. Moyer has recently discussed this and other aspects of Newcomb the scientific activist in *A Scientist's Voice in American Culture: Simon Newcomb and the Rhetoric of Scientific Method* (Berkeley: University of California Press, 1992).

[85]Simon Newcomb to Daniel C. Gilman, 1 October 1893, Gilman Papers.

it, and during this eleven-day voyage he drafted a preliminary report of his findings on mathematics education in the United States.

In this unpublished document, Klein likened America to an untapped, subterranean well full of scientific talent. He stressed that this potential went beyond the proverbial Yankee ingenuity for experimentation and invention, noting that figures like Willard Gibbs, George Hill, and Simon Newcomb had reached high levels of attainment in the theoretical sciences as well. These three hardly represented the typical product of the North American educational system, but Klein argued that "this only shows that the talent for higher scientific achievement is present in the American population,"[86] a statement that echoed Sylvester's earlier assessment of the situation. The question remained, however, how could this hidden resource be tapped? Klein pointed to the secondary schools as the weakest link in the education system. "[I]n striving for well-roundedness," he wrote, "they tend toward superficiality so that the [students'] logical powers and the capability of carrying out concentrated work remain undeveloped."[87] As for the universities, he found that the instruction was often too mechanical and overly dependent on the use of textbooks. This, he implied, resulted from an overemphasis on teaching and too little concern with research.

Klein cautioned that one must view these weaknesses in perspective, by taking into account not just "the present conditions in America" but "the entire development ... especially in this land which is moving incredibly quickly."[88] From this vantage point, he added, "the picture changes completely. *The Americans are completely aware of the above-mentioned flaws, and they are working to remove them by following more than ever the model of the German universities.*"[89] Klein refused to make any predictions about the success of these ventures, although he expressed high hopes for them. He did spell out, in the plainest possible language, what he felt was at stake for Germany. "Without question," he asserted, "*at the present time and for the immediate future, America represents the richest and most promising object [das grösstmöglichste und glücklichste Objekt] for scientific colonization.*"[90] He added that "[a]lready in past years, large numbers of Americans have studied at German universities. Up until now, however, this has happened without any special initiative on our part. My trip evidently represents a change in this policy, and I would like to add immediately that I have already made definite plans and provisional arrangements to return to America in the future."[91]

---

[86]Klein, "Bericht über die Reise nach Chicago," NSUB, Göttingen (our translation).

[87]*Ibid.*

[88]*Ibid.*

[89]*Ibid.* (Klein's emphasis).

[90]*Ibid.* (Klein's emphasis).

[91]Klein had in mind to deliver another cycle of lectures in the United States once the Evanston lectures had had a chance to make an impact. Three years later, while attending Princeton's

Klein certainly returned from the United States singing a very different tune from the one he had voiced before his arrival, for, without a doubt, he wanted to bask in the personal triumph which he identified, predictably enough, with the impressive performance of German science at the Chicago Fair. He described this performance fairly glowingly in a report on the Fair officially prepared by the German government:

> I was altogether surprised by the Universities Exhibit, and I only heard unanimous acclaim for how beautiful and resplendent it appeared. In particular, the mathematics exhibit was excellent. It was very rich, and I myself became acquainted for the first time with a large number of models and apparatus that had not yet come on the market. I have no doubt that they made a great and lasting impression on the experts.[92]

As for the performance of German mathematics, Klein gave this self-congratulatory assessment:

> I must add a few remarks about the Mathematics Congress. Here too there was no lack of pessimistic voices in advance [including his own], and afterward everyone agreed that most of the planned scientific congresses enjoyed but little success, especially since they failed to receive the general support of the American scientific community. In contrast with this, we may take satisfaction that everywhere where representatives of German science were present they [the congresses] met with success ... . It is well known that mathematics in America is still rather modestly developed when compared with her sister sciences, physics and astronomy, which are in their full flowering. I am very hopeful that the exhibit, Congress, and Colloquium will together remain a source of inspiration for the development of American mathematics.[93]

These remarks may perhaps be written off as just so much wanton breast-beating, but, if nothing else, they reflect how earnestly Klein viewed his mission. Like Friedrich Althoff, the powerful ministerial official for higher edu-

---

sesquicentennial celebration in 1896, he gave a series of four lectures on "The Mathematical Theory of the Top" in which he showed how this classical subject could be treated more elegantly by utilizing Riemannian function theory. These lectures were edited by Henry Fine and published as a small booklet: Felix Klein, *The Mathematical Theory of the Top* (New York: Charles Scribner's Sons, 1897). Two days after the Princeton lectures, the American Mathematical Society held a meeting in Princeton, which gave Klein occasion to entertain its members with a supplementary lecture "On the Stability of a Sleeping Top," *Bulletin of the American Mathematical Society* 3 (1897):129–133.

[92] *Weltausstellung in Chicago,* 1893: *Reichskommissar für die amtlicher Bericht über die Weltausstellung in Chicago,* 1893 (Berlin: Reichsdruckerei, 1894), pp. 988–993 (our translation).

[93] *Ibid.*

cation in the Prussian university system, Klein recognized that the time was ripe for German scholars to step down from their ivory towers. Both men believed that German learning represented a potentially powerful asset for conducting diplomacy and fashioning the country's dynamic new Empire.[94] Since the utility of modern mathematics had increasingly come into question at the *Technische Hochschulen*, Klein welcomed the chance to point out how he and other "pure" mathematicians could serve the larger aims of the *Reich.*[95]

The Aftermath

Unfortunately, these rather self-satisfied reflections were interrupted abruptly when Klein's fellow scientific traveler, Helmholtz, fell from a ladder and badly injured himself after about a week at sea. Although he ostensibly recovered from this accident, Helmholtz suffered an apparent paralytic stroke the following July which left him virtually incapacitated.[96] When this second tragedy struck, he had been hard at work on an invited address for the September meeting of the Society of German Natural Scientists and Physicians to be held in Vienna. His death came eighteen days before it convened on 8 September. The man chosen to fill his place on that mournful occasion for German science was Felix Klein, who spoke on "Riemann and his Importance for the Development of Modern Mathematics," a topic that gave him ample opportunity to make more propaganda for one of his favorite causes.[97]

As for the festivities in Chicago, they too ended on a tragic note. Mayor Carter Harrison, a flamboyant, Old-West style figure who had become something of a symbol for the wild times Chicagoans were enjoying while hosting the Fair, was assassinated by a lunatic the day before the Columbian Exposition ended. This not only gave the closing ceremonies a somber air but also set the tone for the events that followed. The winter weather grew particularly harsh, while the economic depression deepened, and the South Park Commission proved unable to care properly for the buildings on the fairgrounds. These were then occupied by indigents and tramps who set small fires and damaged some of the remaining sculpture. On the evening of 5 July 1894 a major fire—presumably set by arsonists—broke out in the Court

[94] Althoff played an instrumental role, for example, in founding exchange professorships with Harvard and Columbia. See Bernhard vom Brocke, "Der deutsch-amerikanischen Professorenaustausch: Preußische Wissenschaftspolitik, internationale Wissenschaftsbeziehungen und die Anfänge einer deutschen auswärtigen Kulturpolitik vor dem ersten Weltkrieg," *Zeitschrift für Kulturaustausch* 31 (1981):128–182.

[95] We have discussed this, as well as the cultural agenda of the mathematicians, in "Embedded in the Culture: Mathematics at the World's Columbian Exposition of 1893," *The Mathematical Intelligencer* 15 (2) (1993):40–45.

[96] Königsberger, pp. 412–415.

[97] Felix Klein, "Riemann und seine Bedeutung für die Entwicklung der modernen Mathematik," *Jahresbericht der deutschen Mathematiker-Vereinigung* 4 (1894–1895):71–87, or *Klein: GMA*, 3:482–498.

of Honor. Henry White and his neighbors in Evanston could see the smoke from twenty miles away as it slowly climbed into the sky.[98] Two hours later, nearly everything lay in a heap of rubble.

The "White City," a magic mirage built on a marshland, was gone just as quickly as it had appeared. Its creators had tried in vain to present a vision of unity amid the growing complexities of modern-day urban life. It was a grand vision that proved illusory, so what did it signify? Henry Adams (writing in the third person) mused:

> Chicago asked in 1893 for the first time the question whether the American people knew where they were driving. Adams answered, for one, that he did not know, but would try to find out. On reflecting sufficiently deeply, under the shadow of Richard Hunt's architecture, he decided that the American people probably knew no more than he did; but that they might still be driving or drifting unconsciously to some point in thought, as their solar system was said to be drifting to some point in space; and that, possibly, if relations enough could be observed, this point might be fixed. Chicago was the first expression of American thought as unity; one must start there.[99]

As for American mathematics, its future course was in many respects already set. Even if the gathering of mathematicians on the shores of Lake Michigan looked more like a regional meeting than an international congress, this event and the Evanston Colloquium that followed on its heels carried repercussions of immense significance for the crystallization and emergence of the American community of research mathematicians. As we shall see in Chapter 9, E. H. Moore quickly capitalized on the success of the Congress by engaging the New York Mathematical Society to subsidize the publication of the Congress proceedings. This arrangement, in turn, helped legitimize the Society as a national organization, an orientation it would officially assume in 1894 when it became the American Mathematical Society. Two years later, inspired by the example of Klein's Evanston Colloquium lectures, Northwestern's Henry White helped found the regular series of Colloquium Lectures sponsored by the AMS. Thus, the events that took place in the Chicago metropolitan area during the late summer of 1893 helped galvanize Midwestern mathematicians, and particularly Moore's department at the University of Chicago, at a time when the American mathematical research community was just beginning to gel.

Furthermore, while few may have noticed it at the time, Klein's visit clearly

---

[98]Henry Seely White to Felix Klein, 11 November 1894, Klein Nachlass XII, NSUB, Göttingen. White was deeply saddened by this inglorious ending and enclosed photographs of the ruined buildings with this letter.

[99]Henry Adams, *The Education of Henry Adams* (Boston: Houghton Mifflin Co., 1961), p. 343.

signaled his desire to bow out gracefully after a decade of involvement in the American mathematical scene. His report to the Prussian Ministry suggests, however, that he still believed Germany could and should exercise considerable influence on the mathematical community in the United States. Having German mathematicians like Bolza and Maschke at a first-class American university like Chicago was certainly one way to maintain such a sphere of influence. Presumably, this circumstance made it easier for Klein to relinquish to the Americans the responsibility for their higher education in mathematics. Over the course of the next few years, the University of Chicago, with E. H. Moore at its mathematical helm, quickly emerged not only as the leading center for mathematics in the United States but also as the first American institution of higher education to offer mathematical training comparable to that available at leading European universities.

Chapter 9
Meeting the Challenge: The University of Chicago and the American Mathematical Research Community

With the close of the Mathematical Congress and the Evanston Colloquium, the mathematicians at the University of Chicago threw themselves back into the task they had begun just one year earlier, namely, the development of a department devoted to mathematical training and original research. At the undergraduate level, they fashioned a program which would not only expose the student body to mathematics as part of general culture but also prepare the mathematically inclined for more advanced work at the graduate level. For those already at this higher stage, they put together a broad range of courses designed to bridge the gap between studying mathematics and doing mathematical research. Finally, at the post-doctoral level, they strove to heighten research activity both within the department, through the regular meetings of their Mathematical Club, and nationally, through their involvement in the American Mathematical Society. Within the conducive and supportive, research-oriented atmosphere of William Rainey Harper's University, E. H. Moore and his colleagues Oskar Bolza and Heinrich Maschke succeeded in realizing their goals.

Running the Chicago Department of Mathematics

When the University opened for instruction on 1 October 1892, almost six hundred young men and women—among them 242 undergraduates and 170 graduate students in arts and sciences—enrolled to take part in Harper's carefully crafted educational institution.[1] Indeed, the President had tried to detail every aspect of the school's operation in drawing up his series of *Official*

[1] *The President's Report: Administration—The Decennial Publications*, 1st ser., vol. 1 (Chicago: University of Chicago Press, 1902), p. 11. In addition to the undergraduates and graduate students in arts and sciences, 84 graduate divinity students and 62 unclassified students were also in attendance during the first quarter of operation. See *ibid.*, p. 175. These figures differ somewhat from those given in Thomas Wakefield Goodspeed, *A History of the University of Chicago Founded by John D. Rockefeller: The First Quarter-Century* (Chicago: University of Chicago Press, 1916), p. 248. The University of Chicago was coeducational from the beginning.

Bulletins, seemingly leaving nothing to chance and little to the discretion of individual faculty members. (Recall from Chapter 6 that it was precisely this ultra-preparedness which had caused Bolza to question, in January of 1893, whether the faculty would enjoy the curricular freedoms he felt were so essential to the success of the German educational system.) Relative to the duties of the Head Professor, or Departmental Chair, for example, Harper stipulated that

> [h]e was to supervise the entire work of his department in general, prepare all entrance examination papers and approve all course examinations prepared by other instructors, arrange course offerings from quarter to quarter, examine all theses offered in the department, determine the textbooks to be used, edit any appropriate papers or journals, conduct a club or seminar, consult with the librarian about needed books and periodicals, consult with the President on appointments of instructors, and countersign the course certificates in the department.[2]

In practice, the Head Professor did not play the dictatorial role suggested by this prescription; in practice, Harper's system was, as he put it, "flexible."[3] Under the Head Professor's direction, each department enjoyed the freedom to devise its own curriculum, choose its own textbooks, and establish its own standards of evaluation. E. H. Moore faithfully implemented Harper's directives in Chicago's Department of Mathematics while setting and maintaining the most exacting standards.

From the start, every participant under Moore's direction—from colleagues to young Assistants and Associates to Fellows and graduate students—had to live up to his expectations or suffer the consequences. This management style never presented a problem for Bolza or Maschke, both of whom shared Moore's vision for the department. For Harris Hancock, however, the situation was quite different and illustrates well both Moore's level of expectation and what could be his feisty personality.

As mentioned earlier, Hancock had been chosen to serve on the first faculty in the post-doctoral rank of Associate. Contrary to what he had led Harper and Moore to believe in the correspondence prior to his appointment, though, he did not ultimately succeed in earning his Berlin doctorate under Lazarus Fuchs in 1892. He thus asked for a leave of absence during Chicago's first

[2]Richard J. Storr, *A History of the University of Chicago: Harper's University The Beginnings* (Chicago: University of Chicago Press, 1966), p. 62. This passage is also quoted in Karen Hunger Parshall, "Eliakim Hastings Moore and the Founding of a Mathematical Community in America, 1892–1902," *Annals of Science* 41 (1984):313–333 on p. 319; reprinted in *A Century of Mathematics in America—Part* II, ed. Peter Duren *et al.* (Providence: American Mathematical Society, 1989), pp. 155–175 on p.160.

[3]Oskar Bolza to Felix Klein, 15 January 1893, Klein Nachlass VIII, Niedersächsische Staats- und Universitätsbibliothek, Göttingen (hereinafter abbreviated NSUB, Göttingen) (our translation). See the conclusion of Chapter 6 for the full context of the quotation.

year in order to complete his work. Moore did not favor this idea, and Harper denied the request. To add insult to injury, in Hancock's view, his rank was lowered from Associate (or Tutor) to Assistant for the 1892–1893 academic year, due to the fact that he did not yet have the required credential. Taking an approved leave the following year instead, Hancock finally took his doctorate in 1894 and resumed his assistantship later that fall.[4] Unfortunately, this first post-graduate year also went badly when he failed to shine as a teacher and when the lectures he subsequently gave on his research failed to impress Moore favorably. Although he was promoted to an associateship in 1895, Moore condemned his work in reports to Harper, calling it "essentially compilation," and refused to promote Hancock at the rate the young man believed he merited.[5] In response to this, Hancock kept up a steady correspondence to Harper over the five-year period from 1895 to 1900, in which he extolled his own virtues, asked for promotions, and denounced Moore's treatment of him. Harris Hancock finally left Chicago in 1900 to take a professorship at the University of Cincinnati where he remained until his retirement in 1937.[6]

As the saga of Hancock's association with the University of Chicago suggests, Moore unwaveringly maintained the standards he had set for his department, and he did not choose to suffer those who failed to meet them. With those who did make the grade, however, he worked collegially and effectively. The ultimate success of his administrative style was well reflected in his coordination of the formulation of a comprehensive mathematics program at Chicago, which not only met the demands of a varied student body but also satisfied the faculty's need for research time.

Following the example set by the mathematicians at the Johns Hopkins and gradually adopted at such established institutions as Harvard and Yale, the Chicago group under Moore put together a serious mathematics curriculum at the elementary and intermediate as well as at the graduate level. To begin with, the University required undergraduates to take a basic "course in Algebra, Plane Trigonometry, and Cöordinate Geometry of the Point, Line and Circle ... , as a means of general culture."[7] Then, the Chicago student wishing to major in mathematics had to take a first course in calculus prior

[4]Hancock apparently did write his dissertation, "Ein Form des Additionstheorem für Hyperelliptische Functionen erster Ordnung," under Fuchs. He also studied under Schwarz and Frobenius while in Berlin. See Harris Hancock, "Short Biographical Sketch of Harris Hancock, with Especial Reference to His Scholastic Attainments," dated 1 May 1896, Special Collections Division, The Milton S. Eisenhower Library, The Johns Hopkins University.

[5]E. H. Moore to W. H. Harper, 31 January 1896, University of Chicago Archives, University Presidents' Papers 1889–1925, Box 17, Folder 2 (hereinafter abbreviated UC Archives, UPP 1889–1925).

[6]Hancock is best remembered today for his two-volume exposition entitled *Foundations of the Theory of Algebraic Numbers*, 2 vols. (New York: Macmillan Co., 1932). For the documents pertaining to Hancock's stay at Chicago, see UC Archives, UPP 1889–1925, Box 37, Folder 12.

[7]*The University of Chicago Annual Register July* 1, 1892–*July* 1, 1893 *with Announcements for* 1893–1894 (Chicago: University Press of Chicago, 1893), p. 77.

THE RYERSON PHYSICAL LABORATORY AT THE UNIVERSITY OF CHICAGO CA. 1905, THE EARLY HOME OF THE DEPARTMENT OF MATHEMATICS

to embarking on the intermediate work. With these prerequisites completed, the major proceeded to "Second courses in Algebra, Analytic Geometry, the Calculus; courses in Differential Equations, Applications of Calculus to Geometry, Analytical Mechanics, Elements of Projective Geometry, Elements of Elliptic Functions, Elements of Theory of Functions" with electives possible in number theory and the mathematical theories of electricity and potential.[8] With few exceptions, the level of mathematical sophistication suggested by this undergraduate course of study was unprecedented in the history of higher education in the United States. Yet, as more universities and colleges fell under the influence of the curricular advances at the undergraduate level made at institutions like Hopkins and Chicago, the educational foundation needed to produce mathematical researchers grew more solid. By the turn of the century, the construction of American research mathematics was well under way.

The strengthening of the undergraduate curriculum represented only one crucial step in this process, however. In logical progression, the next step was the improvement of the graduate offerings. Here, Moore, Bolza, and Maschke faced their most daunting challenge, namely, to reverse the trend that had allowed Göttingen and other German universities to monopolize this final phase in the training of America's best mathematicians. The Chicagoans re-

[8] *Ibid.*

alized that for their program in mathematics to attract the cream of American talent, its offerings had to rival those of the ever-attractive German institutions. Thus, they fashioned a course of studies

> intended to give the student a comprehensive view of modern mathematics, to develop him to scientific maturity, and to enable him to follow, without further guidance, the scientific movement of the day, and, if possible to take an active part in it by original research. To this end general courses on the most important branches of modern mathematics such as: Theory of Functions, Elliptic Functions, Theory of Invariants, Modern Analytical Geometry, Higher Plane Curves, Theory of Substitutions, Theory of Numbers, Synthetic Geometry, Quaternions, Theory of the Potential, are given at least once in two years, while other courses of a more special character and the Seminars are intended to introduce to research work.[9]

At least five students came to Chicago as Fellows to pursue this higher course of studies during the first year of operation: N. B. Heller and John Irwin Hutchinson followed Bolza from Clark to work on topics related to his research; Herbert Ellsworth Slaught relinquished his post as Principal of a New Jersey school to bring his young family to the Midwest; James Archy Smith came and stayed for two years before abandoning his plans for a Chicago Ph.D.; and Mary Francis "May" Winston became an "Honorary Fellow" for one year, leaving her fellowship at Bryn Mawr to take advantage of the coeducational policy and educational promise of the new institution in her native Illinois.[10] Like all of the other participants in Harper's University, these Fellows, too, had a clearly defined role to play. Harper prescribed

---

[9] *Ibid.* As the course listings at Chicago and Göttingen for 1894–1895 reflect, the Chicago curriculum, at least in pure mathematics, compared favorably with the course offerings at Göttingen: (1) Chicago—elliptic modular functions, projective geometry, configurations, groups, theory of functions of a complex variable, quaternions, substitutions, analytic geometry of three dimensions, higher plane curves, advanced integral calculus, Weierstrass' theory of elliptic functions, algebraic surfaces, analytical mechanics, number theory, theory of equations, and calculus of variations (*Bulletin of the New York Mathematical Society* 3 (1894):260), and (2) Göttingen—theory of numbers, introduction to higher mathematics, calculus of variations, algebra, descriptive geometry, introduction to function theory, hydrodynamics, partial differential equations, thermodynamics, applications of potential theory, higher geodesy, determination of orbits of planets and comets (*Bulletin of the American Mathematical Society* 1 (1894):55); Abel's and Riemann's functions, optics, theory of electricity, theoretical astronomy, kinetic theory of gases, elliptic functions, theory of determinants, mechanics, analytical geometry, chapters from Lie's theory of continuous groups (*Bulletin of the American Mathematical Society* 1 (1895):238).

[10] As listed in *The President's Report*, p. 48. Of these first Fellows only Smith appears to have dropped out of mathematics. Heller became Professor of Mathematics at Temple University; Hutchinson had a long career at Cornell; Slaught stayed on and moved through the ranks at Chicago; and Mary Winston (later Newson), after earning her Ph.D. under Klein in Göttingen, taught successively at Kansas State College (1897–1900), Washburn College (1913–1921), and Eureka College (1921–1942).

that they "spend five-sixths of their time in original investigation under the guidance of the Professor, one-sixth being reserved for service in connection with the University."[11] By setting these guidelines, Harper sought to insure that his Fellows not lose sight of the goal in view, namely, the pursuit of their own studies and original researches. Perhaps due to the complete coverage of the department's teaching obligations and the well-balanced sequence of courses that Moore, Bolza, and Maschke had put together for them, the five mathematics Fellows probably had little need for Harper's rather paternalistic prescription. They were already there to devote themselves fully to mathematics.

To that end, they sampled from topics which spanned the discipline. Moore lectured from Heinrich Weber's recently published book, *Elliptische Functionen und algebraische Zahlen*, as well as on general complex function theory.[12] Bolza offered courses in hyperelliptic function theory and in the theory of substitutions as applied to algebraic equations. In pure mathematics, Maschke taught line geometry based on the work of Ernst Eduard Kummer and Julius Plücker as well as the theory of finite groups of linear substitutions based largely on Felix Klein's study of the icosahedron. On the applied side, Maschke utilized his training in electrical engineering to put together courses on potential theory and on electricity. Finally, Young lectured on his specialties, number theory and the theory of invariants.[13]

Slaught's elevation from Fellow to Instructor in 1897 further freed particularly Bolza and Maschke for more graduate-level instruction. By the 1896–1897 academic year, the department's graduate offerings had multiplied to include new courses on mathematical pedagogy (given by Young); differential geometry, higher plane curves, algebraic surfaces, and linear differential equations (taught by Maschke); the calculus of variations (offered by Bolza); and group theory, projective geometry, and number theory (presented by Moore).[14] The initial demand had necessitated that the number of graduate courses start out small just as Moore had predicted in his organizational letter to Harper of 2 March 1892. By 1897, however, graduate offerings had increased more than twofold from nine to twenty-two annually. Furthermore, while the total number of courses leveled out at about two dozen during Chicago's first fifteen years, the special topics changed to reflect

[11] Goodspeed, p. 145.

[12] Heinrich Weber, *Elliptische Functionen und algebraische Zahlen* (Braunschweig: F. Vieweg und Sohn, 1891).

[13] *Annual Register July 1, 1892–July 1, 1893*, pp. 78–79.

[14] *The University of Chicago Annual Register July* 1, 1896–*July* 1, 1897 *with Announcements for* 1897–1898 (Chicago: University of Chicago Press, 1897), pp. 274–276. Written lecture notes from many of the courses given by Moore, Bolza, and Maschke during those early years attest to the mathematical sophistication of the courses taught at Chicago. See University of Chicago Archives, Mathematics Department Lecture Notes 1894–1913 (hereinafter denoted UC Archives, Math. Lecture Notes). Unlike for Klein's courses at Göttingen, however, there were no official sets of notes produced for the courses at Chicago.

recent developments in the discipline. Thus, in 1900–1901, Moore lectured on David Hilbert's 1897 report on algebraic number fields, and just one year later he moved on to cover the latter's *Grundlagen der Geometrie* of 1899.[15]

The latest advances in mathematical research also reached Chicago via its Summer Quarter program. Based most likely on his own deep commitment to summer educational programs like Chautauqua, President Harper had designed a year-long academic calendar divided into four twelve-week quarters separated by week-long recesses. This sort of calendar not only had obvious instructional advantages for the students but also allowed for the fullest utilization of the University's facilities, thereby increasing its overall cost-effectiveness. In addition, the Summer Quarter "[made] it possible for the University to use, besides its own corps of teachers, the best men of other institutions both in this country and in Europe,"[16] a clear echo of Gilman's call at Hopkins for the use of Visiting Professors to supplement course offerings. Beginning in the summer of 1897, the Mathematics Department realized precisely this possibility when it invited Edgar O. Lovett, who had just taken his Leipzig doctorate under Sophus Lie, to lecture on Lie's theory of transformation groups.[17] In fact, the Summer Quarter continued to serve as a special research focus for the mathematicians at Chicago until well into the twentieth century.

As evidenced by both the sequence of regular offerings and the summer series devoted primarily to special topics, the Chicago graduate program in mathematics aimed to bring its students to the threshold of contemporary research. To see them over that threshold, however, the faculty needed to do more than just present the appropriate material. It had to teach that material effectively. Moore, Bolza, and Maschke, each in his own characteristic way, managed to accomplish that task.

Of the three men, Moore had perhaps the most unconventional teaching style, one which tended to appeal to only the most determined and talented students. Like Sylvester before him, Moore often came to his graduate classroom fresh with his latest ideas on the topic under discussion. While this did not generally make for tightly organized and polished lectures, it did instill in his students "a sense of the hunt." With Moore at the blackboard, he and his class would work together to resolve some stubborn detail or to prove some theorem. Sometimes they succeeded, and the class came to a satisfying conclusion; sometimes they failed and dispersed scratching their heads

---

[15]*The University of Chicago Bulletin of Information*: *Annual Register July*, 1900–*July*, 1901 *with Announcements for* 1901–1902 (Chicago: University of Chicago, n.d.), p. 265; and E. H. Moore, "On the Projective Axioms of Geometry," *Transactions of the American Mathematical Society* 3 (1902):142–158 on p. 143, note *. See, also, David Hilbert, "Die Theorie der algebraischen Zahlkörper," *Jahresbericht der deutschen Mathematiker-Vereinigung* 4 (1897):175–546, or David Hilbert, *Gesammelte Abhandlungen*, 3 vols. (Berlin: Julius Springer, 1932–1935), 1:63–363; and *Grundlagen der Geometrie* (Leipzig: B. G. Teubner, 1899).

[16]Goodspeed, p. 141.

[17]See Chapter 5 on Lovett's subsequent career.

in confusion. Yet regardless of their success or failure, Moore demanded the quick attention of his students, and if they fell behind as his thoughts raced, they soon felt the sting of what Mary Winston termed his "peppery temper."[18] Intending nothing personal by these outbursts, Moore simply forgot about accepted social conventions when overwhelmed by the heat of mathematical thought. Students in his classes may have needed thicker skins than those in his colleagues' courses, but, as one student put it, "it was a proud moment when one who was ambitious and interested found himself in the relatively small group of those who could stand the pace."[19]

While the tempo may have been no slower in Chicago's other mathematics lecture rooms, the teaching styles found there differed markedly. Both Bolza and Maschke presented their material in a very thorough and logical fashion. Less intense than Moore before their classes, neither man intimidated his students mathematically, or at least not in the setting of student and teacher. Like Moore, Bolza moved quickly through the material of the day, forcing his auditors to work and think fast, but unlike his senior colleague, he brought with his speed of presentation no stinging intolerance. Finally, Maschke embodied yet a third pedagogical style. He delivered his lectures more slowly and deliberately than either Moore or Bolza to give his students the opportunity to keep apace of the mathematics as they encountered it rather than forcing them to catch up. Their varied approaches to teaching aside, however, the three Chicago mathematicians shared a common sense of mission. They each sought first to bring their students to the boundaries of mathematical research within the contexts of their respective graduate classrooms and then to direct them beyond through their participation in the Mathematical Club.

Not surprisingly, the mandate for departmental clubs designed to inspire an *esprit de corps* as well as to provide a forum for the discussion of research-level work had also come initially from the President of the University himself. In detailing the duties of the Head Professor in that first *Official Bulletin*, Harper had specified that each department would sponsor a club or seminar to stimulate interaction between faculty and advanced students.[20] Of course, the idea was not new in mathematics (or in other areas for that matter). As noted earlier in Chapters 2 and 5, the seminar played crucial—but very different—roles in the graduate teaching both of Sylvester at the Johns Hopkins and of Klein at Göttingen. In Baltimore, Sylvester's semi-

[18]Mary Frances Winston to her mother, 4 March 1895, Smith College Archives, Sophia Smith Collection, M. F. Winston letters, Box 1, Folder 10. Indeed, she wondered, and this as late as 1895, "how those people will get along. Bolza and Maschke are sworn friends, but Moore has such a peppery temper I am afraid he will have trouble in keeping things smooth between himself and his colleagues."

[19]Gilbert A. Bliss, "Eliakim Hastings Moore," *Bulletin of the American Mathematical Society*, 2d. ser., 39 (1933):831–838 on p. 834. We rely on Bliss, in various sources, for these general impressions of the teaching styles of Moore, Bolza, and Maschke.

[20]Storr, p. 66.

nar had functioned as an extension of his classroom, providing yet another venue for the spontaneous airing of new results by both the professor and his students. At Göttingen, on the other hand, Klein's seminar served as a stage for the semester-long development of a given theory. Fully aware of these precedents, the mathematicians at Chicago proposed "[a] bi-weekly meeting throughout the year for the review of memoirs and books, and for the presentation of results of research" which was "open to all graduate students of Mathematics."[21]

The department actually kept up a much more ambitious schedule of weekly meetings during its first year of operation. Perhaps owing to the fact that Bolza's first official vacation coincided with the University's first Fall Quarter, Moore delayed the formation of the Mathematical Club until the start of the Winter Quarter of 1893.[22] After making a "statement of the character and purposes of the organization," he gave the inaugural presentation on Luigi Cremona's researches on "A figure in space from which the properties of Pascal's hexagon in the plane are easily deducible."[23] As the Winter and Spring Quarters of 1893 progressed, the club also heard lectures from Bolza and Maschke, Hancock and Young, Mary Winston and James Smith, N. B. Heller and John Hutchinson. In short, each faculty member and Fellow gave at least one lecture before the group (with the exception of Herbert Slaught who was busily trying to make up mathematically for his years spent in high school teaching), and the topics they presented reflected the breadth and depth of the department's mathematical interests. (See Table 9.1.)

In keeping with the Club's stated objectives, some of the talks, like Maschke's "Paper on the Historical Development of the Theory of Fourier Series," were expository or historical in nature and aimed at putting a particular area in a meaningful perspective (roughly the same objective Klein would soon have in view in his Evanston Colloquium Lectures). Some, like Moore's discussion of Eugen Netto's work on the theory of (Steiner) triple systems, gave a careful presentation of results in some recently published research article with an eye toward using that work to generate new theorems. Others, like Moore's two follow-up talks on triple systems, put original mathematics before the club for scrutiny and criticism. As Moore's lectures that first year exemplified, the Mathematical Club was designed, from its very inception, to serve as a breeding ground for new research.[24]

---

[21] *Annual Register July* 1, 1892–*July* 1, 1893, p. 79.

[22] The faculty got one quarter off each year. Bolza arranged to take his first break immediately in order to return to Germany and take care of various family matters.

[23] UC Archives, Mathematical Club Records 1893–1921, Box 1, Folder 1 (hereinafter denoted Math. Club Records).

[24] For a discussion of Moore's work on (Steiner) triple systems, see below. Another noted example of the influence of the Mathematics Club in the production of a key result involved the work in 1905 of Joseph H. M. Wedderburn and Leonard E. Dickson on the finite division

And the faculty was not the Club's only beneficiary. Gilbert Ames Bliss, whose attendance at the gatherings began in 1898 when he entered the graduate program in mathematics and continued after 1908 when he joined the Chicago faculty permanently, vividly conveyed the intellectual excitement imparted to the graduate student members during the often intense meetings of the group. Over thirty years later, Bliss reminisced that

> [t]hose of us who were students in those early years remember well the tensely alert interest of these three men [Moore, Bolza, and Maschke] in the papers which they themselves and others read before the Club. They were enthusiasts devoted to the study of mathematics, and aggressively acquainted with the activities of the mathematicians in a wide variety of domains. The speaker before the Club knew well that the excellence of his paper would be fully appreciated, but also that its weaknesses would be discovered and thoroughly discussed. Mathematics, as accurate as our powers of logic permit us to make it, came first in the minds of these leaders in the youthful department at Chicago, but it was accompanied by a friendship for others having serious mathematical interests which many who experienced their encouragement will never forget.[25]

Bliss's account also suggests that, in mathematics at least, the departmental club had realized Harper's goals. The discussion of advanced mathematics between the faculty and the graduate students generated a strong sense of camaraderie and common purpose, just as Harper had hoped. But even more, the three Chicago faculty members succeeded in instilling in their students that very important intangible in mathematics, a sense of style and taste. They challenged muddy or ill-presented ideas. They questioned unclear expression. They had high standards, which they conveyed to their students not only in the classroom, in the seminar room, and in the Mathematical Club but also through the example of their own published work.

The Mathematical Interests of Moore and His Students

As researchers, the Chicago mathematicians embraced many of the leading areas of late nineteenth- and early twentieth-century research. Moore, who had started his career working in algebraic geometry, had switched by the 1890s to group theory and by the turn of the century had moved into axiomatics and the foundations of analysis. His colleague Bolza initially fo-

algebra theorem. On this work, see Karen Hunger Parshall, "In Pursuit of the Finite Division Algebra Theorem and Beyond: Joseph H. M. Wedderburn, Leonard E. Dickson, and Oswald Veblen," *Archives internationales d'Histoires des Sciences* 33 (1983):274–299.

[25]Bliss, "Eliakim Hastings Moore," p. 833.

cused on the theories of hyperelliptic and elliptic integrals and functions, spending the nineties in such pursuits before shifting into the calculus of variations after 1901. Finally, Maschke had first devoted his energies to the theory of finite groups of linear substitutions (that is, finite linear groups) before taking up the invariant theory of differential forms from 1900 until his death in 1908. All three mathematicians shared their work and their new ideas with their advanced students and guided them to open problems. Yet, it was E. H. Moore who, sniffing the changing mathematical winds, uncannily and successfully shifted his mathematical course and brought numerous students along in his wake.

As the lone American and Acting Head (until 1896 when he became Head) of the Department, Moore had perhaps the most to prove of the Chicago mathematicians. Although fortunate enough to have studied under Hubert Newton at Yale, Moore suffered nevertheless from an early lack of exposure to much of late nineteenth-century mathematical research. As we saw in Chapter 1, Newton's interests had moved from geometry to astronomy in the 1860s, and while he had tried to keep abreast of subsequent developments in mathematics, he enjoyed only limited range and depth as a mathematical adviser later in his career. Not surprisingly, he steered his student E. H. Moore into the area he knew and could evaluate best, namely, geometry. In light of the exciting work being done in that area on the Continent, and particularly in Germany and Italy, during the early 1880s, Newton would appear to have directed Moore well. The geometry to which he introduced his student, however, was not in the latest German or Italian style but rather in outmoded French and British fashions of earlier decades.

When Moore earned his B.A. in 1883 and decided to pursue graduate studies under Newton, a new book, the *Mathematical Papers by William Kingdon Clifford*, had only recently appeared.[26] There, in one manageable volume lay all of the pearls of Clifford's mathematical—not to mention philosophical—wisdom. Given the paucity of available journals in nineteenth-century American universities, what better way to learn of a major mathematician's ideas than through a handy edition of collected works? With this source in hand, Moore apparently set out to master Clifford's geometrical papers and to isolate a problem for his thesis research. To aid him in this dual quest, he relied both on George Salmon's *Geometry of Three Dimensions*, the standard geometrical treatise in English, and on his adviser's expertise in the French synthetic approach.[27] By 1885, Moore had succeeded in pulling all of these resources together to prove the new results he presented for his doctoral degree. His thesis, entitled "Extensions of Certain Theorems of Clifford and

[26]William Kingdon Clifford, *Mathematical Papers by William Kingdon Clifford*, ed. Robert Tucker (London: Macmillan and Co., 1882; reprint ed., New York: Chelsea Publishing Company, 1968) (hereinafter referred to as *Math. Papers WK*).

[27]George Salmon, *A Treatise on the Analytic Geometry of Three Dimensions*, 4th ed. (Dublin: Hodges, Figgis, and Co., 1882).

Cayley in the Geometry of n Dimensions" and published in September of 1885, opened with rather trivial generalizations of three theorems on skew curves brought to light by Clifford in 1878. From there, Moore moved on to set up the machinery needed to give his generalization of Pascal's theorem for six points on a curve lying on a surface in $(m+1)$-space, and closed by extending an 1872–1873 theorem of Arthur Cayley on quadric surfaces in five dimensions.[28] Although not a truly significant piece of work, Moore's thesis certainly suggested that its author had the potential to become a productive mathematical researcher.

Unfortunately, the next seven years provided Moore scant opportunity to realize that potential. After his year-long study tour of Germany, he returned to a string of teaching-intensive positions in the United States which left him little time for his own research. Despite the odds against him, he did produce three new articles. The first two of these, published in the *American Journal of Mathematics* in 1888, not only extended the lines of investigation he had opened in his dissertation but also reflected his exposure to the German approach to algebraic geometry.[29] The third, which appeared two years later in the newly founded *Rendiconti del Circolo Matematico di Palermo*, marked his first contribution to function theory, an area which would come to occupy his scientific attentions completely by 1910.[30] This work notwithstanding, Moore's mathematical career had by no means taken off when he accepted the University of Chicago's first mathematics professorship in 1892.

As evidenced by his participation in the Mathematical Club (see Table 9.1), at the start of the Winter Quarter of 1893 Moore had not yet abandoned the sort of geometrical questions which had engaged him as a graduate student. By the third week of the term, however, his interests had fixed on ideas of a related but more purely algebraic nature. In fact, Moore hardly returned during the remainder of his career to the area in which he had earned his

---

[28]Eliakim Hastings Moore, "Extensions of Certain Theorems of Clifford and of Cayley in the Geometry of n-Dimensions," *Transactions of the Connecticut Academy of Arts and Sciences* 7 (1885–1886):9–26. See, also, William Kingdon Clifford, "On the Classification of Loci," *Philosophical Transactions of the Royal Society of London* 169 (1878):663–681, or *Math. Papers WK*, pp. 305–329; and Arthur Cayley, "On the Superlines of a Quadric Surface in 5-Dimensional Space," *Quarterly Journal of Pure and Applied Mathematics* 12 (1872–1873):176–180, or Arthur Cayley, *The Collected Mathematical Papers of Arthur Cayley*, ed. Arthur Cayley and A. R. Forsyth, 14 vols. (Cambridge: University Press, 1889–1898), 9:79–83. For a fairly complete survey of Moore's work, see Gilbert A. Bliss, "The Scientific Work of Eliakim Hastings Moore," *Bulletin of the American Mathematical Society* 40 (1934):501–514.

[29]Eliakim Hastings Moore, "Algebraic Surfaces of Which Every Plane Section is Unicursal in the Light of n-Dimensional Geometry," *American Journal of Mathematics* 10 (1888):17–28; and "A Problem Suggested in the Geometry of Nets of Curves and Applied to the Theory of Six Points Having Multiply Perspective Relations," *op. cit.*, pp. 243–257.

[30]Eliakim Hastings Moore, "Note Concerning a Fundamental Theorem of Elliptic Functions as Treated in Halphen's Traité, vol. 1, pp. 39–41," *Rendiconti del Circolo Matematico di Palermo* 4 (1890):186–194. On the historical development and significance of the *Circolo Matematico di Palermo,* see Aldo Brigaglia and Guido Masotto, *Il Circolo Matematico di Palermo* (Bari: Edizioni Dedalo, 1982).

Ph.D. One exception did occur several years later in 1900 when he gave a generalization of the 1898 thesis research of his second student, Herbert Slaught, but even that work was just geometrically motivated group theory.[31] From 19 January 1893 to the end of the century, Moore's focus turned to group theory and away from algebraic geometry.

Several factors associated with the mathematical environment at Chicago may well have contributed to Moore's change in mathematical direction. First, his new colleague Bolza was offering a course during that first Winter Quarter on permutation groups, an area which had become increasingly fashionable following the publication of works by Joseph Serret and Camille Jordan in France and Eugen Netto in Germany.[32] Undoubtedly using Netto's *Substitutionstheorie* as one of his texts, Bolza could easily have provided Moore with the impetus to study and absorb the ideas in that work. Next, his colleague Maschke specialized in the theory of finite linear groups and, more particularly, in the search for complete systems of invariants for such groups. Finally, Jacob Young, the junior member of the Chicago team, had done his dissertation under Bolza at Clark on finite groups of order p^n, p a prime.[33] In such a group-theoretic milieu it is not surprising that the mathematically isolated Moore would have moved in a more algebraic direction, if only for the sake of mathematical companionship. Furthermore, he had a sixth sense for the hot and fast-breaking fields which manifested itself for the first time in this opportune shift into group theory. At any rate, as the Mathematical Club records indicate, by 19 January Moore had begun looking at the theory of finite groups, and by 13 April he had picked up on a recent paper by Netto in the *Mathematische Annalen* on triple systems.

The notion of a triple system as an abstract algebraic entity worthy of study had grown out of work in algebraic geometry by Max Noether, as well as out of research by Netto and others aimed at describing the Galois group associated with various polynomials of certain degrees.[34] Suppose $p(x) = 0$ is an

[31]Slaught published his thesis, entitled "The Cross-Ratio Group of 120 Quadratic Cremona Transformations of the Plane," in the *American Journal of Mathematics* 22 (1900):343–388. Moore generalized his student's results in "The Cross Ratio Group of $n!$ Cremona Transformations of Order $n-3$ in Flat Space of $n-3$ Dimensions" also in the *American Journal of Mathematics* 22 (1900):279–291. Moore followed this up with the not unrelated paper, "Concerning Klein's Group of $(n+1)!$ n-ary Collineations," *American Journal of Mathematics* 22 (1900):336–342.

[32]See Joseph Serret, *Cours d'Algèbre supérieure*, 2 vols., 3d ed. (Paris: Gauthier-Villars, 1866); Camille Jordan, *Traité des Substitutions et des Équations algébriques* (Paris: Gauthier-Villars, 1870); and Eugen Netto, *Substitutionstheorie und ihre Anwendungen auf die Algebra* (Leipzig: B. G. Teubner, 1882) or its English translation, *The Theory of Substitutions and Its Applications to Algebra,* trans. Frank Nelson Cole (Ann Arbor: G. Wahr, 1892; reprint ed., Chelsea Publishing Co., n.d.). As mentioned in Chapter 6, Bolza had already taught such a course at the Johns Hopkins and at Clark prior to joining the Chicago faculty. He had published a polished version of his course "On the Theory of Substitution Groups and Its Applications to Algebraic Equations" in the *American Journal of Mathematics* 13 (1891):59–144.

[33]Jacob W. A. Young, "On the Determination of Groups Whose Order Is a Power of a Prime," *American Journal of Mathematics* 15 (1893):124–178.

[34]Actually, these appear to have arisen even earlier in different contexts in Thomas Kirkman, "On a Problem in Combination," *Cambridge and Dublin Mathematical Journal* 2 (1847):191–

irreducible polynomial equation of degree n (over a field of characteristic zero). For Noether, this was a *triad equation* if the roots can be arranged in disjoint triples (or triads) $\{x_i, x_j, x_k\}$ such that any two roots of the triple uniquely determine the third root rationally.[35] If so, this yields a *triad* $\{x_i, x_j, x_k\}$ *of roots*. Since there are $\frac{n(n-1)}{2}$ possible pairs of the n roots and since any pair of roots determines a third root uniquely, there must be $\frac{n(n-1)}{6}$ triads associated to any triad equation. Since this number must be an integer, and since it is easily seen that n must be odd, it follows that n has the form $6m+1$ or $6m+3$, for $0 \leq m \in \mathbb{Z}^+$. Netto used this construct to calculate the Galois groups of the irreducible triad equations of degrees 7 and 9 in his *Substitutionstheorie* of 1882.[36]

By 1893, Netto had generalized the notion of a triad equation to the more abstract concept of the purely combinatorial (*Steiner*) *triple system*, that is, a collection Δ_t consisting of triples chosen from a set of t elements in such a way that any pair of elements (with no repetitions allowed) appears in one and only one triple. In his "Zur Theorie der Tripelsysteme," Netto posed two questions about these entities: first, if $t \in \mathbb{Z}^+$ and $t = 6m+1$ or $6m+3$ for $0 \leq m \in \mathbb{Z}^+$, must a triple system Δ_t necessarily exist?; and second, given a t of the appropriate form, are all triple systems Δ_t isomorphic?[37] By providing the necessary algorithms, Netto answered the first question affirmatively for all $t < 100$, except $t = 25$ and 85. His methods proved less effective relative to the second query, however. For each $t \in \{3, 7, 9\}$, he showed that all triple systems Δ_t are isomorphic, but definitive conclusions about higher values of t eluded him. Moore used Netto's partial result as a springboard in his first research foray into combinatorial algebra and showed that, in general, all triple systems Δ_t are *not* isomorphic for a general t. In his article "Concerning Triple Systems" and published, like Netto's work, in the *Mathematische Annalen*, Moore gave an explicit construction which produced at least two nonisomorphic triple systems Δ_t for every $t > 13$ of the form $6m+1$ or $6m+3$.[38] Dated 29 April 1893, Moore's published paper brought

204, and independently in Jacob Steiner, "Combinatorische Aufgabe," *Journal für die reine und angewandte Mathematik* 45 (1853):181–182. See Richard A. Brualdi, *Introductory Combinatorics*, 2nd ed. (New York: North-Holland, 1992), p. 363. Today, they are known as *Steiner* triple systems, but Netto, Moore, and others referred to them merely as "triple systems."

[35]Max Noether, "Ueber die Gleichungen achten Grades und ihr Auftreten in der Theorie der Curven vierter Ordnung," *Mathematische Annalen* 15 (1879):89–110 on pp. 101–103; and Netto, *The Theory of Substitutions*, p. 229.

[36]Netto, *The Theory of Substitutions*, pp. 229–239.

[37]Eugen Netto, "Zur Theorie der Tripelsysteme," *Mathematische Annalen* 43 (1893):143–152. Actually, unbeknownst to Netto, Kirkman had shown that the answer to the first of these questions was in the affirmative in his paper of 1847. See note 34 above, and Brualdi, p. 364.

[38]Eliakim Hastings Moore, "Concerning Triple Systems," *Mathematische Annalen* 43 (1893): 271–285.

together the results he had presented to Chicago's Mathematical Club on 20 and 29 April. Thus, it would seem that the fledgling club served not only as a forum for Moore's new results but also as a stimulus for his research efforts. The timing, just before the Mathematical Congress at the World's Columbian Exposition, could not have been better.

In his contribution to the Congress (discussed in Chapter 7 above), Moore took up some specific questions in the theory of finite simple groups.[39] Following the 1870 work of Jordan on the classical groups over prime fields, and the later research of Otto Hölder on the noncyclic simple groups of order less than 200,[40] this more specialized area of finite group theory had increasingly sparked the interest of mathematicians. In the United States, in fact, Michigan's Frank Nelson Cole published on finite simple groups as early as 1892 in the *American Journal.* The theory of finite simple groups thus represented an active area of research at home as well as abroad; as such it determined an obvious topic for discussion in those first meetings of Chicago's Mathematical Club during the Winter and Spring Quarters of 1893. Young gave lectures at two of the meetings on Hölder's recent work. Likewise, Moore spoke on the construction (most likely the one due to Netto) of a certain simple group of order 168 as a permutation group on seven letters (in modern notation, the group $PSL_2(7) \cong PSL_3(2)$).[41] With this as background, Moore's definition and development of a class of simple groups in his Chicago Congress paper followed quite naturally.

Moore opened that work with a complete list of the general formulas for the orders of the *classes* of finite simple groups then known.[42] His list included, besides the cyclic groups of prime order p and the alternating groups A_m ($m > 4$), the particular class of groups now denoted $PSL_m(p)$ of order

$$\frac{(p^m-1)p^{m-1}(p^{m-1}-1)p^{m-2}\cdots(p^2-1)p}{\delta}$$

where $(p, m) \neq (2, 2), (3, 2)$ and $\delta = \gcd(m, p-1)$.[43] As noted in Chapter 7, Moore focused in his paper on the special case $m = 2$, that is, on the class of groups $PSL_2(p) \cong SL_2(p)/(\pm 1)$.

[39] Eliakim Hastings Moore, "A Doubly-Infinite System of Simple Groups," in *Mathematical Papers Read at the International Mathematics Congress Held in Connection with the World's Columbian Exposition: Chicago* 1893, ed. E. H. Moore, Oskar Bolza, Heinrich Maschke, and Henry S. White, (New York: Macmillan & Co., 1896), pp. 208–242 (hereinafter referred to as *Congress Papers*). Moore published a lengthy abstract of this paper under the same title in *Bulletin of the New York Mathematical Society* 3 (1893):73–78.

[40] Otto Hölder, "Die einfachen Gruppen im ersten und zweiten Hundert der Ordnungszahlen," *Mathematische Annalen* 40 (1892):55–88. For a discussion of Hölder's work in historical context, consult Julia Nicholson, "Otto Hölder and the Development of Group Theory and Galois Theory" (unpublished Dissertation, Oxford University, 1993).

[41] See Netto, *The Theory of Substitutions*, pp. 232–234.

[42] Moore, "A Doubly-Infinite System of Simple Groups," pp. 208–209.

[43] Jordan had done significant work on these and the other finite groups of substitutions over the prime field $\mathbb{Z}/p\mathbb{Z}$ in his *Traité des Substitutions* of 1870. See note 32 above.

But Moore had also noticed two key examples of finite simple groups, which, although they resembled groups which Klein and others had studied, did not have orders of the appropriate form. Moore had analyzed the first such example, a group of order 360, as early as the fall of 1892 and had announced his findings in the *Proceedings of the American Association for the Advancement of Science.*[44] Then, in the spring of 1893, Cole had discovered another simple group, this one of order 504, which also failed to fit into one of the six known classes.[45] In fact, Cole, like Moore, had lectured on his new group at the Chicago Congress. As Moore later explained in the written version of his Congress talk, these two examples had led him to generalize the simple groups $PSL_2(p)$ to what he termed the *doubly-infinite* (or two-parameter) groups of order $\frac{p^n(p^{2n}-1)}{\delta}$, for $(p, n) \neq (2, 1), (3, 1)$. He proceeded to show that these new groups (the $PSL_2(p^n)$) are actually simple. Preliminary to his main result, though, Moore needed to get a better grip on the fields with p^n elements which underlay his new groups. To that end, he proved the theorem for which his Congress-volume paper has been best remembered, namely, "[e]very existent [finite] field $F[s]$ is the abstract form of a Galois-field, $GF[p^n]$, where $s = p^n$."[46] To do this, Moore adopted the traditional definition of a Galois field: given an indeterminate X, take an irreducible monic polynomial $f(X) \in \mathbb{Z}_p[X]$ of degree n over the prime field $\mathbb{Z}_p = \mathbb{Z}/p\mathbb{Z}$. Then the Galois field $GF[p^n]$ is the collection of p^n equivalence classes of $\mathbb{Z}_p[X]/(f(X))$.[47] The recognition of finite fields as Galois fields, structures whose properties had been thoroughly explored in the readily accessible works of Serret and Jordan, allowed Moore to apply this well-developed theory in the analysis of his new group. Tucked away as it was in a group-theoretic paper, Moore's theorem on the classification of finite fields appeared somewhat out of context. It would finally find a more natural place in the literature in 1905 when Chicago's young visitor Joseph H. M. Wedderburn would use it as a stepping-stone to his celebrated finite division algebra theorem; that is, the finite fields are the *only* finite division algebras.[48] Furthermore, the study of finite linear groups—or, more generally, the structure of reductive groups over finite fields—would develop into a major theme in twentieth-century algebraic research, particularly in the United States.

---

[44] E. H. Moore, "Concerning a Congruence Group of Order 360 Contained in the Group of Linear Fractional Substitutions," *Proceedings of the American Association for the Advancement of Science* 41 (1892):62.

[45] F. N. Cole, "On a Certain Simple Group," in *Congress Papers*, pp. 40–43; and "Simple Groups As Far As Order 660," *American Journal of Mathematics* 15 (1893):303–315.

[46] Moore, "A Doubly-Infinite System of Simple Groups," p. 211.

[47] *Ibid.*

[48] Joseph H. M. Wedderburn, "A Theorem on Finite Algebras," *Transactions of the American Mathematical Society* 6 (1905):349–352. For the history of this theorem, see Parshall, "In Pursuit of the Finite Division Algebra Theorem."

These results on triple systems, finite simple groups, and finite fields effectively ended Moore's days as an algebraic geometer and heralded the beginning of the second, and most creative, phase of his career. The fifteen years from 1893 to 1908 witnessed not only Moore's continued progress on triple systems and in group theory generally but also his total immersion in yet a third area of interest, namely, foundational questions raised in such works as David Hilbert's *Grundlagen der Geometrie*.[49] As with those of Sylvester, Klein, and Lie, Moore's students during this fruitful period tended to pursue research problems reflective of their adviser's interests. (See Table 9.2.)

Coming to Chicago in 1894 with a B.S. and an M.A. from the University of Texas, Moore's first doctoral student, Leonard Eugene Dickson, quickly embraced and extended the group-theoretic research which Moore had presented in his 1893 Congress lecture. By 1896, Dickson had earned one of Chicago's first Ph.D.s in mathematics for a thesis on "The Analytic Representation of Substitutions on a Power of a Prime Number of Letters with a Discussion of the Linear Group."[50] In the first of this two-part work, he focused, like his adviser, on a finite field $F = GF[p^n]$ for p a prime and $n \in \mathbb{Z}^+$. For a polynomial $\phi(X)$ of degree $k \leq p^n$ (with coefficients in F), Dickson called the associated mapping $\phi : F \to F$, $\xi \mapsto \phi(\xi)$ a *substitution quantic* $SQ[k; p^n]$ of degree k on p^n letters, provided it was bijective. He then devoted the first part of his dissertation to a "complete determination of all quantics up to as high a degree as practicable which are suitable to represent substitutions on p^n letters," and he obtained complete results for degrees $k < 7$ with partial results for degrees 7 and 11.[51]

In his dissertation's second part, Dickson turned to a study of the general linear group $GL_m(F)$ where, as in part one, $F = GF[p^n]$. For the finite fields $F = GF[p]$, these groups had already been studied thoroughly by Jordan in his *Traité des Substitutions*, but as Dickson explained, he aimed to generalize Jordan's work.[52] To this end, he proved that his $GL_m(F)$ was a group, determined its order, and examined its composition series. The latter analysis led him to one of his main results; namely, if Z denotes the center of $SL_m(F)$, then $SL_m(F)/Z$ is simple provided $(m, n, p) \neq (2, 1, 2)$ or

[49] See, also, David Hilbert, *The Foundations of Geometry*, trans. E. J. Townsend, 2d ed. (Chicago: The Open Court Publishing Company, 1910). All subsequent citations are from this translation. Edgar Townsend, who taught at the University of Illinois from 1893 to 1929, was one of David Hilbert's doctoral students. He earned his Göttingen degree in 1900 for a thesis entitled "Über den Begriff und die Anwendung des Doppellinies." His first English translation of Hilbert's book appeared in 1902 and was based on the original German edition.

[50] Leonard E. Dickson, "The Analytic Representation of Substitutions on a Power of a Prime Number of Letters with a Discussion of the Linear Group," *Annals of Mathematics* 11 (1897):65–143, or *The Collected Mathematical Papers of Leonard Eugene Dickson*, ed. A. Adrian Albert, 6 vols. (New York: Chelsea Publishing Company, 1975, 1983), 2:651–729.

[51] *Ibid.*, pp. 68–120 on p. 66, or pp. 652–706 on p. 652. He developed his notation and the general theory on pp. 68–77, or pp. 654–663. The rest of the first part is devoted to a case-by-case study for degrees up to and including six.

[52] *Ibid.*, p. 67, or p. 653.

LEONARD EUGENE DICKSON (1874–1954)

$(2, 1, 3)$. Dickson's theorem thus generalized his adviser's research of 1893 to *triply-infinite* systems of simple groups (in the three parameters m, n, and p), and Dickson himself acknowledged his indebtedness to Moore for the suggestion to pursue such a generalization.[53] Dickson's analysis also yielded a previously unknown class of finite simple groups, the groups $SL_m(F)/Z$ for $m \geq 3$ and $n > 1$.[54]

After earning his degree, Dickson continued to pursue the lines of investigation suggested in his doctoral dissertation, first as a postdoctoral student at the Universities of Leipzig and Paris during the academic year 1896–1897, next on the faculties of the University of California (1897–1899) and the University of Texas (1899–1900), and finally as an Assistant Professor back at Chicago (1900–1907).[55] Over the course of these eleven years, he published one book and almost eighty articles on the theory of linear groups, thereby establishing his reputation as one of the rising stars of American mathematics. In particular, his book, *Linear Groups with an Exposition of the Galois Field Theory* published by B. G. Teubner in 1901, extended the theory of linear groups he had explored in his thesis, examined its connections with the problem of finding all finite simple groups, and presented a complete list of all the finite simple groups then known.[56] Besides inspiring further research of his own, Dickson's book generated much interest nationally in finite group theory, an area in which American mathematicians would continue to excel for decades to come. His work also suggested a wealth of research problems and attracted the attention of the Chicago graduate students. As a result, Dickson had already supervised the first of his more than sixty Ph.D. students by the age of twenty-seven.[57] (See Table 9.3.) Thanks to a career on the Chicago faculty which ended only in 1939, E. H. Moore's student and later colleague L. E. Dickson helped to perpetuate his adviser's brand of algebraic research and to establish the Chicago Mathematics Department as a world center in the study of groups, algebras, and rings, a distinction it would enjoy throughout most of the twentieth century. (See Chapter 10.)

[53] *Ibid.*

[54] *Ibid.*, pp. 128–138, or pp. 714–724. Jordan had treated the case of $n = 1$ and m arbitrary but finite, and Moore had dealt with the case n arbitrary but finite and $m = 2$.

[55] For the details of Dickson's life, see, among other articles, Raymond C. Archibald, *A Semicentennial History of the American Mathematical Society* 1888–1938 (New York: American Mathematical Society, 1938; reprint ed., New York: Arno Press, 1980), pp.183–194; and A. Adrian Albert, "Leonard Eugene Dickson 1874–1954," *Bulletin of the American Mathematical Society* 61 (1955):331–345. At the time of Dickson's foreign study tour, Sophus Lie was at Leipzig and Camille Jordan was at Paris.

[56] Leonard Eugene Dickson, *Linear Groups with an Exposition of the Galois Field Theory* (Leipzig: B. G. Teubner, 1901; reprint ed., New York: Dover Publications, Inc., 1958). See, also, Karen V. H. Parshall, "A Study in Group Theory: Leonard Eugene Dickson's *Linear Groups*," *The Mathematical Intelligencer* 13 (1) (1991):7–11. In particular, he studied the symplectic and orthogonal groups over arbitrary finite fields, further extending work of Jordan.

[57] For an almost complete list of Dickson's doctoral students (up to 1938), see Archibald, *Semicentennial History*, p. 185.

While most mathematicians would consider themselves lucky to have just one student during their career as talented as Dickson, E. H. Moore oversaw the doctoral research of three more of similar stature during Chicago's first fifteen years. Their doctoral work, like Dickson's, bore the strong imprint of their adviser's changing research focus. Whereas Dickson had come under Moore's guidance during his group-theoretic second phase, Oswald Veblen and Robert L. Moore arrived in Chicago at the height of his foundational, third period, and George Birkhoff followed after his interests had made their final shift into functional analysis. The wide variety and incisiveness of these students' results attested to their talent as mathematicians as well as to Moore's unique gifts as an adviser and to his ability to move swiftly and effectively into promising areas of mathematical research.

In 1899, for example, David Hilbert published his *Grundlagen der Geometrie*, a book based on a course he had given during the Winter Semester of 1898–1899 and had written for the occasion of the dedication of the Gauss-Weber memorial in Göttingen. Furthering the investigations of mathematicians such as Moritz Pasch in Germany and Guiseppe Peano, Guiseppe Veronese, and others in Italy, Hilbert attempted in his new book "to choose for geometry a *simple* and *complete* set of *independent* axioms and to deduce from these the most important geometrical theorems in such a manner as to bring out as clearly as possible the significance of the different groups of axioms and the scope of the conclusions to be derived from the individual axioms."[58] In the spirit of Euclid but in the light of nineteenth-century developments in set theory and non-Euclidean geometry, Hilbert sought a redundancy-free, axiomatic approach to geometry that brought its structural features into the sharpest possible relief.

To this end, he postulated three fundamental elements of geometry—points, straight lines, and planes—and characterized their mutual interrelations by means of five groups of axioms. There were seven axioms of connection, five of order, a (Euclidean) axiom of parallels, six of congruence, and an (Archimedean) axiom of continuity. With twenty axioms in all, Hilbert's system may have greatly exceeded Euclid's in size, but it also greatly surpassed it in logical completeness. Following Pasch's lead, Hilbert included axioms of order, thereby formalizing the traditionally implicit notion of betweenness, and he explicitly postulated an axiom of continuity, thereby assuring the intuitively—if not logically—clear existence of points of intersection. The true novelty of Hilbert's approach, however, lay in his methodical construction of a system of axioms that was both consistent (that is, not mutually contradictory) and independent. By exhibiting various models satisfying only some of the axioms, he proved (or almost proved) their mutual independence as well as their necessity for deriving Euclidean geom-

[58]Hilbert, p. 1 (his emphasis). See, also, Michael-Markus Toepell, *Über die Entstehung von David Hilberts "Grundlagen der Geometrie"* (Göttingen: Vandenhoeck & Ruprecht, 1986) for a discussion of the background that led to Hilbert's work.

etry.[59] It was these ideas that lured E. H. Moore away from algebraic and into foundational studies at the turn of the century.

In the Chicago department's continuing efforts to introduce its graduate students to contemporaneous research developments, Moore offered a seminar during the Fall Quarter of 1901 on the foundations of geometry and analysis. As any good seminar leader should, he exposed his students to the latest literature, and so, given the topic at hand, he lectured on Hilbert's *Grundlagen* and on the new work it had inspired. Of especial interest was a recent paper by Friedrich Schur which contested Hilbert's claim of the independence of the axioms of connection and order.[60] In his book, Hilbert had stated, without proof, that relative to the axioms of connection, order, and congruence "it is easy to show that the axioms of these groups are each independent of the others of the same group."[61] Hilbert contended that, for example, no one axiom of connection was deducible from the other six. While this may have been the case, it did not insure that the axioms were, in fact, mutually independent. Hilbert failed to show that no one axiom of connection, say, was deducible from the other axioms of connection and order. Focusing on this gap, Schur argued that three of Hilbert's axioms of connection followed from the other four axioms of connection together with the five axioms of order.

During the course of his Chicago seminar, Moore discovered that, although correct in principle, Schur had incorrectly identified the redundancies in Hilbert's two sets of axioms. Thinking out loud before his class in that style which his students came to view as so typical, Moore showed that while redundancies existed, they involved only one axiom of connection and one of order.[62] This time, Moore's unconventional teaching techniques succeeded not only in generating another new mathematical result but also in enticing one of the students in his class to pursue foundational studies.

Oswald Veblen had come to Chicago as a graduate student in 1900, fresh

[59] Hilbert, pp. 3–26. Pasch had already given a set of axioms of order as early as 1882 in *Vorlesungen über neuere Geometrie* (Leipzig: B. G. Teubner, 1882), and Hilbert acknowledged his indebtedness to Pasch's work. Hilbert's proof of consistency, however, depended on showing that his system of axioms for the real numbers was consistent, a problem that led into very deep waters. See G. Kreisel, "What Have We Learnt from Hilbert's Second Problem?," *Mathematical Developments Arising from Hilbert's Problems*, ed. Felix Browder, 2 vols., Proceedings of Symposia in Pure Mathematics, vol. 28 (Providence: American Mathematical Society, 1976), 1:93–130. Hilbert did not add the so-called axiom of completeness or *Vollständigkeit* until the French edition of his work in 1901.

[60] Friedrich Schur, "Über die Grundlagen der Geometrie," *Mathematische Annalen* 55 (1901): 265–292.

[61] Hilbert, p. 30. It is important to realize that Hilbert's principal goal was to demonstrate the logical relations between his five groups of axioms as these pertain to the foundations of classical plane and solid Euclidean geometry. The status of the individual axioms within these groups was only of secondary concern. The desire to preserve a clear, intuitive content thus took precedence over a stringent effort to economize the set of axioms. For more on this, see Arnold Schmidt, "Zu Hilberts Grundlegung der Geometrie," *Hilbert GA*, 2: 404–414, pp. 406–407.

[62] Eliakim Hastings Moore, "On the Projective Axioms of Geometry."

from earning an A.B. at Harvard. Taking courses under each of the Chicago mathematicians, he found himself in Moore's seminar during the fall of 1901 and immediately distinguished himself in that setting. So impressed was Moore by the young Veblen's abilities that he singled him out for special praise in the published version of the seminar results, writing that "queries and remarks of members of this course, in particular of Mr. O. Veblen, have been a source of much stimulus."[63] Just a year and a half later in the Spring Quarter of 1903, this bright, young student completed a doctoral dissertation under Moore's direction, which presented an axiom system based not on the undefined notions of point, line, and plane but rather on point and order. With this as his starting point, Veblen set down twelve axioms for Euclidean geometry and proved their completeness as well as their independence.[64]

Veblen stayed on at Chicago after earning his doctorate, spending the academic years of 1903–1904 and 1904–1905 in the capacity of Associate.[65] He occupied the first of these years by honing his thesis into publishable form and discussing his evolving ideas with two of Chicago's geometrically inclined graduate students, his friends Nels J. Lennes and Robert Lee Moore.[66] Of these two, R. L. Moore most completely embraced the axiomatic approach which E. H. Moore taught and which characterized Veblen's early research.

As an undergraduate and then Master's student at the University of Texas at Austin, R. L. Moore had already made somewhat of a splash in geometry and axiomatics thanks to his professor there, Sylvester's former student George Bruce Halsted. In response to a question Halsted had raised in April of 1902, the twenty-year-old R. L. Moore proved, independently of E. H. Moore's published result, the same redundancy in Hilbert's axioms of order. Moore saw the young man's proof in the *American Mathematical Monthly* and wrote to him at once to offer both his congratulations on a "delightfully simple proof" and his best wishes for continued progress in mathematics.[67] Barely a year later in 1903, the elder Moore got the chance to assure this

[63] *Ibid.*, p. 143. On Veblen's life, see, for example, Archibald, *Semicentennial History*, pp. 206–211; and Deane Montgomery, "Oswald Veblen," *Bulletin of the American Mathematical Society* 69 (1963):26–36.

[64] Oswald Veblen, "A System of Axioms for Geometry," *Transactions of the American Mathematical Society* 5 (1904):343–384.

[65] For a glimpse at Veblen's subsequent career, see Chapter 10.

[66] Veblen, "A System of Axioms for Geometry," p. 344. After earning his Chicago Ph.D. in 1907 under E. H. Moore (see Table 9.2), Lennes served as an Instructor first at MIT (1907–1910) and then at Columbia (1910–1913). In 1913, he accepted the professorship at the University of Montana which he held until his retirement on 1944. On Lennes, see Deane Montgomery and Oswald Veblen, "Nels Johann Lennes," *Bulletin of the American Mathematical Society* 60 (1954):264–265.

[67] Eliakim Hastings Moore, "The Betweenness Assumptions," *American Mathematical Monthly* 9 (1902):152–153. R. L. Moore's proof had appeared somewhat earlier under Halsted's name. See George Bruce Halsted, "The Betweenness Assumptions," *American Mathematical Monthly* 9 (1902):98–101.

OSWALD VEBLEN (1880–1960)

progress when R. L. Moore entered the Ph.D. program at Chicago on Halsted's strongest recommendation.[68]

Befriended almost immediately by Veblen, R. L. Moore found himself in the thick of Chicago's axiomatic research and used this exposure to his best advantage. By 1905, he had earned his doctorate for a dissertation, entitled "Sets of Metrical Hypotheses for Geometry," which Veblen directed in concert with E. H. Moore.[69] In his thesis, R. L. Moore, like Veblen, experimented with an alternative axiom system for geometry. Taking point, order, and congruence as his underlying undefined notions, Moore first constructed a complete set of redundancy-free axioms for Euclidean geometry and then studied the geometries resulting from carefully altered versions of that set. In particular, he showed that all of his various axiomatizations defined geometries in which theories of proportion as well as rigid motion are derivable.[70]

While Veblen and R. L. Moore continued to work on the foundations of geometry during the century's opening decade, their mentor turned his attentions to other foundational matters. In 1902, on the heels of his work on Hilbert's axioms, E. H. Moore joined Heinrich Weber, Harvard's E. V. Huntington,[71] and others in formulating a set of formal axioms for a group.

[68]Much has been written on R. L. Moore's life, but see, in particular, Archibald, *Semicentennial History*, pp. 240–244; R. L. Wilder, "Robert Lee Moore, 1882–1974," *Bulletin of the American Mathematical Society* 82 (1976):417–427; Charles C. Gillespie, ed. *Dictionary of Scientific Biography*, 16 vols. 2 supps. (New York: Charles Scribner's Sons, 1970–1990), supp. II, s.v. "Moore, Robert Lee," by Albert C. Lewis; Robert E. Greenwood, R. H. Bing, W. T. Guy, Jr, and R. C. Osborn, "In Memoriam: Robert Lee Moore," Faculty Resolution, University of Texas at Austin; D. Reginald Traylor with William Bane and Madeline Jones, *Creative Teaching: Heritage of R. L. Moore* (Houston: University of Houston Press, 1972); and Robert E. Greenwood, "The Kinship of E. H. Moore and R. L. Moore," *Historia Mathematica* 4 (1977):153–155. (In the latter article, Greenwood concluded that E. H. and R. L. were seventh cousins.) Relative to his recommendation in support of R. L. Moore's candidacy for a Chicago fellowship, Halsted wrote to R. L. on 16 March 1903 to tell him that "I have just written a tremendously strong letter to Prof. Bolza in your favor for a Fellowship at the University of Chicago, and I think you may surely count on receiving the appointment." A fair amount of correspondence relating to R. L. Moore is housed with the above-quoted letter in the R. L. Moore Papers, Archives of American Mathematics, University of Texas at Austin.

[69]A certain amount of confusion surrounds the question of who served officially as R. L. Moore's adviser. In *Semicentennial History*, Raymond C. Archibald listed R. L. Moore as a student of Veblen (p. 209). R. L. Moore himself thanked both E. H. Moore and Veblen in the published version of his dissertation "for suggestions and criticisms," while acknowledging that "Professor Veblen, who suggested the undertaking of this investigation, has not only made numerous suggestions and criticisms, but has given me much help in the actual way of collaboration." See Robert L. Moore, "Sets of Metrical Hypotheses for Geometry," *Transactions of the American Mathematical Society* 9 (1908):487–512 on p. 488. As far as we have been able to determine, the documentation needed to settle this question categorically no longer survives at the University of Chicago.

[70]R. L. Moore, "Sets of Metrical Hypotheses for Geometry," pp. 487–512.

[71]Edward Vermilye Huntington took A.B. (1895) and A.M. (1897) degrees at Harvard before earning a Ph.D. from the University of Strassburg in 1901. Prior to joining the Harvard faculty permanently in 1901, he taught at Williams College. He served as Editor of the *Annals of Mathematics* after its move to Harvard and worked primarily in axiomatics during the first decade of the twentieth century. He died in 1952 at the age of seventy-eight.

Moore compared four possible axiom systems—one proposed by Weber, two put forth by Huntington, and one favored by himself—according to explicitly articulated criteria for the "best" system upon which to base further group-theoretic investigations. Whereas, for example, one of Huntington's formulations called for four axioms—closure under multiplication, associativity, and the existence of left and right inverses—Moore believed in the superiority of his own version, one with six axioms: closure under multiplication, associativity, the existence of an idempotent element, the requirement that every idempotent be both a left and a right identity element, and the existence of a lefthand inverse for every element relative to every idempotent. He felt that the latter "definition seems ... to be an advantageous one both from the group-theoretic and the logical standpoints," but few of his contemporaries or successors apparently agreed.[72] After this brief foray into the axiomatization of algebra, Moore returned to analysis, this time to examine *its* foundational underpinnings. In so doing, he entered one of the fastest-breaking areas of turn-of-the-century mathematical research.

As we have seen, Moore had a longstanding interest in analysis dating back at least to 1890. During Chicago's first fifteen years, however, his own research had centered on algebra while mathematicians such as Thomas Stieltjes, Vito Volterra, Erik Fredholm, David Hilbert, and Henri Lebesgue had made great strides in developing functional analysis, operator theory, and integration theory.[73] By 1906, Moore had completed work preliminary to the linkage of these special theories under one theoretical umbrella which he dubbed "general analysis." He presented his new ideas to the American Mathematical Society in the form of his New Haven Colloquium lectures of 1906 and to the International Congress of Mathematicians at its meeting in Rome two years later.[74]

[72]Eliakim Hastings Moore, "A Definition of Abstract Groups," *Transactions of the American Mathematical Society* 3 (1902):485–492; and "On a Definition of Abstract Groups," *op. cit.*, 6 (1905):179–180. The quote appears on p. 491 of the former paper. Moore reacted principally to the definitions given in Heinrich Weber, *Lehrbuch der Algebra*, 2nd ed., 3 vol. (Braunschweig: F. Vieweg & Sohn, 1899–1912), 2:3–4; and E. V. Huntington, "Simplified Definition of a Group," *Bulletin of the American Mathematical Society* 8 (1902):296–300; and "A Second Definition of a Group," *op. cit.*, pp. 388–391. Moore's student and colleague Dickson also wrote on this question, but several years later, in "Definitions of a Group and a Field by Independent Postulates," *Transactions of the American Mathematical Society* 6 (1905):198–204.

[73]For the history of some of these new developments, see Thomas Hawkins, *Lebesgue's Theory of Integration: Its Origins and Development* (New York: Chelsea Publishing Company, 1970). As Hawkins pointed out (pp. 143–145), Moore's work of 1901 on Harnack's theory of integration influenced the theory of integration which Lebesgue subsequently presented in his *Leçons sur l'Intégration et la Recherche des Fonctions primitifs* (Paris: Gauthier-Villars, 1904). See E. H. Moore, "Concerning Harnack's Theory of Improper Definite Integrals," *Transactions of the American Mathematical Society* 2 (1901):296–330; and "On the Theory of Improper Definite Integrals," *op. cit.*, pp. 459–475.

[74]E. H. Moore, "Introduction to a Form of General Analysis," *The New Haven Colloquium* (New Haven: Yale University Press, 1910); and "On a Form of General Analysis with Application to Linear Differential and Integral Equations," in *Atti del* 4 *Congresso internazionale dei Matematici Rome* 1908, 3 vols. (Rome: Accademia dei Lincei, 1909), 2:98–114.

Robert Lee Moore (1882–1974)

In these various lectures, Moore outlined a "theory of systems of classes of functions, functional operations, etc., involving at least one general variable on a general range," which aimed to bring the theory of functions under one extremely general and all-encompassing postulational system.[75] He based this theory on several key structures: a system A of real or complex numbers, a general range[76] B, a class M of functions μ from B to A, a class K of functions κ from $B \times B$ to A, and a functional operator $J : K \times M \to M$. Moore then subjected the basis, namely, $\Sigma_1 = (A, B, M, K, J)$ "to a set of postulates sufficiently stringent in character to provide a theory of the equation $\mu_1 = \mu_2 - zJ\kappa\mu_2$ [z a variable] which would have as instances the theory of the Fredholm integral equation and certain other theories."[77] Within this general framework and employing a highly idiosyncratic notation, Moore succeeded in giving a unified presentation of such special cases as the theory of linear equations over a Hilbert space and Fredholm's theory of linear integral equations. As a reviewer of his New Haven Colloquium Lectures characterized Moore's theory:

> There can be no doubt that the principal mathematical results of these lectures are of a simple character, admitting of very brief proof, ... But Professor Moore has broken up his treatment into its component abstract parts, and at the same time has employed a sufficiently extensive technical notation to distinguish numerous special cases by their abstract properties; in this way a complex mathematical situation has arisen. The reason which led Professor Moore to adopt this form of treatment lies of course in the indisputable fact that the whole of mathematics needs to be presented from a standpoint which recognizes common elements of thoughts in diverse fields. It is Professor Moore who has most consistently advanced this important thesis.[78]

In fact, Moore clearly articulated this thesis—a principle which he stressed to the graduate students in his analysis courses back at Chicago—at the start of his New Haven lectures. As he put it, "[t]he existence of analogies between central features of various theories implies the existence of a general theory which underlies the particular theories and unifies them with respect to

[75]E. H. Moore, "Introduction to a Form of General Analysis," p. 9.

[76]As Moore conceived it, the elements in a general range were completely arbitrary. They could be points, functions, curves, etc. Furthermore, the elements in a given range need not even all be of the same type.

[77]Eliakim Hastings Moore with Raymond Walter Barnard, *General Analysis: Part* 1 (Philadelphia: The American Philosophical Society, 1935), p. 2. We have used the more modern notation of maps here, although Moore did not.

[78]George D. Birkhoff, Review of "The New Haven Colloquium. By Eliakim Hastings Moore, Ernest Julius Wilczynski, Max Mason. Yale University Press, 1910. x + 222 p." *Bulletin of the American Mathematical Society* 17 (1911):414–428 on p. 422.

those central features."[79] This viewpoint, so reminiscent of the philosophical orientation of Klein and Hilbert, reveals how Moore, too, was imbued with the "modern" spirit of mathematics. One of the Chicago students who was exposed to and who took heed of this modernist philosophy was the above-mentioned reviewer, George David Birkhoff.[80]

A Michigan native, Birkhoff began his college career at the University of Chicago before transferring to Harvard in 1903 after one year. He came under the mathematical sway of Osgood and especially Bôcher while in Cambridge, but returned to Chicago to do his graduate work under E. H. Moore after earning a Harvard A.B. in 1905. As a student of mathematics, Birkhoff had thus sampled the classical approach to analysis of his Harvard teachers as well as the more avant-garde and abstract slant of his Chicago adviser. His thesis, submitted in 1907 and published in two separate articles in the *Transactions of the American Mathematical Society* in 1908, successfully fused these two points of view.[81] In its treatment of asymptotic expansions of ordinary linear differential equations, boundary-value problems, and Sturm-

[79]Moore with Barnard, p. 1. Actually, Moore presented two distinct versions of his so-called general analysis. The first, which we have described somewhat here, was developed by Moore between 1906 and 1915. Bolza gave a cogent account of Moore's first version of general analysis in his article, "Einführung in E. H. Moore's 'General Analysis' und deren Anwendung auf die Verallgemeinerung der Theorie der linearen Integralgleichungen," *Jahresbericht der deutschen Mathematiker-Vereinigung* 23 (1914):248–303. From 1915 until his death in 1932, however, Moore worked on reformulating his theory from a simpler, constructive point of view. Although Moore published very few of these later ideas, they came to light posthumously in Moore with Barnard. Among the ideas which emerged from his reformulation and which he did publish during his lifetime is the "Moore-Smith limit" and the associated notion of "Moore-Smith convergence." See E. H. Moore, "Definition of Limit in General Integral Analysis," *Proceedings of the National Academy of Sciences* 1 (1915):628–632; and E. H. Moore and H. L. Smith, "A General Theory of Limits," *American Journal of Mathematics* 44 (1922):102–121. For a classic, modern treatment of Moore-Smith convergence, see John L. Kelley, *General Topology* (New York: Springer-Verlag, 1955), pp. 62–83.

[80]On this modernist philosophy as manifested at Göttingen, see David E. Rowe, "Felix Klein, David Hilbert, and the Göttingen Mathematical Tradition," in *Science in Germany: The Intersection of Institutional and Intellectual Issues*, ed. Kathryn M. Olesko, *Osiris*, 2nd series, 5 (1989): 189–213. In the spirit of give-and-take so characteristic of the Chicago Department, Moore also piqued Bolza's interest in general analysis. See, in particular, Oskar Bolza, "An Application of the Notions of 'General Analysis' to a Problem in the Calculus of Variations," *Bulletin of the American Mathematical Society* 16 (1910):107–110. As remarked in the previous note, Bolza also tried to bring Moore's ideas on general analysis before a broader mathematical public.

[81]George D. Birkhoff, "On the Asymptotic Character of the Solutions of Certain Linear Differential Equations Containing a Parameter," *Transactions of the American Mathematical Society* 9 (1908):219–231; and "Boundary Value and Expansion Problems of Ordinary Linear Differential Equations," *op. cit.*, pp. 373–395. These papers have also been published in George David Birkhoff, *Collected Mathematical Papers*, 3 vols. (New York: American Mathematical Society, 1950), 1:1–13 and 14–36, respectively. A copy of the lecture notes from Moore's 1908–1909 course on "General Analysis—With Applications to Differential Equations" is housed in the UC Archives. For more on Birkhoff's life, see Archibald, *Semicentennial History*, pp. 212–218; Marston Morse, "George David Birkhoff and His Mathematical Work," *Bulletin of the American Mathematical Society* 52 (1946):357–391; and Oswald Veblen, "George David Birkhoff (1884–1944)," *Yearbook of the American Philosophical Society* (1946):279–285.

GEORGE DAVID BIRKHOFF (1884–1944)

Liouville theory, Birkhoff's thesis research clearly mirrored Bôcher's interests, but its reliance on more modern and abstract methods such as those of Fredholm just as clearly reflected the influence of Moore.[82] In the first of his dissertation-related papers, Birkhoff developed the asymptotic representations of the solutions of differential equations of the form

$$\frac{d^n z}{dx^n} + \rho a_{n-1}(x, \rho)\frac{d^{n-1} z}{dx^{n-1}} + \cdots + \rho^n a_0(x, \rho) z = 0,$$

where the functions $a_i(x, \rho)$ along with all of their derivatives were assumed continuous on the interval $a \leq x \leq b$ and where ρ was a complex parameter with $|\rho|$ assumed to be large.[83] He then applied this result in the second paper to boundary-value and expansion problems. For the analytic functions $p_2(x), p_3(x), \ldots, p_n(x)$ of a real variable $x \in [a, b]$, Birkhoff defined the operator

$$L(z) = \frac{d^n z}{dx^n} + p_2(x)\frac{d^{n-2} z}{dx^{n-2}} + \cdots + p_n(x) z$$

and its adjoint

$$M(z) = (-1)^n \frac{d^n z}{dx^n} + (-1)^{n-2} \frac{d^{n-2}}{dx^{n-2}}[p_2(x) z] + \cdots + p_n(x) z.$$

He next constructed the nth-order differential equation $L(u) + \lambda u = 0$ in u and took n linearly independent boundary conditions $W_i(u) = 0$, $1 \leq i \leq n$, such that the W_i's are linear and homogeneous in u as well as in the first $n - 1$ derivatives of u evaluated at $x = a$ and $x = b$. With this set-up, Birkhoff was the first to associate properly the adjoint differential equation $M(v) + \lambda v = 0$ and n adjoint boundary conditions $V_i(v) = 0$, $1 \leq i \leq n$. He then proceeded to examine the solutions $u(x)$ and $v(x)$ associated with the characteristic values of the complex parameter λ.[84] Birkhoff's work on these problems appeared prior to and independently of similar results of the Russian mathematical emigré J. D. Tamarkin.[85]

---

[82]Birkhoff thanks Moore at several points in this work for his suggestions and also cites the research of Bôcher.

[83]Birkhoff, "On the Asymptotic Character of the Solution of Certain Linear Differential Equations," pp. 225–226.

[84]Birkhoff, "Boundary Value and Expansion Problems," pp. 380–382. For more details on this and related works by Birkhoff, consult Morse, "George David Birkhoff and His Mathematical Work," pp. 360–364.

[85]Morse, "George David Birkhoff and His Mathematical Work," p. 362; and Jacob D. Tamarkin, "Sur quelques Pointes de la Théorie des Équations différentielles linéaires ordinaires et sur la Généralisation de la Série de Fourier," *Rendiconti del Circolo Matematico di Palermo* 24 (1912): 345–382; "Sur un Problème de la Théorie des Équations différentielles linéaires ordinaires," *op. cit.*, 27 (1914):376–378. Tamarkin emigrated to the United States in 1925, serving as an Assistant Professor at Dartmouth College from 1925 to 1927 before taking a position at Brown University. He served as an editor of the *Mathematical Reviews* and the *Transactions of the AMS* in addition to holding the Vice Presidency of the AMS in 1942. We discuss Birkhoff's subsequent career in Chapter 10 below.

As indicated by the succession of students—Dickson, Veblen, R. L. Moore, Birkhoff—the program in mathematics at the University of Chicago succeeded, during its first fifteen years, in attracting some of the best of America's next mathematical generation. Quite naturally, many of these students came under the spell of the program's dynamic leader and eventually chose to do their doctoral research under his guidance. Moore's ability first to focus on and then to stay abreast of and contribute to the latest developments in mathematics—finite group theory in the 1890s where he made his most significant contributions, axiomatics and the foundations of geometry at the turn of the century, and the foundations of functional analysis by 1907—drew his students into areas then currently of interest to leading European mathematicians. Moreover, his very modern interests, which, as should now be evident, closely mirrored those of David Hilbert, gave his students—and through them a significant cross-section of the American mathematical research community of the succeeding generation—a decidedly purist thrust which would propel them into the early decades of the twentieth century. (See Chapter 10 for a discussion of the impact of this purist tendency on American mathematics in the period from 1900 to the 1930s.)

The Mathematical Interests of Bolza, Maschke, and Their Students

Regardless of Moore's attractions as an adviser, many students in the Chicago program preferred the work or the style of his two colleagues, Bolza and Maschke, and so opted to study under one of them. The more successful adviser of the two, Bolza directed nine dissertations between 1896 and 1910, the year of his return to Germany, while Maschke supervised five students between 1899 and his untimely death in 1908. (See Tables 9.4 and 9.5.) Like Moore, each of these men guided his students along lines of investigation consonant with his own evolving research interests.

When Bolza joined the Chicago faculty in 1892, his work continued to revolve around the questions which had defined the topic of his dissertation under Felix Klein, namely, the interrelations between hyperelliptic and elliptic integrals.[86] As his contribution to the Chicago Mathematical Congress of 1893 shows, Bolza's research in this area aimed primarily to link Klein's invariant-theoretic point of view with Karl Weierstrass' approach to the theory.[87] One of the major fields of nineteenth-century mathematical inquiry, elliptic and hyperelliptic function theory attracted three of Bolza's first four Chicago students: John Hutchinson, William Gillespie, and John McDonald.[88] In fact, Hutchinson, who, along with Dickson, earned the first Chicago

[86]For a brief indication of Bolza's doctoral research, see Chapter 5 above.

[87]On Bolza's contribution to the Mathematical Congress in Chicago, recall Chapter 7.

[88]After earning his Ph.D., Hutchinson took a position at Cornell where he remained until his death in 1935. Hutchinson also served mathematics at the national level as the Vice President

Ph.D.s in mathematics in 1896, had followed Bolza from Clark in order to continue his research in this area. By 1901, however, Bolza's own mathematical interests had shifted to the calculus of variations, and the dissertation work of his subsequent students reflected that change.[89]

Bolza had encountered the calculus of variations in Weierstrass' lecture course of 1879. After coming to Chicago, he himself had given graduate-level instruction in the area, but his teaching did not redirect his research efforts. That crucial turn occurred in February of 1901 when the organizers of the third AMS Colloquium invited him to serve as one of the keynote speakers in Ithaca, New York in August. The series of AMS Colloquia (begun in 1896) aimed, like the precedent-setting Evanston Colloquium Lectures given by Klein in 1893, to survey broad expanses of the mathematical terrain before a general mathematical audience and thereby to suggest paths for further research. With these goals in mind, the organizers asked Bolza to speak on the theory of hyperelliptic functions. Bolza opted not to review that well-worked topic, however, choosing instead to formulate a series of lectures on advances in the calculus of variations: the ideas of Weierstrass and others aimed at rigorously establishing old, eighteenth- and early nineteenth-century results, the sufficiency proofs of both Weierstrass and Hilbert, Hilbert's existence proof, and the then recent transversal theorem of Adolf Kneser.[90] Bolza's literature review uncovered a whole range of unanswered questions which refocused his research attention for the remainder of his career and which immediately yielded new results in his hands. At the Ithaca meeting of the Society, in addition to giving his series of Colloquium lectures, Bolza also presented the first fruits of his new labors, a "New Proof of a Theorem of Osgood's in the Calculus of Variations."[91]

On 27 April 1901 four months prior to the Ithaca meeting, William Osgood had presented a paper to the Society in which he proved a key prop-

---

of the AMS in 1910 and as an Associate Editor of its *Transactions*. Gillispie, who had been an instructor at Princeton, returned there after earning his doctorate and entered the ranks of Woodrow Wilson's preceptors in 1905. Finally, McDonald took his expertise in the theory of hyperelliptic functions to the West Coast when he accepted a position at the University of California (Berkeley) in 1901.

[89]One of Bolza's post-1901 students, however, did not work on the calculus of variations. Buz M. Walker chose to do his dissertation research on the singularities of algebraic curves. On earning his doctorate in 1906, he returned to the Mississippi Agricultural and Mechanical College where he had held the rank of Professor since 1888.

[90]Oskar Bolza, *Aus meinem Leben* (München: Verlag Ernst Reinhardt, 1936), p. 33. For an historical discussion of these turn-of-the-century advances in the calculus of variations, see Herman H. Goldstine, *A History of the Calculus of Variations from the 17th through the 19th Century* (New York: Springer-Verlag, 1980), pp. 314–389.

[91]Oskar Bolza, "New Proof of a Theorem of Osgood's in the Calculus of Variations," *Transactions of the American Mathematical Society* 2 (1901):422–427. Édouard Goursat communicated yet another simple proof of this theorem in a letter to Osgood dated 21 June 1903, which appeared as Édouard Goursat, "A Simple Proof of a Theorem in the Calculus of Variations (Extract of a Letter to Mr. W. F. Osgood)," *Transactions of the American Mathematical Society* 5 (1904):110–112.

erty characteristic of the strong minimum of an integral of the form $I = \int_{t_0}^{t_1} F(x, y, x', y')dt$.[92] (Let (K) be a specified class of curves $C : x = \phi(t)$, $y = \psi(t)$, $t_0 \leq t \leq t_1$. Then a particular $C \in (K)$ *minimizes* I if $J := \int_{t_0}^{t_1} F(\phi(t), \psi(t), \phi'(t), \psi'(t))dt$ is minimal among all such integrals for $\bar{C} \in (K)$. (See the quote below.)) In his Berlin lectures of 1879, Weierstrass had given a condition sufficient for C to minimize I, but Osgood pushed Weierstrass' work somewhat further. He asked:

> Assuming that Weierstrass's sufficient condition is fulfilled, so that no curve of the class (K) distinct from C will give the integral I so small a value as J, may we not still have a set of curves belonging to this class, $\bar{C}_1$, $\bar{C}_2$, ..., which do not cluster around C as their limit and which have the property that, if I_1, I_2, . . . denote respectively the values of the integral I formed for these curves,
>
> $$\lim_{n=\infty} I_n = J?$$
>
> In particular, it is conceivable that these curves might approach a curve $\bar{C}$ of the class (K) as their limit, distinct from C.[93]

In his paper, Osgood showed that this cannot, in fact, happen, but his proof involved a tedious examination of the definite integral $H = I - J$ transformed into Kneser's curvilinear coordinates.[94] Instead of initially casting the problem in terms of the rectangular coordinates x and y and then translating into curvilinear coordinates as Osgood had done, Bolza set things up in terms of curvilinear coordinates from the start and succeeded not only in simplifying Osgood's proof but also in rendering it more conceptual.[95]

Following this début, Bolza went on to write almost thirty papers and two book-length manuscripts on the calculus of variations. His first monograph on the subject, written at E. H. Moore's urging in celebration of the tenth anniversary of the founding of the University of Chicago, elaborated on his Colloquium lectures and appeared in 1904 in Chicago's series of Decennial Publications.[96] By 1909, he had greatly expanded upon this treatment in his highly influential *Vorlesungen über Variationsrechnung*.[97] There, in a characteristically clear and engaging style, he detailed and systematized the approaches and results in the calculus of variations of Weierstrass, Kneser,

[92] William F. Osgood, "On a Fundamental Property of a Minimum in the Calculus of Variations and the Proof of a Theorem of Weierstrass's," *Transactions of the American Mathematical Society* 2 (1901):273–295. See pp. 273–276 for the full notational set-up.

[93] *Ibid.*, p. 274. In his text, Osgood writes $n = \infty$ instead of the now-standard $n \to \infty$.

[94] *Ibid.*, pp. 276–286.

[95] Bolza, "New Proof of a Theorem of Osgood's in the Calculus of Variations," pp. 422–427.

[96] Oskar Bolza, *Lectures on the Calculus of Variations* (Chicago: University of Chicago Press, 1904).

[97] Oskar Bolza, *Vorlesungen über Variationsrechnung* (Leipzig: B. G. Teubner, 1909).

and Hilbert, among others. In particular, he treated then recent developments on Lagrange's problem with fixed end points and on the calculus of variations for double integrals.[98] By combining his teaching at Chicago with his exposition of this book, Bolza attracted several students into dissertation work in this area.

Bolza's first—and most prominent—student in the calculus of variations actually graduated the year before the Ithaca Colloquium lectures took place. Gilbert Ames Bliss earned his doctorate in 1900 for a thesis on "The Geodesic Lines on the Anchor Ring."[99] In this work, Bliss extended previous results of Jacobi and Hans von Mangoldt on the geodesics of surfaces of either negative (Jacobi) or positive (Mangoldt) curvature by focusing on the geodesics of a torus, a figure with curvature neither always positive nor always negative. He showed that only points on the inner equator of the torus were of what he termed the *first kind,* that is, only the geodesics passing through points on the inner equator remain a minimum throughout their entire lengths. All other points on the torus were therefore of the *second kind* by definition. This result nicely complemented the research of Jacobi and Mangoldt, who had found all points on the special surfaces they examined to be exclusively of the first kind or exclusively of the second kind, respectively.[100]

During the preparation of this thesis, Bolza had yet to add his own original research contributions to the calculus of variations. Whether or not Bliss's work in this area served as a catalyst in redefining Bolza's research interests, by the time Bliss returned permanently to Chicago in 1908 to replace the lamented Maschke, he and his adviser were both deeply immersed in the field. On Bolza's return to Germany two years later, Bliss not only took up his mentor's teaching duties but also maintained the University of Chicago's status as the American center for the study of the calculus of variations.[101] (See Chapter 10.) The more geometrically oriented line of teaching which Maschke had worked to establish also enjoyed a certain continuity with the

---

[98]Gilbert A. Bliss, "Oskar Bolza—In Memoriam," *Bulletin of the American Mathematical Society* 50 (1944):478–489 on p. 484. For a statement of Lagrange's problem, see *Variationsrechnung*, pp. 543–546 and 566–569; or Goldstine, pp. 132–136 and 150. Today, Bolza is best remembered for "Bolza's Problem" in the calculus of variations. For statements of the problem, see Oskar Bolza, "Über den 'Anormalen Fall' beim Lagrangeschen und Mayerschen Problem mit gemischten Bedingungen und variablen Endpunkten," *Mathematische Annalen* 74 (1913):430–446; and Goldstine, pp. 373–383.

[99]Gilbert A. Bliss, "The Geodesic Lines on the Anchor Ring," *Annals of Mathematics* 4 (1902–1903):1–21.

[100]*Ibid.*, p. 1.

[101]On Bliss's life, see Archibald, *Semicentennial History*, pp. 201–206; and Lawrence M. Graves, "Gilbert Ames Bliss 1876–1951," *Bulletin of the American Mathematical Society* 58 (1952):251–264. Bliss had several prominent students, among them: William L. Duren, Jr. (Ph.D. 1930), Edward J. "Jimmy" McShane (Ph.D. 1930), and Magnus R. Hestenes (Ph.D. 1932). See Graves, pp. 261–264 for a complete list of Bliss's students together with the years and titles of their dissertations.

GILBERT AMES BLISS (1876–1951)

appointment in 1910 of the American projective differential geometer Ernest Julius Wilczynski.[102]

When Maschke arrived to take up his position at Chicago in the fall of 1892, he had been away from his work in pure mathematics for two years while pursuing his new career in electrical engineering in Newark, New Jersey. This hiatus, combined with the fact that he had no experience in university-level teaching, placed him at a disadvantage early on. Although Maschke wrote up a result from 1888 on "The Invariants of a Group of $2 \cdot 168$ Linear Quaternary Substitutions" for his contribution to the Chicago Congress in 1893, he did not really hit his stride until the latter half of the decade. The records of the Chicago Mathematical Club reveal that, by the spring of 1897, he had once again taken up his old, pet theory of finite linear groups and had produced a quick succession of incisive general results which contrasted markedly with his very special computational work of the 1880s.

On 28 May 1897, Maschke delivered a survey lecture before the club on "The History and Present State of the Theory of Linear Homogeneous Substitution-Groups."[103] At that time, he announced a key, new result which would appear in the *Mathematische Annalen* the following year. Maschke had proved the so-called cyclotomic theorem, namely, "if a finite linear substitution group contains at least one substitution with distinct characteristic roots, then the group can be transformed in such a way that the coefficients of every substitution in it are cyclotomic, that is, rationally expressible in terms of roots of unity."[104] His proof proceeded constructively. By means of successive base changes on the finite substitution group $G \subset GL_n(\mathbb{C})$, he systematically showed that any $A \in G$ could be represented by an $n \times n$ matrix $[a_{ik}]$ with cyclotomic entries a_{ik}.[105] This theorem soon attracted the attention of William Burnside and Issai Schur, both of whom generalized it to broader classes of groups—Burnside to transitive permutation groups and Schur to all solvable groups.[106]

---

[102]Wilczynski took a Berlin Ph.D. in 1897 and served successively as a computer at the Nautical Almanac Office (1898), on the faculty at the University of California, Berkeley (1898–1906), and at the University of Illinois (1907–1910), before joining the Chicago staff.

[103]UC Archives, Math. Club Records, Box 1, Folder 3, p. 43.

[104]"Ueber den arithmetischen Charakter der Coefficienten der Substitutionen endlicher linearer Substitutionsgruppen," *Mathematische Annalen* 50 (1898):492–498 on p. 498 (our translation). Also, see Oskar Bolza, "Heinrich Maschke: His Life and Work," *Bulletin of the American Mathematical Society* 15 (1908):85–95, for an account of Maschke's major research contributions.

[105]Maschke, "Ueber die Coefficienten linearer Substitutionsgruppen," p. 493.

[106]See William Burnside, "On the Complete Reduction of Any Transitive Permutation Group and on the Arithmetic Nature of the Coefficients in Its Irreducible Components," *Proceedings of the London Mathematical Society* 3 (1905):239–252; and Issai Schur, "Arithmetische Untersuchungen über endliche Gruppen linearer Substitutionen," *Sitzungsberichte der Preussischer Akademie der Wissenschaften* (1906):164–184. In 1945, Richard Brauer presented the result in its fullest generality, showing that "[i]f G is a group of finite order g, then every irreducible representation L of G can be written in the field Ω of the g-th roots of unity (that is, L is similar to a representation with coefficients in Ω)." See Richard Brauer, "On the Represen-

In the form of a lemma ancillary to the published proof of the cyclotomic theorem, Maschke also proved a special case of the theorem which has since come to bear his name. By December of 1898, he had obtained a more general version of Maschke's Theorem, namely, "every finite linear substitution group all of whose substitutions contain (in the same place) one, off-diagonal zero coefficient, is intransitive."[107] (For Maschke, a linear substitution group was *intransitive* if it could be transformed in such a way that the new variables split up into a number of sets where the variables of each set are transformed among themselves.) His proof followed virtually as a corollary of a critical result which E. H. Moore had announced and proven in the Mathematical Club at Chicago on 26 February 1897: "[a] finite group of n-ary linear homogeneous substitutions leave[s] absolutely invariant an n-ary positive Hermitian form."[108] In his talk of 28 May, Maschke had specifically singled out Moore's result as one of three important "[g]eneral theorems concerning groups of any number of variables," thereby underscoring not only his familiarity with and appreciation of the result but also the positive effect of the Mathematical Club on the research environment at Chicago.[109] A year and a half later, Moore's theorem would give Maschke exactly what he needed to prove the intransitivity, or, to use terminology popularized by Burnside, to establish "complete reducibility."[110]

Moore's theorem on Hermitian forms also served as the key tool in Maschke's complete determination of all (complex) ternary and quaternary collineation groups isomorphic to the symmetric or alternating groups. Whereas in the 1880s his research had focused primarily on ternary and quaternary groups of specific orders, Maschke noted in his Mathematics Club lecture of 28 May 1897 that "[n]o attempt has been made to determine all quaternary

tation of a Group of Order g in the Field of the g-th Roots of Unity," *American Journal of Mathematics* 67 (1945):461–471; or Richard Brauer, *Richard Brauer: Collected Papers*, ed. Paul Fong and Warren J. Wong, 3 vols. (Cambridge: MIT Press, 1980), 1:518–528. For a modern treatment of this theorem using Schur indices, see Charles W. Curtis and Irving Reiner, *Representation Theory of Finite Groups and Associative Algebras* (New York: Interscience Publishers, 1962), pp. 292–295.

[107]Heinrich Maschke, "Beweis des Satzes, dass diejenigen endlichen linearen Substitutionsgruppen, in welchen einige durchgehends verschwindende Coefficienten auftreten, intransitiv sind," *Mathematische Annalen* 52 (1899):363–368 on p. 363 (our translation).

[108]UC Archives, Math. Club Records, Box 1, Folder 2, p. 66. Moore subsequently published this result as "An Universal Invariant for Finite Groups of Linear Substitutions: With Application in the Theory of the Canonical Form of a Linear Substitution of Finite Period," *Mathematische Annalen* 50 (1898):213–219. Alfred Loewy also announced this result independently—but without proof—in 1896.

[109]UC Archives, Math. Club Records, Box 1, Folder 3, p. 43.

[110]William Burnside, *Theory of Groups of Finite Order*, 2nd ed. (Cambridge: University Press, 1911; reprint ed., n.p.: Dover Publications, Inc. 1955), pp. 258–263. There are other well-known equivalent statements of Maschke's Theorem. For example, if G is a finite group of order $o(G)$ and if F is a field of characteristic zero or of characteristic p such that p does not divide $o(G)$, then the group algebra $F[G]$ is semisimple. See, for example, I. N. Herstein, *Noncommutative Rings* (n.p.: Mathematical Association of America, 1968), pp. 26–27.

groups" although "[t]hree are known so far; they are all due to Klein."[111] In a paper dated November 1897 but only published in 1899, Maschke made significant progress toward the solution of this classification problem by rather painlessly determining (up to isomorphism) the seven ternary and the twenty-five quaternary collineation groups.[112] He managed this using both Moore's result on Hermitian forms and his characterization by generators and relations of the symmetric and alternating groups. In 1897, Moore had given, for the first time, the now-usual presentations of these groups. In particular, he showed that

> [t]he abstract group $G(k)$ (≥ 2) defined by the $k-1$ generators
>
> $$B_d \qquad (d = 1, 2, \ldots k-1),$$
>
> with the generational relations
>
> $$B_d^2 = 1 \qquad (d = 1, 2, \ldots, k-1),$$
>
> $$(B_d B_{d+1})^3 = 1 \qquad (d = 1, 2, \ldots, k-2),$$
>
> $$(B_d B_e)^2 = 1 \qquad (d = 1, 2, \ldots, k-3), \quad (e = d+2, d+3, \ldots, k+1)$$
>
> ... has the order $O(k) = k!$, and is holoedrically isomorphic with the symmetric substitution-group on k letters.[113]

The results on collineation groups which Maschke derived based on Moore's work followed closely on the heels of the dissertation research of one of Maschke's own students, George Lincoln Brown. At Moore's suggestion, Brown had undertaken an analysis of the structure of the ternary

[111] UC Archives, Math. Club Records, Box 1, Folder 3, p. 46.

[112] Note, that Maschke was still concerned here with ternary and quaternary groups isomorphic to the symmetric or alternating groups only. Heinrich Maschke, "Bestimmung aller ternären und quaternären Collineationsgruppen, welche mit symmetrischen und alternierenden Buchstabenvertauschungsgruppen holoedrisch isomorph sind," *Mathematische Annalen* 51 (1897):253–298.

[113] E. H. Moore, "Concerning the Abstract Groups of Order $k!$ and $\frac{k!}{2}$ Holoedrically Isomorphic with the Symmetric and the Alternating Substitution-Groups on k Letters," *Proceedings of the London Mathematical Society* 28 (1897):357–366. In the 1930s, H. S. M. Coxeter systematically studied the now so-called *Coxeter groups*, generated by elements R_i, $1 \leq i \leq n$, subject to relations $R_i^2 = (R_i R_j)^{k_{ij}} = 1$, where $2 \leq k_{ij} = k_{ji} \leq \infty$ for $i \neq j$. See H. S. M. Coxeter, "Discrete Groups Generated by Reflections," *Annals of Mathematics* 35 (1934):588–621; "The Complete Enumeration of Finite Groups of the Form $R_i^2 = (R_i R_j)^{k_{ij}} = 1$," *Journal of the London Mathematical Society* 10 (1935):21–25. Although Coxeter credited Moore with the initial discovery of this presentation of the symmetric group in his book, *Regular Polytopes* (New York: The Macmillan Company, 1963; reprint ed., New York: Dover Publications, Inc., 1973), p. 199, this fact seems little known. For example, Bourbaki made no mention of Moore's work in the historical notes of his *Groupes et Algèbres de Lie: Chapîtres* IV, V, *et* VI (Paris: Hermann, 1968), pp. 234–240 on p. 237.

collineation group of order 360 sometime prior to the summer of 1896. Writing to Klein on 28 July, Maschke proudly reported that Brown had not only established the isomorphism of this group with the alternating group on six letters but also, under Maschke's guidance, determined its complete invariant system.[114] So pleased was he with his student's results that he asked Klein whether it might not be possible for him to write up an abstract of the work for inclusion in either the *Mathematische Annalen* or the *Göttinger Nachrichten.* Unfortunately, Maschke's hopes for the public announcement of Brown's research achievements proved short-lived. When the most recent number of the *Mathematische Annalen* arrived in Chicago, it contained an article by Wiman which duplicated Brown's results. As Maschke rather bitterly reported to Klein: "Brown spoke on G_{360} last Friday, that is, before the number of the *Annalen* arrived. Naturally, my question concerning publication has now become superfluous. Our Brown, however, will not have done his work in vain; he will publish it in one of the American journals with reference to Wiman's work."[115]

Although atypical in that Brown experienced firsthand the dangers of working in an active research area against seasoned competition, his case nevertheless exemplified the close working relationship both among the Chicago faculty members and among faculty and graduate students. Moore proposed Brown's research problem; Maschke helped him push his results; both men took a keen interest in his progress. Over and over again, this kind of cooperation and interaction enriched the research not only of the graduate students but also of the faculty members themselves. At Chicago, Moore, Bolza, and Maschke had created a lively, vibrant atmosphere in which department members shared a sense of common cause, a commitment to the production of original mathematical results. They worked together; they kept abreast of each other's research; they commented upon and criticized each other's ideas and presentations. In so doing, they created a whole which was greater than the sum of its parts. The early Chicago department owed its research successes as much to this dynamic working environment as to its unique pool of mathematical talent.

Mathematical Activism: The Role of the Chicagoans in the American Mathematical Society

This glimpse of the research record of Chicago's mathematical fathers and their first generation of offspring unequivocally confirms Bliss 's characterization of Moore, Bolza, and Maschke. Mathematics did indeed come "first in

[114]Heinrich Maschke to Felix Klein, 28 July 1896, Klein Nachlass, NSUB, Göttingen. Compare Moore's work on the finite simple group of order 360 discussed above, and see note 44.

[115]Heinrich Maschke to Felix Klein, 24 August 1896, Klein Nachlass, NSUB, Göttingen (our translation).. Brown apparently never published this work. See Anders Wiman, "Ueber eine einfache Gruppe von 360 ebenen Collineationen," *Mathematische Annalen* 47 (1896):531–556. On earning his doctorate, Brown returned to his professorship at South Dakota State College. He became President of the institution in 1940.

the minds of these leaders in the youthful department at Chicago."[116] Their shared sense of the goal of the mathematician, together with the stated mission of their university, dictated their vigorous approach to graduate teaching as well as their dedication to the production of new results. But if the development of mathematics came first in their minds, the development of a mathematical community followed a close second. The Chicago mathematicians, and especially E. H. Moore, recognized that only rarely was mathematics created in a vacuum. Communication through direct contact and conversation as well as through publication enhanced mathematical productivity. Thus, for the emergent community of American research mathematicians to prosper, it needed an organization to link its widely dispersed members. As noted in Chapter 6, the seed of just such an organization, the New York Mathematical Society, had sprouted in New York in 1888 and had grown steadily in the years following.

Despite this relative success, the Society had nevertheless managed to enjoy only limited effectiveness. Although its *Bulletin* provided a forum for the exchange of mathematical ideas, its monthly meetings (initially held only during the academic year) served only those in and around New York City. To be sure, some of the Society's more geographically remote members like E. H. Moore in Chicago or F. N. Cole in Ann Arbor occasionally attended a meeting while passing through New York, but given the vagaries of late nineteenth-century travel, not to mention the problems of time and money, regular attendance and so regular interaction was impossible for all but a few of the members. The New York Mathematical Society had national aspirations, but, as its name implied, it functioned primarily at a local level only. By the mid-nineties, several events had taken place which helped liberate the Society from its provinciality.

As we have seen, in August of 1893, the Mathematical Congress of the World's Columbian Exposition brought four mathematicians from abroad and forty-one from the United States together for six days of lectures in Chicago. The first mathematics meeting of note held in the United States, the Congress drew participants from nineteen of the forty-four states then in the Union, with participants traveling from the East, Northeast, and Midwest, and from as far west as California and Colorado, from as far south as Texas and Alabama. With Felix Klein as the main mathematical attraction, most of the Americans came no doubt to see and hear this near-legendary figure. Yet the Congress unquestionably encompassed more than the work and persona of Felix Klein. In response to the invitations sent out by Moore, Bolza, Maschke, and White in their capacity as the local organizing committee,

---

[116] Bliss, "E. H. Moore," p. 833.

thirteen American and twenty-six foreign mathematicians submitted papers to be read before the assemblage.[117]

Charged with securing a publisher for these proceedings, the local committee set to work immediately to complete its final task. This publication venture proved more difficult than initially supposed, however, for "[n]either the management of the Exposition nor the government of the United States had made provision for the publishing of the proceedings of any of the Chicago Congresses. No publisher was found willing to issue the papers at his own risk."[118] Faced with this clash between the economics of the publishing industry and the desire to disseminate technical mathematical knowledge,[119] the editorial committee turned first to the University of Chicago and then to the New York Mathematical Society for help.

Moore, as chair of the committee, was well aware of President Harper's insistence upon the desirability of departmental publications, and confidently approached his university for financial support. He presented an ostensibly airtight argument in a letter to Harper and the University's Executive Committee on 30 April 1894:

> In accordance with the policy of The Trustees of The University of Chicago, each department has *as one of its functions* the publication of a departmental journal or of a series of scientific monographs. Hence it was my own earnest desire from the beginning that this publication of the collected Congress Memoirs should be undertaken by our, let me say, *your* Department of Mathematics. In the scientific world—and that is the world in which every department of the University really *lives*—the University would gain much credit by undertaking this publication. The volume would certainly be one of the few *very important* mathematical books issued in the United States, while the contributions of the Chicago staff add materially to the value of the book .... .
>
> I ask of the Executive Committee, on behalf of the Department of Mathematics an appropriation of $700 *for publishing*. With this (which may eventually not all be needed) as a fixed basis, I am confident that the New York Society will make some contribution, and that the publication can be consummated.

[117]Here, those mathematicians affiliated with American institutions or having American addresses have been counted as "American." For a sketch of the broader activities of the members of the early Chicago Mathematics Department, see Parshall, "Eliakim Hastings Moore and the Founding of a Mathematical Community in America, 1892–1902."

[118]Moore, Bolza, Maschke, White, ed., *Congress Papers*, p. v.

[119]Henry White had also recently dealt with this dilemma in his efforts to secure the publication of Klein's Evanston Colloquium lectures. Macmillan & Co., which ultimately did publish the volume, viewed the project apathetically. See Henry White to Felix Klein, 5 October 1893, Klein Nachlass XII, NSUB, Göttingen.

> *Is this too much to ask*, when you consider the great service thereby done to Mathematics in the United States, and the prominent position o[ur University] will hold in the volume?[120]

Apparently so. Whether due to the absence of the needed technical type in Chicago or to the fact that this did not represent an ongoing publication venture, the University declined to underwrite the project.[121]

If Moore's confidence in his university's backing ultimately proved unjustified, his faith in the New York Mathematical Society was well founded. In June of 1894, the New York organization pledged $600 of a $1000 guaranty fund needed to secure the publishing services of Macmillan and Co.[122] In the long run, in fact, the University of Chicago's failure to act favorably on Moore's request forced the developing American mathematical research community to take charge of its own affairs. As the historian of the Society's first fifty years, Raymond C. Archibald, saw matters, "[t]his major publication enterprise, transcending local considerations and sentiment, quickened the desire of the Society for a name indicative of its national or continental character."[123] As a result, on 1 July 1894, barely one month after the *New York* Society appropriated funds to Moore and his committee, the *American* Mathematical Society emerged to assume its broadened leadership role.[124] It made its début at another significant event, its first summer meeting.

Capitalizing on the occasion of the summer meeting of the American Association for the Advancement of Science (AAAS) at the Brooklyn Polytechnic Institute in August of 1894, the new American Mathematical Society arranged

---

[120]E. H. Moore to W. R. Harper and the Executive Committee, 30 April 1894, UC Archives, UPP 1889–1925, Box 17, Folder 2 (Moore's emphasis).

[121]The University may also have rejected Moore's appeal because it could not afford to take on the role of Congress publisher. Given that Congresses in many disciplines were held at the Exposition, each associated department within the University could have argued for University support. The President and his Executive Committee may simply have decided that they could not fund mathematics if they could not afford to fund all other similar requests. Their denial of Moore's request followed closely on the heels of Henry White's solicitation of support for the publication of the Evanston Colloquium lectures. The University Press had actually agreed to put up $200.00 for the publication of 500 copies of the Klein work, but White judged the offer too low. See Henry White to Felix Klein, 5 October 1893, Klein Nachlass XII, NSUB, Göttingen. White's own university, Northwestern, flatly refused to assist in the project. White related his thoughts on this to Klein: "Not to take it was a mistake on their part which I could not understand at first. It is less surprising when I learn that of the Executive Committee, to whom the question was submitted, only *one* has enjoyed a college education." Henry White to Felix Klein, 14 January 1894, Klein Nachlass XII, NSUB, Göttingen (White's emphasis; our translation).

[122]The American Mathematical Society ultimately paid Macmillan $773.76. See Archibald, *Semicentennial History*, p. 7.

[123]*Ibid.*

[124]It is interesting to note that the London Mathematical Society also eventually became a national—as opposed to a local—organization, but it did not change its name to reflect its broader mission.

to meet concurrently with the membership of the AAAS's Section A. In all, roughly two dozen AMS members attended ten predominantly astronomical lectures, but they also heard purely mathematical talks given by E. H. Moore and by one of his mentors, Ormond Stone (who by then had moved from the Cincinnati Observatory to the Leander McCormick Observatory at the University of Virginia).[125] Even though this first summer meeting took place in New York City like the regular gatherings during the academic year, its initiators had more widely reaching goals in mind: by instituting a series of summer meetings, they felt that "members from all over the country would often be free to attend," and the Society would be *American* in more than name only.[126]

As at all of the Society's meetings, however, virtually anyone who wished to present a paper could do so. This resulted quite naturally in conferences with no overall focal point, where those in attendance heard about snippets of research ranging over the whole of mathematics. Mathematicians in the United States of the last decade of the nineteenth century undoubtedly needed forums for the presentation of the greatest amount of mathematics in the shortest span of time—forums which these Society meetings provided—but they lacked opportunities for more broadly accessible, seminar-like presentations. In Chicago or New Haven or Cambridge or Baltimore, mathematicians could attend university seminars designed to give the listener a systematic and in-depth look at a given area of mathematics, but this was a privilege not accorded to the geographically remote rank and file.

Northwestern's Henry White, one of the organizers and beneficiaries of Felix Klein's series of Evanston Colloquium lectures, understood full well the value of more intensive and focused lectures. Writing to AMS founder, Thomas Fiske on 23 February 1896, White offered the opinion that

> Our summer meeting last year was profitable in various ways; but not specially, perhaps, as a stimulus to mathematical thought. One would likely derive more direct advantage from an hour in any one of several lecture rooms in this country. Yet each one found two or three papers out of the whole program of high interest. Now, why would it not be possible to combine with this miscellaneous program (which ought by all means to be kept up), something more akin to university models? Would not a series of three or six lectures on nearly related topics, if well chosen, prove attractive and useful to larger numbers?[127]

With Fiske's endorsement and the AMS Council's permission, White wrote

[125] Archibald, *Semicentennial History*, p. 86.

[126] *Ibid.*, p. 85.

[127] *Ibid.*, p. 67.

a circular to all members of the Society describing his idea for a colloquium to supplement the regular summer meeting.[128] White's proposal—supported by the signatures of Fiske, E. H. Moore, W. F. Osgood of Harvard, F. N. Cole of Columbia, Alexander Ziwet of Michigan, and Frank Morley then of Haverford—garnered a favorable response, and the first Colloquium of the AMS took place on 2–5 September 1896 in Buffalo, New York. As its kick-off speakers, the AMS invited James Pierpont of Yale, who spoke on his work in the Galois theory of equations, and Maxime Bôcher of Harvard, who talked on second-order linear differential equations.[129] Unlike White's handling of the Evanston Colloquium lectures, however, the AMS did not publish the proceedings of this first colloquium.[130] Its first publication of this type would only appear in 1903, when, as chance would have it, White was one of the three chosen speakers.[131] The published volume of Evanston Colloquium lectures—delivered by Klein ten years earlier—thus served as a forerunner of the AMS Colloquium Publication series.

This was not the only way that that historic colloquium affected the development of the AMS. As a result of its success, the two departments at Chicago and Northwestern resolved to maintain a regular, interactive schedule of seminars. As early as June of 1893, in fact, White had already come down from Evanston to give the final Chicago Mathematics Club presentation of the academic year on "Hesse's Enumeration of the Bitangents of the Plane Quartic Curve."[132] The two departments understood the value of sharing mathematical perspectives and strengths, and after the Evanston Colloquium, they realized the importance of diversity of approach and interest. After all, Klein's lectures at Northwestern had brought together geographically dispersed mathematicians and had given them occasion to talk not only to Felix Klein but also to one another. Yet the Evanston Colloquium was an apparently once-in-a-lifetime occurrence for the Midwest. As we have seen, the regular meetings of the Society continued to be held in New York City following the official name change in 1894. Even the newly instituted summer

---

[128]See Thomas S. Fiske, "The Buffalo Colloquium," *Bulletin of the American Mathematical Society* 3 (1896):49–59 for the complete text of the proposal White put forth before the Society. This communication also includes abstracts of the lectures given by Pierpont and Bôcher.

[129]Archibald, *Semicentennial History*, pp. 66–67.

[130]The talks were published, however, either in whole or in part. See James Pierpont, "Galois Theory of Equations," *Annals of Mathematics*, 2nd ser., 1 (1900):22–56; and Maxime Bôcher, *Regular Points of Linear Differential Equations of the Second Order* (Cambridge, MA: Harvard University, 1896), and "Notes on Some Points in the Theory of Linear Differential Equations," *Annals of Mathematics* 12 (1898):45–53.

[131]White spoke on "Linear Systems of Curves on Algebraic Surfaces." The other two lecturers were F. S. Woods, speaking on "Forms of Non-Euclidean Space" and E. B. Van Vleck, talking on "The Theory of Divergent Series and Continued Fractions." See *AMS Colloquium Publications*, vol. 1 (New York: Macmillan and Co., 1905).

[132]UC Archives, Math. Club Records, Box 1, Folder 1. See, also, Table 9.1. This talk covered a topic closely related to the seminar lectures White delivered in Göttingen (recall Chapter 5).

meetings suffered from a geographical bias toward the Northeast.[133] Thus, only an occasion like a Congress and Colloquium with a keynote speaker like Klein seemed momentous enough to motivate the Northeastern mathematicians, comfortable with the level of activity within their own environment, to undertake the trek to a city as seemingly far-removed as Chicago. (And, as we saw in Chapter 7, for some, like Osgood and Bôcher, not even this was enough.)

The New Yorker's, or more generally the Easterner's, view of the United States was, to be sure, even more exaggerated then than it is purported to be now. Whatever lay west of the Hudson River, much less the Ohio or Mississippi, was raw, untamed, and uncultured frontier. In spite of the spectacular architectural advances made after the equally spectacular fire of 1871, in spite of the lasting impression of progress created by the World's Fair of 1893, many Easterners still shared a vision of Chicago reminiscent of that recorded by the Rhode Island botanist Edward L. Peckham on a visit there in June 1857. "Chicago, the world renowned Chicago," he wrote, "is as mean a spot as I ever was in, yet."[134] It was characterized as a filthy, sooty, crime-ridden, rough-and-tumble, meatpacking, railroading town with little culture and much corruption.

To many who lived there and to some shrewd observers, however, Chicago was the city destined to set the trends of the twentieth century. Writing in 1900, the Scottish journalist William Archer exclaimed

> What a wonderful city Chicago will be! That is the ever-recurring burden of one's cognitions. For Chicago is awake, and intelligently awake, to her destinies; so much one perceives even in the reiterated complaints that she is asleep. Discontent is the condition of progress, and Chicago is not in the slightest danger of relapsing into a condition of inert self-complacency. Her sons love her, but they chasten her. They are never tired of urging her on, sometimes (it must be owned) with the most unfilial objurations; and she, a quite unwearied Titan, is bracing up her sinews for the great task of the coming century. ... Nowhere in the world, ... does the "to be continued in our next" interest take hold on one with such a compulsive grip.[135]

Although Archer's path undoubtedly had not crossed that of the Chicago mathematicians, his sense of the broader Chicago could not have captured

[133]By December 1896, three summer meetings had been held: the first in Brooklyn in 1894, the second in Springfield, Massachusetts in 1895, and the third in Buffalo, New York in 1896.

[134]Edward L. Peckham, "My Journey Out West," *Journal of American History* 17 (1923):227–230, as quoted in Bessie Louise Peirce and Joe L. Norris, ed., *As Others See Chicago: Impressions of Visitors*, 1673-1933 (Chicago: University of Chicago Press, 1933), p. 166.

[135]William Archer, *America To-Day* (London: William Heinemann, 1900), pp. 87–98, as quoted in Peirce and Norris, ed., pp. 411–412.

better their sense of the city mathematically. Moore, Bolza, and Maschke together with White and his younger colleague Thomas F. Holgate shared a strong desire to make their city, as much as New York, a focal point of the developing mathematical community. This was a group of men driven by Moore's ambition to make mathematics in the United States something over which they had influence.[136]

With still vivid memories of the Evanston Colloquium, they spearheaded a movement to establish an official Midwestern or Chicago section of the AMS. Receiving the go-ahead from the New York establishment, Moore and his colleagues sounded a "Call to a Conference in Chicago" in a circular mailed to Society members who lived within striking distance of the city. The Chicago group presented the case in this way:

> Our Society represents the organized mathematical interests of this country. Its function is to promote those interests in all possible ways.
>
> Do we not need most of all frequent meetings? Those who have attended the summer meetings know the keen stimulus and inspiration resulting from personal contact—inside and outside the stated meeting—with colleagues from other institutions. The regular monthly meetings of the Society afford similar opportunities to those who live in the vicinity of New York.
>
> By the organization of *sections* of the Society can similar advantages be secured for other parts of the country? Shall, for instance, a Chicago section be organized? Obviously only if the members of the Society residing in the vicinity of Chicago wish the section organized and are willing to support sectional meetings by attendance and by the contribution of papers. How shall the sections be related to the Society?
>
> Those members of the Society who may be interested in the consideration of these and cognate questions we invite to meet in conference in Chicago ... . The deliberations of the conference may result in recommendations to the Society or to the Council of the Society.[137]

When the two-day-long meeting convened on the morning of 31 December 1896, seventeen Society members along with a number of unaffiliated mathematicians had opted to celebrate New Year's Eve by pursuing the cause of mathematics in the Midwest.

They came from Wisconsin, Nebraska, Kansas, Ohio, Indiana, Michigan,

---

[136]This objective had also partly motivated the Chicago mathematicians in their organization of the Congress in 1893. See our article "Embedded in the Culture: Mathematics at the World's Columbian Exposition of 1893," *The Mathematical Intelligencer* 15 (2) (1993):40–45.

[137]Archibald, *Semicentennial History*, p. 75 (their emphasis).

Illinois, and even Louisiana to listen to the fourteen presented papers but, more importantly, to discuss the issues raised in Moore's circular.[138] Their deliberations translated into an addition to the Bylaws of the AMS allowing for "Sections" of the Society, and on 24 April 1897, they met officially as the "Chicago Section of the American Mathematical Society." At that meeting, the participants elected E. H. Moore as Chair of the new group, Thomas Holgate as its Secretary, and Holgate, Alexander Ziwet, and Arthur S. Hathaway as its Program Committee. The thirteen Society members in attendance on this precedent-setting occasion heard ten papers on topics ranging from the theory of matrices to the mathematical theory of fluid flow and began what would ultimately be the slow process of making the AMS an organization of truly national proportion and influence.[139]

During its first twenty-five years, in fact, the Chicago Section profited from ever-increasing levels of mathematical intensity at its gatherings. Starting out small, the Section never drew more than thirty people to its biannual sessions over the first decade of its existence, but the next fifteen years saw attendance steadily rise to an average of fifty participants per meeting. Furthermore, from 1897 to 1922, those who came heard 1,102 lectures from a total of 278 different speakers. Of these presentations, 649 or 59% eventually appeared in print in a wide variety of journals both foreign and domestic. In his capacity as final Secretary and chronicler of the Section's quarter-century history, the University of Wisconsin's Arnold Dresden put all of these statistics in a meaningful context when he expressed his feeling:

> that this aspect of our meetings [i.e., the presented papers] is perhaps most widely useful. For a mathematician to know that he can find a group of more or less informed colleagues who will at least listen to what he has to say about his own work, and perhaps comment on it with a measure of understanding of what he is working for, encourage him if he deserves it, discourage him if it be otherwise, must be for him a stimulus of which each one of us can best appreciate the value from his own experience.[140]

By providing a forum for this necessary and fruitful sort of interchange, the Chicago Section of the AMS functioned as a seedbed under the cultivation—especially during those early years—of the mathematicians at the University

[138] For a roster of those Society members in attendance as well as the schedule of lectures, see the "Notes" in the *Bulletin of the American Mathematical Society* 3 (1897):199–200.

[139] Moore served in this capacity for the five years from 1897 to 1902, and Holgate remained secretary until 1905. For the complete list of officers and their years of service, see Archibald, *Semicentennial History*, p. 76. The program of this first official meeting appears in the "Notes" of the *Bulletin of the American Mathematical Society* 3 (1897):353–354. Programs for subsequent meetings also appear in this source.

[140] Arnold Dresden, "A Report on the Scientific Work of the Chicago Section, 1897–1922," *Bulletin of the American Mathematical Society* 28 (1922):303–307 on p. 305. All of the above statistics come from this source.

of Chicago for the mathematical growth of the Midwest.

Mathematicians in other regions of the country also recognized the obvious success of the Chicago experiment and worked to create additional Sections of the AMS. In May of 1902, twenty West Coast Society members—including Irving Stringham and Mellen Haskell—met and founded a San Francisco Section. By December of 1906, Hilbert's student Earle Raymond Hedrick, the Russian-born Alexander Chessin, and Hopkins Ph.D. Ellery W. Davis, among others, had met in Columbia, Missouri and had formed a Southwestern Section initially numbering some thirty-five strong. The sustained activity of all of these sections slowly resulted in the extension of the organized mathematical community and in the diffusion of power away from the Northeastern establishment. This gradual movement from strict centralization was sanctioned in 1929 when the AMS began to recognize sectional meetings as regular meetings of the Society.[141] Thus, only in 1929 did geographical snobbery end officially—if not always in practice—in the AMS.

As this late date suggests, regional tensions persisted within the Society in spite of the various sectional initiatives. As early as 1911, they came dramatically to a head over the site for the national meeting at which Harvard's Maxime Bôcher would give his presidential address. "At that time," according to John Hasbrouck Van Vleck, son of then AMS Council Member E. B. Van Vleck,

> there was bitter feeling, especially by the Chicago group, that no national, rather than sectional, meetings were held west of the Alleghenies. ... [T]he Harvard group was particularly adamant: Osgood, I remember father said, claimed he suffered from insomnia if he left Cambridge. ... Cole, the secretary of the Society, was apparently willing to settle for a plan to more or less Balkanize it, whereas my father believed that the organization should be national in scope.[142]

Fully cognizant of the depth of the ill feelings of his Midwestern colleagues, E. B. Van Vleck secured Cole's permission to appeal directly to Bôcher in the matter of scheduling. On 27 February 1911, Bôcher informed Van Vleck of his decision: "I wrote this morning to Cole ... saying I would give the presidential address in Chicago. In doing this I yield my own judgment, which tells me that no useful purpose will be accomplished, to yours. You are on the

[141] For the full reports of these first meetings, see Ernest J. Wilczynski, "The First Meeting of the San Francisco Section of the American Mathematical Society," *Bulletin of the American Mathematical Society* 8 (1902):429–437; and Alexander S. Chessin, "The Preliminary Meeting of the Southwestern Section," *Bulletin of the American Mathematical Society* 13 (1907):213–223. For complete lists of the officers of the San Francisco and Southwestern Sections, see Archibald, *Semicentennial History*, pp. 8–9.

[142] John Hasbrouck Van Vleck, "Edward Burr Van Vleck, My Father," Program from the Dedication Dinner for Edward Burr Van Vleck Hall, University of Wisconsin, 13 May 1963, pp. 19–28 on p. 20.

ground, and ought to know."[143] The national meeting took place in Chicago; an important precedent had been set. Recalling this incident later in life, E. B. Van Vleck confided to his son "that he believed he [had] averted an open break in the American Mathematical Society."[144] In view of the key role that particularly the Chicago group played in the subsequent development of the American mathematical research community, their alienation at this stage could definitely have retarded the process.

As in the movement toward true nationalization of the AMS, the Chicagoans also formed the vanguard of the movement to provide better publication opportunities for those conducting original mathematical research in the United States. By the close of the nineteenth century, aspiring contributors could effectively choose from only four American journals when submitting their work: the *American Journal of Mathematics*, which, as we have seen, was founded at the Johns Hopkins University by J. J. Sylvester in 1878; the *Annals of Mathematics*, which Ormond Stone started at his home base at the University of Virginia in 1884; the *Bulletin of the AMS* begun in 1891 by Thomas Fiske as the *Bulletin of the NYMS*; and the *American Mathematical Monthly*, which débuted in 1894 under the joint editorship of Benjamin F. Finkel of Drury College in Springfield, Missouri (from 1895–1937) and John M. Colaw, a high school teacher in Monterey, Virginia.[145] Despite this seemingly ample number of publication outlets, the growing community of American mathematicians still sensed an unfulfilled need. None of these four journals, it was held, specialized in first-rate, original research from American contributors.

While the avowedly *American Journal*, the oldest and most venerable mathematical publication in the United States, had maintained the high standards of publishable research established by Sylvester between 1878 and 1884, its subsequent editorial staff under mathematical astronomer Simon Newcomb had also upheld Sylvester's European bias. In 1878 when the journal first appeared in print, this bias was not at all unreasonable. As mentioned in Chapter 2, Sylvester had feared that, initially at least, high-level contributions from the United States would not come in numbers sufficient to sustain the venture. For this reason, he actively solicited papers from abroad as well as from the members of his own department to assure that the issues had the requisite amount of material. Sylvester also realized fully that part of the mission of a research-level mathematics journal in America at that time was to provide American mathematicians with ready access to the latest developments abroad. This would then facilitate their entry into what was considered the mainstream of mathematical research by allowing them to

[143] *Ibid.*, p. 21.

[144] *Ibid.*, p. 20.

[145] Recall Table 1.2. Finkel earned a B.S. from Ohio Northern in 1888 and an A. M. and Ph.D. from the University of Pennsylvania in 1904 and 1906, respectively. Colaw took his A. B. and A. M. from Dickinson College in 1882 and 1892, respectively.

isolate interesting and timely research problems. Moreover, a relatively high percentage of distinguished foreign contributors would tend to give credibility to the American mathematical enterprise and would serve to introduce mathematicians abroad to the work of their counterparts on the other side of the Atlantic. In the six volumes issued under Sylvester's charge, foreign contributions comprised on average 25% of the papers published. What characterized the *American Journal* even more than its foreign contributors, however, was its high percentage of Hopkins or Hopkins-connected authors. Also during Sylvester's six years at the journal's helm, an average of 45% of each volume came from those who taught at, were Fellows of, or had received Ph.D.s from the Johns Hopkins University. With only thirty percent of the journal's pages available for other American work, the *American Journal* did not provide a particularly rich publication source.

By contrast, the *Annals of Mathematics* published only three papers by foreign contributors during its first fifteen years of existence. With each of its issues under Stone's direction featuring a section of problems and solutions, it, like the *American Mathematical Monthly*, catered to a more popular audience than the purely research-oriented *American Journal.* Furthermore, the papers which did appear were, for the most part, derivative, of only moderate depth, and far from the cutting edge of mathematical research. The latter characterization gradually began to change during the 1890s as the Chicago mathematicians, together with those fresh from studies abroad, opted to use the journal as an outlet for their new ideas. In 1898–1899, for example, the *Annals* served as the American mathematical community's first introduction to the dissertation research on the general linear group Leonard Dickson would subsequently expand into his book *Linear Groups.* Although this trend toward high quality, original research and away from recreational problem-solving continued after the *Annals* moved to Harvard in 1899 and was complete by the time it made its final move to Princeton in 1911, the leading mathematicians of the late 1890s felt that their Society should officially support research by sponsoring its own research-level journal. The *Bulletin*—intended, as stated on its title page, as an "Historical and Critical Review of Mathematical Science"—simply did not suffice. By 1898 moves were afoot to get AMS control over a high-class publication, and the Society had its sights set on the by then financially troubled *American Journal.*

In August of 1898, the Council mandated the appointment of a committee to draft a plan of cooperation between the AMS and the Johns Hopkins University. The committee, with Thomas Fiske presiding over E. H. Moore, Simon Newcomb, Maxime Bôcher, and James Pierpont, made its written recommendation to the Council in October and went ahead with its negotiations with the university. Since committee member Newcomb was President of the Society as well as an editor of the *American Journal* and a distinguished member of the Hopkins faculty, these negotiations should have proceeded effortlessly. Contrary to first appearances, though, the University and the Society ultimately failed to come to terms. As William Osgood later recalled the chain of events, "Professor Newcomb said that any plan which should place the acceptance of papers in the hands of a board of editors not controlled by The Johns Hopkins, would be unacceptable, and that the appearance of the name of the American Mathematical Society on the title page, in any form, would be distasteful. ... But the question of a journal refused to retire into the background."[146] By February of 1899, the AMS had resolved to publish a wholly new journal which would meet the needs of its constituency. To be called the *Transactions of the AMS*, it would officially record that mathematical research which had been presented before a sanctioned meeting of the Society and would, symbolically at least, compete less directly with the *American Journal.*[147] Later in the year, Moore, Fiske, and Ernest W. Brown of Haverford[148] accepted appointments as the *Transactions*' first Editors, and Moore was elected Editor-in-Chief.

Under Moore's direction, the *Transactions of the AMS* developed into the kind of journal its proponents had envisioned. In view of its transactional nature, it could only accept those articles which had previously been read (in person or *in absentia*) before a meeting of the Society or one of its Sections. Operating under this constraint, it would favor neither foreign contributors nor authors from or associated with Moore's own University of Chicago. Unlike Sylvester's *American Journal,* then, Moore's *Transactions* accepted only slightly more than 12% of its papers from foreign sources, somewhat more than 22% from the Chicago contingent, and fully 65% from other American authors.[149] Like Sylvester, though, Moore set and maintained publication standards which produced volumes of consistently high quality. Ernest

---

[146]Archibald, *Semicentennial History*, p. 58. For Newcomb's side of the story, see Simon Newcomb to Thomas S. Fiske, 23 January 1899, quoted in *ibid.*, p. 57.

[147]This last issue had particularly concerned some of the older members of the Society, like Emory McClintock, who had helped to insure the success of the *American Journal* through their faithful contributions and who did not want to see its viability jeopardized. They felt that starting a new *journal* would represent a hostile move toward the venerable Baltimore publication but ultimately agreed that any society had the right to publish its *transactions.* As William Osgood recounted the debate, "[w]ith that formula, the one word 'Transactions' replacing an opprobrious word like 'Journal' or 'Annals' or 'Acta,' the whole opposition collapsed and good relations were established. The power of a word!" For the quotation, see Archibald, *Semicentennial History*, p. 58.

[148]In Parshall, "E. H. Moore and the Founding of a Mathematical Community in America," p. 526, Brown's academic affiliation was incorrectly given as Yale University. While Brown did serve actively on the Yale faculty from 1907 to 1932, he was at Haverford College from 1891–1907, and so had a Haverford affiliation at the time of his appointment to an editorship of the *Transactions.* Brown served as the President of the AMS from 1915 to 1916.

[149]These figures are based on data taken from the first eight volumes of the *Transactions*, under Moore as Editor-in-Chief. The relative percentages of foreign to American contributors remained fairly constant over the first thirty volumes of the journal (1900–1928). The statistics as compiled by AMS Secretary (from 1921 to 1940), Roland G. D. Richardson, showed a total of 322 contributors of which thirty-five, or slightly more than 10%, were foreign. See R. G. D. Richardson to members of the AMS in UC Archives, E. H. Moore Papers, Box 1, Folder 2.

Brown reflected on this aspect of their early editorial work together, when he described "the immense amount of trouble we all took—and especially Moore—to get the best information, the best printing, the best editing and the best papers before the first number appeared. And the work did not stop there. We wrestled with our younger contributors to try to get them to put their ideas into good form. The refereeing was a very serious business .... Most of it in those days was, I believe, done by Moore himself though he sought outside assistance whenever possible."[150]

The publication which resulted from these efforts contained the work of such notable foreign mathematicians as Paul Gordan, Édouard Goursat, David Hilbert, Jacques Hadamard, Maurice Fréchet, and Henri Poincaré and such talented Americans as Leonard Dickson, Oswald Veblen, Maxime Bôcher, George Hill, Max Mason, and Luther Eisenhart. Whether or not reflective of a conscious editorial decision, the early issues of the *Transactions* showcased the best American work against the research of some of the best Europeans and in so doing demonstrated not only the potentiality but also the actuality of the growing American mathematical research community. The readiness of these and other prominent Europeans to submit their work to the American Mathematical Society also indicated that Europe was coming to recognize the United States as a legitimate member-nation in the world of mathematics.[151]

As Editor of the *Transactions* E. H. Moore by no means effected such attitudinal changes single-handedly. It is nevertheless undeniable that his capable leadership of this new publication enterprise as well as his role in the steady improvement of the American mathematical community in the 1890s did much to foster them both at home and abroad. Dramatic changes were clearly taking place. From September 1894 to January 1901, the AMS's membership had increased by one third from 244 to 357. The number of papers read annually before the Society had multiplied almost fivefold from 24 to 112 in the same time period. This last increase was due in large part to the establishment of the Chicago Section in 1897 with its regular schedule of talk-filled meetings and to the institution of the AMS summer meetings in 1896. Finally, between 1891 and 1900, the Society had created two successful journals and had subsidized the publication of the Chicago Congress volume.[152]

---

[150]Archibald, *Semicentennial History*, p. 60.

[151]During the fifteen years from 1891–1906, over one thousand people participated in the American mathematical research community. Of these over 10% were foreign. For these and other data descriptive of the early American mathematical research community, see Della Dumbaugh Fenster and Karen Hunger Parshall, "A Profile of the American Mathematical Research Community: 1891–1906," and "Women in the American Mathematical Research Community," both in *The History of Modern Mathematics*, vol. 3, ed. Eberhard Knobloch and David E. Rowe (Boston: Academic Press, Inc., 1994), to appear.

[152]For these statistics, see F. N. Cole, "The Seventh Annual Meeting of the American Mathematical Society," *Bulletin of the American Mathematical Society* 7 (1901):199–210 on pp. 199–200.

In recognition of his accomplishments, the AMS membership elected E. H. Moore to serve as its sixth President from January 1901 through December 1902.

At the time of his election, Moore was only thirty-eight, twelve years younger than any of his predecessors. He was also the first pure mathematician to hold the office, but he was certainly no stranger to the inner sanctum of AMS politics. Moore came to the presidency on the heels of three years in the vice presidency (1898–1900), and had found himself thrust into the thick of the AMS machinery by his editorial duties on behalf of the *Transactions*. He understood how the organization worked, and he succeeded in keeping it running smoothly. During his two years in office, the membership grew by another 10% to 399, and the amount of mathematics presented at Society meetings jumped by almost 40% from 112 papers in 1900 to 156 in 1902.[153] Of greater importance than these statistics, Moore had loosened the Northeastern hold on American mathematics by becoming the Society's first Midwestern President. His five predecessors—John van Amringe, Emory McClintock, George Hill, Simon Newcomb, and Robert Woodward—together with the indefatigable F. N. Cole and T. S. Fiske, had also already built a firm foundation for the Society and its publications during their successive regimes, so he was free to use his term in office in the pursuit of potentially more far-reaching goals.[154] Rather than concentrating on the mechanics of building the Society, Moore took the opportunity provided by his national office to champion the advancement of mathematics education at all levels of the curriculum nationwide.

Like Klein in his draft to the Prussian Ministry of Education in 1893 on the state of mathematics education in the United States, Moore had realized that the American educational system needed reform at the lower levels in order to facilitate mathematical education at the research level. As early as 1894, in fact, he had encouraged active discussion of and fresh research on mathematical pedagogy at the University of Chicago. During the Summer Quarter of that year, Chicago's Mathematics Club sponsored a series of lectures "On Teaching Mathematics" given by Ernest Skinner, then an instructor at Wisconsin; Henry Benner, a teacher at the Chicago Manual Training School; T. M. Blakslee of Des Moines College; and Chicago mathematics graduate student Herbert Slaught. By 1897, the University had promoted Jacob Young to an assistant professorship specifically in mathematical pedagogy. He then lectured regularly before the Club on such topics as "The Presentation of the First Principles of the Calculus" in an effort to help his colleagues improve their teaching effectiveness.[155] More generally, the Chicago Section of

[153]F. N. Cole, "The Ninth Annual Meeting of the American Mathematical Society," *Bulletin of the American Mathematical Society* 9 (1903):281–295 on p. 282.

[154]For biographical information on these early Presidents, see Archibald, *Semicentennial History*, pp. 110–144.

[155]See UC Archives, Math. Club Records, Box 1, Folders 1 and 7.

the AMS became increasingly involved in issues of mathematics education during Moore's tenure in the chair from 1897 through March of 1902. At his instigation, for example, a three-member committee was formed at the Section's winter meeting in January 1902 to settle on a uniform standard for colleges and universities to follow in granting the Master's degree in mathematics.[156] Quite naturally, Moore's efforts in connection with the Chicago Section carried over into his concurrent vice presidency and then presidency of the AMS.

During the closing months of his vice presidency, Moore participated in a session on the undergraduate mathematics curriculum at the summer meeting in 1900. There, the question which he addressed along with James Harkness of Bryn Mawr, Osgood of Harvard, Morley by then at Hopkins, and his own Chicago colleague Young, was "What courses in mathematics shall be offered to the student who desires to devote one-half, one-third, or one-fourth of his undergraduate time to preparation for graduate work in mathematics?"[157] Representing at least a certain segment of the AMS, these speakers recognized that strong graduate students came from well-prepared undergraduates.[158] In their view, the Society needed to tackle more than the problem of organizing and governing established mathematicians in order to assure the continual growth of American mathematics. It needed to widen its focus by devising a strategy for promoting mathematics at various levels of the curriculum. In the summer of 1900, the undergraduate curriculum defined the principal area of educational concern. Two years later at the ninth summer meeting held in Evanston, Illinois, then President Moore directed his Society to consider college entrance standards. On this occasion, he appointed a committee of five "to report on standard definitions of requirements in mathematical subjects for admission to colleges and scientific schools."[159] Master's requirements, undergraduate major requirements, college entrance requirements, where should the official interests of the AMS stop in this educational hierarchy? Should it continue in this decreasing progression to

---

[156]The members of the committee were Clarence Waldo of Purdue University, Edgar Townsend of the University of Illinois, and Oskar Bolza. Their findings appeared in their "Report on the Requirements for the Master's Degree," *Bulletin of the American Mathematical Society* 10 (1904):380-385.

[157]W. H. Maltbie, "The Undergraduate Mathematical Curriculum: Report of the Discussion at the Seventh Summer Meeting of the American Mathematical Society," *Bulletin of the American Mathematical Society* 7 (1900):14–24. Moore's abstract appeared on pp. 15–16.

[158]Furthermore, as John Servos has cogently argued, adequate preparation at all levels of mathematics was crucial for the development of the other sciences—in addition to mathematics —in the United States. See his paper, "Mathematics and the Physical Sciences in America, 1880–1930," *Isis* 77 (1986):611–629.

[159]See Edward Kasner, "The Ninth Summer Meeting of the American Mathematical Society," *Bulletin of the American Mathematical Society* 9 (1902):73–94; and H. W. Tyler, T. S. Fiske. W. F. Osgood, Alexander Ziwet, and J. W. A. Young, "Report of the Committee of the American Mathematical Society on Definitions of College Entrance Requirements in Mathematics," *Bulletin of the American Mathematical Society* 10 (1903):74–77.

address the problems of high school and then elementary school mathematics curricula? Just how far should its efforts toward standardization extend? E. H. Moore used his presidential address in December of 1902 to make his opinions known on precisely these questions.

In his lecture "On the Foundations of Mathematics," Moore exploited the *double entendre* of his title to compare and contrast the turn-of-the-twentieth-century passion for axiomatization with the actual acquisition of mathematical knowledge. As Moore saw it, works like Hilbert's *Grundlagen der Geometrie* had come to set the tone of mathematical research. While he found this research intensely interesting—even going so far, as we have seen, as to run a seminar on the subject to encourage his graduate students to undertake similar studies—Moore wondered "whether the abstract mathematicians in making precise the metes and bounds of logic and the special deductive sciences are not losing sight of the evolutionary character of all life-processes."[160]

The establishment of a field like geometry upon a consistent and non-redundant set of axioms and undefined terms created a theoretically and æsthetically pleasing research framework, but it masked the trial-and-error element of experimentation inherent in the discipline.[161] More often than not, experience, not some mere logical deduction, leads the mathematician to the statement of what the theorem should be. Examples are studied. Conjectures are tested. A theorem is proposed and, if valid, is logically deducible within the axiomatic setting. In Moore's view:

> In the ultimate analysis for any epoch, we have general logic, the mathematical sciences, that is, all special formally and abstractly deductive self-consistent sciences, and the natural sciences, which are inductive and informally deductive. While this classification may be satisfactory as an ideal one, it fails to recognize the fact that in mathematical research one by no means confines himself to processes which are mathematical according to this definition; and if this is true with respect to the research of professional mathematicians, how much more is it true with respect to the study, which should throughout be conducted in the spirit of research, on the part of students of mathematics in the elementary schools and colleges and universities.[162]

---

[160] E. H. Moore, "On the Foundations of Mathematics," *Science* 17 (1903):401–416 on p. 403.

[161] Recall that Sylvester, too, had highlighted precisely this aspect of mathematical research in his Exeter address of 1869 and through his teaching style at Hopkins in the 1870s and 1880s. Klein also often expressed similar views. See, for example, the discussion of his sixth Evanston Colloquium lecture in Chapter 8. Moore's ideas were undoubtedly influenced by John Dewey, who served on the Chicago faculty from 1894 to 1904. Compare, for example, John Dewey, *Lectures in the Philosophy of Education: 1899*, ed. Reginald D. Archimbault (New York: Random House, 1966).

[162] Moore, "On the Foundations of Mathematics," p. 404.

Obviously, he thought a more experiential approach was possible, and he proposed to accomplish it through the widespread implementation of the so-called laboratory method of teaching championed by his one-time Chicago colleague, John Dewey.[163] Without going into the details of Dewey's educational philosophy, suffice it to say that he argued for "learning by doing." He held that children learned best when ideas were presented to them in a concrete, experientially oriented form. Moore contended that this idea could be applied effectively to mathematics teaching not merely at the elementary level but at the secondary and university levels as well. In the high schools, it would involve a complete integration of algebra, geometry, and physics, teaching them essentially as one, undifferentiated subject aimed at the solution of practical problems. At the university level, the usual lecture method of presenting theorems and proofs would also be motivated by experience:

> The teacher should lead up to an important theorem gradually in such a way that the precise meaning of the statement in question, and further the practical—i.e., computational or graphical or experimental—truth of the theorem is fully appreciated; and, furthermore, the importance of the theorem is understood, and, indeed, the desire for the formal proof of the proposition is awakened, before the formal proof itself is developed. Indeed, in most cases, much of the proof should be secured by the research work of the students themselves.[164]

In his presidential address, Moore called on the AMS to work vigorously for the adoption of these pedagogical reforms at all levels of the curriculum, and he urged it to broaden its membership base to include "the strongest teachers in the secondary schools."[165]

Although the Society continued to take a tangential interest in general educational issues after Moore stepped down from the presidency in 1902, it never embraced these causes to the extent he had hoped for, and it never actively solicited high school teachers for membership. In Chicago, however, Moore himself implemented the laboratory method with some limited success in his own department during the 1903–1904 academic year.[166] His commitment to mathematical pedagogy also encouraged one of his students, Herbert Slaught, to pursue his own interests along those same lines. (Of course, another of Moore's students, R. L. Moore, subsequently became famous—or infamous—for his own novel teaching methods. See Chapter 10.)

In 1907, Slaught replaced Leonard Dickson on the editorial staff of the

[163] This, of course, is not to be confused with what we termed Sylvester's "laboratory" in Chapters 2 and 3 above. Dewey conceived of his laboratory method as part of a highly articulated philosophy of education at all levels and in all areas, while Sylvester "directed" his laboratory only in the more limited graduate-level setting. In the final analysis, though, Sylvester and Dewey did share some of the same ideas.

[164] Moore, "On the Foundations of Mathematics," pp. 412–413.

[165] *Ibid.*, p. 414. Membership, however, was essentially open to all who paid their dues.

[166] Bliss, "E. H. Moore," p. 834.

American Mathematical Monthly and propounded his pedagogical ideas there in articles like "The Teaching of Mathematics in the College."[167] By 1914, he had revived Moore's idea of expanding the AMS to meet the needs of secondary school instructors and had instigated a study into the matter by the Chicago Section. When the committee put forth the proposition in December of that year that the AMS assume the sponsorship of the *Monthly* to reach this constituency, the proposal met with the Section's unanimous approval. In spite of this support, the AMS Council resolved in April 1915 that "[i]t is deemed unwise for the American Mathematical Society to enter into the activities of the special field now covered by the *American Mathematical Monthly*; but the Council desires to express its realization of the importance of the work in this field and its value to mathematical science, and to say that should an organization be formed to deal specifically with this work the Society would entertain toward such an organization only feelings of hearty good will and encouragement."[168] By the end of 1915, Slaught had laid the foundation for just such an organization, the Mathematical Association of America, and had partially filled the need Moore had sensed as early as 1902. The AMS's reluctance to widen its purview to include mathematics instructors at the secondary level clearly underscored the emphasis it placed on research—as opposed to pedagogy—in its self-definition as well as the underlying and growing tension between teaching and research.[169]

Through their ability to train a next generation of mathematicians, through their own mathematical accomplishments, and through their more broadly directed interests, the Chicago mathematicians under the leadership of E. H. Moore played a singular role in the establishment of a strong and productive mathematical research community in the United States. Archibald summed up their contributions well in his *Semicentennial History* when he wrote that "[t]he Chicago group (including those from Urbana, Ann Arbor, Madison, and Minneapolis, who were also members of the Section) founded the Mathematical Association of America; originated the ideas of Colloquium, Section, and Symposium; hurried up the changing of the Society's name; and contributed four presidents to the Society, six of the nine managing editors for the *Transactions*, and ten of the thirty-three colloquium lecturers."[170] "Perhaps," as he concluded, "enough has been indicated, faintly to suggest how vital and far-reaching have been the contributions of members of the Chicago group, in the development of the Society to its present position of eminence and in the establishment of a basis for the American Mathematical School of the present day."[171]

[167]Herbert Slaught, "The Teaching of Mathematics in the Colleges," *American Mathematical Monthly* 16 (1909):173–177.

[168]See W. D. Cairns, "Herbert Ellsworth Slaught—Editor and Organizer," *American Mathematical Monthly* 45 (1938):1–4 on p. 2.

[169]This same tension also manifested itself in the physics community of the early twentieth century. See Daniel J. Kevles, *The Physicists: The History of a Scientific Community in Modern America* (Cambridge, MA: Harvard University Press, 1971), pp. 34–36, 71–72, and 181–182.

[170]Archibald, *Semicentennial History*, p. 81.

[171]*Ibid.*

TABLE 9.1
MATHEMATICAL CLUB LECTURES: WINTER AND SPRING 1893

Meeting 1—5 January 1893:

Moore—"Statement of the Character and Purposes of the Organization"
Moore—"Cremona: A Figure in Space from Which the Properties of Pascal's Hexagon in the Plane Are Easily Deducible"

Meeting 2—12 January 1893:

Maschke—"The Complete Form-System of the Hessian Group of Ternary Linear Homogeneous Substitutions"
Hancock—"Note on the Divisibility of Numbers"

Meeting 3—19 January 1893:

Young—"On Hölder's Enumeration of All Simple Groups Whose Order Is Not Greater Than 200"
Moore—"An Existence-Proof of the Group of Order 168 as a Group of Substitutions on 7 Letters"

Meeting 4—2 February 1983:

Bolza—"Weierstrass, Zur Theorie der aus n Haupteinheiten gebildeten complexen Grössen"
Maschke—"A Remark of *Eisenstein* on Invariants"
Young—"*Hölder's* Proof That a Simple Group of Order 180 Does Not Exist"

Meeting 5—9 February 1893:

Hancock—"*Fuchs'* Normal Form for Linear Differential Equations of Second Order All of Whose Integrals Are Regular"
Moore—"A Note on the Theory of Numbers"

Meeting 6—16 February 1893:

Hutchinson—"The Transformation of Hyperelliptic Integrals to Elliptic Integrals"
Young—"*Kronecker's* Determination of All Commutative Groups"

Meeting 7—23 February 1893:

Winston—"A Theorem Concerning Linear Differential Equations with Constant Coefficients"
Hancock—"*Fermat's* Theorem"

Meeting 8—2 March 1893:

Bolza—"Gamma Functions of a Complex Variable"

Meeting 9—9 March 1893:

Moore—"Galois' Theory of Imaginaries in the Theory of Numbers"
Bolza—"Gamma Functions of a Complex Variable (Second Paper)"

Meeting 10—16 March 1893:

Maschke—"A Ternary Algebraic Problem"

Meeting 11—23 March 1893:

T. J. J. See (Department of Astronomy)—"The Secular Action of Tidal Friction"

Meeting 12—6 April 1893:

Bolza—"An Invariantive Problem in the Theory of Linear Differential Eq[uation]s"

Meeting 13—13 April 1893:

Hancock—"Concerning a Linear Differential Equation Which Is Connected with the Hyperelliptic Integral of the First Order & of the First Kind"
Moore—"Netto: Zur Theorie der Tripelsysteme"

Meeting 14—20 April 1893:

Smith—"A Study of the Equilateral Hyperbola Based on a Correspondence Between the Ellipse and Hyperbola with the Same Axis"
Moore—"A Triple System in 25 Elements"

Meeting 15—27 April 1893:

Moore—"A Method of Construction of Triple Systems in Any Number t of Elements, t Being of the Form $6m + 1$ or $6m + 3$

Meeting 16—4 May 1893:

Maschke—"Paper on the Historical Development of the Theory of Fourier Series"

Meeting 17— 11 May 1893:

Hutchinson—"The Transformation of Hyperelliptic Integrals into Elliptic Integrals (Second Paper)"

Meeting 18—18 May 1893:

Young—"On the Determination of Groups Whose Order Is a Power of a Prime"

Meeting 19—25 May 1893:

Heller—"The Fundamental Theorem of Algebra: Gauss' First Proof"
Bolza—"Sketches of the Proofs by Gauss (the Third), Argand, Cauchy, Weierstrass"

Meeting 20—1 June 1893:

Moore—"Concerning Binary Matrices"

Meeting 21—8 June 1893:

H. S. White (Northwestern)—"Hesse's Enumeration of the Bitangents of the Plane Quartic Curve"

From the University of Chicago Archives, Mathematical Club Records 1893–1913, Box 1, Folder 1.

TABLE 9.2
E. H. MOORE'S EARLY CHICAGO STUDENTS

1. Leonard Eugene Dickson (1896)—"The Analytical Representation of Substitutions on a Power of a Prime Number of Letters; with a Discussion of the Linear Group"

2. Herbert Ellsworth Slaught (1898)—"The Cross-Ratio Group of 120 Cremona Transformations of the Plane"

3. Derrick Norman Lehmer (1900)—"Asymptotic Evaluation of Certain Totient Sums"

4. William Findlay (1901)—"The Sylow Subgroups of the Symmetric Group on K Letters"

5. Oswald Veblen (1903)—"A System of Axioms for Geometry"

6. Thomas Emery McKinney (1905)—"Concerning a Certain Type of Continued Fractions Depending Upon a Variable Parameter"

7. Robert Lee Moore (1905 and with Oswald Veblen)—"Sets of Metrical Hypotheses for Geometry"

8. George David Birkhoff (1907)—"Asymptotic Properties of Certain Ordinary Differential Equations with Applications to Boundary Value and Expansion Problems"

9. Nels Johann Lennes (1907)—"Curves in Non-Metrical Analysis Situs, with Applications to the Calculus of Variations and Differential Equations"

10. Frederick William Owens (1907)—"The Introduction of Ideal Elements and Construction of Projective n-space in Terms of a Plane System of Points Involving Order and Desargues's Theorem"

11. Harry Franklin McNeish (1909)—"Linear Polars of the k-hedron in n-space"

12. R. P. Baker (1910)—"The Problem of the Angle Bisectors"

TABLE 9.3
DICKSON'S EARLY CHICAGO STUDENTS

1. Thomas Milton Putnam (1901)—"Concerning the Linear Fractional Group on Three Variables with Coefficients in the Galois Field of Order p^n"

2. William Henry Bussey (1904)—"Generational Relations for the Abstract Group Simply Isomorphic with the Linear Fractional Group"

3. Herbert Edwin Jordan (1904)—"Group Characters of Various Types of Linear Groups"

4. Arthur Ranum (1907)—"A New Kind of Congruence Groups"

5. R. L. Börger (1908)—"On the Determination of Ternary Linear Groups in the Galois Field of Order p^2"

TABLE 9.4
BOLZA'S CHICAGO STUDENTS

1. John Irwin Hutchinson (1896)—"On the Reduction of a Hyperelliptic Function to Elliptic Functions by a Transformation of the Second Degree"

2. Gilbert Ames Bliss (1900)—"The Geodesic Lines on the Anchor Ring"

3. William Gillespie (1900)—"On the Reduction of Hyper-Elliptic Integrals ($p = 3$) to Elliptic Integrals by Transformation of the Second and Third Degrees"

4. John Hector McDonald (1900)—"Concerning the System of the Binary Cubic and Quadratic, with Application to the Reduction of Hyperelliptic Integrals by a Transformation of Order 4"

5. Buz M. Walker (1906)—"On the Resolutions of Higher Singularities of Algebraic Curves in Ordinary Double Points"

6. Anthony Lispenard Underhill (1906)—"Invariants Under Point Transformations in the Calculus of Variations"

7. Mary Emily Sinclair (1908)—"On a Compound Discontinuous Solution Connected with the Surface of Revolution of a Minimum Area"

8. Norman Richard Wilson (1908)—"Isoperimetric Problems Which Are Reducible to Non-Isoperimetric Problems"

9. Arnold Dresden (1909)—"The Second Derivatives of the External Integral"

TABLE 9.5
MASCHKE'S CHICAGO STUDENTS

1. John Anthony Miller (1899)—"Concerning Certain Elliptic Modular Functions of Square Rank"

2. George Lincoln Brown (1900)—"Ternary Linear Transformation Group $G3360$ and the Complete Invariant System"

3. Ernest Brown Skinner (1900)—"On Ternary Monomial Substitution Groups of Finite Order with Determinant $+1$"

4. A. W. Smith (1905)—"The Symbolic Treatment of Differential Geometry"

5. Louis Ingold (1908)—"Vector Interpretation of Symbolic Differential Parameters"

Chapter 10
Epilogue: Beyond the Threshold: The American Mathematical Research Community, 1900–1933

In surveying American mathematics during the hundred years that followed the signing of the Declaration of Independence, Judith Grabiner suggested that this earlier epoch had made possible the outburst of mathematical activity in the United States that commenced after 1876. Summarizing these developments, Grabiner wrote that

> this explosion in American mathematics was not a creation out of nothing, not a sudden flowering out of previously barren soil. Its roots lie in the influx of French mathematics teaching in the 1820s; it was nurtured by government support for applied mathematics throughout the century, and by the increase in science education which began in the 1850s; and it came to fruition in the universities of the 1870s, 1880s, and 1890s. We may, then, proudly exhibit the institutions and the people that produced the flowering of mathematics at the end of the nineteenth century as the major achievement of American mathematics in its first hundred years.[1]

While this assessment is certainly accurate in substance, it clearly stresses the continuity of the events before and after 1876. As should now be evident, however, the present study centers on the quarter-century from 1876 to 1900 as a crucial *period* in the history of research mathematics in the United States. In so doing, it emphasizes periodization—not continuity—in an effort to provide a more meaningful and potent historical analysis.

Arguably, four reasonably distinct and well-defined periods have shaped mathematics in the United States thus far: the hundred years from 1776 to 1876, during which mathematics developed not separately but within the

[1]Judith V. Grabiner, "Mathematics in America: The First Hundred Years," in *The Bicentennial Tribute to American Mathematics*, 1776–1976, ed. Dalton Tarwater (n.p.: The Mathematical Association of America, 1977), pp. 9–24 on pp. 22–23.

context of the general structure-building of American—as opposed to colonial—science as a whole; the quarter-century from 1876 to 1900, which witnessed the emergence of the true research community; the first third of the twentieth century, during which the institutions and research traditions largely established in the previous era consolidated and grew; and the period from 1933 to roughly 1960, which saw the massive influx of European mathematicians beginning in the mid-1930s, the development of various areas of applied mathematics, and the institutionalization of large-scale governmental funding of basic research both during and after the Second World War.[2] By detailing the distinguishing features of the second of these periods, the present study has implicitly delimited parameters that define the first and third periods as well.

As documented in Chapter 1, developments within American higher education and within the broader context of American science significantly affected American mathematics during the period prior to 1876. Americans could point to the work of Nathaniel Bowditch; the scientific accomplishments at the United States Coast Survey of Superintendents Ferdinand Hassler, Alexander Dallas Bache, and Benjamin Peirce; the influence of engineering mathematics at West Point with its curriculum modeled on that of the *École polytechnique*; and the astronomical research of Simon Newcomb and George William Hill. However, few of the achievements associated with these institutions and individuals had any lasting effect upon the generation that followed in the last quarter of the nineteenth century. Quite simply, prior to 1876, nothing even remotely resembling a mathematical research community existed in the United States, nor did the time appear ripe for its imminent emergence. Rather, the century from 1776 to 1876 witnessed the formation of an American scientific profession intimately related to the country's institutions of higher education and to the federal government. While general scientific societies and their publications, like the American Academy of Arts and Sciences and its *Proceedings*, served to define the scientific community, critical numbers of practitioners did not exist in most fields to sustain specialized societies or journals. Colleges broke from the confines of the colonial era by expanding their faculties with scientists and their curricula within the sciences. Concomitantly, the traditional mathematics curriculum, which had largely been restricted to the first few books of Euclid's *Elements*, incorporated pedagogical innovations issuing mostly from France and began

[2]For information on the activities of the AMS during this latter period, consult Everett Pitcher, *A History of the Second Fifty Years: American Mathematical Society* 1939–1988, American Mathematical Society Centennial Publications, vol. 1 (Providence: American Mathematical Society, 1988). As noted in the preface (see note 13), Thomas Fiske suggested a different periodization of the history of American mathematics in his paper, "Mathematical Progress in America," *Bulletin of the American Mathematical Society* 11 (1905):238–246, reprinted in *A Century of Mathematics in America—Part* I, ed. Peter Duren *et al.* (Providence: American Mathematical Society, 1988), pp. 3–11.

to include the calculus, among other topics. These changes within higher education, however, did not imply the existence of institutional support for or an encouragement of basic scientific research. In fact, the general lack of support for research within the institutions of higher education fundamentally distinguishes the periods before and immediately after 1876.

Prior to 1876, when it was fostered at all, research was promoted primarily within the federal government—in agencies like the Coast Survey and the Nautical Almanac Office—but only exceptionally within the colleges. As a result, the kind of research done before 1876 had, by and large, an applied flavor. The impressive accomplishments of Hill and others notwithstanding, American mathematics, as it unfolded after 1876, had little in common with the research "tradition" of the previous era. The next generation—associated with institutions of higher education in which departmental structures discouraged the kind of cooperation generally needed for applied research—focused its attention almost exclusively on the pure side of the mathematical spectrum, rather than pursuing celestial mechanics as had Bowditch, Benjamin Peirce, and Hill. Moreover, its leading figures reinforced their mathematical predilections by forging a viable community during this period, which successfully incorporated research-level mathematics into the intellectual fabric of the country. Three factors, which had been altogether absent prior to 1876, decisively influenced this community's emergence.

First, the founding of research-oriented universities—beginning with the Johns Hopkins in 1876, followed by Clark University in 1889, and culminating with the University of Chicago in 1892—reflected a fundamental change in the American academic climate. Within these new institutions, research in pure mathematics as well as other fields attained a degree of credibility and support that beforehand had been lacking. Moreover, the existence and example of these schools prompted some older institutions—Harvard, Columbia, Yale, and Princeton—to develop viable graduate programs during this era or soon thereafter. It had quickly become clear to leading American educators, many of them inspired by the model of German universities, that graduate training and research went hand in hand, and those institutions that pursued this policy soon established a competitive edge within a fast-changing academic environment.

As a measure of the rapidity of the transformation that ensued, consider the following statistics on the number of doctoral degrees in mathematics earned by Americans between 1875 and 1900. Before 1875, American universities had conferred a total of only six degrees in the field. During the next fifteen years, thirty-nine Americans took doctorates in the United States, and another fifteen earned their degrees abroad. These figures were dwarfed again by those of the final decade of the century, which witnessed a total of 107 new Ph.D.s in mathematics, eighty-four of them earned at home.[3]

---

[3]R. G. D. Richardson, "The Ph.D. Degree and Mathematical Research," *American Mathe-*

A second factor crucial in the emergence of the American mathematical research community during the last quarter of the nineteenth century was the founding of the New York Mathematical Society in 1888. As we saw in Chapter 6, its beginnings were inauspicious enough: by the end of 1889 it had attracted only sixteen members. Within two years, however, membership had ballooned to 210, and the Society had begun issuing its *Bulletin*. In the wake of the Chicago Congress and the prototypic Evanston Colloquium, the organization assumed national dimensions and changed its name to the American Mathematical Society in 1894. Following the initiative of E. H. Moore and Henry White, a group of Midwestern mathematicians established the Chicago Section of the American Mathematical Society in 1897. Thus, from the mid-1890s onward, the AMS served as the principal organizational vehicle for meetings, colloquia, and other activities of interest to the budding community of research mathematicians. By the turn of the century, it had also initiated its *Transactions* as a complement both to its *Bulletin* and to the two older, research-oriented periodicals, the *American Journal of Mathematics* and the *Annals of Mathematics*. Thus, by the end of this pivotal period, the United States already supported sufficiently many researchers to sustain four respectable mathematical journals.

While it would be difficult to overestimate the role the American Mathematical Society played in promoting mathematical research in the United States, its early success depended on a third ingredient: the coming of age of a generation of Americans not merely interested in mathematics but who possessed the requisite knowledge *and* the institutional support to educate the next generation of researchers. From 1876 to 1883, Sylvester had instructed a number of students at Hopkins who had sincere mathematical interests and a certain amount of real mathematical talent. As discussed at the close of Chapter 3, however, if they managed, by and large, to raise mathematical standards at the schools where they ultimately found positions, they tended not to succeed in passing the research torch on to students of their own. This state of affairs changed dramatically after the mid-1880s when the first wave of aspiring American mathematicians reached Europe. Whether they studied with Felix Klein in Göttingen, with Sophus Lie in Leipzig, or with other prominent mathematicians elsewhere, this generation went on to introduce research-level mathematics at numerous colleges and universities throughout the United States. Led by E. H. Moore's Chicago school, a few select institutions also developed solid doctoral programs whose graduates shaped and directed American mathematics in the decades that followed.

As the preceding chapters show, these three factors interacted to transform what amounted to a few scattered pockets of mathematical expertise into a cohesive and extensive mathematical community, that is, an interac-

---

matical Monthly 43 (1936):199–215; reprinted in *A Century of Mathematics in America—Part II*, ed. Peter Duren *et al.* (Providence: American Mathematical Society, 1989), pp. 361–378 on p. 366.

tive group of individuals linked by common interests. Although the account presented here has focused on certain qualitative features of this community, it is further enhanced and supported by a recent quantitative analysis undertaken by Della Dumbaugh Fenster and Karen Hunger Parshall.[4] Drawing on a wide variety of information recorded in the first fifteen volumes of the *Bulletin of the New York* (later *American*) *Mathematical Society*, Fenster and Parshall document the extent, depth, and principal research interests of this community around the turn of the century. Their work uncovered over 1,000 geographically dispersed participants at various levels of interest and activity who formed a pyramidal community broadly based upon more than 500 interested—if not active—participants and tapering up to an apex of some sixty highly active and productive researchers. Within this community, those most active in research spanned the landscape of pure mathematics from algebra to analysis to geometry, while tending to favor the subdisciplines of group theory, the theory of automorphic functions, the calculus of variations, real and complex function theory, classical projective, algebraic, and differential geometry, and foundational studies. (Applied areas, although represented, held less attraction for turn-of-the-century American research mathematicians, due, in part, to the fact that they drew their inspiration from pure mathematicians like Sylvester, Klein, and E. H. Moore.) This broad support of pure mathematics at the research level that solidified during the pivotal period from 1876 to 1900 undoubtedly contributed to the sustained growth in American mathematical activity that characterized the ensuing period from 1900 to 1933. The final quarter of the nineteenth century thus marks a true watershed in the history of American mathematics. The major institutional structures and research traditions of American mathematics stood firmly in place by the end of the first decade of the twentieth century, when the United States began to make its presence felt on the international mathematical scene. No dramatic new qualitative changes affected the community's overall contours until the influx of European refugees began in the mid-1930s.

The remainder of this chapter provides an indication of the principal links and interrelations between the second and third periods in the historical development of American mathematics. It outlines the most prominent features of the American mathematical research community during the first third of the twentieth century by looking not only at how the institutions and research traditions established between 1876 and 1900 were perpetuated and strengthened after the turn of the century but also at how the momentum of the earlier period continued to propel the mathematical community forward. Furthermore, it suggests a number of larger themes and open questions pertaining to this community as it developed during the period that followed.

---

[4]Della Dumbaugh Fenster and Karen Hunger Parshall, "A Profile of the American Mathematical Research Community, 1891–1906," in *The History of Modern Mathematics*, vol. 3, ed. Eberhard Knobloch and David E. Rowe (Boston: Academic Press, Inc., 1994), forthcoming.

What was the state of the research community during the first decades of the twentieth century, and how did it compare both quantitatively and qualitatively with its counterparts in other leading nations? How effective were the new American graduate programs in training productive researchers? To what extent did the research ethic take hold at less élite institutions of higher education in the United States? Which fields of research did American mathematicians prefer and when? What do these preferences indicate about the balance between pure and applied fields? How did the overall picture of mathematical activity change in these decades? Did mathematics develop an indigenous tradition in the United States and, if so, what were its distinguishing features? Clearly, any attempt to provide detailed answers to these and other questions would require an extensive investigation of subsequent developments that lie beyond the scope of the present study. Nevertheless, the following sketch of the contours of this period may offer some tentative conclusions and suggestions of promising avenues for future historical research.

Mathematics at the Johns Hopkins and at Yale

By the turn of the century, the United States had at least seven established graduate programs in mathematics. Three resided at Ivy League institutions: Columbia, Harvard, and Yale. Three others—at Chicago, Clark, and the Johns Hopkins—had burst onto the scene with the founding of private, research-oriented universities. The seventh program, located at Cornell, was supported by the only land-grant institution with a strong commitment to higher mathematics at this time.[5] Twenty years later, a number of other schools had also entered the fray, in particular, California, Illinois, Michigan, Pennsylvania, and Wisconsin. In fact, out of a total of 406 doctoral degrees in mathematics awarded by American universities during the period from 1900 to 1920, 354 were conferred by these twelve schools.[6]

It should be emphasized, however, that only a few of these doctoral programs consistently turned out strong research mathematicians. By the 1890s, when the program at the Johns Hopkins had entered a sort of holding pattern and Clark's had begun its swift decline, Chicago and Harvard stood poised to establish solid research traditions and strong graduate programs that would

[5]This is confirmed by Fenster and Parshall's quantitative study. In the period from 1891–1906, these top seven schools were followed by the University of Pennsylvania, Princeton, and Syracuse, which had granted at least eleven, five, and four doctorates in mathematics, respectively. Bryn Mawr, Kansas, Michigan, and Nebraska had awarded at least two doctorates during this time. See *ibid.*, Table 9. Since the data in Fenster and Parshall's study result from reports in and communications to the *Bulletin*, they reflect lower bounds. Wisconsin, for example, actually gave at least three doctorates in mathematics over the period 1891–1906 although only one was reported in the *Bulletin*. Recall that Cornell was not a pure land-grant institution, as it relied on private financing as well.

[6]Richardson, p. 366.

attract the best and the brightest young talent in the United States for years to come. Princeton operated in a rather different academic environment and made a series of judicious appointments that enabled its faculty to emerge quietly as the third premier center for mathematical research in the United States.

Although mathematics at the Johns Hopkins University did not regain the prominent place it had enjoyed under Sylvester, the presence of men like Thomas Craig and, for a time, Fabian Franklin at least insured that its program remained sound and respectable. The man most responsible for maintaining these standards—as well as a distinctly nineteenth-century British style of mathematics—during the early decades of the new century was the English-born geometer Frank Morley.[7] After taking degrees at King's College, Cambridge in the mid-1880s, Morley accepted a position at the Quaker college in Haverford, Pennsylvania, where he taught for twelve unusually productive years. An inveterate problem solver, Morley submitted numerous solutions to geometrical problems posed in the *Educational Times.* His accomplishments earned him an appointment as Full Professor at Hopkins in 1900, where, over the course of the next thirty years, he guided the research of some forty-eight doctoral students. His most influential student, Arthur Byron Coble, later recalled how his mentor made it "a cardinal point to have on hand a sufficient variety of thesis problems to accommodate particular tastes and capacities."[8] Morley's approach faintly echoed the venerable Cambridge tradition in which coaches honed the skills of their future Wranglers by posing tough problems aimed at preparing them for the rigors they would encounter in the Tripos examinations. In this sense, his work in the rich, but fast-fading areas of nineteenth-century algebraic and projective geometry complemented that of two other leading representatives of this discipline in the United States, Klein's former pupils, Virgil Snyder and Henry Seely White. In sum, Morley's style reflected the overall state of the Hopkins graduate program, which continued to have both feet firmly planted in older mathematical traditions of only marginal importance by the early decades of the twentieth century.

Whereas Morley might well be considered an intellectual heir of Cayley and Salmon, a second Englishman, Ernest William Brown, came to the United States from Christ's College, Cambridge with a very different mathematical pedigree.[9] Brown not only reflected the British tradition of his mentor G. H. Darwin (as well as that of Stokes and Maxwell), but he also walked in the footsteps of America's distinguished mathematical astronomer George William Hill. In fact, as a post-graduate student, Brown had been counseled by

[7] On Morley, see Raymond Clare Archibald, *A Semicentennial History of the American Mathematical Society*, 1888–1938 (New York: American Mathematical Society, 1938), pp. 194–201.

[8] *Ibid.*, p. 196.

[9] *Ibid.*, pp. 173–183.

Darwin to study Hill's "Researches in the Lunar Theory,"[10] advice that proved crucial for the young man's future career. Joining Morley at Haverford College in 1891, Brown began a lifelong quest to analyze the mutual influences of all known forces affecting the moon's motion. His work in this field prompted the Royal Society of London to elect the thirty-two-year-old to a fellowship in 1898, and nine years later the Royal Astronomical Society awarded him its seventh gold medal. Also in 1907, Brown accepted an appointment as Professor of Mathematics at Yale where he subsequently became the first occupant of the Josiah Willard Gibbs Chair. Prodigiously productive throughout his career, Brown was generally regarded as the world's leading expert on lunar theory;[11] yet, within the context of the American *mathematical* community of the day, he, like Gibbs before him, was clearly a fringe figure. Although he served as the thirteenth President of the AMS, he embodied a centuries-old research tradition in celestial mechanics that would soon sever ties with its former practitioners, the mathematicians.

Yale's other leading mathematicians during this period, James Pierpont and Percey F. Smith, had both studied in Europe during the mid-1890s, gaining solid training that enabled them to maintain a productive graduate program in New Haven for many years.[12] Pierpont had taken his doctorate in Vienna before joining the Yale faculty in 1894, whereas Smith had spent two years abroad in Göttingen, Berlin, and Paris before returning to his *alma mater* in 1896. A respected authority on real and complex analysis, Pierpont wrote several research articles and textbooks in these fields. He also joined Harvard's Maxime Bôcher in delivering the first series of AMS Colloquium lectures in 1896. As further signs of his stature within the mathematical community, the AMS chose him as its third Gibbs Lecturer in 1925, and G. D. Birkhoff, in his survey of American mathematics up to 1938, praised Pierpont as one of those "who have shown the rare quality of leadership."[13] He might have added the name of Pierpont's student R. G. D. Richardson,

[10]George William Hill, "Researches in the Lunar Theory," *American Journal of Mathematics* 1 (1878):5–26, 129–147, and 245–260. About this work, Brown later wrote: "This memoir of but fifty quarto pages has become fundamental for the development of celestial mechanics in three different directions. It would be difficult to say as much for any other publication of its length in the whole range of modern mathematics, pure or applied. Poincaré's remark that in it we may perceive the germ of all the progress which has been made in celestial mechanics since its publication is doubtless fully justified." See Archibald, *Semicentennial History*, p. 119.

[11]Another sign of this recognition came from Felix Klein, who chose Brown to write an article on lunar theory for the *Encyklopädie*. See Ernest W. Brown, "Theorie des Erdmondes," *Encyklopädie der mathematischen Wissenschaften* VI-2 (1915):667–728.

[12]See Harold Dorwart, "Mathematics and Yale in the Nineteen Twenties," in *A Century of Mathematics in America—Part* II, pp. 87–98.

[13]The others Birkhoff mentioned in this connection were: E. H. Moore, G. B. Halsted, L. E. Dickson, R. L. Moore, O. Veblen, G. A. Bliss, G. C. Evans, S. Lefschetz, M. Morse, J. F. Ritt, M. H. Stone, and N. Wiener; George David Birkhoff, "Fifty Years of American Mathematics," in *Semicentennial Addresses of the American Mathematical Society* (New York: American Mathematical Society, 1938), pp. 270–315 on p. 276.

who succeeded Frank Nelson Cole as AMS Secretary in 1921. As Dean of the Graduate School at Brown University, Richardson played an instrumental role in strengthening Brown's mathematics program during the 1930s.

While mathematicians like Yale's Pierpont, Morley of the Johns Hopkins, and Virgil Snyder at Cornell inspired scores of aspiring American mathematicians during the early decades of the twentieth century, few of their numerous doctoral students went on to become productive researchers. Perhaps even more telling, among those who did engage in research, none attained a level of achievement comparable with his or her respective mentor. The question thus arises: why did the programs at these schools produce so few strong mathematicians? Did the faculties at Hopkins, Yale, Columbia, Clark, and Cornell simply lack the necessary depth and/or breadth to train solid researchers? These departments clearly could not match the talented faculties at Chicago and Harvard. Moreover, even their strongest individual members—Pierpont, Morley, and Snyder—could not be described as major mathematicians working in pivotal areas of research. Much of the mathematics they had learned in the 1880s and 1890s had either fallen out of fashion or become literally obsolete after the turn of the century. Even so, one could still train gifted, young mathematicians to do old-fashioned mathematics given the right conditions. Morley's student Arthur Coble, for example, worked in the very same tradition as his adviser and ultimately supervised the thesis research of twenty-two doctoral students at the Johns Hopkins and Illinois between 1914 and 1938.[14]

Considerations such as these make it clear that qualitative differences beyond the research strengths of their respective faculties distinguished the dynamic departments at Chicago, Harvard, and (somewhat later) Princeton from those a notch below. A variety of factors—from institutional support and financial integrity to location and student talent—combined to determine the relative strengths and weaknesses of the mathematics programs at the major universities. In the final analysis, however, one feature differentiated the former three schools from all the rest: by the early decades of the twentieth century they alone had established solid research traditions in particular mathematical fields. These traditions not only served to channel the resources of these three dominant centers for mathematical research in the United States but they also acted as catalysts that directed the growth of American mathematics as a whole throughout much of the next century.

The Emergence of "The Big Three"

As we saw in Chapter 9, the success of the mathematics program at the University of Chicago depended on E. H. Moore's vision and drive as well as on favorable circumstances and, sometimes, sheer luck. More than any other contemporary American mathematician, Moore had a bloodhound's nose for

[14]The list of Coble's students appears in Archibald, *Semicentennial History*, p. 235.

important new developments in mathematical research, and once he found the right scent, he doggedly pursued his game. This uncanny hunter's instinct made him ideally suited for the task of pointing young students down promising new paths of research. His gifts, both as a teacher and an administrator, thus gave the Chicago program a decisive edge over its competitors and enhanced its ability to attract the critical mass of talented students necessary to sustain an ongoing graduate program. Between 1890 and 1915, in fact, Chicago graduated over sixty doctoral students in mathematics, far more than any other school in the nation.[15] The five most significant among these—Dickson, Bliss, Veblen, R. L. Moore, and Birkhoff—went on to exert a profound influence on the American mathematical scene.[16]

These five matured mathematically in the vibrant research environment which Moore together with Bolza and Maschke had already established at Chicago by the dawn of the new century. This triumvirate had also sown the seeds for an incipient research tradition in two nearly unrelated, yet important fields: modern algebra and the calculus of variations. While 1908 brought Maschke's death, Bolza's subsequent departure for Germany in 1910, and the end of a dynamic era of dominance at Chicago, it also heralded the firm entrenchment of the University's bipartite research tradition. The calculus of variations continued to flourish under Bolza's successor and leading student, G. A. Bliss, while Leonard Dickson, a member of the Chicago faculty since 1900, assumed the mantel of algebra once worn by Moore and Maschke.[17] Dickson remained the leading algebraist in the country for another two decades, guiding the work of more than sixty doctoral students during his career at Chicago.

By 1900, the beginnings of Harvard's reputation as a major center for research in analysis could also be clearly discerned. Led by Klein's two most distinguished American students, Osgood and Bôcher, the institution soon overtook the Johns Hopkins and established itself as the dominant center for higher mathematics on the eastern seaboard. These two principals "spearheaded a revolution in mathematics at Harvard,"[18] restructuring and modernizing their department's graduate program with a series of courses that stressed clearly stated definitions and air-tight proofs. In keeping with this new critical spirit in analysis, Osgood published the first truly rigorous proof

[15]Richardson, p. 366.

[16]Another important member of this group of doctoral students, Theophil Henry Hildebrandt, took his degree in 1910. Hildebrandt spent his entire career at the University of Michigan, where he served as Chair of the Department of Mathematics from 1935 to 1957. He was the President of the AMS from 1945 to 1946.

[17]Maschke's successor was the Berlin-trained geometer Ernst Julius Wilczynski, an important figure in the field of projective differential geometry. See Saunders Mac Lane, "Mathematics at the University of Chicago: A Brief History," in *A Century of Mathematics in America—Part* II, pp. 127–154.

[18]Garrett Birkhoff, "Mathematics at Harvard, 1836–1944," in *ibid.*, pp. 3–58 on p. 15.

of the Riemann mapping theorem in 1900.[19] One year later, his seminal article on the theory of functions of a single complex variable appeared in Klein's *Encyklopädie der mathematischen Wissenschaften*. In the words of G. D. Birkhoff, this article "represents the first careful and systematic presentation of the Riemannian point of view ... as against the earlier Weierstrassian approach, based on the use of power series."[20] Osgood expanded his masterful treatment of this subject in the 1907 treatise *Funktionentheorie*, which became a standard international reference work for the next three decades.[21]

Bôcher's mathematical interests covered a somewhat broader terrain than his colleague's, but they still bore a close relationship to Klein's innovative research program in potential theory. In a series of three papers published in the *Bulletin of the American Mathematical Society*, Bôcher pushed this program forward by providing the first analytic proof of Klein's oscillation theorem.[22] He also gave the first adequate treatment of the "Gibbs phenomenon," a term he coined for the well-known blurring effect that arises at boundary points in certain Fourier series expansions.[23]

Although Osgood's reputation as a mathematical scholar was unsurpassed in the United States, he was certainly no match for Bôcher as a teacher. In fact, Osgood supervised only four doctoral theses over more than forty years at Harvard, whereas his colleague guided the work of seventeen doctoral students up until his death in 1918.[24] Two of Bôcher's students, Lester R. Ford and Griffith C. Evans, became colleagues at the Rice Institute in Houston. There, Ford wrote his authoritative text, *Automorphic Functions*, which is still a standard reference today.[25] Evans became Chair of the Department at the University of California, Berkeley in 1934, and soon afterward, he succeeded in propelling it into world prominence.[26] Bôcher also inspired G. D. Birkhoff during his undergraduate years. As noted in Chapter 9, Birkhoff's early research centered on asymptotic expansions, boundary-value problems, and Sturm-Liouville theory, topics he had begun to master under Bôcher's tutelage.[27]

---

[19]William Fogg Osgood, "On the Existence of the Green's Function for the Most General Simply Connected Plane Region," *Transactions of the American Mathematical Society* 1 (1900):310–314; and 2 (1901):484–485.

[20]G. D. Birkhoff, "Fifty Years of American Mathematics," p. 293.

[21]William Fogg Osgood, *Lehrbuch der Funktionentheorie*, vol. 1 (Leipzig: Verlag B.G. Teubner, 1907).

[22]Maxime Bôcher, "The Theorems of Oscillation of Sturm and Klein," *Bulletin of the American Mathematical Society* 4 (1898):295–313 and 365–376; and 5 (1898):22–43.

[23]Maxime Bôcher, "Introduction to Fourier Series," *Annals of Mathematics* 7 (1906):81–152.

[24]See Archibald, *Semicentennial History*, pp. 153 and 163.

[25]Lester B. Ford, *Automorphic Functions* (New York: McGraw-Hill, 1929).

[26]See Robin Rider, "An Opportune Time: Griffith C. Evans and Mathematics at Berkeley," in *A Century of Mathematics in America—Part* II, pp. 283-302.

[27]Birkhoff paid homage to his former teacher in George David Birkhoff, "The Scientific

Besides analysis, Harvard's Mathematics Department also soon developed particular strength in geometry, thanks primarily to the presence there of Julian Lowell Coolidge.[28] After joining the Harvard faculty in 1902, Coolidge, a former Harvard undergraduate, set off on a two-year study tour that took him to Paris, Greifswald, Turin (to attend lectures of Corrado Segre), and finally Bonn. He took his doctorate at Bonn under his former teacher in Griefswald, Eduard Study, writing a thesis on non-Euclidean line geometry. (Study had presumably suggested that Coolidge hear Segre's lectures on this subject before embarking on his dissertation research. Recall that Study had also provided Henry Fine with his dissertation topic while working as a *Privatdozent* in Leipzig.)

Coolidge's reputation as an authority on classical algebraic and projective geometry rested largely on the numerous textbooks he wrote on these subjects.[29] Their rambling style and rich content were, in many ways, reminiscent of the better-known works of Felix Klein, and this resemblance was perhaps not accidental. Coolidge's admiration for Klein came through clearly in one of his later books the widely read *History of Geometrical Methods*. There, Coolidge referred to Klein as "the greatest synthesist that geometry has ever known" and further claimed exuberantly that the German's *Erlanger Programm* had "probably influenced geometrical thinking more than any other work since the time of Euclid, with the exception of Gauss and Riemann."[30]

As these brief discussions of the strengths of the Departments of Mathematics at Chicago and Harvard show, both of these programs had already proven their mettle by the first decade of the century. The years from 1900 to 1910, however, witnessed the emergence of a third major mathematical center, at Princeton University. The school's rise attests to the viability of the by now extant American mathematical research community, for it was from the first generation of that community that Princeton drew much of its fresh, young talent. Partly owing to its relative youth, Princeton's program, unlike those of Chicago and Harvard, did not play a major role in training graduate students until the 1930s. It earned its reputation instead by virtue of the quality of the research produced by its faculty. The man largely responsible for assembling the original core group that went on to launch Princeton's

Work of Maxime Bôcher," *Bulletin of the American Mathematical Society* 25 (1919):197–215; reprinted in *A Century of Mathematics in America—Part* II, pp. 59–78.

[28]See Dirk J. Struik, "Julian Lowell Coolidge in Memoriam," *American Mathematical Monthly* 62 (1955):669–682.

[29]Julian Lowell Coolidge, *The Elements of Non-Euclidean Geometry* (Oxford: Clarendon Press, 1909); *A Treatise on the Circle and the Sphere* (Oxford: Clarendon Press, 1916); *The Geometry of the Complex Domain* (Oxford: Clarendon Press, 1924); *A Treatise on Algebraic Plane Curves* (Oxford: Clarendon Press, 1931); *A History of Geometrical Methods* (Oxford: Clarendon Press, 1940); *A History of the Conic Sections and Quadric Surfaces* (Oxford: Clarendon Press, 1945); and *The Mathematics of Great Amateurs* (Oxford: Clarendon Press, 1949).

[30]Coolidge, *History of Geometrical Methods*, p. 293.

mathematical tradition was Klein's former student, Henry Burchard Fine.[31]

Fine had done well nurturing his Princeton ties, including the friendship he had maintained with Woodrow Wilson since their undergraduate days. When Wilson returned to his *alma mater* as its President in 1903, he appointed his friend, Fine, to the deanship of the faculty. Within the span of a dozen years, Fine attracted some superb mathematical talent to Princeton. In 1905, he appointed Luther Pfahler Eisenhart (who had taken his doctorate at Hopkins in 1900), Oswald Veblen, G. A. Bliss, and John Wesley Young to preceptorships.[32] That same year, he also hired the distinguished British mathematician, James Jeans, as Professor of Applied Mathematics. After Bliss returned to Chicago and Young left for Illinois in 1908, Fine found two more distinguished replacements, the Scottish algebraist Joseph H. M. Wedderburn and E. H. Moore's latest stellar student, G. D. Birkhoff. This faculty would solidify during the interwar years under the leadership of Veblen and Eisenhart to establish key research foci in differential geometry and topology. (See below.)

Thus, by the opening decades of the century, the mathematics departments at three universities—Chicago, Harvard, and Princeton—had, to a considerable extent, established their respective fields of expertise and staked out their positions as the élite centers for mathematical research in the United States. With these three mathematics departments and several other reasonably sound doctoral programs around the country to choose from, aspiring young American mathematicians no longer felt compelled to go abroad for their training as they had throughout much of the period 1876–1900. Almost without exception, the leading figures of the following period had either studied or taught at one or more of the "Big Three" institutions. These schools were clearly filling the institutional void that had existed prior to 1876.

Hilbert's Göttingen and the Impact of World War I

Despite the dramatic turn of events for mathematics in the United States during the final quarter of the nineteenth century, German universities continued to attract large numbers of American mathematicians, and, as before,

[31]See William Aspray, "The Emergence of Princeton as a World Center for Mathematical Research, 1896–1939," in *History and Philosophy of Modern Mathematics*, ed. William Aspray and Philip Kitcher (Minneapolis: University of Minnesota Press, 1988), pp. 346–366; reprinted in *A Century of Mathematics in America—Part* II, pp. 195–215.

[32]Young had taken his Bachelor's degree at Ohio State University in 1899 and his A.M. (1901) and Ph.D. (1904) at Cornell. Prior to joining the Princeton faculty, he had served as an Instructor at Northwestern. He stayed at Princeton for the three years from 1905 to 1908, moved on to an assistant professorship at the University of Illinois from 1908 to 1910, and accepted a professorship and the departmental headship at the University of Kansas in 1910. His final move was to Dartmouth College in 1911, where he remained until his death in 1932. Young's main areas of expertise were geometry and algebra, but he also devoted much of his time and energy to mathematics education. For more on his life and works, consult, for example, K. D. Beetle and C. E. Wilder, "John Wesley Young: In Memoriam," *Bulletin of the American Mathematical Society* 38 (1932):603–610.

most of them flocked to Göttingen. Their principal mentor, however, was no longer Felix Klein but David Hilbert, who supervised the research of some sixty doctoral students (ten of whom were Americans) between 1898 and the outbreak of World War I.[33] Moreover, the mathematical atmosphere the younger Americans found on their arrival in Göttingen differed markedly from that in which White, Osgood, Bôcher, and their contemporaries had moved a little more than a decade earlier. As we saw in Chapter 5, the American contingent formed a significant portion of the entire corps of mathematics students in Göttingen during the late 1880s and early 1890s. Consequently, these Americans, like other foreign students, enjoyed fairly regular contact with faculty members and considerable camaraderie with their peers.

By early in the new century, however, Klein had guided Göttingen mathematics to a position of dominance over even its traditional rival, Berlin University. After reaching a low of less than 100 during the early 1890s, enrollments in mathematics and the natural sciences at Göttingen skyrocketed after 1895 to 300 by the turn of the century and nearly 800 by 1913.[34] Besides Klein and Hilbert, its faculty included such luminaries as Hermann Minkowski, Carl Runge, Ludwig Prandtl, and Edmund Landau. Klein, having by now fully abdicated his former role as master teacher, focused most of his attention on organizational and administrative matters. This left the bulk of the responsibility for training doctoral students squarely on Hilbert's shoulders. How he fared may be judged easily enough from some of the names that adorn the long list of his students: Otto Blumenthal, Max Dehn, Felix Bernstein, Erhard Schmidt, Ernst Hellinger, Hermann Weyl, Alfred Haar, Richard Courant, Erich Hecke, Hugo Steinhaus, and Hellmuth Kneser.[35] Clearly, the Americans who took their degrees in Göttingen during this period faced extraordinarily keen competition. Unlike their more fortunate predecessors, who for over a decade had been among the prime beneficiaries of Klein's teaching activities, this new group found they were just so many more faces in the talented crowd surrounding Hilbert and his colleagues.

As recounted in Chapter 9, Hilbert's work on the foundations of geometry had exerted a strong influence on E. H. Moore and through him on Oswald

[33]The ten were Anne Lucy Bosworth (1899), Edgar Jerome Townsend (1900), Earle Raymond Hedrick (1901), Charles Albert Noble (1901), Oliver Dimon Kellogg (1902), Charles Max Mason (1903), David Clinton Gillespie (1906), Arthur Robert Crathorne (1907), Charles Haseman (1907), and William Deweese Cairns (1907). Hilbert also supervised the dissertation of Haskell Brooks Curry in 1929. The complete list can be found in David Hilbert, *Gesammelte Abhandlungen*, 3 vols. (Berlin: Springer-Verlag, 1932-1935), 3:431–433 (hereinafter cited as *Hilbert GA*). On Hilbert's life, consult Constance Reid, *Hilbert* (New York: Springer-Verlag, 1970).

[34]Wilhelm Lorey, *Das Studium der Mathematik an den deutschen Universitäten seit Anfang des* 19. *Jahrhunderts*, Abhandlungen über den mathematischen Unterricht in Deutschland, Band III, Heft 9 (Leipzig: Verlag B. G. Teubner, 1916).

[35]*Hilbert GA*, 3:431–433.

DAVID HILBERT (1862–1943)

Veblen and R. L. Moore. (Even more influential for American mathematical research in the long run was Hilbert's famous Paris lecture of 1900, which highlighted twenty-three unsolved problems for the attention of future investigators.[36]) By the time Minkowski arrived in Göttingen in 1902, Hilbert had already entered a new and very different research phase dominated by a wide range of problems arising from integral equation theory, potential theory, and the mathematical foundations of physics.[37] This period, up until the outbreak of World War I, marks the pinnacle of achievement for Göttingen mathematics, in general, and for the Hilbert school, in particular. Hilbert and his coworkers created a powerful array of mathematical concepts and techniques for physics which ultimately found expression in the instant classic of Courant and Hilbert, *Methods of Mathematical Physics*.[38] One of these techniques, Hilbert space theory, motivated the development of an entirely new field of mathematical analysis.

Two early contributions stemming from Hilbert's work in analysis came from his American students, Oliver Dimon Kellogg and Max Mason. In fact, Kellogg completed his 1902 dissertation, "Zur Theorie der Integralgleichungen und des Dirichlet'schen Prinzips," two years before the publication of Hilbert's first papers in this field. Mason's thesis research also dealt with boundary-value problems in ordinary differential equations.[39] Both men later assumed leading positions in the American scientific community: Kellogg, as a member of the Harvard faculty, and Mason through his administrative work as President of the University of Chicago and head of the Rockefeller Foundation.

Another Hilbert student who went on to become a motive force within the American mathematical community was Earl Raymond Hedrick.[40] After taking his A. B. degree at Michigan in 1896, Hedrick studied with Bôcher and Osgood at Harvard for three years before arriving in Göttingen as a Parker Fellow in 1899. With solid training in mechanics and analysis, he had little difficulty completing his 1901 dissertation on the analytic character

[36]David Hilbert, "Mathematische Probleme," *Archiv für Mathematik und Physik* 1 (1901):44–63 and 213–237, or *Hilbert GA*, 3:290–329; and David Hilbert, "Mathematical Problems: Lecture Delivered Before the International Congress of Mathematicians at Paris in 1900," trans. Mary F. Winston, *Bulletin of the American Mathematical Society* 8 (1902):437–479; reprinted in *Mathematical Developments Arising from Hilbert's Problems*, ed. Felix Browder, Symposia in Pure Mathematics, vol. 28 (Providence: American Mathematical Society, 1976), 1:1–34.

[37]This should not be taken to mean that Hilbert had lost interest in foundational questions. For a detailed study of his ongoing interests in this area, see Volker Peckhaus, *Hilbertprogramm und kritische Philosophie*, Studien zur Wissenschafts-, Sozial-, und Bildungsgeschichte der Mathematik, vol. 7 (Göttingen: Vandenhoeck & Ruprecht, 1990).

[38]Richard Courant and David Hilbert, *Methoden der mathematischen Physik* (Berlin: Springer-Verlag, 1924).

[39]For a discussion of the general context that motivated the dissertations of Kellogg and Mason, see Ernst Hellinger, "Hilbert's Arbeiten über Integralgleichungen und unendliche Gleichungssysteme," in *Hilbert GA*, 3:94–145, especially on p. 100.

[40]On Hedrick, see Archibald, *Semicentennial History*, pp. 223–228.

of solutions to certain differential equations.[41] Besides Hilbert's lectures, he and another American, Charles Noble, also attended Klein's classes. Later, in the 1930s, these two would help propagate the Kleinian teaching legacy by collaborating on the English translation of his popular textbooks for teachers, *Elementary Mathematics from an Advanced Standpoint.*[42] After leaving Göttingen, Hedrick journeyed to Paris, where he met Édouard Goursat, Émile Picard, Jacques Hadamard, and Paul Appell. This experience intensified his interest in pedagogical problems and spurred him to translate Goursat's *Cours d'Analyse.*[43]

Although clearly a capable mathematician, Hedrick left his mark primarily as an editor and organizational leader. For some thirty years, he edited two series of textbooks: one dealing with advanced mathematical subjects for secondary schools and colleges and the other covering nearly every facet of the engineering sciences.[44] As a strong proponent of higher standards for mathematics education, Hedrick helped to launch the Mathematical Association of America and served as its first President in 1916. From 1921 to 1937, he edited the *Bulletin of the American Mathematical Society*, and he was elected AMS President for the 1929–1930 term. After teaching at the University of Missouri for about twenty years, he assumed the Chair of the Mathematics Department at the University of California at Los Angeles in 1924, and in 1937 he became UCLA's Provost. The 1930s also found him representing the interests of the mathematical community as Secretary of Section A of the American Association for the Advancement of Science.

The careers of Hilbert's early students—Hedrick, Kellogg, Mason, and others—amply illustrate the major role these young Americans played in building up their country's mathematical institutions. As we have emphasized, however, the main features of American mathematics had already been shaped by their predecessors. The previous generation, many of whom had studied under Klein, had served as the real architects and engineers of research-level mathematics in the United States. Their like-minded compatriots, those who studied abroad in the early years of the twentieth century, still found plenty of opportunity to strengthen and augment the edifice on their return, but with the decided advantage that the underlying foundation firmly supported their efforts.

---

[41]Earl Raymond Hedrick, *Ueber den analytischen Character des Lösungen von Differentialgleichungen* (Göttingen: n.p., 1901).

[42]Felix Klein, *Elementary Mathematics from an Advanced Standpoint. Arithmetic, Algebra, Analysis*, vol. 1, trans. E. R. Hedrick and C. A. Noble (New York: Dover Publications, Inc., 1932); and *Elementary Mathematics from an Advanced Standpoint. Geometry*, vol. 2, trans. E. R. Hedrick and C. A. Noble (New York: Dover Publications, Inc., 1939).

[43]Édouard Goursat, *A Course in Mathematical Analysis*, trans. E. R. Hedrick, 2 vol. in 3 (Boston: Ginn and Co., 1904–1917).

[44]*A Series of Mathematical Texts* and *Engineering Science Series*; a complete list of the volumes that had appeared up to 1938 can be found in Archibald, *Semicentennial History*, pp. 227–228.

While Hilbert continued to attract many American students until 1907, the numbers fell off precipitously thereafter. When the guns sounded in August of 1914 marking the opening of World War I, the once vibrant "American Colony" in Göttingen had already largely vanished from the scene. As with so much else, the Great War placed a vast gulf between formerly friendly mathematicians who now found it more difficult to remain on cordial terms with colleagues from enemy nations. American scholars like the public at large tried, for the most part, to maintain a neutralist stance, but some felt that Germany's scientific, literary, and artistic leaders, by rallying to defend their nation's aggression, had forever tarnished the ideals of *Wissenschaft*.[45] By the time Woodrow Wilson's government decided to send American troops to Europe, Germany had few friends left to defend its policies. Although the German defeat brought an end to the fighting, the harsh terms of the Versailles Treaty hindered reconciliation. Europe's leading nations long remained embittered enemies, and the issue of nationalist loyalties seriously undermined international scientific cooperation.[46]

The war years also marked an interlude during which many American mathematicians took a furlough from their formerly purist endeavors. According to information compiled by David A. Rothrock, at least two hundred mathematicians engaged in some form of national service.[47] Max Mason, for example, made important progress on submarine detection as part of a secret project sponsored by the newly established National Research Council. The largest mathematical contingent consisted of some thirty mathematicians who analyzed ordnance tests conducted at the Army's Aberdeen Proving Grounds. The Chicago mathematical astronomer and Army Major, Francis R. Moulton, led another group of eight mathematicians—including Princeton's J. W. Alexander, Dunham Jackson of Harvard, and Joseph F. Ritt of Columbia—who worked on similar tests there. The Aberdeen researchers included such key figures as Veblen, Griffith C. Evans, Marston Morse, Warren Weaver, Norbert Wiener, Hans F. Blichfeldt, and G. A. Bliss, the first four of whom played significant roles in mobilizing the country's mathematical expertise during World War II. Be this as it may, this hands-on research appears to

---

[45] As one of the ninety-three signatories of the infamous manifesto "An die Kulturwelt! Ein Aufruf" issued during the early stages of World War I, Klein was summarily dismissed from the French Academy. Unlike Max Planck, Klein refused later to distance himself publicly from its more chauvinistic statements, despite his disagreement with them. On Planck's response, see John L. Heilbron, *The Dilemmas of an Upright Man: Max Planck as Spokesman for German Science* (Berkeley: University of California Press, 1986), p. 7.

[46] Germany, for example, was barred from participating in the International Congresses held in Strassburg (1920) and Toronto (1924). See Donald J. Albers, Gerald L. Alexanderson, and Constance Reid, *International Mathematical Congresses: An Illustrated History* (New York: Springer-Verlag, 1987), pp. 16–21.

[47] See David A. Rothrock, "Mathematicians in War Service," *American Mathematical Monthly* 26 (1919):40–44; reprinted in *A Century of Mathematics in America—Part* I, ed. Peter Duren *et al.* (Providence: American Mathematical Society, 1988), pp. 269–273.

have given American mathematicians little inclination to pursue applications after their return to civilian life.[48]

The Interwar Years: Sustaining the Momentum

The 1920s and early 1930s witnessed further growth and consolidation within American mathematics, as many established programs in the East and throughout the Midwest expanded into solid departments. Besides Chicago, Harvard, and Princeton, publicly funded universities like Michigan, Illinois, Wisconsin, California, and Ohio State developed programs competitive with the older ones at institutions like Columbia, Yale, the Johns Hopkins, and Cornell. Meanwhile, the mathematics faculty at the Massachusetts Institute of Technology vied for a place alongside the nation's "Big Three" schools. Outside of this more élite coterie, however, teaching occupied twelve to eighteen hours per week, making it difficult for students trained in these strong programs to maintain their mathematical momentum once they entered the academic work force. Only on rare occasions might a mathematician at one of the more teaching-intensive institutions be given a lighter load as an inducement to do research. According to data collected by R. G. D. Richardson for the period 1862–1933, in fact, some sixty individuals (about 5% of the Ph.D.s granted during this time) accounted for half of the published papers of the entire American mathematical community.[49] And newly graduated Ph.D.s were not the only ones who experienced difficulty in maintaining their research momentum.

By the 1920s, the dynamism of E. H. Moore's Chicago school had largely run its course, whereas the departments at Harvard and Princeton continued to gain strength during the interwar years. At Chicago, Bliss perpetuated the tradition in the calculus of variations as Department Chair from 1927 to 1941, while Dickson and, after 1931, his student A. Adrian Albert carried the torch of algebra.[50] These two disciplines remained at the core of the Chicago school, flourishing in the good times and surviving in the bad, throughout the first half of the century. (Even after the Second World War, Chicago's mathematical reputation has partly been due to its contributions to abstract algebra.) During Bliss's regime, the Chicago program continued to turn out new mathematicians in record numbers, graduating some ten doctoral candidates each year on average. A handful of them—including

[48]Veblen, however, did push for a greater emphasis on certain applications of mathematics to physics which he, unlike the physicists, viewed as significant. See Loren Butler, "Mathematical Physics and the American Mathematical Community: Disciplinary Values, Professional Interests, and the Place of Borderland Research 1880–1940," (unpublished Ph.D. dissertation, University of Chicago, 1992).

[49]Richardson, p. 373. His data also revealed that slightly less than one-third of those who taught mathematics at the college level held doctoral degrees, and only about one-fifth of these Ph.D.s had been consistently productive researchers.

[50]Irving Kaplansky, "Abraham Adrian Albert," *Biographical Memoirs of the National Academy of Sciences* 51 (1980):3–22, reprinted in *A Century of Mathematics in America—Part* I, pp. 244–264.

E. J. McShane, M. R. Hestenes, Mina Rees, and Herman Goldstine—went on to become important figures in the American mathematical community, but many abandoned research after taking their degrees.[51] As Saunders Mac Lane expressed it, Chicago had become something of a "Ph.D. mill" during the interwar years.[52]

As Chicago's fortunes fell, Harvard's stock rose, thanks largely to the presence there after 1912 of G. D. Birkhoff. One of the most distinguished American mathematicians of his generation, Birkhoff broke fertile new ground with work that won recognition internationally. His most important breakthroughs came in two related fields, dynamical systems and ergodic theory, that faintly echoed the work from America's scientific past of Hill and Gibbs.[53] In 1912, the year he came to Harvard from Princeton, Birkhoff proved an outstanding conjecture of Poincaré which states that any one-to-one area-preserving transformation of an annulus that moves the two boundary circles in opposite directions must have two fixed points.[54] This result implied the existence of periodic orbits in a restricted three-body problem. Twenty years later, he established the individual ergodic theorem, inspired by the work of John von Neumann and others.[55]

Harvard's dominance in analysis was further strengthened by the acquisition of O. D. Kellogg, an expert in mechanics and potential theory. Both Kellogg and the geometer William C. Graustein joined the Harvard faculty in 1920.[56] Together with G. D. Birkhoff, Kellogg proved the first fixed-point theorem for a function space about a decade before the emergence of Leray-Schauder theory. Furthermore, Kellogg published his noteworthy textbook, *Foundations of Potential Theory*, in 1929.[57] In the meantime, Birkhoff showed himself to be a worthy spiritual successor to E. H. Moore by supervising the dissertations of a remarkable group of young American mathematicians, including Marston Morse, Joseph L. Walsh, Marshall Stone, C. B.

[51]Mac Lane, pp. 138–143.

[52]*Ibid.*, p. 138.

[53]See Marston Morse, "George David Birkhoff and his Mathematical Work," *Bulletin of the American Mathematical Society* 52 (1946):357–391; Archibald, *Semicentennial History*, pp. 212–218; and Garrett Birkhoff's discussion of his father's work in G. Birkhoff, pp. 39–42 and 56–57.

[54]George D. Birkhoff, "Proof of Poincaré's Geometric Theorem," *Transactions of the American Mathematical Society* 14 (1913):14–22.

[55]George D. Birkhoff, "Proof of the Ergodic Theorem," *Proceedings of the National Academy of Sciences* 17 (1931):656–660. See, also, G. D. Birkhoff, "Fifty Years of American Mathematics," pp. 314–315.

[56]Graustein, like his elder colleague J. L. Coolidge, maintained the longstanding interest in complex geometry he had imbibed as a 1913 doctoral student of Eduard Study in Bonn. Graustein also did significant work on differential invariants in the tradition of Ricci and Levi-Civita. See William C. Graustein, "Invariant Methods in Classical Differential Geometry," *Bulletin of the American Mathematical Society* 36 (1930):489–521.

[57]Oliver Dimon Kellogg, *Foundations of Potential Theory* (New York: Dover Publications, Inc., 1929).

Morrey, and Hassler Whitney. (The first four later served as AMS Presidents.) All but Morrey, who enjoyed a long career in analysis at Berkeley, subsequently became members of the Harvard faculty.[58]

As Harvard continued to strengthen its research tradition in analysis, a second major mathematical center emerged in Cambridge—the Massachusetts Institute of Technology. Under the leadership of Harry W. Tyler, MIT quickly established its claim during the 1920s as one of the nation's preeminent mathematics departments.[59] Like Princeton's Henry Fine, Tyler never developed into a serious research mathematician, but both men brought back from Germany an appreciation of what it meant to do mathematics. Tyler also possessed the administrative skill and vision necessary to run a major department.

Most of MIT's active researchers during this period—F. S. Woods, Philip Franklin, Dirk Struik, and Virgil Snyder's student, Clarence L. E. Moore—worked in geometry. The star of the show, however, was Norbert Wiener.[60] Wiener had taken his doctorate in philosophy from Harvard in 1913 at the age of eighteen and had gone on to study logic and philosophy with Bertrand Russell at Cambridge. There, he also took courses with G. H. Hardy that effectively sowed the seeds for his late-budding interest in mathematics. He joined the MIT faculty as an Instructor in 1919. Ten years later, he published a lengthy memoir on generalized harmonic analysis, and three years after this another on Tauberian theorems, a subject that Hardy and Littlewood had pioneered. For these works, the AMS awarded Wiener its Bôcher Prize in 1933.[61] This work notwithstanding, Wiener went on to do influential research

[58]Walsh joined the department in 1921. His work on interpolation and approximation theory made him known as one of the few American experts in a field that had grown too classical for the abstract tastes of his countrymen. Morse arrived five years later, having already begun his seminal work on the calculus of variations in the large. In 1935 he went to Princeton as one of the original members of the faculty of the School of Mathematics at the Institute for Advanced Study (IAS). Whitney, one of the most influential figures in American topology, taught at Harvard from 1934 to 1952, when he, too, joined the faculty at the IAS. Finally, Stone spent the period from 1927 to 1946 at Harvard (except for a two-year stint at Yale) working on Fourier analysis, Hilbert space theory, and topological Boolean algebras. He left to become Chair at the University of Chicago, where he ushered in a new era for this once dominant research school. See Marshall Stone, "Men and Institutions in American Mathematics," in *Men and Institutions in American Mathematics*, ed. J. Dalton Talwater, John T. White, and John D. Miller, Graduate Studies Texas Tech University, no. 13 (Lubbock: Texas Tech Press, 1976), pp. 7–29.

[59]See Dirk Struik, "The MIT Department of Mathematics during its First Seventy-Five Years: Some Recollections," in *A Century of Mathematics in America–Part* III, ed. Peter Duren *et al.* (Providence: American Mathematical Society, 1989), pp. 163-178.

[60]On Wiener, see Pesi R. Masani, *Norbert Wiener*, 1894–1964, Vita Mathematica, vol. 5 (Basel: Birkhäuser Verlag, 1990). See, also, the insightful study by Steve J. Heims, *John von Neumann and Norbert Wiener* (Cambridge, MA: MIT Press, 1980).

[61]Wiener shared the Bôcher Prize with Marston Morse. The two major works for which he was recognized were Norbert Wiener, "Generalized Harmonic Analysis," *Acta Mathematica* 55 (1930):117–258; and "Tauberian Theorems," *Annals of Mathematics* 33 (1932):1–100; both works were reprinted in *Selected Papers of Norbert Wiener* (Cambridge, MA: MIT Press, 1964); and in a paperback edition, *Generalized Harmonic Analysis and Tauberian Theorems*

in applied mathematics, and today he is best remembered as the founder of the field of cybernetics.

Sparked by Birkhoff and Wiener, the Cambridge area emerged as America's foremost center for mathematical research by the mid-1920s. Trailing but a half-step behind was Princeton, where the hopes of Henry Fine found fulfillment during the interwar years. Led by Oswald Veblen and Luther Eisenhart, Princeton mathematicians began making their mark in two areas that became the cornerstones of their program: differential geometry and topology.

By the outset of the 1920s, Veblen and Eisenhart had already established their reputations as the nation's two premier differential geometers. Eisenhart had worked primarily in the classical tradition of Darboux and Bianchi, but, following Veblen, he began to focus on the intrinsic geometry of manifolds. In 1922, the two Princeton colleagues collaborated on an important paper dealing with the non-Riemannian geometry of paths in a manifold.[62] Three years later, Eisenhart delivered the AMS Colloquium lectures on "The New Differential Geometry," which he followed with the publication of the two volumes *Riemannian Geometry* and *Non-Riemannian Geometry*.[63] Veblen's contributions to this field culminated in three major monographs that appeared between 1927 and 1933.[64]

Veblen also served to solidify the Princeton program through the training of graduate students. Although Princeton graduated only thirty doctoral students in mathematics prior to 1930, their quality, when measured against the products of more prolific programs, like those at the Johns Hopkins and Yale, was especially high. With the notable exception of James W. Alexander (who joined the Princeton faculty in 1915 and was a student of Veblen in every other respect), most of the more distinguished Princeton graduates wrote their dissertations under Veblen's guidance. These included the geometers Philip Franklin and T. Y. Thomas, the logician Alonzo Church, the topologist J. H. C. Whitehead, and the physicist Banesh Hoffmann.[65] Church later joined the Princeton faculty in 1929 and soon thereafter inaugurated the school's on-going program in mathematical logic.

(Cambridge, MA: MIT Press, 1964).

[62]Luther P. Eisenhart and Oswald Veblen, "The Riemannian Geometry and its Generalization," *Proceedings of the National Academy of Sciences* 8 (1922):19–23.

[63]Luther P. Eisenhart, *Riemannian Geometry* (Princeton: University Press, 1926); and *Non-Riemannian Geometry* (New York: American Mathematical Society, 1927).

[64]The pertinent works are: Oswald Veblen, *Invariants of Quadratic Differential Forms*, Cambridge Tracts in Mathematics and Mathematical Physics, vol. 24 (Cambridge: University Press, 1927); Oswald Veblen and J. H. C. Whitehead, *The Foundations of Differential Geometry*, Cambridge Tracts in Mathematics and Mathematical Physics, vol. 29 (Cambridge: University Press, 1932); and Oswald Veblen, *Projektive Relativitätstheorie*, Ergebnisse der Mathematik und ihrer Grenzgebiete, vol. 2, no. 1 (Berlin: Springer-Verlag, 1933).

[65]A list of Veblen's students up to 1936 is given in Archibald, *Semicentennial History*, p. 209. Alexander actually wrote his doctoral thesis on univalent functions under the Swedish analyst Thomas Gronwall.

As Veblen and Eisenhart institutionalized differential geometry at Princeton, their younger colleague James Alexander demonstrated his knack for proving important theorems in the equally fashionable field of topology. An early breakthrough came in 1915, when he showed how to bypass the problematic *Hauptvermutung* and still prove the invariance of the Betti numbers and torsion coefficients of a manifold. Then, in 1922, he found a generalization of the Jordan curve to higher dimensions, a result known today as Alexander duality. Two years later, he gave the first example of a "wild embedding," the Alexander horned sphere, and four years after that, he introduced one of the principal invariants in knot theory, the so-called Alexander polynomial.[66] All of these results proved fundamental to the development of algebraic and geometric topology in the decades ahead.

If Alexander may be regarded as the founder of Princeton's research tradition in topology, though, its moving spirit was unquestionably Russian-born Solomon Lefschetz.[67] Originally trained as an engineer at the *École centrale* in Paris, Lefschetz came to the United States in 1905 at the age of twenty-one. In 1907, while working as an engineer for Westinghouse in Pittsburgh, he lost both of his hands in an industrial accident. Three years later, he took up graduate studies in mathematics at Clark where he earned his doctorate under William Story.[68] Thereafter, he taught briefly at Nebraska and eleven years at Kansas (where he did some of his best work) before eventually coming to Princeton in 1924 on the recommendation of Alexander. An intuitive thinker with little patience (or gift) for details, Lefschetz initially worked in algebraic geometry. His novel use of topological methods in connection with algebraic varieties not only set the stage for his work in algebraic topology but also profoundly influenced the research of algebraic geometers like the Englishman W. V. D. Hodge.[69]

---

[66]See James W. Alexander, "A Proof of the Invariance of Certain Constants of Analysis Situs," *Transactions of the American Mathematical Society* 16 (1915):148–154; "A Proof and Extension of the Jordan-Brouwer Separation Theorem," *Transactions of the American Mathematical Society* 23 (1922):333–349; "An Example of a Simply Connected Surface Bounding a Region Which Is Not Simply Connected," *Proceedings of the National Academy of Sciences* 10 (1924):8–10; and "Topological Invariants of Knots and Links," *Transactions of the American Mathematical Society* 30 (1928):275–306, respectively. Jean Dieudonné also gives numerous references to Alexander's work in *A History of Algebraic and Differential Topology*, 1900-1960 (Boston: Birkhäuser, 1989). Alexander later became one of the original faculty members at the Institute for Advanced Study. See Armand Borel, "The School of Mathematics at the Institute for Advanced Study," in *A Century of Mathematics in America—Part* III, pp. 119–148.

[67]See Solomon Lefschetz, "Reminiscences of a Mathematical Immigrant in the United States," in *A Century of Mathematics in America—Part* I, pp. 201–207; and Archibald, *Semicentennial History*, pp. 236–240.

[68]See Roger Cooke and V. Frederick Rickey, "W.E. Story of Hopkins and Clark," in *A Century of Mathematics in America—Part* III, pp. 29–76 on pp. 63–65.

[69]Most of Lefschetz's subsequent investigations concerned fixed-point theorems, a subject which first arose in topological research around 1910 with the classical fixed-point theorem of L. E. J. Brouwer. His first major paper on fixed point theorems was Solomon Lefschetz, "Intersections and Transformations of Complexes and Manifolds," *Transactions of the American*

A teacher famous for his sarcasm, Lefschetz trained a number of important American mathematicians: Paul A. Smith and Norman Steenrod taught topology for many years at Columbia and Princeton, respectively, and Albert W. Tucker became a pioneering figure in game theory and linear programming at Princeton.[70] Other noteworthy Lefschetz students included the knot theorist Ralph Fox, the statistician John Tukey, and the topologist Henry Wallman. Lefschetz also left his mark on the *Annals of Mathematics*, which evolved into one of the world's foremost mathematics journals under his editorship. Finally, he was the author of two notoriously difficult, yet influential textbooks, *Topology* and *Algebraic Topology*.[71]

While algebraic topology flourished at Princeton through the work of Alexander, Lefschetz, Steenrod, and their students, general topology and point set topology became the nearly exclusive province of an expanding school that developed around R. L. Moore, who taught at the University of Pennsylvania from 1911–1920 and at the University of Texas for several decades thereafter.[72] Although well known for his work in general topology, Moore also derived his reputation from his influence as an educator. His so-called "Moore Method" represented a new pedagogical approach which required students to rely on their own ingenuity in devising proofs of the problems presented by the instructor in class. In true Socratic fashion, the professor then posed questions which forced the students to clarify their reasoning and mathematical argumentation.[73] Moore's success as a teacher may have had more to do with his personality than with the efficacy of his highly acclaimed method for training students. The merits or demerits of his approach aside, there was no denying his success. At Pennsylvania, Moore's best student was J. R. Kline, who served for many years as Secretary of the AMS. After Moore's arrival at the University of Texas, however, he came into his own as the sage of point set topology. There, he supervised the dissertations of such American topologists as R. L. Wilder (University of Michigan), G. T. Whyburn (University of Virginia), R. H. Bing (University of Texas), E. E. Moise (City University of New York), Gail Young (Tulane University), Mary Ellen Rudin (University of Wisconsin), and R. D. Anderson (Louisiana State University).[74]

Mathematical Society 28 (1926):1–49.

[70]See the interview with Tucker in *Mathematical People*, ed. Donald J. Albers and Gerald L. Alexanderson (Chicago: Contemporary Books, Inc., 1985), pp. 337–348; and Albert W. Tucker, "Solomon Lefschetz: A Reminiscence," in *ibid.*, pp. 349–352.

[71]Solomon Lefschetz, *Topology* (New York: American Mathematical Society, 1930); and *Algebraic Topology* (New York: American Mathematical Society, 1942).

[72]Raymond L. Wilder, "The Mathematical Work of R.L. Moore: Its Background, Nature, and Influence," *Archive for History of Exact Sciences* 26 (1982): 73–97; reprinted in *A Century of Mathematics in America—Part* I, pp. 265–291.

[73]For a discussion of the influential "Moore Method," see D. Reginald Traylor, *et al.*, *Creative Teaching: Heritage of R. L. Moore* (Houston: University of Houston Press, 1972).

[74]A lengthy list of Moore's academic progeny can be found in Traylor.

As this brief outline indicates, the graduate programs and research traditions that prospered at America's leading institutions during the interwar years had clearly discernible roots that led back to the pivotal period of 1876 to 1900. In the cases of both Chicago and Harvard, the evidence for this could hardly be more compelling, and yet the same can be said about the newer programs at Princeton and MIT as well. The administrative groundwork for their respective successes had largely been laid by Henry Fine and Harry Tyler, who had studied under Felix Klein in Germany during the 1880s.

Looking beyond these local centers to the national scene, and especially to the activities of the American Mathematical Society, this same pattern reappears even more strikingly. With the notable exception of Solomon Lefschetz, practically all of those who assumed leadership positions in American mathematics during the 1920s and 1930s—Dickson, Bliss, Veblen, R. L. Moore, and G. D. Birkhoff—were either students of E. H. Moore at Chicago or descendents of the Chicago school. The mathematicians of this generation, building on the foundation laid by their predecessors, largely shaped and defined those mathematical interests that gave the research issuing from the United States its own distinctive contours. Moreover, they succeeded in pushing American achievements a step beyond those of the preceding generation. Unlike such distinguished figures as Moore, Osgood, and Bôcher, whose influence remained largely confined to the national scene, Birkhoff, Dickson, Wiener, Veblen, Lefschetz, and Alexander attained a level of achievement of international proportion. This group succeeded in clearing new paths of its own, whereas earlier Americans had largely followed in the tracks of the likes of Klein and Hilbert. By the mid-1930s, American mathematicians had shown not only that they could make significant contributions to established areas of research but also, and more importantly, that they could create new fields and isolate new problems that enriched mathematics while attracting researchers, both at home and abroad.

The Flight from Europe and the Dawn of a New Era

While the American mathematical community expanded rapidly over the first third of the century, its development followed a pattern the main features of which were, for the most part, discernible at the outset. Consonant with the research strengths of the country's strongest programs, Americans made notable contributions to group theory (particularly finite groups), algebra, differential geometry, topology, logic, and certain aspects of analysis. Several figures—among them E. H. Moore, Veblen, R. L. Moore, and Harvard's E. V. Huntington—had played important roles in promoting the axiomatization of modern mathematics.[75] At the same time, other major fields—number

[75]See Michael Scanlon, "Who Were the American Postulate Theorists?" *The Journal of Symbolic Logic* 56 (1991):981–1002; and G. D. Birkhoff, "Fifty Years of American Mathematics," pp. 280–286.

theory, most of classical and modern functional analysis, differential equations, and mathematical physics—had received only scant attention. American researchers had begun to define the scope of their enterprise, but even in the 1920s their work (with several notable exceptions) still manifested a rather provincial character.

External events soon intervened that changed the course of American mathematics dramatically, however. Shortly after Hitler was elevated to the office of Chancellor of Germany in January of 1933, the Nazis managed to destroy the once-mighty Mathematics Institute at Göttingen. Had it not been for a few resolute figures within the German mathematical community, the whole enterprise might have fallen prey to the political machinations of Ludwig Bieberbach, who sought to "purify" German mathematics by ideological means.[76] Although Bieberbach and his followers ultimately failed, the ever-worsening political climate triggered a "brain drain" of truly staggering proportions. In the end, the United States emerged as the prime beneficiary.

Four of Hilbert's former students—Dehn, Bernstein, Weyl, and Courant—eventually emigrated to the United States. Weyl and the most famous of the émigrés, Albert Einstein, accepted positions on the original faculty of the Institute for Advanced Study, where they were later joined by the number theorist Carl Ludwig Siegel. Although less prominent displaced scholars sometimes faced cooler receptions owing in part to anti-Semitic sentiments within the American academic community,[77] many of the problems were mitigated through the efforts of a few energetic and well-connected individuals, most notably Oswald Veblen and R. G. D. Richardson.[78]

The successful placement and eventual integration of this new mathematical talent affected the character of American mathematics in several ways. One of the most dramatic changes, however, involved the nation's orientation toward applied mathematics. Following the example of their predecessors in the period from 1876 to 1900, mathematicians in the United States had been almost exclusively involved in pure research during the first third of the century. In fact, mathematical research all over the world was moving in a very purist direction during America's period of expansion through the 1930s. It was thus hardly an accident that the number of those engaged in significant applications had diminished to such a point that, in 1938, G. D. Birkhoff could name only six Americans with serious interests in traditional applied

[76]See Norbert Schappacher, "Fachverband–Institut–Staat," in *Ein Jahrhundert Mathematik, 1890–1990. Festschrift zum Jubiläum der DMV*, ed. Gerd Fischer *et al.* (Braunschweig: Vieweg, 1990), pp. 1–82; and Herbert Mehrtens, "Ludwig Bieberbach and 'Deutsche Mathematik,' " in *Studies in the History of Mathematics*, ed. Esther Phillips, MAA Studies in Mathematics, vol. 26 (Washington, D.C.: Mathematical Association of America, 1987), pp. 195–241.

[77]See Lipman Bers, "The European Mathematicians' Migration to America," in *A Century of Mathematics in America—Part* I, pp. 231–243; and Nathan Reingold, "Refugee Mathematicians in the United States of America, 1933–1941: Reception and Reaction," *Annals of Science* 38 (1981):313–338; reprinted in *A Century of Mathematics in America—Part* I, pp. 175–200.

[78]The Oswald Veblen Papers, located in the Library of Congress, contain much fascinating correspondence documenting his efforts to aid displaced scholars.

mathematics, and four of these six had received their training in Britain.[79] The pendulum of research interests had swung, in the course of about fifty years, to the extreme opposite of the position it had occupied in the days of Benjamin Peirce, G. W. Hill, and Simon Newcomb.

The ensuing period from 1933 to 1960 marked a gradual swing back in the other direction (although the overall orientation of American mathematical research would remain largely purist when compared, for example, with work done in the Soviet Union). This era saw the creation of major centers for applied mathematics led by Richardson at Brown, Courant at New York University, Theodor von Kármán at California Institute of Technology, and Jerzy Neyman at the University of California at Berkeley.[80] At the same time, it witnessed the emergence of Princeton's Institute for Advanced Study as a new type of research environment, an intensely interactive community reminiscent of Göttingen during Hilbert's heyday. What we have styled the "Big Three" continued as major centers, but they now shared the limelight with several newer (and some older) rivals. The American research community lost its provincial innocence and entered a new era characterized by government grants, the advent of computer technology, and expanding educational opportunities. Not only had the cast of leading players changed dramatically but so had the institutional substructure that guided their efforts. At the same time, the character of American research became more eclectic, reflecting the diverse roots of its highly internationalized community.

In certain respects, the dynamic events that shaped American mathematics during this fourth period, from 1933 to 1960, paralleled those that took place during the final quarter of the nineteenth century. In fact, just as internal institutional and external international influences affected the shift in historical developments that ended the third period of quiet consolidation and expansion and led to the tumultuous era that followed, so similar forces had shaped the crucial second period from 1876 to 1900 during which a self-sustaining American mathematical research community had emerged. Individuals both at home and abroad, educational institutions both domestic and foreign, general developments in science and its social and cultural status, broader philosophies of education, political rivalries, and the encroachment of modernity in its several guises, these were among the factors that formed the matrix in which research-level mathematics evolved in the United States during the last quarter of the nineteenth century. By isolating and analyzing the dynamic interrelations between these various factors during this period, we hope not only to have provided insight into the overall historical development of American mathematics but also to have awakened interest in an area which, we believe, has much light to shed on the history of American science.

[79]G. D. Birkhoff, Fifty Years of American Mathematics," p. 313.

[80]On the lives of Courant and Neyman, see Constance Reid, *Courant* (New York: Springer-Verlag, 1976) and *Neyman—From Life* (New York: Springer-Verlag, 1982).

Bibliography

Adams, Henry, *The Education of Henry Adams*, Boston: Houghton Mifflin Co., 1961.

Albers, Donald J. and Alexanderson, Gerald L., ed., *Mathematical People*, Chicago: Contemporary Books, Inc., 1985.

Albers, Donald J.; Alexanderson, Gerald L.; and Reid, Constance, *International Mathematical Congresses: An Illustrated History*, New York: Springer-Verlag, 1987.

Albert, A. Adrian, "Leonard Eugene Dickson 1874-1954," *Bulletin of the American Mathematical Society* **61** (1955):331–345.

Alexandrov, P. S., ed., *Die Hilbertschen Probleme*, Ostwalds Klassiker der exakten Wissenschaften, vol. 252, Leipzig: Akademische Verlagsgesellschaft Geest & Portig, 1979.

Andrews, George E., "Generalizations of the Durfee Square," *Journal of the London Mathematical Society* **3** (1971):563–570.

______, "On a Partition Problem of J. J. Sylvester," *Journal of the London Mathematical Society* **2** (1970):571–576.

______, *The Theory of Partitions*, Encyclopedia of Mathematics and its Applications, vol. 2, Reading: Addison-Wesley Publishing Co., 1976.

Appelbaum, Stanley, *The Chicago World's Fair of* 1893: *A Photographic Record*, New York: Dover Publications, Inc., 1980.

Archer, William, *America To-Day*, London: William Heinemann, 1900.

Archibald, Raymond Clare, "Material Concerning James Joseph Sylvester," *Studies and Essays in the History of Science and Learning Offered in Homage to George Sarton on the Occasion of His Sixtieth Birthday,* 31 *August* 1944, New York: Schuman, n.d., pp. 209–217.

______, *A Semicentennial History of the American Mathematical Society,* 1888–1938, New York: American Mathematical Society, 1938, reprint ed., New York: Arno Press, 1980.

______, "Unpublished Letters of James Joseph Sylvester and Other New Information Concerning His Life and Work," *Osiris* **1** (1936):85–154.

Archibald, Raymond Clare, ed., *Semicentennial Addresses of the American Mathematical Society*, New York: American Mathematical Society, 1938.

Archibald, Raymond Clare *et al.*, "Benjamin Peirce," *American Mathematical Monthly* **32** (1925):1–30.

Aris, R.; David, H. T.; and Stuewer, R. H., ed., *Springs of Scientific Creativity: Essays on Founders of Modern Science*, Minneapolis: University of Minnesota Press, 1983.

Aspray, William, "The Emergence of Princeton as a World Center for Mathematical Research, 1896–1939," in *History and Philosophy of Modern Mathematics,* ed. William Aspray and Philip Kitcher, Minneapolis: University of Minnesota Press, 1988, pp. 346–366.

Aspray, William and Kitcher, Philip, ed., *History and Philosophy of Modern Mathematics*, Minneapolis: University of Minnesota Press, 1988.

Badger, Reid, *The Great American Fair: The World's Columbian Exposition and American Culture*, Chicago: Nelson Hall, 1979.

Baker, H. F., "Biographical Notice," in *The Collected Mathematical Papers of James Joseph Sylvester,* ed. H. F. Baker, 4 vols., Cambridge: University Press, 1904–1912, reprint ed., New York: Chelsea Publishing Co., 1973, 4:xv-xxxvii.

Beach, Mark, "Was There a Scientific Lazzaroni?," in *Nineteenth-Century American Science: A Reappraisal,* ed. George H. Daniels, Evanston: Northwestern University Press, 1972, pp. 115–132.

Beardsley, Edward A., *The Rise of the American Chemical Profession,* 1850–1900, Gainesville: University of Florida Press, 1964.

Beckert, Herbert and Schumann, Horst, ed., 100 *Jahre Mathematisches Seminar der Karl-Marx-Universität Leipzig,* Berlin: VEB Deutscher Verlag der Wissenschaften, 1981.

Bedini, Silvio A., *The Life of Benjamin Banneker*, New York: Charles S. Scribner's Sons, 1971.

Beetle, K. D. and Wilder, C. E., "John Wesley Young: In Memoriam," *Bulletin of the American Mathematical Society* **38** (1932):603–610.

Belhoste, Bruno, *Augustin-Louis Cauchy: A Biography*, trans. Frank Ragland, New York: Springer-Verlag, 1991.

———, *Cauchy*: Un Mathématicien légitimiste au XIXe Siècle, Paris: Librairie Classique Eugène Belin, 1985.

Bell, Eric Temple, *The Development of the History of Mathematics*, New York: McGraw-Hill, 1945.

———, "Fifty Years of Algebra in America, 1888–1938," in *Semicentennial Addresses of the American Mathematical Society,* ed. Raymond C. Archibald, New York: American Mathematical Society, 1938, pp. 1–34.

Bellot, H. Hale, *University College London* 1826–1926, London: University of London Press, Ltd., 1929.

Benson, Keith R., "American Morphology in the Late Nineteenth Century: The Biology Department at Johns Hopkins University," *Journal of the History of Biology* **18** (1985):163–205.

Bers, Lipman, "The European Mathematicians' Migration to America," in *A Century of Mathematics in America—Part* I, ed. Peter Duren *et al.*, Providence: American Mathematical Society, 1988, pp. 231–243.

Betsch, Gerhard, "Alexander von Brill (1842-1935)," *Bausteine zur Tübingen Universitätsgeschichte* **3** (1989):71–90.

Biermann, Kurt-R., "Johann Peter Gustav Lejeune Dirichlet, Dokumente für sein Leben und Wirken," *Abhandlungen der deutschen Akademie der Wissenschaften zu Berlin, Klasse für Mathematik, Physik, und Technik* **2** (1959):2–88.

———, *Die Mathematik und ihre Dozenten an der Berliner Universität,* 1810–1933, Berlin: Akademie-Verlag, 1988.

"Biographical Sketch of the President of the Association [Hubert Anson Newton]," *Science* **6** (1885):161–162.

Birkhoff, Garrett, "Mathematics at Harvard, 1836-1944," in *A Century of Mathematics in American—Part* II, ed. Peter Duren *et al.*, Providence: American Mathematical Society, 1989, pp. 3–58.

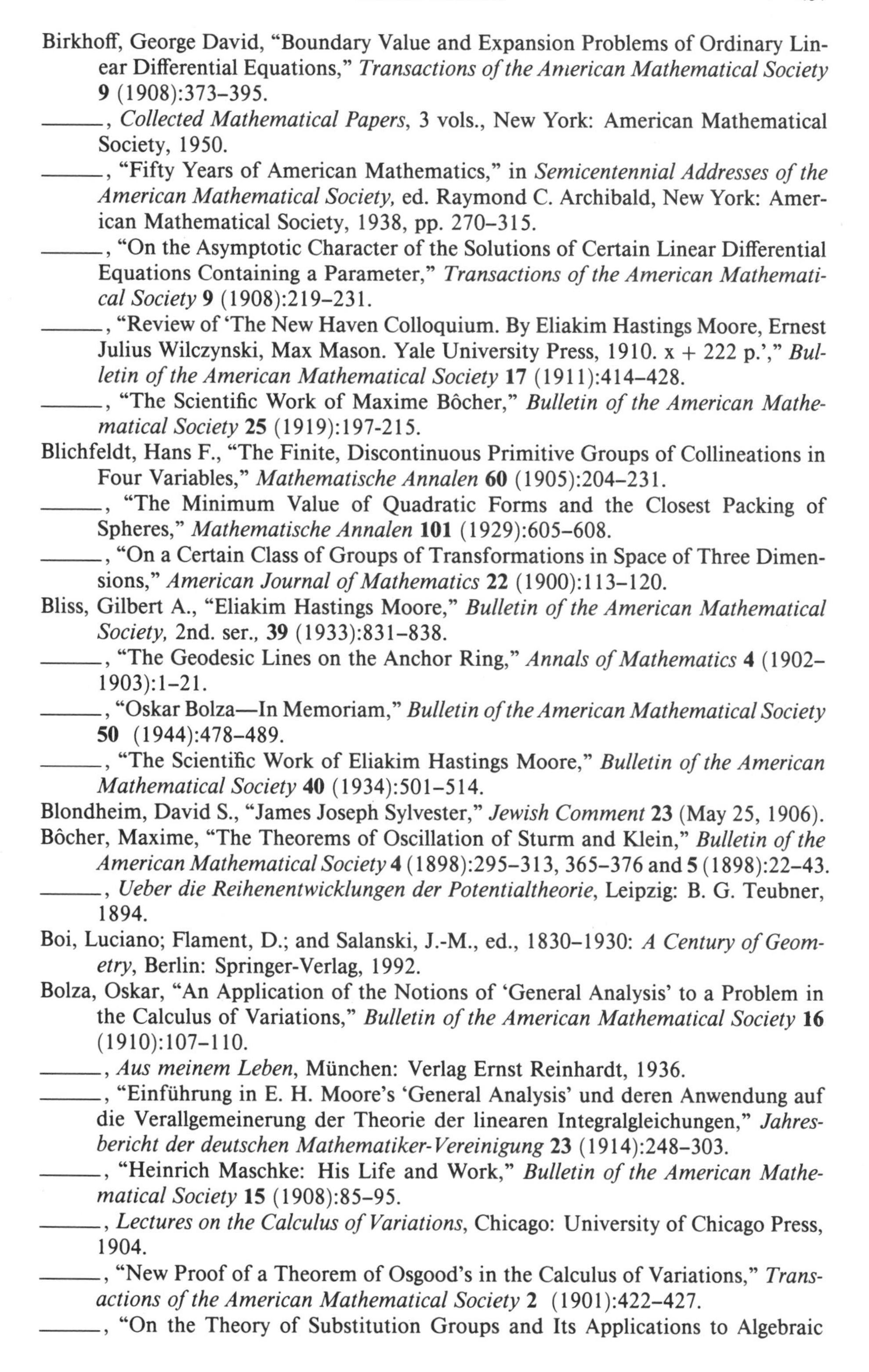

Birkhoff, George David, "Boundary Value and Expansion Problems of Ordinary Linear Differential Equations," *Transactions of the American Mathematical Society* **9** (1908):373–395.

———, *Collected Mathematical Papers*, 3 vols., New York: American Mathematical Society, 1950.

———, "Fifty Years of American Mathematics," in *Semicentennial Addresses of the American Mathematical Society*, ed. Raymond C. Archibald, New York: American Mathematical Society, 1938, pp. 270–315.

———, "On the Asymptotic Character of the Solutions of Certain Linear Differential Equations Containing a Parameter," *Transactions of the American Mathematical Society* **9** (1908):219–231.

———, "Review of 'The New Haven Colloquium. By Eliakim Hastings Moore, Ernest Julius Wilczynski, Max Mason. Yale University Press, 1910. x + 222 p.'," *Bulletin of the American Mathematical Society* **17** (1911):414–428.

———, "The Scientific Work of Maxime Bôcher," *Bulletin of the American Mathematical Society* **25** (1919):197-215.

Blichfeldt, Hans F., "The Finite, Discontinuous Primitive Groups of Collineations in Four Variables," *Mathematische Annalen* **60** (1905):204–231.

———, "The Minimum Value of Quadratic Forms and the Closest Packing of Spheres," *Mathematische Annalen* **101** (1929):605–608.

———, "On a Certain Class of Groups of Transformations in Space of Three Dimensions," *American Journal of Mathematics* **22** (1900):113–120.

Bliss, Gilbert A., "Eliakim Hastings Moore," *Bulletin of the American Mathematical Society,* 2nd. ser., **39** (1933):831-838.

———, "The Geodesic Lines on the Anchor Ring," *Annals of Mathematics* **4** (1902–1903):1-21.

———, "Oskar Bolza—In Memoriam," *Bulletin of the American Mathematical Society* **50** (1944):478–489.

———, "The Scientific Work of Eliakim Hastings Moore," *Bulletin of the American Mathematical Society* **40** (1934):501-514.

Blondheim, David S., "James Joseph Sylvester," *Jewish Comment* **23** (May 25, 1906).

Bôcher, Maxime, "The Theorems of Oscillation of Sturm and Klein," *Bulletin of the American Mathematical Society* **4** (1898):295–313, 365–376 and **5** (1898):22–43.

———, *Ueber die Reihenentwicklungen der Potentialtheorie*, Leipzig: B. G. Teubner, 1894.

Boi, Luciano; Flament, D.; and Salanski, J.-M., ed., 1830–1930: *A Century of Geometry*, Berlin: Springer-Verlag, 1992.

Bolza, Oskar, "An Application of the Notions of 'General Analysis' to a Problem in the Calculus of Variations," *Bulletin of the American Mathematical Society* **16** (1910):107–110.

———, *Aus meinem Leben*, München: Verlag Ernst Reinhardt, 1936.

———, "Einführung in E. H. Moore's 'General Analysis' und deren Anwendung auf die Verallgemeinerung der Theorie der linearen Integralgleichungen," *Jahresbericht der deutschen Mathematiker-Vereinigung* **23** (1914):248–303.

———, "Heinrich Maschke: His Life and Work," *Bulletin of the American Mathematical Society* **15** (1908):85–95.

———, *Lectures on the Calculus of Variations*, Chicago: University of Chicago Press, 1904.

———, "New Proof of a Theorem of Osgood's in the Calculus of Variations," *Transactions of the American Mathematical Society* **2** (1901):422–427.

———, "On the Theory of Substitution Groups and Its Applications to Algebraic

Equations," *American Journal of Mathematics* **13** (1891):59–144.

_____, "On Weierstrass' Systems of Hyperelliptic Integrals of the First and Second Kind," in *Mathematical Papers Read at the International Mathematical Congress Held in Connection with the World's Columbian Exposition Chicago* 1893, ed. E. H. Moore *et al.*, New York: Macmillan & Co., 1896, pp. 1–12.

_____, "Über den 'Anormalen Fall' beim Lagrangeschen und Mayerschen Problem mit gemischten Bedingungen und variablen Endpunkten," *Mathematische Annalen* **74** (1913):430–446.

_____, *Vorlesungen über Variationsrechnung*, Leipzig: B. G. Teubner, 1909.

Boole, George, "Exposition of a General Theory of Linear Transformations," *Cambridge Mathematical Journal* **3** (1841–1842):1–20 and 106–119.

Borel, Armand, "The School of Mathematics at the Institute for Advanced Study," in *A Century of Mathematics in America—Part* III, ed. Peter Duren *et al.*, Providence: American Mathematical Society, 1989, pp. 119–148.

Bottazzini, Umberto, *The Higher Calculus: A History of Real and Complex Analysis from Euler to Weierstrass*, trans. Warren Van Egmond, New York: Springer-Verlag, 1986.

_____, "Ursprünge der Riemannschen Theorie der Funktionen einer veränderlichen komplexen Grösse," unpublished manuscript.

Bourbaki, Nicolas, *Éléments d'Historire des mathématiques*, Paris: Masson, 1984.

_____, *Groupes et Algèbres de Lie: Chapîtres* IV, V, *et* VI, Paris: Hermann, 1968.

Bouton, Charles L., "Invariants of the General Linear Differential Equation and Their Relation to the Theory of Continuous Groups," *American Journal of Mathematics* **21** (1899):25–84.

Bowditch, Henry Ingersoll, "Memoir," in Pierre Simon de la Place, *Celestial Mechanics*, trans. Nathaniel Bowditch, 4 vols., New York: Chelsea Publishing Co., 1966.

Bowditch, Nathaniel, *American Practical Navigator*, Washington: Government Printing Office, 1802, reprint ed., n. p.: Defense Mapping Agency Hydrographic/ Topographic Center, 1984.

Brent, Joseph, *Charles Sanders Peirce: A Life*, Bloomington: Indiana University Press, 1993.

Brieskorn, Egbert and Knörrer, Horst, *Plane Algebraic Curves*, trans. John Stillwell, Boston: Birkhäuser Verlag, 1986.

Brill, Alexander, "Max Noether," *Jahresbericht der deutschen Mathematiker-Vereinigung* **32** (1923):211–233.

Brill, Alexander and Noether, Max, "Die Entwicklung der Theorie der algebraischen Functionen in alterer und neuerer Zeit," *Jahresbericht der deutschen Mathematiker-Vereinigung* **3** (1892):107–566.

Brocke, Bernhard vom, "Der deutsch-amerikanischen Professorenaustausch. Preußische Wissenschaftspolitik, internationale Wissenschaftsbeziehungen und die Anfänge einer deutschen auswärtigen Kulturpolitik vor dem ersten Weltkrieg," *Zeitschrift für Kulturaustausch* **31** (1981):128–182.

Browder, Felix, ed., *Mathematical Developments Arising from Hilbert's Problems*, Symposia in Pure Mathematics, vol. 28, Providence: American Mathematical Society, 1976.

Brown, Ernest W., "Theorie des Erdmondes," *Encyklopädie der mathematischen Wissenschaften* vol. VI-2 (1915):667–728.

Brualdi, Richard A., *Introductory Combinatorics*, 2nd ed., New York: North-Holland, 1992.

Bruce, Robert V., *The Launching of Modern American Science*: 1846–1876, New York: Alfred A. Knopf, 1987.

———, "A Statistical Profile of American Scientists, 1846–1876," in *Nineteenth-Century American Science*: *A Reappraisal,* ed. George H. Daniels, Evanston: Northwestern University Press, 1972, pp. 63–94.

Buel, James William, *The Magic City*: *A Massive Portfolio of Original Photographic Views of the Great World's Fair and Its Treasures of Art, Including a Vivid Presentation of the Famous Midway Plaisance*, St. Louis: Historical Publishing Co., 1894, reprint ed., New York: Arno Press, 1974.

Bumstead, H. A., "Josiah Willard Gibbs," *The Scientific Papers of Josiah Willard Gibbs* ... , 2 vols., New York: Longmans, Green, & Co., 1906, reprint ed., New York: Dover Publications, Inc., 1961, 1:xi–xxvi.

———, "Josiah Willard Gibbs," *American Journal of Science,* ser. 4, **16** (1903):187–202.

Burkhardt, Heinrich, "Grundzüge einer allgemeinen Systematik der hyperelliptischen Funktionen erster Ordnung: Nach Vorlesungen von F. Klein," *Mathematische Annalen* **35** (1890):198–296.

———, "Ueber einige mathematische Resultaten astronomischer Untersuchungen, insbesondere über irreguläre Integrale linearer Differentialgleichungen," in *Mathematical Papers Read at the International Mathematical Congress Held in Connection with the World's Columbian Exposition Chicago* 1893, ed. E. H. Moore *et al.*, New York: Macmillan & Co., 1896, pp. 350–366.

Burckhardt, J. J., "Der Briefwechsel von E. S. von Federow und F. Klein, 1893," *Archive for History of Exact Sciences* **9** (1972):85–93.

———, "Der Briefwechsel von E. S. von Federow und A. Schoenflies," *Archive for History of Exact Sciences* **7** (1971):91–141.

Burke, Colin B., *American Collegiate Population*: *A Test of the Traditional View*, New York and London: New York University Press, 1982.

Butler, Loren, "Mathematical Physics and the American Mathematical Community: Disciplinary Values, Professional Interests, and the Place of Borderland Research 1880–1940," unpublished Ph.D. dissertation, University of Chicago, 1992.

Cairns, W. D., "Herbert Ellsworth Slaught—Editor and Organizer," *American Mathematical Monthly* **45** (1938):1–4.

Cajori, Florian, *The Teaching and History of Mathematics in the United States*, Washington: Government Printing Office, 1890.

Caldi, D. G. and Mostow, G. D., ed., *Proceedings of the Gibbs Symposium*: *Yale University, May* 15–17, 1989, Providence: American Mathematical Society, 1990.

Cartan, Élie, *Oeuvres complètes*, 3 vols. in 6 pts., Paris: Gauthier-Villars, 1952–1955.

Cattell, James McKeen, *American Men of Science*: *A Biographical Directory*, 1st ed., New York: Science Press, 1906.

———, *American Men of Science*: *A Biographical Directory*, 2nd ed., New York: Science Press, 1910.

Cayley, Arthur, *The Collected Mathematical Papers of Arthur Cayley*, ed. Arthur Cayley and A. R. Forsyth, 14 vols., Cambridge: University Press, 1889-1898.

———, "An Introductory Memoir Upon Quantics," *Philosophical Transactions of the Royal Society of London* **144** (1854):244–258.

———, "A Ninth Memoir on Quantics," *Philosophical Transactions of the Royal Society of London* **161** (1871):17–50.

———, "On Linear Transformations," *Cambridge and Dublin Mathematical Journal* **1** (1846):104–122.

———, "On the Theory of Linear Transformations," *Cambridge Mathematical Journal* **4** (1845):193–209.

———, "A Second Memoir Upon Quantics," *Philosophical Transactions of the Royal Society of London* **146** (1856):101–126.

———, "A Sixth Memoir Upon Quantics," *Philosophical Transactions of the Royal Society of London* **149** (1859):61–90.

———, "A Third Memoir Upon Quantics," *Philosophical Transactions of the Royal Society of London* **146** (1856):627–647.

Chandrasekharan, Komaravolu, *Elliptic Functions*, New York: Springer-Verlag, 1985.

Chauvenet, William, *A Treatise on Elementary Geometry*, Philadelphia: J. B. Lippincott, 1870.

———, *Treatise on Plane and Spherical Trigonometry*, Philadelphia: Hogan, Perkins & Co., 1850.

Chessin, Alexander S., "The Preliminary Meeting of the Southwestern Section," *Bulletin of the American Mathematical Society* **13** (1907):213–223.

Chittenden, Russell H., *History of the Sheffield Scientific School of Yale University: 1846–1922*, 2 vols., New Haven: Yale University Press, 1928.

Clebsch, Alfred, "Über die Anwendung der Abel'schen Functionen in der Geometrie," *Journal für die reine und angewandte Mathematik* **63** (1864):189–243.

———, *Vorlesungen über Geometrie*, ed. Ferdinand Lindemann, Leipzig: B. G. Teubner, 1876.

Clebsch, Alfred and Gordan, Paul, *Theorie der Abelschen Functionen*, Leipzig: B. G. Teubner, 1866.

Clifford, William Kingdon, *Mathematical Papers by William Kingdon Clifford*, ed. Robert Tucker, London: Macmillan and Co., 1882, reprint ed., New York: Chelsea Publishing Company, 1968.

Coble, Arthur B., "Virgil Snyder, 1869–1950," *Bulletin of the American Mathematical Society* **56** (1950):468–471.

Cochrane, Rexmond C., *The National Academy of Sciences: The First Hundred Years, 1863–1963*, Washington: The Academy, 1978.

Cole, Frank N., "A Contribution to the Theory of the General Equation of the Sixth Degree," *American Journal of Mathematics* **8** (1886):265–286.

———, "Klein's Ikosaeder," *American Journal of Mathematics* **9** (1887):45–61.

———, "On a Certain Simple Group," in *Mathematical Papers Read at the International Mathematics Congress Held in Connection with the World's Columbian Exposition Chicago* 1893, ed. E. H. Moore *et al.*, New York: Macmillan & Co., 1896, pp. 40–43.

———, "Simple Groups As Far As Order 660," *American Journal of Mathematics* **15** (1893):303-315.

Condit, Carl W., *The Chicago School of Architecture: A History of Commercial and Public Building in the Chicago Area*, 1875–1925, Chicago: University of Chicago Press, 1964.

Cooke, Roger, "Abel's Theorem," in *The History of Modern Mathematics*, ed. David E. Rowe and John McCleary, 2 vols., Boston: Academic Press, Inc., 1989, 1:389–421.

———, *The Mathematics of Sonya Kovalevskaya*, New York: Springer-Verlag, 1984.

Cooke, Roger and Rickey, V. Frederick, "W.E. Story of Hopkins and Clark," in *A Century of Mathematics in America–Part* III, ed. Peter Duren *et al.*, Providence: American Mathematical Society, 1989, pp. 29–76.

Coolidge, Julian Lowell, *A History of Geometrical Methods*, Oxford: Clarendon Press, 1940.

———, *A History of the Conic Sections and Quadric Surfaces*, Oxford: Clarendon Press, 1945.

———, *The Mathematics of Great Amateurs*, Oxford: Clarendon Press, 1949.

———, "Robert Adrain, and the Beginnings of American Mathematics," *American Mathematical Monthly* **33** (1926):61–76.

———, *A Treatise on Algebraic Plane Curves*, Oxford: Clarendon Press, 1931.

Cordasco, Francesco, *Daniel Coit Gilman and the Protean Ph.D.: The Shaping of American Graduate Education*, Leiden: E. J. Brill, 1960.

Courant, Richard, "Felix Klein," *Die Naturwissenschaften* **37** (1925):765–772.

Courant, Richard and Hilbert, David, *Methoden der mathematischen Physik*, Berlin: Springer-Verlag, 1924.

Coxeter, H. S. M., "The Complete Enumeration of Finite Groups of the Form $R_i^2 = (R_iR_j)^{k_{ij}} = 1$," *Journal of the London Mathematical Society* **10** (1935):21–25.

———, "Discrete Groups Generated by Reflections," *Annals of Mathematics* **35** (1934):588–621.

———, *Regular Complex Polytopes*, Cambridge: University Press, 1974.

———, *Regular Polytopes*, New York: The Macmillan Company, 1963, reprint ed., New York: Dover Publications, Inc., 1973.

Craig, Gordan, *The Germans*, New York: New American Library, 1983.

Craig, Thomas, "The Counter-Pedal Surface of the Ellipsoid," *American Journal of Mathematics* **4** (1881):358–378.

———, "On the Parallel Surface to an Ellipsoid," *Journal für die reine und angewandte Mathematik* **93** (1882):251–270.

———, "Orthomorphic Projection of an Ellipsoid on a Sphere," *American Journal of Mathematics* **3** (1880):114–127.

———, "Some Elliptic Function Formulae," *American Journal of Mathematics* **5** (1882):62–75.

———, *A Treatise on Projections*, Washington: United States Coast and Geodetic Survey, 1882.

Cremin, Lawrence A., *The Transformation of the School: Progressivism in American Education* 1876–1957, New York: Alfred A. Knopf, 1961.

Crilly, Anthony James, "The Mathematics of Arthur Cayley with Particular Reference to Linear Algebra", unpublished Ph.D. dissertation, Middlesex Polytechnic, June 1981.

———, "The Decline of Cayley's Invariant Theory (1863–1895)," *Historia Mathematica* **15** (1988):332–347.

———, "The Rise of Cayley's Invariant Theory (1841–1862)," *Historia Mathematica* **13** (1986):241–254.

Crowe, Michael J., *A History of Vector Analysis: The Evolution of the Idea of a Vectorial System*, Notre Dame: University of Notre Dame Press, 1967.

Crozet, Claude, *A Treatise on Descriptive Geometry for the Use of the Cadets of the United States Military Academy*, New York: A. T. Goodrich & Co., 1821.

Currey, J. Seymour, *A Century of Marvelous Growth*, 5 vols., Chicago: Clarke Publishing Co., 1912, especially volume 3, *Chicago and Its History and Builders.*

Curtis, Charles W. and Reiner, Irving, *Representation Theory of Finite Groups and Associative Algebras*, New York: Interscience Publishers, 1962.

Dalmedico, Amy Dahan, *Mathématisations: Augustin-Louis Cauchy et l'École française*, Argenteuil: Éditions du Choix and Paris: Albert Blanchard, 1993.

Daniels, George H., *American Science in the Age of Jackson*, New York: Columbia University Press, 1968.

______, "The Process of Professionalization in American Science: The Emergent Period, 1820–1860," *Isis* **58** (1967):151–166.

Daniels, George H., ed., *Nineteenth-Century American Science: A Reappraisal*, Evanston: Northwestern University Press, 1972.

Davies, Charles, *Elementary Algebra: Embracing the First Principles of the Science*, New York: A. S. Barnes & Burr, 1862.

Day, Jeremiah, *An Introduction to Algebra, Being the First Part of a Course of Mathematics, Adapted to the Method of Instruction in the American Colleges*, New Haven: Howe & Deforest, 1814.

______, *The Mathematical Principles of Navigation and Surveying, with the Mensuration of Heights and Distances, Being the Fourth Part of a Course of Mathematics, Adapted to the Method of Instruction in the American Colleges*, New Haven: Steele & Gray, 1817.

______, *A Practical Application of the Principles of Geometry to the Mensuration of Superficies and Solids, Being the Third Part of a Course of Mathematics, Adapted to the Method of Instruction in the American College*, New Haven: Oliver Steele, 1814.

______, *A Treatise of Plane Trigonometry to Which is Prefixed a Summary View of the Nature and Use of Logarithms, Being the Second Part of a Course of Mathematics, Adapted to the Method of Instruction in the American Colleges*, New Haven: Howe & Deforest, 1815.

Dedekind, Richard, "Schreiben an Herrn Borchardt über die Theorie der elliptischen Modulfunktionen," *Journal für die reine und angewandte Mathematik* **83** (1877):265–292.

Delambre, Jean-Baptiste, *Rapport historique sur les Progrès des Sciences mathématiques depuis* 1789, *et sur leur État actuel*, Paris: Imprimerie impériale, 1810, reprint ed., Amsterdam: N. V. Boekhandel & Antiquariaat B. M. Israël, 1966.

Demidov, Sergei S.; Folkerts, Menso; Rowe, David E.; Scriba, Christoph, eds., *Amphora: Festschrift für Hans Wussing zu seinem* 65. *Geburtstag*, Basel: Birkhäuser Verlag, 1992.

Dewey, John, *Lectures in the Philosophy of Mathematics: 1899*, ed. Reginald D. Archimbault, New York: Random House, 1966.

Dickson, Leonard E., "The Analytic Representation of Substitutions on a Power of a Prime Number of Letters with a Discussion of the Linear Group," *Annals of Mathematics* **11** (1897):65–143.

______, *The Collected Mathematical Papers of Leonard Eugene Dickson*, ed. A. Adrian Albert, 6 vols., New York: Chelsea Publishing Company, 1975 and 1983.

______, "Definitions of a Group and a Field by Independent Postulates," *Transactions of the American Mathematical Society* **6** (1905):198–204.

______, "Eliakim Hastings Moore," *Science* **77** (1933):79–80.

______, *Linear Groups with an Exposition of the Galois Field Theory*, Leipzig: B. G. Teubner, 1901, reprint ed., New York: Dover Publications, Inc., 1958.

Dieudonné, Jean, *A History of Algebraic and Differential Topology*, 1900–1960, Boston: Birkhäuser Verlag, 1989.

______, "The Tragedy of Grassmann," *Linear and Multilinear Algebra* **8** (1) (1979–1980):1–14.

Dieudonné, Jean and Carrell, James B., *Invariant Theory, Old and New*, New York: Academic Press, Inc., 1971.

Dirichlet, Peter Lejeune, *Vorlesungen über Zahlentheorie*, ed. Richard Dedekind, 1st ed., Braunschweig: F. Vieweg und Sohn, 1863.

Donaldson, James A., "Black Americans in Mathematics," in *A Century of Math-*

ematics in America—Part III, ed. Peter Duren *et al.*, Providence: American Mathematical Society, 1989, pp. 449–469.

Dorwart, Harold, "Mathematics and Yale in the Nineteen Twenties," in *A Century of Mathematics in America—Part* II, ed. Peter Duren *et al.*, Providence: American Mathematical Society, 1989, pp. 87-98.

Douglas, Jesse, "Edward Kasner", *Biographical Memoirs of the National Academy of Sciences*, 31 1958, pp. 180–209.

Dresden, Arnold, "A Report on the Scientific Work of the Chicago Section, 1897–1922," *Bulletin of the American Mathematical Society* **28** (1922):303–307.

Dronke, Adolf, *Julius Plücker, Professor der Mathematik und Physik an der Rheinischen Friedrich Wilhelms-Universität in Bonn*, Bonn: Adolf Marcus, 1871.

Dugac, Pierre, *Richard Dedekind et les Fondements des Mathématiques*, Collection des Travaux de l'Académie internationale d'Histoire des Sciences, vol. 24, Paris: Librairie philosophique J. Vrin, 1976.

Dunnington, G. Waldo, *Carl Friedrich Gauss: Titan of Science*, New York: Exposition Press, 1955.

Dupree, A. Hunter, *Science in the Federal Government: A History of Policies and Activities*, Baltimore: Johns Hopkins University Press, 1986.

Duren, Peter *et al.*, ed., *A Century of Mathematics in America—Part* I, Providence: American Mathematical Society, 1988.

——, *A Century of Mathematics in America—Part* II, Providence: American Mathematical Society, 1989.

——, *A Century of Mathematics in America—Part* III, Providence: American Mathematical Society, 1989.

Durfee, William Pitt, "Tables of the Symmetric Functions of the Twelfthic," *American Journal of Mathematics* **5** (1882):45–61.

——, "The Tabulation of Symmetric Functions," *American Journal of Mathematics* **5** (1882):348–349.

Dyck, Walther von, ed., *Amtlicher Bericht über die Weltausstellungen in Chicago* 1893, *erstattet von Reichskommisar*, 2 vols., Berlin: Reichsdruckerei, 1894.

——, *Deutsche Unterrichtsausstellung in Chicago,* 1893: *Special-Katalog der mathematischen Ausstellung*, Berlin: n.p., 1893.

——, *Katalog mathematischer und mathematische-physikalischer Modelle, Apparate und Instrumente*, Munich: C. Wolf, 1892.

Echeverria, Javier *et al.*, ed., *The Space of Mathematics: Philosophical, Epistemological, and Historical Explorations*, Berlin: Walter de Gruyter, 1992.

Eddy, Henry T., *Neue Constructionen aus der graphischen Statik*, Leipzig: B. G. Teubner, 1880.

——, "Modern Graphical Developments," in *Mathematical Papers Read at the International Mathematical Congress Held in Conjunction with the World's Columbian Exposition Chicago* 1893, ed. E. H. Moore *et al.*, New York: Macmillan & Co., 1896, pp. 58–71.

——, *Researches in Graphical Statics*, New York: D. Van Nostrand, 1878.

Edwards, Harold M., *Fermat's Last Theorem: A Genetic Approach to Algebraic Number Theory*, New York: Springer-Verlag, 1977.

Eisenhart, Luther P., *Non-Riemannian Geometry*, New York: American Mathematical Society, 1927.

——, *Riemannian Geometry*, Princeton: University Press, 1926.

Elliott, Clark A., *Biographical Dictionary of American Science: The Seventeenth through the Nineteenth Centuries*, Westport: Greenwood Press, 1979.

Elliott, Edwin Bailey, *An Introduction to the Algebra of Quantics*, Oxford: University

Press, 1895, reprint ed., Bronx, N. Y.: Chelsea Publishing Co., 1964.

Ely, George S., "Bibliography of Bernoulli Numbers," *American Journal of Mathematics* **5** (1882):228–235.

Engel, Friedrich, "Zur Erinnerung an Sophus Lie," *Berichte über die Verhandlungen der königlichen sächsischen Gesellschaft der Wissenschaften zu Leipzig* **51** (1899):11–61.

Enros, Philip C., "The Analytical Society: Mathematics at Cambridge in the Early Nineteenth Century", doctoral dissertation, University of Toronto, 1979.

———, "The Analytical Society (1812–1813): Precursor of the Revival of Cambridge Mathematics," *Historia Mathematica* **10** (1983):24–47.

———, "Cambridge University and the Adoption of Analytics in Early Nineteenth-Century England," in *Social History of Nineteenth Century Mathematics,* ed. Herbert Mehrtens, Henk Bos, and Ivo Schneider, Boston: Birkhäuser Verlag, 1981, pp. 135–148.

Ernst, Wilhelm, *Julius Plücker: Eine zusammenfassende Darstellung seines Lebens und Wirkens als Mathematiker und Physiker auf Grund unveröffentlicher Briefe und Urkunden*, Bonn: Universitäts-Buchdruckerei, 1933.

Faà de Bruno, Francesco, *Théorie des Formes binaires*, Turin: P. Marietti, 1876.

Farrar, John and Emerson, George, *First Principles of Differential and Integral Calculus, or The Doctrine of Fluxions ... Taken Chiefly from the Mathematics of Bézout*, Cambridge: Hilliard & Metcalf, 1824.

Fenster, Della Dumbaugh and Parshall, Karen Hunger, "A Profile of the American Mathematical Research Community, 1891–1906," in *The History of Modern Mathematics,* vol. 3, ed. Eberhard Knobloch and David E. Rowe, Boston: Academic Press, Inc., 1994, forthcoming.

———, "Women in the American Mathematical Research Community," in *The History of Modern Mathematics,* vol. 3, ed. Eberhard Knobloch and David E. Rowe, Boston: Academic Press, Inc., 1994, forthcoming.

Feuer, Lewis S., "America's First Jewish Professor: James Joseph Sylvester at the University of Virginia," *American Jewish Archives* **36** (1984):151–201.

———, "Sylvester in Virginia," *The Mathematical Intelligencer* **9** (2) (1987):13–19.

Fine, Henry B., "Kronecker and his Arithmetical Theory of Algebraic Equations," *Bulletin of the New York Mathematical Society* **1** (1892):173–184.

Fischer, Gerd, ed., *Mathematische Modelle*, 2 vols., Berlin: Akademie Verlag, 1986.

Fischer, Gert *et al.*, ed., *Ein Jahrhundert Mathematik,* 1890–1990. *Festschrift zum Jubiläum der DMV*, Braunschweig: Vieweg, 1990.

Fiske, Thomas S., "Frank Nelson Cole," *Bulletin of the American Mathematical Society* **33** (1927):773–777.

———, "Mathematical Progress in America," *Bulletin of the American Mathematical Society* **11** (1905):238–246.

Forman, Paul; Heilbron, John L.; and Weart, Spencer, "Physics circa 1900: Personnel and Productivity of the Academic Establishments," *Historical Studies of the Physical Sciences* **5** (1975):1-185.

Franklin, Fabian, *The Life of Daniel Coit Gilman*, New York: Dodd, Mead and Co., 1910.

———, "On the Calculation of the Generating Functions and Tables of Groundforms for Binary Quantics," *American Journal of Mathematics* **3** (1880):128–153.

———, "On a Problem of Isomerism," *American Journal of Mathematics* **1** (1878): 365–368.

———, "Sur le Développement du Produit infini $(1-x)(1-x^2)(1-x^3)(1-x^4)\cdots$," *Comptes rendus* **82** (1881):448–450.

Frei, Günther, ed., *Der Briefwechsel David Hilbert–Felix Klein* (1886–1918), Arbeiten aus der niedersächsischen Staats- und Universitätsbibliothek Göttingen, vol. 19, Göttingen: Vandenhoeck & Ruprecht, 1985.

French, John C., *History of the University Founded by Johns Hopkins*, Baltimore: Johns Hopkins University Press, 1946.

Frobenius, Ferdinand Georg, *Ferdinand Georg Frobenius: Gesammelte Abhandlungen*, ed. Jean-Pierre Serre, 3 vols., Berlin: Springer-Verlag, 1968.

_____, "Ueber lineare Substitutionen und bilineare Formen," *Journal für die reine und angewandte Mathematik* **84** (1878):1–63.

Gauss, Carl Friedrich, *Disquisitiones arithmeticæ*, trans. Arthur A. Clarke, rev. William C. Waterhouse *et al.*, New York: Springer-Verlag, 1966.

Gibbs, Josiah Willard, *Elementary Principles of Statistical Mechanics*, New York: C. Scribner's Sons, 1902.

_____, "Hubert Anson Newton," *American Journal of Science*, 4th ser. **3** (1897):358–378.

_____, *The Scientific Papers of J. Willard Gibbs, Ph.D., LL.D. formerly Professor of Mathematical Physics in Yale University*, 2 vols., New York: Longmans, Green, & Co., 1906, reprint ed., New York: Dover Publications, Inc., 1961.

Giedion, Sigfried, *Space, Time, and Architecture*, 3rd ed., Cambridge, MA: Harvard University Press, 1954.

Gillies, Donald, ed., *Revolutions in Mathematics*, Oxford: Clarendon Press, 1992.

Gillispie, Charles C., ed., *Dictionary of Scientific Biography*, 16 vols., 2 supps., New York: Charles Scribner's Sons, 1970–1990.

Gilman, Daniel Coit, *The Launching of a University and Other Papers: A Sheaf of Remembrances*, New York: Dodd, Mead & Co., 1906.

Goldstine, Herman H., *A History of the Calculus of Variations from the 17th through the 19th Century*, New York: Springer-Verlag, 1980.

Goodspeed, Thomas Wakefield, *A History of the University of Chicago Founded by John D. Rockefeller: The First Quarter-Century*, Chicago: University of Chicago Press, 1916.

Gordan, Paul, "Beweis, dass jede Covariante und Invariante einer binären Form eine ganze Function mit numerische Coefficienten einer endlichen Anzahl solchen Formen ist," *Journal für die reine und angewandte Mathematik* **69** (1868):323–354.

_____, *Vorlesungen über Invariantentheorie*, ed. Georg Kerschensteiner, 2 vols., Leipzig: B. G. Teubner, 1885–1887.

Gorenstein, Daniel, "The Classification of the Finite Simple Groups. A Personal Journey: The Early Years," in *A Century of Mathematics in America—Part I*, ed. Peter Duren *et al.*, Providence: American Mathematical Society, 1988, pp. 447–476.

Gould, Joseph E., *The Chautauqua Movement: An Episode in the Continuing American Revolution*, n.p.: State University of New York, 1961.

Goursat, Édouard, "A Simple Proof of a Theorem in the Calculus of Variations (Extract of a Letter to Mr. W. F. Osgood)," *Transactions of the American Mathematical Society* **5** (1904):110–112.

Grabiner, Judith V., "Mathematics in America: The First Hundred Years," in *The Bicentennial Tribute to American Mathematics*, 1776–1976, ed. Dalton Tarwater, n.p.: The Mathematical Association of America, 1977, pp. 9–24.

_____, *The Origins of Cauchy's Rigorous Calculus*, Cambridge, MA: MIT Press, 1981.

Grassmann, Hermann, *Die lineale Ausdehnungslehre, ein neuer Zweig der Mathematik dargestellt und durch Anwendungen auf die übrigen Zweigen der Mathematik, wie auch auf die Statik, Mechanik, die Lehre vom Magnetismus und die Krystallonomie erläutert*, Leipzig: O. Wigand, 1844.

Grattan-Guinness, Ivor, *Convolutions in French Mathematics,* 1800–1840: *From the Calculus and Mechanics to Mathematical Analysis and Mathematical Physics*, 3 vols., Basel: Birkhäuser Verlag, 1990.

——, *The Development of the Foundations of Mathematical Analysis from Euler to Riemann*, Cambridge, MA: MIT Press, 1970.

——, "Does History of Science Treat of the History of Science? The Case of Mathematics," *History of Science* **28** (1990):149–173.

——, "A Mathematical Union: William Henry and Grace Chisholm Young," *Annals of Science* **29** (1972):105–183.

Graustein, William C., "Invariant Methods in Classical Differential Geometry," *Bulletin of the American Mathematical Society* **36** (1930):489–521.

Graves, Lawrence M., "Gilbert Ames Bliss 1876–1951," *Bulletin of the American Mathematical Society* **58** (1952):251–264.

Gray, Jeremy, "Algebraic Geometry in the late Nineteenth Century," in *The History of Modern Mathematics,* ed. David E. Rowe and John McCleary, 2 vols., Boston: Academic Press, Inc., 1989, 1:361–385.

——, *Linear Differential Equations and Group Theory from Riemann to Poincaré*, Boston: Birkhäuser Verlag, 1986.

Greene, John C., *American Science in the Age of Jefferson*, Ames: The Iowa State University Press, 1984.

Green, Judy, "Christine Ladd-Franklin (1847-1930)," in *Women of Mathematics: A Biobibliographic Sourcebook,* ed. Louise S. Grinstein and Paul J. Campbell, New York: Greenwood Press, 1987, pp. 121–128.

Green, Judy and LaDuke, Jeanne, "Women in the American Mathematical Community: The Pre-1940 Ph.D.'s," *The Mathematical Intelligencer* **9** (1) (1987):11–23.

Gregory, Frederick, "Kant, Schelling, and the Administration of Science in the Romantic Era," in *Science in Germany: The Intersection of Institutional and Intellectual Issues*, ed. Kathryn Olesko, *Osiris*, 2nd ser., 5 (1989):17–35.

Grinstein, Louise S. and Campbell, Paul J., ed. *Women of Mathematics: A Biobibliographic Sourcebook*, New York: Greenwood Press, 1987.

Guralnick, Stanley M., "The American Scientist in Higher Education: 1820–1910," in *The Sciences in the American Context: New Perspectives,* ed. Nathan Reingold, Washington: Smithsonian Institution Press, 1979, pp. 99–141.

Haas, Arthur, ed., *A Commentary on the Scientific Writings of J. Willard Gibbs Ph.D., LL.D. formerly Professor of Mathematical Physics in Yale University*, 2 vols., New Haven: Yale University Press, 1936.

Hall, G. Stanley, *Life and Confessions of a Psychologist*, New York: Appleton-Century-Crofts, Inc., 1923.

Halphen, Georges, *Traité des Fonctions élliptiques et de leurs Applications*, 3 vols., Paris: Gauthier-Villars, 1886–1891.

Halsted, George Bruce, "The Betweenness Assumptions," *American Mathematical Monthly* **9** (1902):98–101.

——, "Bibliography of Hyper-Space and Non-Euclidean Geometry," *American Journal of Mathematics* **1** (1878):261–276 and 384–385; **2** (1879):65–70.

Hamilton, William Rowan, "On Quaternions, or on a New System of Imaginaries in Algebra," *Philosophical Magazine,* 3rd ser., **29** (1846):26–31.

Hancock, Harris, *Foundations of the Theory of Algebraic Numbers*, 2 vols., New York: Macmillan & Co., 1932.

Hankins, Thomas, *Sir William Rowan Hamilton*, Baltimore: Johns Hopkins University Press, 1980.

Harmon, P. M., ed., *Wranglers and Physicists: Studies on Cambridge Physics in the Nineteenth Century*, Manchester: University Press, 1985.

Harper, William Rainey, *The President's Report: Administration—The Decennial Publications*, 1st ser., vol. 1, Chicago: University of Chicago Press, 1902.

Haskell, Mellen W., "Über die zu der Kurve $\lambda^3\mu+\mu^3\nu+\nu^3\lambda = 0$ im projektiven Sinne gehörende mehrfache Überdeckung der Ebene," *American Journal of Mathematics* **13** (1890):1–52.

Hathaway, Arthur S., "Some Papers on the Theory of Numbers," *American Journal of Mathematics* **6** (1884):316–330.

Hawkins, Hugh, *Pioneer: A History of the Johns Hopkins University,* 1874–1889, Ithaca: Cornell University Press, 1960.

Hawkins, Thomas, "Another Look at Cayley and the Theory of Matrices," *Archives internationales d'Histoire des Sciences* **26** (1977):82–112.

_____, "Hypercomplex Numbers, Lie Groups, and the Creation of Group Representation Theory," *Archive for History of Exact Sciences* **8** (1972):243–287.

_____, *Lebesgue's Theory of Integration: Its Origins and Development*, New York: Chelsea Publishing Company, 1970.

_____, "Line Geometry, Differential Equations, and the Birth of Lie's Theory of Groups," in *The History of Modern Mathematics,* ed. David E. Rowe and John McCleary, 2 vols., Boston: Academic Press, Inc., 1989, 1:275–329.

_____, "The Theory of Matrices in the 19th Century," in *Proceedings of the International Congress of Mathematicians: Vancouver,* 1974, 2 vols., n.p.: Canadian Mathematical Congress, 1975, 2:56–70.

_____, "Weierstrass and the Theory of Matrices," *Archive for History of Exact Sciences* **17** (1977):119–163.

Hedrick, Earl Raymond, *Ueber den analytischen Character des Lösungen von Differentialgleichungen*, Göttingen: n.p., 1901.

Heilbron, John L., *The Dilemmas of an Upright Man: Max Planck as Spokesman for German Science*, Berkeley: University of California Press, 1986.

Heims, Steve J., *John von Neumann and Norbert Wiener*, Cambridge, MA: MIT Press, 1980.

Hellinger, Ernst, "Hilbert's Arbeiten über Integralgleichungen und unendliche Gleichungssysteme," in David Hilbert, *Gesammelte Abhandlungen,* 3 vols., Berlin: Springer-Verlag, 1932–1935, 3:94–145.

Hensel, Susann, "Die Auseinandersetzungen um die mathematische Ausbildung der Ingenieure an den Technischen Hochschulen in Deutschland Ende des 19. Jahrhunderts," in S. Hensel, K.-N. Ihmig, and M. Otte, *Mathematik und Technik im* 19. *Jahrhundert in Deutschland*, Studien zur Wissenschafts-, Sozial-, und Bildungs geschichte der Mathematik, vol. 6, Göttingen: Vandenhoeck & Ruprecht, 1989, pp. 1–111.

Hensel, Susann; Ihmig, K. -N.; and Otte, M., *Mathematik und Technik im* 19. *Jahrhundert in Deutschland*, Studien zur Wissenschafts-, Sozial-, und Bildungsgeschichte der Mathematik, vol. 6, Göttingen: Vandenhoeck & Ruprecht, 1989.

Herstein, I. N., *Noncommutative Rings*, n.p.: Mathematical Association of America, 1968.

Hilbert, David, *The Foundations of Geometry*, trans. E. J. Townsend, 2d ed., Chicago: The Open Court Publishing Company, 1910.

_____, *Gesammelte Abhandlungen*, 3 vols., Berlin: Springer-Verlag, 1932–1935.

———, *Grundlagen der Geometrie*, Leipzig: B. G. Teubner, 1899.

———, "Mathematical Problems: Lecture Delivered Before the International Congress of Mathematicians at Paris in 1900," trans. Mary F. Winston, *Bulletin of the American Mathematical Society* **8** (1902):437–479.

———, "Mathematische Probleme," *Archiv für Mathematik und Physik* **1** (1901): 44–63.

———, *Natur und mathematisches Erkennen*, ed. David E. Rowe, Basel: Birkhäuser, 1991.

———, "Die Theorie der algebraischen Zahlkörper," *Jahresbericht der deutschen Mathematiker-Vereinigung* **4** (1897):175–546.

———, "Ueber die Theorie der algebraischen Formen," *Mathematische Annalen* **36** (1890):473–534.

———, "Ueber die Theorie der algebraischen Invarianten," in *Mathematical Papers Read at the International Mathematical Congress Held in Conjunction with the World's Columbian Exposition Chicago* 1893, ed. E. H. Moore *et al.*, New York: Macmillan & Co., 1896, pp. 116–124.

Hill, George William, *Collected Mathematical Works of George William Hill*, 4 vols., Washington: Carnegie Institution, 1905–1907.

Hindle, Brooke, *The Pursuit of Science in Revolutionary America*, 1735–1789, Chapel Hill: University of North Carolina Press, 1956, reprint ed., New York: Norton, 1974.

Hines, Thomas S., *Burnham of Chicago, Architect and Planner*, Chicago: University of Chicago Press, 1979.

Hofstadter, Richard, "The Revolution in Higher Education," in *Paths of American Thought*, ed. Arthur M. Schlesinger, Jr. and Morton White, Boston: Houghton Mifflin Company, 1970, pp. 269–290.

Hofstadter, Richard and Smith, Wilson, ed., *American Higher Education: A Documentary History*, 2 vols., Chicago: University of Chicago Press, 1961.

Hogan, Edward R., "George Baron and the *Mathematical Correspondent*," *Historia Mathematica* **3** (1976):403–415.

———, "The Mathematical Miscellany (1836–1839)," *Historia Mathematica* **12** (1985):245– 257.

———, "Robert Adrain: American Mathematician," *Historia Mathematica* **4** (1977): 157–172.

———, "Theodore Strong and Ante-Bellum American Mathematics," *Historia Mathematica* **8** (1981):439–455.

Hölder, Otto, "Die einfachen Gruppen im ersten und zweiten Hundert der Ordnungszahlen," *Mathematische Annalen* **40** (1892):55–88.

———, "n Die Gruppen der Ordnungen p^3, pq^2, pqr, p^4," *Mathematische Annalen* **43** (1893):301–412.

———, "Ueber einfache Gruppen," *Mathematische Annalen* **40** (1892):55–88.

Howard, Michael, *The Franco-Prussian War: The German Invasion of France*, 1870–1871, London: Rupert Hart-Davis, 1961.

Huntington, Edward V., "A Second Definition of a Group," *Bulletin of the American Mathematical Society* **8** (1902):388–391.

———, "Simplified Definition of a Group," *Bulletin of the American Mathematical Society* **8** (1902):296–300.

Husemöller, Dale, *Elliptic Curves*, New York: Springer-Verlag, 1986.

Jacobi, Carl G. J., *C. G. J. Jacobi's Gesammelte Werke*, ed. Carl W. Borchardt and Karl Weierstrass, 7 vols., Berlin: Verlag von G. Reimer, 1881–1891.

Johns Hopkins University Celebration of the Twenty-Fifth Anniversary of the Founding of the University and Inauguration of Ira Remsen, LL.D. As President of the University, Baltimore: Johns Hopkins University Press, 1902.

Jordan, Camille, *Traité des Substitutions et des Équations algébriques*, Paris: Gauthier-Villars, 1870.

Jungnickel, Christa and McCormach, Russell, *Intellectual Mastery of Nature: Theoretical Physics from Ohm to Einstein*, 2 vols., Chicago: University of Chicago Press, 1986.

Kaplansky, Irving, "Abraham Adrian Albert," *Biographical Memoirs of the National Academy of Sciences* **51** (1980):3–22.

Kaufmann-Bühler, Walter, *Gauss: A Biographical Study*, New York: Springer-Verlag, 1981.

Kellogg, Oliver Dimon, *Foundations of Potential Theory*, New York: Dover Publishing Co., 1929.

Kelley, Brooks Mather, *Yale: A History*, New Haven: Yale University Press, 1974.

Kenschaft, Patricia C., "Charlotte Angas Scott, 1858–1931," *College Mathematics Journal* **18** (1987):98–110.

_____, "Charlotte Angas Scott (1858–1931)," in *Women of Mathematics: A Biobibliographic Sourcebook*, ed. Louise S. Grinstein and Paul J. Campbell, New York: Greenwood Press, 1987, pp. 193–203.

Kevles, Daniel J., *The Physicists: The History of a Scientific Community in Modern America*, Cambridge, MA: Harvard University Press, 1971.

Klein, Felix, "A Comparative Review of Recent Researches in Geometry," trans. M. W. Haskell, *Bulletin of the New York Mathematical Society* **2** (1893):215–249.

_____, *Einleitung in die höhere Geometrie, Vorlesungen gehalten im Wintersemester* 1892–93 *und Sommersemester* 1893, 2nd ed., Leipzig: B. G. Teubner, 1907.

_____, *Elementarmathematik vom höheren Standpunkte aus*, Berlin: Springer-Verlag, 1928.

_____, *Elementary Mathematics from an Advanced Standpoint. Arithmetic, Algebra, Analysis*, vol. 1, trans. E. R. Hedrick and C. A. Noble, New York: Dover Publishing Co., 1932.

_____, *Elementary Mathematics from an Advanced Standpoint. Geometry*, vol. 2, trans. E. R. Hedrick and C. A. Noble, New York: Dover Publishing Co., 1939.

_____, *The Erlangen Program, Evanston Colloquium Lectures, and Other Selected Works*, ed. Jeremy Gray and David E. Rowe, New York: Springer-Verlag, forthcoming.

_____, "Ernst Ritter," *Jahresbericht der deutschen Mathematiker-Vereinigung* **4** (1894–1895):52–54.

_____, *The Evanston Colloquium Lectures on Mathematics*, New York: Macmillan & Co., 1893, reprint ed., New York: American Mathematical Society, 1911.

_____, *Gesammelte Mathematische Abhandlungen*, 3 vols., Berlin: J. Springer, 1921–1923.

_____, "Göttinger Professoren. Lebensbilder von eigener Hand," *Mitteilungen des Universitätsbundes Göttingen* **5** (1923):11–36.

_____, *Lectures on the Icosahedron and the Solution of Equations of the Fifth Degree*, trans. George Gavin Morrice, London: Kegan Paul, Trench, Truber & Co., 1888.

_____, *The Mathematical Theory of the Top*, New York: Charles Scribner's Sons, 1897.

_____, "Neue Beiträge zur Riemannschen Funktionentheorie," *Mathematische Annalen* **21** (1882–1883):141–218.

_____, "The Present State of Mathematics," in *Mathematical Papers Read at the*

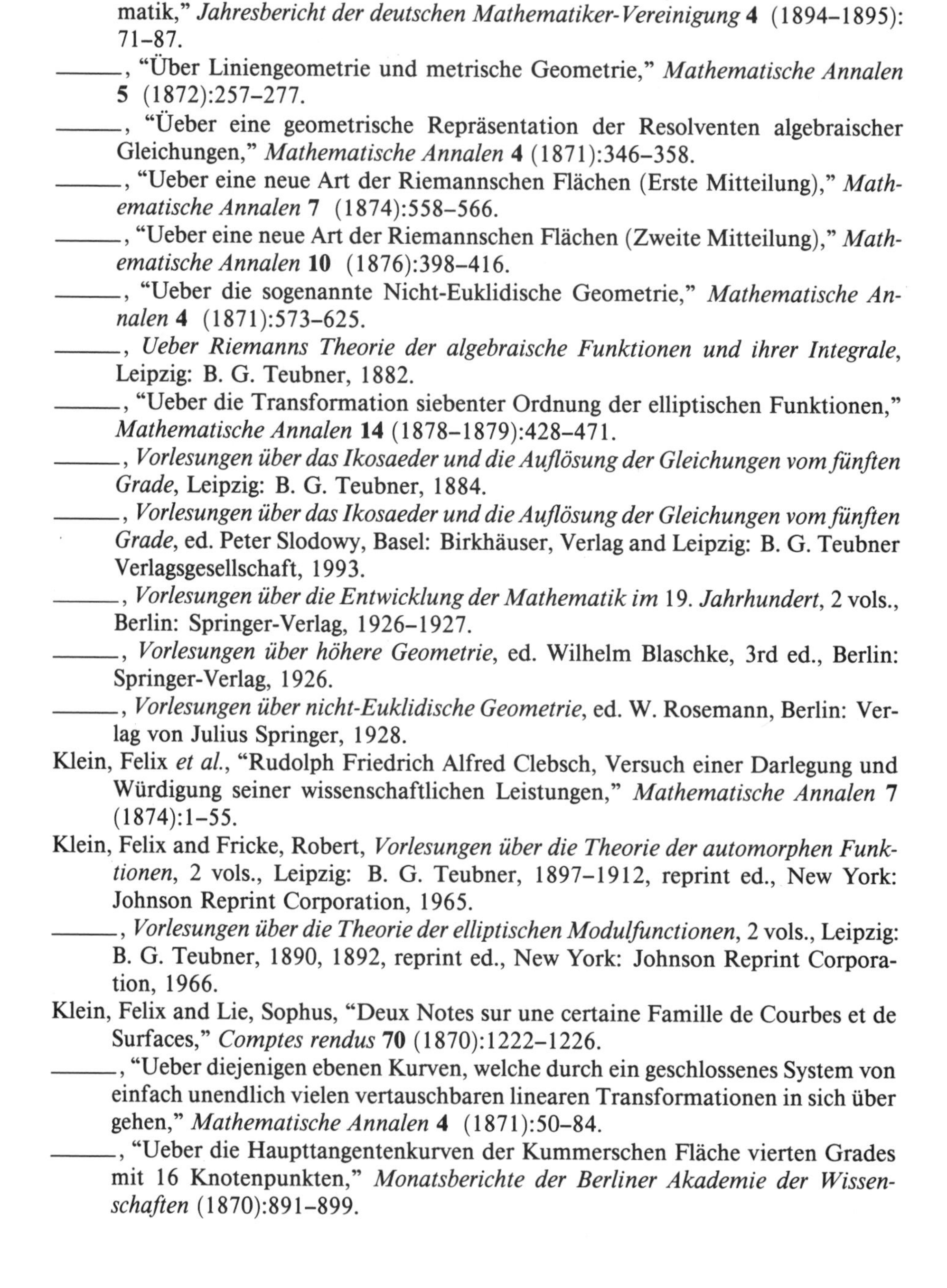

International Mathematical Congress Held in Connection with the World's Columbian Exposition Chicago 1893, ed. E. H. Moore *et al.*, New York: Macmillan & Co., 1896, pp. 133–135.

———, *Riemannsche Flächen. Vorlesungen, gehalten in Göttingen 1891/92*, ed. Günther Eisenreich and Walter Purkert, Teubner-Archiv zur Mathematik, vol. 5, Leipzig: B. G. Teubner, 1986.

———, "Riemann und seine Bedeutung für die Entwicklung der modernen Mathematik," *Jahresbericht der deutschen Mathematiker-Vereinigung* **4** (1894–1895): 71–87.

———, "Über Liniengeometrie und metrische Geometrie," *Mathematische Annalen* **5** (1872):257–277.

———, "Üeber eine geometrische Repräsentation der Resolventen algebraischer Gleichungen," *Mathematische Annalen* **4** (1871):346–358.

———, "Ueber eine neue Art der Riemannschen Flächen (Erste Mitteilung)," *Mathematische Annalen* **7** (1874):558–566.

———, "Ueber eine neue Art der Riemannschen Flächen (Zweite Mitteilung)," *Mathematische Annalen* **10** (1876):398–416.

———, "Ueber die sogenannte Nicht-Euklidische Geometrie," *Mathematische Annalen* **4** (1871):573–625.

———, *Ueber Riemanns Theorie der algebraische Funktionen und ihrer Integrale*, Leipzig: B. G. Teubner, 1882.

———, "Ueber die Transformation siebenter Ordnung der elliptischen Funktionen," *Mathematische Annalen* **14** (1878–1879):428–471.

———, *Vorlesungen über das Ikosaeder und die Auflösung der Gleichungen vom fünften Grade*, Leipzig: B. G. Teubner, 1884.

———, *Vorlesungen über das Ikosaeder und die Auflösung der Gleichungen vom fünften Grade*, ed. Peter Slodowy, Basel: Birkhäuser, Verlag and Leipzig: B. G. Teubner Verlagsgesellschaft, 1993.

———, *Vorlesungen über die Entwicklung der Mathematik im* 19. *Jahrhundert*, 2 vols., Berlin: Springer-Verlag, 1926–1927.

———, *Vorlesungen über höhere Geometrie*, ed. Wilhelm Blaschke, 3rd ed., Berlin: Springer-Verlag, 1926.

———, *Vorlesungen über nicht-Euklidische Geometrie*, ed. W. Rosemann, Berlin: Verlag von Julius Springer, 1928.

Klein, Felix *et al.*, "Rudolph Friedrich Alfred Clebsch, Versuch einer Darlegung und Würdigung seiner wissenschaftlichen Leistungen," *Mathematische Annalen* **7** (1874):1–55.

Klein, Felix and Fricke, Robert, *Vorlesungen über die Theorie der automorphen Funktionen*, 2 vols., Leipzig: B. G. Teubner, 1897–1912, reprint ed., New York: Johnson Reprint Corporation, 1965.

———, *Vorlesungen über die Theorie der elliptischen Modulfunctionen*, 2 vols., Leipzig: B. G. Teubner, 1890, 1892, reprint ed., New York: Johnson Reprint Corporation, 1966.

Klein, Felix and Lie, Sophus, "Deux Notes sur une certaine Famille de Courbes et de Surfaces," *Comptes rendus* **70** (1870):1222–1226.

———, "Ueber diejenigen ebenen Kurven, welche durch ein geschlossenes System von einfach unendlich vielen vertauschbaren linearen Transformationen in sich über gehen," *Mathematische Annalen* **4** (1871):50–84.

———, "Ueber die Haupttangentenkurven der Kummerschen Fläche vierten Grades mit 16 Knotenpunkten," *Monatsberichte der Berliner Akademie der Wissenschaften* (1870):891–899.

Klein, Felix and Riecke, Eduard, *Neue Beiträge zur Frage des mathematischen und physikalischen Unterrichts an den höheren Schulen*, Leipzig: B. G. Teubner, 1904.

Klein, Martin J., "The Scientific Work of Josiah Willard Gibbs," in *Springs of Scientific Creativity: Essays on Founders of Modern Science,* ed. R. Aris, H. T. Davis, and R. H. Stuewer, Minneapolis: University of Minnesota Press, 1983, pp. 142–162.

Kline, Morris, *Mathematical Thought from Ancient to Modern Times*, New York: Oxford University Press, 1972.

Knobloch, Eberhard and Rowe, David E., ed., *The History of Modern Mathematics*, vol. 3, Boston: Academic Press, Inc., 1994, forthcoming.

Knorr, Wilbur R., *The Evolution of the Euclidean Elements*, Dordrecht: Reidel, 1975.

Koblitz, Ann Hibner, *A Convergence of Lives: Sofia Kovalevskaia, Scientist, Writer, Revolutionary*, Boston: Birkhäuser Verlag, 1983.

Koebe, Paul, "Ueber die Uniformisierung beliebigen analytischen Kurven," *Journal für die reine und angewandte Mathematik* **138** (1910):192–253.

Koelsch, William A., *Clark University* 1887–1987: *A Narrative History*, Worcester: Clark University Press, 1987.

Kohlstedt, Sally Gregory, *The Formation of the American Scientific Community: The American Association for the Advancement of Science* 1848–60, Urbana: University of Illinois Press, 1976.

Kohlstedt, Sally Gregory and Rossiter, Margaret W., *Historical Writing on American Science: Perspectives and Prospects*, Baltimore: Johns Hopkins University Press, 1986.

Königsberger, Leo, *Hermann von Helmholtz*, trans. Frances A. Welby, New York: Dover Publications, Inc., 1965.

Kowalewski, Gerhard, *Bestand und Wandel: Meine Lebenserinnerungen zugleich ein Beitrag zur neueren Geschichte der Mathematik*, Munich: Verlag R. Oldenbourg, 1950.

"The Krupp Exhibit at the Great Fair," *Scientific American* **69** (July–Dec. 1893):33 and 40.

Ladd (-Franklin), Christine, "On the Algebra of Logic," in *Studies in Logic by Members of the Johns Hopkins University,* ed. Charles S. Peirce, Boston: Little and Co., 1883, pp. 17–71.

Ladd(-Franklin), Christine, "On De Morgan's Extension of the Algebraic Processes," *American Journal of Mathematics* **3** (1880):210–225.

_____, "The Pascal Hexagram," *American Journal of Mathematics* **2** (1879):1–12.

Langer, Rudolf E., "Josiah Willard Gibbs," *American Mathematical Monthly* **46** (1939):75–84.

Langer, Rudolf E. and Ingraham, Mark H., "Edward Burr Van Vleck, 1863-1943," *Biographical Memoirs of the National Academy of Sciences* **30** (1957):399–409.

Layton, Edwin, "Mirror-Image Twins: The Communities of Science and Technology in 19th-Century America," *Technology and Culture* **12** (1971):562–580.

Lefschetz, Solomon, *Algebraic Topology*, New York: American Mathematical Society, 1942.

_____, "Reminiscences of a Mathematical Immigrant in the United States," in *A Century of Mathematics in America—Part* I, ed. Peter Duren *et al.*, Providence: American Mathematical Society, 1988, pp. 201–207.

_____, *Topology*, New York: American Mathematical Society, 1930.

Lehner, Joseph, *Discontinuous Groups and Automorphic Functions*, Mathematical Surveys, vol. 8, Providence: American Mathematical Society, 1964.

Lewis, Albert C., "The Building of the University of Texas Mathematics Faculty, 1883–1938," in *A Century of American Mathematics—Part* III, ed. Peter Duren *et al.*, Providence: American Mathematical Society, 1989, pp. 205–239.

———, "George Bruce Halsted and the Development of American Mathematics," in *Men and Institutions in American Mathematics,* ed. J. Dalton Tarwater, John T. White, and John D. Miller, Graduate Studies, Texas Tech University, No. 13, Lubbock: Texas Tech Press, 1976, pp. 123–129.

Lie, Sophus, *Gesammelte Abhandlungen*, ed. Friedrich Engel and Poul Heegaard, 7 vols., Oslo: H. Aschehoug & Co. and Leipzig: B. G. Teubner, 1922–1960.

———, *Theorie der Transformationsgruppen*, 3 vols., Leipzig: B. G. Teubner, 1888–1893.

———, "Ueber die Grundlagen der Geometrie," *Berichte über die Verhandlungen der königlichen sächsischen Gesellschaft der Wissenschaften zu Leipzig,* **3** (1890): 355–418.

———, "Über Komplexe, insbesondere Linien- und Kugelkomplexe, mit Anwendung auf die Theorie partieller Differentialgleichungen," *Mathematische Annalen* **5** (1872):145–256.

———, *Vorlesungen über Differentialgleichungen mit bekannten infinitesimalen Transformation*, ed. Georg Scheffers, Leipzig: B. G. Teubner, 1891.

Lie, Sophus and Scheffers, Georg, *Geometrie der Berührungstransformationen*, Leipzig: B. G. Teubner, 1896.

Lindemann, Ferdinand, "Über die Zahl π," *Mathematische Annalen* **20** (1882): 213–225.

Lorey, Wilhelm, *Das Studium der Mathematik an den deutschen Universitäten seit Anfang des* 19. *Jahrhunderts*, Abhandlungen über den mathematischen Unterricht in Deutschland, vol. 3, no. 9, Leipzig: Verlag B. G. Teubner, 1916.

Lovett, Edgar O., "The Theory of Perturbation and Lie's Theory of Contact Transformations," *The Quarterly Journal of Pure and Applied Mathematics* **30** (1898): 47–149.

Lützen, Jesper, "The Solution of Partial Differential Equations by Separation of Variables," in *Studies in the History of Mathematics,* ed. Esther Phillips, MAA Studies in Mathematics, vol. 26, Washington, D.C.: Mathematical Association of America, 1987, pp. 242–277.

Lützen, Jesper and Purkert, Walter, "Conflicting Tendencies in the Historiography of Mathematics: Moritz Cantor and H. G. Zeuthen," in *The History of Modern Mathematics*, vol. 3, ed. Eberhard Knobloch and David E. Rowe, Boston: Academic Press, Inc., 1994, forthcoming.

Mac Lane, Saunders, "Mathematics at the University of Chicago: A Brief History," in *A Century of Mathematics in America—Part* II, ed. Peter Duren *et al.*, Providence: American Mathematical Society, 1989, pp. 127–154.

Magnus, Wilhelm and Chandler, Bruce, *The History of Combinatorial Group Theory*, New York: Springer-Verlag, 1982.

Maltbie, W. H., "The Undergraduate Mathematical Curriculum: Report of the Discussion at the Seventh Summer Meeting of the American Mathematical Society," *Bulletin of the American Mathematical Society* **7** (1900):14–24.

Manchester, William, *The Arms of Krupp,* 1587–1968, New York: Bantam Books, 1981.

Manegold, Karl-Heinz, *Universität, Technische Hochschule und Industrie. Ein Beitrag zur Emanzipation der Technik im* 19. *Jahrhundert unter besonderer Berücksichtigung der Bestrebungen Felix Kleins,* Berlin: Duncker & Humblot, 1970.

Manin, Yu. I., *Cubic Forms: Algebra, Geometry, Arithmetic*, Amsterdam: North-Holland Publishing Company, 1974.

Masani, Pesi R., *Norbert Wiener,* 1894–1964, Vita Mathematica, vol. 5, Basel: Birkhäuser Verlag, 1990.

Maschke, Heinrich, "Bestimmung aller ternären und quaternären Collineationsgruppen, welche mit symmetrischen und alternierenden Buchstabenvertau schungsgruppen holoedrisch isomorph sind," *Mathematische Annalen* **51** (1899): 253–298.

———, "Beweis des Satzes, dass diejenigen endlichen linearen Substitutionsgruppen, in welchen einige durchgehends verschwindende Coefficienten auftreten, intransitiv sind," *Mathematische Annalen* **52** (1899):363–368.

———, "The Invariants of a Group of 2 · 168 Linear Quaternary Substitutions," in *Mathematical Papers Read at the International Mathematical Congress Held in Connection with the World's Columbian Exposition Chicago* 1893, ed. E. H. Moore *et al.*, New York: Macmillan & Co., 1896, pp. 175-186.

———, "Ueber den arithmetischen Charakter der Coefficienten der Substitutionen endlicher linearer Substitutionsgruppen," *Mathematische Annalen* **50** (1898): 492–498.

Matz, F. P., "Professor Thomas Craig, C.E., Ph.D.," *The American Mathematical Monthly* **8** (1901):183–187.

McClelland, Charles, *State, Society, and University in Germany,* 1700–1914, Cambridge: Cambridge University Press, 1980.

Mehrtens, Herbert, "Ludwig Bieberbach and 'Deutsche Mathematik'," in *Studies in the History of Mathematics,* ed. Esther Phillips, MAA Studies in Mathematics, vol. 26, Washington, D.C.: Mathematical Association of America, 1987, pp. 195–241.

———, *Moderne—Sprache—Mathematik*, Frankfurt am Main: Suhrkamp, 1990.

Mehrtens, Herbert; Bos, Henk; and Schneider, Ivo, ed., *Social History of Nineteenth Century Mathematics*, Boston: Birkhauser Verlag, 1981.

Meinhardt, Günther, *Die Universität Göttingen, Ihre Entwicklung und Geschichte von* 1734–1974, Göttingen: Musterschmidt, 1977.

Merrill, George P., *The First One Hundred Years of American Geology*, New Haven: Yale University Press, 1924.

Merzbach, Uta, "The Study of the History of Mathematics in America: A Centennial Sketch," in *A Century of Mathematics in America—Part* III, ed. Peter Duren *et al.*, Providence: American Mathematical Society, 1989, pp. 639–666.

Meyer, Franz, "Tabellen von endlichen continuierlichen Transformationsgruppen," in *Mathematical Papers Read at the International Mathematical Congress Held in Connection with the World's Columbian Exposition Chicago* 1893, ed. E. H. Moore *et al.*, New York: Macmillan & Co., 1896, pp. 187–200.

Miller, George A., *The Collected Works of George Abram Miller*, 3 vols., Urbana, Ill.: University of Illinois Press, 1935–1946.

Miller, George A.; Blichfeldt, Hans F.; and Dickson, Leonard E., *Theory and Applications of Finite Groups*, New York: John Wiley, 1916.

Minkowski, Hermann, *Geometrie der Zahlen*, Leipzig: B. G. Teubner, 1896.

———, "Ueber Eigenschaften von ganzen Zahlen, die durch räumliche Anschauung erschlossen sind," in *Mathematical Papers Read at the International Mathematical Congress Held in Connection with the World's Columbian Exposition Chicago* 1893, ed. E. H. Moore *et al.*, New York: Macmillan & Co., 1896, pp. 350–366.

Mitchell, Oscar H., "On Binomial Congruences; Comprising an Extension of Fermat's and Wilson's Theorems, and a Theorem of Which Both Are Special Cases," *American Journal of Mathematics* **3** (1880):294–315.

______, "Some Theorems in Numbers," *American Journal of Mathematics* **4** (1881): 25–38.

Möbius, August Ferdinand, *Gesammelte Werke*, 4 vols., Leipzig: B. G. Teubner, 1886.

Monna, A. F., *Dirichlet's Principle: A Mathematical Comedy of Errors and Its Influences on the Development of Analysis*, Utrecht: Oosthoeck, Scheltema, and Holkema, 1975.

Montgomery, Deane, "Oswald Veblen," *Bulletin of the American Mathematical Society* **69** (1963):26–36.

Montgomery, Deane and Veblen, Oswald, "Nels Johann Lennes," *Bulletin of the American Mathematical Society* **60** (1954):264–265.

Moore, Eliakim Hastings, "The Betweenness Assumptions," *American Mathematical Monthly* **9** (1902):152–153.

______, "Concerning the Abstract Groups of Order $k!$ and $\frac{k!}{2}$ Holoedrically Isomorphic with the Symmetric and the Alternating Substitution-Groups on k Letters," *Proceedings of the London Mathematical Society* **28** (1897):357–366.

______, "Concerning Harnack's Theory of Improper Definite Integrals," *Transactions of the American Mathematical Society* **2** (1901):296–330.

______, "Concerning Triple Systems," *Mathematische Annalen* **43** (1893):271–285.

______, "A Definition of Abstract Groups," *Transactions of the American Mathematical Society* **3** (1902):485–492.

______, "Definition of Limit in General Integral Analysis," *Proceedings of the National Academy of Sciences* **1** (1915):628–632.

______, "A Doubly-Infinite System of Simple Groups," in *Mathematical Papers Read at the International Mathematics Congress Held in Connection with the World's Columbian Exposition: Chicago* 1893, ed. E. H. Moore *et al.*, New York: Macmillan & Co., 1896, pp. 208–242.

______, "Extensions of Certain Theorems of Clifford and of Cayley in the Geometry of n-Dimensions," *Transactions of the Connecticut Academy of Arts and Sciences* **7** (1885–1886):9–26.

______, "Introduction to a Form of General Analysis," *The New Haven Colloquium*, New Haven: Yale University Press, 1910.

______, "On a Definition of Abstract Groups," *Transactions of the American Mathematical Society* **6** (1905):179–180.

______, "On a Form of General Analysis with Application to Linear Differential and Integral Equations," in *Atti del* 4 *Congresso internazionale dei Matematici Rome* 1908, 3 vols., Rome: Accademia dei Lincei, 1909, 2:98–114.

______, "On the Foundations of Mathematics," *Science* **17** (1903):401–416.

______, "On the Projective Axioms of Geometry," *Transactions of the American Mathematical Society* **3** (1902):142–158.

______, "On the Theory of Improper Definite Integrals," *Transactions of the American Mathematical Society* **2** (1901):459-475.

Moore, Eliakim Hastings with Barnard, Raymond Walter, *General Analysis:: Part* 1, Philadelphia: The American Philosophical Society, 1935.

Moore, Eliakim Hastings; Bolza, Oskar; Maschke, Heinrich; and White, Henry S., ed., *Mathematical Papers Read at the International Mathematics Congress Held in Connection with the World's Columbian Exposition Chicago* 1893, New York: Macmillan & Co., 1896.

Moore, Eliakim Hastings and Smith, H. L., "A General Theory of Limits," *American Journal of Mathematics* **44** (1922):102–121.

Moore, Robert L., "Sets of Metrical Hypotheses for Geometry," *Transactions of the American Mathematical Society* **9** (1908):487–512.

Morrison, Samuel Eliot, ed., *The Development of Harvard University Since the Inauguration of President Eliot* 1869–1929, Cambridge: Harvard University Press, 1930.

Morse, Marston, "George David Birkhoff and his Mathematical Work," *Bulletin of the American Mathematical Society* **52** (1946):357–391.

Moyer, Albert E., *A Scientist's Voice in American Culture: Simon Newcomb and the Rhetoric of Scientific Method*, Berkeley: University of California Press, 1992.

Muir, Thomas, *The Theory of Determinants in the Historical Order of Development*, 4 vols., London: Macmillan & Co., Ltd., 1906–1923, reprint ed., New York: Dover Publications, Inc., 1960.

Netto, Eugen, *Substitutionstheorie und ihre Anwendungen auf die Algebra*, Leipzig: B. G. Teubner, 1882.

———, *The Theory of Substitutions and its Applications to Algebra,*, trans. Frank Nelson Cole, Ann Arbor: G. Wahr, 1892, reprint ed., Chelsea Publishing Co., n.d.

Newton, Hubert, "On the Transcendental Curves Whose Equation Is $\sin y \sin my = a \sin x \sin nx + b$, With 24 Plates," *Transactions of the Connecticut Academy of Arts and Sciences* **3** (1875):97–107.

Nicholson, Julia, "Otto Hölder and the Development of Group Theory and Galois Theory," unpublished doctoral dissertation, Oxford University, 1993.

Novick, Peter, *That Noble Dream: The "Objectivity Question" and the American Historical Profession*, New York: Cambridge University Press, 1988.

Olesko, Kathryn M., *Physics as a Calling: Discipline and Practice in the Königsberg Seminar for Physics*, Ithaca: Cornell University Press, 1991.

Olesko, Kathryn, ed., *Science in Germany: The Intersection of Institutional and Intellectual Issues, Osiris*, 2nd ser., vol. 5 (1989).

Oleson, Alexandra and Brown, Sanborn C., ed., *The Pursuit of Knowledge in the Early American Republic: American Scientific and Learned Societies from Colonial Times to the Civil War*, Baltimore: Johns Hopkins University Press, 1976.

Oliver, James E.; Wait, Lucien; and Jones, George, *A Treatise on Algebra*, 2nd ed., New York: Dudley F. Finch, 1887.

———, *A Treatise on Trigonometry*, 4th ed., Ithaca: G. W. Jones, 1890.

Olivier, Charles P., "Ormond Stone," *Popular Astronomy* **41** (1933):294–298.

Osgood, William Fogg, "The Life and Services of Maxime Bôcher," *Bulletin of the American Mathematical Society* **25** (1919):337–350.

———, "The Scientific Work of Maxime Bôcher," *Bulletin of the American Mathematical Society* **25** (1919):197–215.

———, *Lehrbuch der Funktionentheorie*, vol. 1, Leipzig: B. G. Teubner, 1907.

Osgood, William F.; Coolidge, Julian L.; and Chase, George H., "Charles Leonard Bouton," *Bulletin of the American Mathematical Society* **28** (1922):123–124.

Page, James M., "On the Primitive Groups of Transformations in Space of Four Dimensions," *American Journal of Mathematics* **10** (1888):293–346.

———, *Ordinary Differential Equations: An Elementary Text Book with an Introduction to Lie's Theory of the Group of One Parameter*, London: Macmillan & Co., 1897.

Parshall, Karen Hunger, "America's First School of Mathematical Research: James Joseph Sylvester at The Johns Hopkins University 1876–1883," *Archive for History of Exact Sciences* **38** (1988):153–196.

———, "A Century-Old Snapshot of American Mathematics," *The Mathematical Intelligencer* **12** (3) (1990):7–11.

———, "Chemistry through Invariant Theory?" James Joseph Sylvester's Mathematization of the Atomic Theory," to appear.

———, "Eliakim Hastings Moore and the Founding of a Mathematical Community in America, 1892–1902," *Annals of Science* **41** (1984):313–333, reprinted in *A Century of Mathematics in America—Part* II, ed. Peter Duren *et al.* Providence: American Mathematical Society, 1989, pp. 155–175.

———, "In Pursuit of the Finite Division Algebra Theorem and Beyond: Joseph H. M. Wedderburn, Leonard E. Dickson, and Oswald Veblen," *Archives internationales d'Histoires des Sciences* **33** (1983):274–299.

———, "Joseph H. M. Wedderburn and the Structure Theory of Algebras," *Archive for History of Exact Sciences* **32** (1985):223–349.

———, "The One-Hundredth Anniversary of the Death of Invariant Theory?," *The Mathematical Intelligencer* **12** (4) (1990):10–16.

———, "The One-Hundredth Anniversary of Mathematics at the University of Chicago," *The Mathematical Intelligencer* **14** (2) (1992):39–44.

———, *Selected Correspondence of James Joseph Sylvester and His Circle*, Teubner-Archiv zur Mathematik, Stuttgart/Leipzig: B. G. Teubner, forthcoming.

———, "A Study in Group Theory: Leonard Eugene Dickson's *Linear Groups*," *The Mathematical Intelligencer* **13** (1) (1991):7–11.

———, "Toward a History of Nineteenth-Century Invariant Theory," in *The History of Modern Mathematics*, ed. David E. Rowe and John McCleary, 2 vols., Boston: Academic Press, 1989, 1:157–206.

Parshall, Karen Hunger and Rowe, David E., "Embedded in the Culture: Mathematics at the World's Columbian Exposition of 1893," *The Mathematical Intelligencer* **15** (2) (1993):40–45.

Peckhaus, Volker, *Hilbertprogramm und kritische Philosophie*, Studien zur Wissenschafts-, Sozial-, und Bildungsgeschichte der Mathematik, vol. 7, Göttingen: Vandenhoeck & Ruprecht, 1990.

Peirce, Benjamin, *An Elementary Treatise on Algebra: To Which Are Added Exponential Equations and Logarithms*, Boston: J. Munroe & Co., 1837.

———, *An Elementary Treatise on Plane and Solid Geometry*, Boston: J. Munroe & Co., 1837.

———, *An Elementary Treatise on Plane and Spherical Trigonometry with its Applications to Heights and Distances, Navigation and Surveying*, Boston: J. Munroe & Co., 1835.

———, "Linear Associative Algebra with Notes and Addenda, by C. S. Peirce," *American Journal of Mathematics* **4** (1881):97–229.

———, "On the Uses and Transformations of Linear Algebra," *Proceedings of the American Academy of Arts and Sciences*, n.s. **2** (1875):395–400.

Peirce, Bessie Louise and Norris, Joe L., ed., *As Others See Chicago: Impressions of Visitors*, 1673–1933, Chicago: University of Chicago Press, 1933.

Peirce, Charles S., *The New Elements of Mathematics*, ed. Carolyn Eisele, 4 vols. in 5 pts., The Hague: Mouton Publishers and Atlantic Highlands, N. J.: Humanities Press, 1976.

———, "On the Algebras In Which Division Is Unambiguous," *American Journal of Mathematics* **4** (1881):225–229.

———, "On the Application of Logical Analysis to Multiple Algebra," *Proceedings of the American Academy of Arts and Sciences* **10** (1875):392–394.

Peirce, Charles S., ed., *Studies in Logic by Members of the Johns Hopkins University*, Boston: Little and Co., 1883.

Phillips, Esther, ed., *Studies in the History of Mathematics*, MAA Studies in Mathematics, vol. 26, Washington, D.C.: Mathematical Association of America, 1987.

Pierpont, James, "Galois Theory of Equations," *Annals of Mathematics,* 2nd ser., **1** (1900):22–56.

Pitcher, Everett, *A History of the Second Fifty Years: American Mathematical Society* 1939–1988, American Mathematical Society Centennial Publications, vol. 1, Providence: American Mathematical Society, 1988.

Poincaré, Henri, *Oeuvres de Henri Poincaré*, ed. Paul Appell *et al.*, 11 vols., Paris: Gauthier-Villars et Cie., 1916–1956.

——, *Papers on Fuchsian Functions*, trans. John Stillwell, New York: Springer-Verlag, 1985.

——, *Science et Méthode*, Paris: Flammarion, 1908.

——, "Sur le Déterminant de Hill," *Bulletin astronomique* **17** (1900):134–143.

——, "Sur les Fonctions uniformes qui se reproduisent par des Substitutions linéaires," *Mathematische Annalen* **19** (1882):553–564.

——, "Sur l'Uniformisation des Fonctions analytiques," *Acta Mathematica* **31** (1907):1–63.

Post, Robert, "Science, Public Policy, and Popular Precepts: Alexander Dallas Bache and Alfred Beach as Symbolic Adversaries," in *The Sciences in the American Context: New Perspectives,* ed. Nathan Reingold, Washington: Smithsonian Institution Press, 1979, pp. 77–98.

Purkert, Walter, "Die Mathematik an der Universität Leipzig von ihrer Gründung bis zum zweiten Drittel des 19. Jahrhunderts," *in 100 Jahre Mathematisches Seminar der Karl-Marx-Universität Leipzig,* ed. Herbert Beckert and Horst Schumann, Berlin: VEB Deutscher Verlag der Wissenschaften, 1981, pp. 9–40.

Pycior, Helena M., "Benjamin Peirce's 'Linear Associative Algebra'," *Isis* **70** (1979): 537–551.

——, "British Synthetic vs. French Analytic Styles of Algebra in the Early American Republic," in *The History of Modern Mathematics,* ed. David E. Rowe and John McCleary, 2 vols., Boston: Academic Press, Inc., 1989, 1:125–154.

Pyenson, Lewis, *Neohumanism and the Persistence of Pure Mathematics in Wilhelmian Germany*, Philadelphia: American Philosophical Society, 1983.

Reid, Constance, *Courant*, New York: Springer-Verlag, 1976.

——, *Hilbert*, New York: Springer-Verlag, 1970.

——, *Neyman—From Life*, New York: Springer-Verlag, 1982.

——, "The Road Not Taken: A Footnote in the History of Mathematics," *The Mathematical Intelligencer* **1** (1) (1978):21–23.

Reingold, Nathan, "Alexander Dallas Bache: Science and Technology in the American Idiom," *Technology and Culture* **11** (1970):163–177.

——, "Definitions and Speculations: The Professionalization of Science in America in the Nineteenth Century," in *The Pursuit of Knowledge in the Early American Republic: American Scientific and Learned Societies from Colonial Times to the Civil War,* ed. Alexandra Oleson and Sanborn C. Brown, Baltimore: Johns Hopkins University Press, 1976, pp. 33–69.

——, "Refugee Mathematicians in the United States of America, 1933–1941: Reception and Reaction," *Annals of Science* **38** (1981):313–338.

Reingold, Nathan, ed., *The Sciences in the American Context: New Perspectives,* Washington: Smithsonian Institution Press, 1979.

——*Science in Nineteenth-Century America: A Documentary History,* Chicago: University of Chicago Press, 1985.

Richards, Joan, *Mathematical Visions: The Pursuit of Geometry in Victorian England*, Boston: Academic Press, Inc., 1988.

Richardson, R. G. D., "The Ph.D. Degree and Mathematical Research," *American Mathematical Monthly* **43** (1936):199–215.

Richenhagen, Gottfried, *Carl Runge* (1856–1927): *Von der reinen Mathematik zur Numerik*, Göttingen: Vandenhoeck & Ruprecht, 1985.

Rider, Robin, "An Opportune Time: Griffith C. Evans and Mathematics at Berkeley," in *A Century of Mathematics in America—Part* II, ed. Peter Duren *et al.*, Providence: American Mathematical Society, 1989, pp. 283–302.

Riemann, Bernhard, *Bernhard Riemann's Gesammelte Mathematische Werke*, ed. Heinrich Weber, 2nd ed., Leipzig: B. G. Teubner, 1892.

——, "Theorie der Abel'schen Funktionen," *Journal für die reine und angewandte Mathematik* **54** (1857):115–155.

——, "Ueber die Hypothesen welche der Geometrie zu Grunde liegen," *Abhandlungen der Königlichen Gesellschaft der Wissenschaften zu Göttingen* **13** (1868): 132-152.

Ringer, Fritz, *The Decline of the German Mandarins: The German Academic Community*, Cambridge, MA: Harvard University Press, 1969.

Rohn, Karl, "Die verschiedenen Gestalten der Kummer'schen Fläche," *Mathematische Annalen* **18** (1881):99–159.

Rosenbaum, Robert A., "There were Giants in those Days: Van Vleck and his Boys," *Wesleyan University Alumnus* (November 1956):2–3.

Rosenfeld, Boris A., *A History of Non-Euclidean Geometry*, trans. Abe Shenitzer, New York: Springer-Verlag, 1988.

Rosenstein, George M., Jr., "The Best Method. American Calculus Textbooks of the Nineteenth Century," in *A Century of American Mathematics—Part* III, ed. Peter Duren *et al.*, Providence: American Mathematical Society, 1989, pp. 77–109.

Ross, Dorothy, *G. Stanley Hall: The Psychologist as Prophet*, Chicago: University of Chicago Press, 1972.

——, *The Origins of American Social Science*, New York: Cambridge University Press, 1991.

Rossiter, Margaret W., *Women Scientists in America: Struggles and Strategies to 1940*, Baltimore: Johns Hopkins University Press, 1982.

Rothrock, David A., "Mathematicians in War Service," *American Mathematical Monthly* **26** (1919):40–44.

Rowe, David E., "Der Briefwechsel Sophus Lie-Felix Klein, eine Einsicht in ihre persönlichen und wissenschaflicthen Beziehungen," *NTM. Schriftenreihe für Geschichte der Naturwissenschaften, Technik und Medizin* **25** (1988):37–47.

——, "The Early Geometrical Works of Sophus Lie and Felix Klein," in *The History of Modern Mathematics,* ed. David E. Rowe and John McCleary, 2 vols., Boston: Academic Press, Inc., 1989, 1:209–273.

——, "Felix Klein's 'Erlanger Antrittsrede.' A Transcription with English Translation and Commentary," *Historia Mathematica* **12** (1985):123–141.

——, "A Forgotten Chapter in the History of Felix Klein's *Erlanger Programm*," *Historia Mathematica* **10** (1983):448-454.

——, "'Jewish Mathematics at Göttingen in the Era of Felix Klein," *Isis* **77** (1986): 422–449.

——, "Klein, Hilbert, and the Göttingen Mathematical Tradition," in *Science in Germany: The Intersection of Institutional and Intellectual Issues*, ed. Kathryn M. Olesko, *Osiris*, vol. 5 (1989), pp. 189–213.

———, "Klein, Lie, and the 'Erlanger Programm,'" *in* 1830–1930: *A Century of Geometry,* ed. Luciano Boi, D. Flament, and J.-M. Salanski, Berlin: Springer-Verlag, 1992, pp. 45–54.

———, "Klein, Mittag-Leffler, and the Klein-Poincaré Correspondence of 1881–1882," in *Amphora*: *Festschrift für Hans Wussing zu seinem* 65. *Geburtstag,* ed. Sergei S. Demidov, Menso Folkerts, David E. Rowe, and Christoph Scriba, Basel: Birkhäuser Verlag, 1992, pp. 597–618.

———, "The Philosophical Views of Klein and Hilbert," in *Proceedings of the* 1990 *Tokyo Symposium on the History of Mathematics,* ed. Chikara Sasaki, Basel: Birkhäuser Verlag, to appear.

Rowe, David E. and McCleary, John, ed., *The History of Modern Mathematics,* 2 vols., Boston: Academic Press, Inc., 1989.

Royden, Halsey, "A History of Mathematics at Stanford," in *A Century of Mathematics in America—Part* II, ed. Peter Duren *et al.*, Providence: American Mathematical Society, 1989, pp. 237–277.

Rudolph, Frederick, *The American College and University*: *A History,* New York: Alfred A. Knopf, 1962.

———, *Curriculum*: *A History of the American Undergraduate Course of Study since* 1636, San Francisco: Jossey-Bass Publishers, 1978.

Rukeyser, Muriel, *Willard Gibbs,* Garden City: Doubleday, Doran & Company, Inc., 1942.

Ryan, W. Carson, *Studies in Early Graduate Education*: *The Johns Hopkins, Clark University, The University of Chicago,* New York: Carnegie Foundation for the Advancement of Teaching, 1939.

Rydell, Robert W., *All the World's a Fair*: *Visions of Empire at American International Expositions,* 1876–1916, Chicago: University of Chicago Press, 1984.

Salmon, George, *Lessons Introductory to the Modern Higher Algebra,* 3rd ed., Dublin: Hodges, Foster, & Co., 1876.

———, *A Treatise on the Analytic Geometry of Three Dimensions,* 4th ed., Dublin: Hodges, Figgis, and Co., 1882.

———, *A Treatise on Higher Plane Curves,* 2nd ed., Dublin: Hodges, Foster, and Co., 1873.

Sanford, Edmund C., *A Sketch of the History of Clark University,* Worcester: Clark University Press, 1923.

Sarton, George, *The Study of the History of Mathematics,* Cambridge, MA: Harvard University Press, 1937; reprint ed., New York: Dover Publications, Inc., 1957.

Sasaki, Chikara, ed., *Proceedings of the* 1990 *Tokyo Symposium on the History of Mathematics,* Basel: Birkhäuser Verlag, to appear.

Scanlon, Michael, "Who Were the American Postulate Theorists," *The Journal of Symbolic Logic* **56** (1991):981–1002.

Schappacher, Norbert, "Fachverband–Institut–Staat," in *Ein Jahrhundert Mathematik,* 1890–1990. *Festschrift zum Jubiläum der DMV,* ed. Gerd Fischer *et al.*, Braunschweig: Vieweg, 1990, pp. 1-82.

Scharlau, Winfried, ed., *Richard Dedekind*: 1831-1981. *Eine Würdigung zu seinem* 150. *Geburtstag,* Wiesbaden: Vieweg, 1981.

Scheffers, Georg, "Ueber die Berechnung von Zahlensystemen," *Berichte über die Verhandlungen der königlichen sächsischen Gesellschaft der Wissenschaften zu Leipzig* **41** (1889):400–457.

———, "Zur Theorie der aus *n* Haupteinheiten gebildeten complexen Grössen," *Berichte über die Verhandlungen der königlichen sächsischen Gesellschaft der Wissenschaften zu Leipzig* **41** (1889):290–307.

Schlesinger, Arthur M., Jr. and White, Morton, ed., *Paths of American Thought*, Boston: Houghton Mifflin Company, 1970.

Schoenflies, Arthur, "Gruppentheorie und Krystallographie," in *Mathematical Papers Read at the International Mathematics Congress Held in Connection with the World's Columbian Exposition Chicago* 1893, ed. E. H. Moore *et al.*, New York: Macmillan & Co., 1896, pp. 341–349.

Scholz, Erhard, "Crystallographic Symmetry Concepts and Group Theory (1850–1880)," in *The History of Modern Mathematics,* ed. David E. Rowe and John McCleary, 2 vols., Boston: Academic Press, Inc., 1989, 2:3–27.

——, *Geschichte des Mannigfaltigkeitsbegriffs von Riemann bis Poincaré*, Basel: Birkhäuser Verlag, 1980.

——, *Symmetrie—Gruppe—Dualität: Zur Beziehung zwischen theoretischer Mathematik und Anwendungen in Kristallographie und Baustatik des* 19. *Jahrhunderts*, Science Networks—Historical Studies, vol. 1, Basel: Birkhäuser Verlag, 1989.

Schubring, Gert, "Pure and Applied Mathematics in Divergent Institutional Settings in Germany: The Role and Impact of Felix Klein," in *The History of Modern Mathematics,* ed. David E. Rowe and John McCleary, 2 vols., Boston: Academic Press, Inc., 1989, 2:171–220.

——, "The Rise and Decline of the Bonn Naturwissenschaften Seminar," in *Science in Germany: The Intersection of Institutional and Intellectual Issues*, ed. Kathryn M. Olesko, *Osiris*, vol. 5 (1989), pp. 56–93.

Schwarz, Hermann Amandus, *Gesammelte Mathematische Abhandlungen*, 2 vols., Berlin: Akademie der Wissenschaften, 1890.

——, "Ueber diejenigen Fälle, in welchen die Gaussische hypergeometrische Reihe eine algebraische Function ihres vierten Elementes darstellt," *Journal für die reine und angewandte Mathematik* **75** (1872):292–335.

Serret, Joseph, *Cours d'Algèbre supérieure*, 2 vols., 3rd ed., Paris: Gauthier-Villars, 1866.

Servos, John, "Mathematics and the Physical Sciences in America, 1880–1930," *Isis* **77** (1986):611–629.

——, *Physical Chemistry from Ostwald to Pauling: The Making of a Science in America*, Princeton: University Press, 1990.

Shafarevich, I. R., "Zum 150. Geburtstag von Alfred Clebsch," *Mathematische Annalen* **266** (1983):135–140.

Simons, Lao G., "The Influence of French Mathematicians at the End of the Eighteenth Century upon the Teaching of Mathematics in American Colleges," *Isis* **15** (1931):104–123.

Slaught, Herbert, "The Cross-Ratio Group of 120 Quadratic Cremona Transformations of the Plane," *American Journal of Mathematics* **22** (1900):343–388.

——, "The Teaching of Mathematics in the Colleges," *American Mathematical Monthly* **16** (1909):173–177.

Slaught, Herbert E. and Bliss, Gilbert A., "The Department of Mathematics," *The University Record,* n.s. **15** (1929):97–104.

Smith, David Eugene and Ginsburg, Jekuthiel, *A History of Mathematics in America Before* 1900, Chicago: The Mathematical Association of America, 1934.

Smith, Percey F., "Josiah Willard Gibbs, Ph.D., LL.D.: A Short Sketch and Appreciation of His Mathematics," *Bulletin of the American Mathematical Society,* 2nd. ser. **10** (1903–1904):34–39.

——, "On Sophus Lie's Representation of Imaginaries in Plane Geometry," *American Journal of Mathematics* **25** (1903):165–179.

Snyder, Virgil, "Review of F. Klein, *Gesammelte Mathematische Abhandlungen*, vol. 1," *Bulletin of the American Mathematical Society* **28** (1922):125–129.

———, "Ueber die lineare Complexe der Lie'schen Kugelgeometrie," doctoral dissertation, University of Göttingen, 1895.
Spivak, Michael, *Differential Geometry*, 5 vols., Boston: Publish or Perish Press, 1970–1975.
Springer, George, *Introduction to Riemann Surfaces*, Reading, MA: Addison-Wesley Publishing Co., 1957.
Springer, Tony A., *Invariant Theory*, Berlin: Springer-Verlag, 1977.
Stigler, Stephen, ed., *American Contributions to Mathematical Statistics in the Nineteenth Century*, 2 vols., New York: Arno Press, 1980.
———, "Mathematical Statistics in the Early States," *Annals of Statistics* **6** (1978): 239-265.
Stone, Marshall, "Men and Institutions in American Mathematics," in *Men and Institutions in American Mathematics*, ed. J. Dalton Tarwater, John T. White, and John D. Miller, Graduate Studies Texas Tech University, no. 13, Lubbock: Texas Tech Press, 1976, pp. 7–29.
Storr, Richard J., *The Beginnings of Graduate Education in America*, Chicago: University of Chicago Press, 1953.
———, *A History of the University of Chicago: Harper's University, The Beginnings*, Chicago: University of Chicago Press, 1966.
Story, William E., "On the Non-Euclidean Geometry," *American Journal of Mathematics* **5** (1882):180–211.
———, "On Non-Euclidean Properties of Conics," *American Journal of Mathematics* **5** (1882):358–381.
———, "On the Non-Euclidean Trigonometry," *American Journal of Mathematics* **4** (1881):332–340.
———, "On the Theory of Rational Derivation on a Cubic Curve," *American Journal of Mathematics* **3** (1880):356–387.
Stringham, W. Irving, "Determination of the Finite Quaternion Groups," *American Journal of Mathematics* **4** (1881):345-357.
———, "A Formulary for an Introduction to Elliptic Functions," in *Mathematical Papers Read at the International Mathematical Congress Held in Connection with the World's Columbian Exposition Chicago* 1893, ed. E. H. Moore *et al.*, New York: Macmillan & Co., 1896, pp. 350–366.
———, "Regular Figures in *n*-Dimensional Space," *American Journal of Mathematics* **3** (1880):1–14.
Strong, Theodore, "On Analytical Trigonometry," *Proceedings of the American Philosophical Society* **1** (1843):49–50.
———, "Theory and Variation of the Arbitrary Constants in Elliptic Motion," *American Journal of Science* **30** (1836):248–266.
Struik, Dirk J., *A Concise History of Mathematics*, 4th rev. ed., New York: Dover Publications, Inc., 1987.
———, "Julian Lowell Coolidge in Memoriam," *American Mathematical Monthly* **62** (1955):669–682.
———, *Lectures on Classical Differential Geometry*, 2nd ed., Reading, MA: Addison-Wesley Publishing Co., Inc., 1961, reprint ed., New York: Dover Publications, Inc., 1988.
———, "The MIT Department of Mathematics during its First Seventy-Five Years: Some Recollections," in *A Century of Mathematics in America—Part* III, ed. Peter Duren *et al.*, Providence: American Mathematical Society, 1989, pp. 163–178.
———, *Yankee Science in the Making*, Boston: Little Brown, 1948.

Study, Eduard, "Ältere und neuere Untersuchungen über Systeme complexer Zahlen," in *Mathematical Papers Read at the International Mathematical Congress Held in Connection with the World's Columbian Exposition Chicago* 1893, ed. E. H. Moore *et al.*, New York: Macmillan & Co., 1896, pp. 350–366.

———, "Über Systeme von complexen Zahlen," *Nachrichten der Gesellschaft der Wissenschaften zu Göttingen, Mathematisch-Physicalische Abteilung* (1889):237–268.

Sullivan, Louis, *The Autobiography of an Idea*, New York: Press of the American Institute of Architects, Inc., 1926.

Sylvester, James Joseph, "Address on Commemoration Day at Johns Hopkins University 22 February, 1877," in *The Collected Mathematical Papers of James Joseph Sylvester*, ed. H. F. Baker, 4 vols, Cambridge: University Press, 1904–1912, reprint ed., New York: Chelsea Publishing Co., 1973, 3:72–87.

———, "Algebraical Researches, Containing a Disquisition on Newton's Rule for the Discovery of Imaginary Roots, and an Allied Rule Applicable to a Particular Class of Equations, Together with a Complete Invariantive Determination of the Character of the Roots of the General Equation of the Fifth Degree, &c," *Philosophical Transactions of the Royal Society of London* **154** (1864):579–666.

———, *The Collected Mathematical Papers of James Joseph Sylvester*, 4 vols., Cambridge: University Press, 1904–1912, reprint ed., New York: Chelsea Publishing Co., 1973.

———, "A Constructive Theory of Partitions, Arranged in Three Acts, an Interact, and an Exodion," *American Journal of Mathematics* **5** (1882):251–330.

———, "Examples of the Dialytic Method of Elimination as Applied to Ternary Systems of Equations," *Cambridge Mathematical Journal* **2** (1841):232–236.

———, *Fliegende Blätter: Supplement to the Laws of Verse*, privately printed in London by Grant & Co., 1876.

———, *The Laws of Verse: or, Principles of Versification Exemplified in Metrical Translations: Together with an Annotated Reprint of the Inaugural Presidential Address to the Mathematical and Physical Sections of the British Association at Exeter*, London: Longmans, Green, & Co., 1870.

———, "Lectures on the Principle of Universal Algebra," *American Journal of Mathematics* **6** (1884):270–286.

———, "Note on the Algebraical Theory of Derivative Points of Curves of the Third Degree," *Philosophical Magazine* **16** (1858):116-119.

———, "On an Application of the New Atomic Theory to the Graphical Representation of the Invariants and Covariants of Binary Quantics,—With Three Appendices," *American Journal of Mathematics* **1** (1878):64–125.

———, "On Certain Ternary Cubic-Form Equations," *American Journal of Mathematics* **2** (1879):280–285 and 357–393; **3** (1880):58–88 and 179–189.

———, "On the General Theory of Associated Algebraical Forms," *Cambridge and Dublin Mathematical Journal* **6** (1851):289–293.

———, "On the Principles of the Calculus of Forms," *Cambridge and Dublin Mathematical Journal* **7** (1852):52–97 and 179–217.

———, "On Quaternions, Nonions, Sedenions, etc.," *Johns Hopkins University Circulars* **2** (1883):7–9.

———, "On a Rule for Abbreviating the Calculation of the Number of In- and Covariants of a Given Order and Weight in the Coefficients of a Binary Quantic of a Given Degree," *Messenger of Mathematics* **8** (1879):1–8.

———, "On a Theory of the Syzygetic Relations of Two Rational Integral Functions, Comprising an Application to the Theory of Sturm's Functions, and That of the Greatest Algebraical Common Measure," *Philosophical Transactions of the Royal Society of London* **143** (1853):407–548.

______, "Presidential Address to Section 'A' of the British Association," *The Collected Mathematical Papers of James Joseph Sylvester,* ed. H. F. Baker, 4 vols., Cambridge: University Press, 1904–1912, reprint ed., New York: Chelsea Publishing Co., 1973, 2:650–661.

______, "Proof of a Hitherto Undemonstrated Fundamental Theorem of Invariants," *Philosophical Magazine* **5** (1878):178–188.

______, "Sur les Covariants fondamentaux d'un Système cubo-biquadratique binaire," *Comptes rendus* **87** (1878):242–244 and 287–289.

______, "Tables of Generating Functions and Ground-Forms for the Binary Quantics of the First Ten Orders," *American Journal of Mathematics* **2** (1879):223–251.

______, "Tables of the Generating Functions and Ground-Forms for Simultaneous Binary Quantics of the First Four Orders Taken Two and Two Together," *American Journal of Mathematics* **2** (1879):293–306 and 324–329.

Taber, Henry, "On the Theory of Matrices," *American Journal of Mathematics* **11** (1889):337–395.

Tait, Peter Guthrie, *Elementary Treatise on Quaternions*, 3rd ed., Cambridge: University Press, 1890.

Tarwater, J. Dalton; White, John T.; Miller, John D., ed., *Men and Institutions in American Mathematics*, Graduate Studies Texas Tech University, no. 13, Lubbock: Texas Tech Press, 1976.

Thompson, Henry Dallas, "Hyperelliptische Schnittsysteme und Zusammenordnung der algebraischen und transcendentalen Thetacharakteristiken," *American Journal of Mathematics* **15** (1893):91–123.

Tobies, Renate, *Felix Klein*, Biographien hervorragender Naturwissenschaftler, Techniker und Mediziner **50** Leipzig: BSB B. G. Teubner Verlagsgesellschaft, 1981.

Tobies, Renate and Rowe, David E., ed., *Korrespondenz Felix Klein–Adolf Mayer, Auswahl aus den Jahren* 1871–1907, Teubner-Archiv zur Mathematik, vol. 14, Leipzig: BSB B. G. Teubner Verlagsgesellschaft, 1990.

Toepell, Michael-Markus, *Über die Entstehung von David Hilberts "Grundlagen der Geometrie"*, Göttingen: Vandenhoeck & Ruprecht, 1986.

Traylor, D. Reginald *et al.*, *Creative Teaching*: *Heritage of R. L. Moore*, Houston: University of Houston Press, 1972.

Truesdell, Clifford, *An Idiot's Fugitive Essays on Science*: *Methods, Criticisms, Training, Circumstances*, New York: Springer-Verlag, 1984.

Tucker, Albert W., "Solomon Lefschetz: A Reminiscence," in *Mathematical People,* ed. Donald J. Albers and Gerald L. Alexanderson, Chicago: Contemporary Books, Inc., 1985, pp. 349–352.

Turner, R. Steven, "The Growth of Professional Research in Prussia, 1818–1848—Causes and Context," *Historical Studies in the Physical Sciences* **3** (1971):137–182.

van der Waerden, B. L., *Algebra,* 2 vols., New York: Springer-Verlag, 1991.

______, *A History of Algebra from al-Khwārizmī to Emmy Noether*, Berlin: Springer-Verlag, 1985.

______, "Topologische Begründung des Kalküls der abzählende Geometrie," *Mathematische Annalen* **102** (1930):337–362.

Van Vleck, Edward Burr, "Zur Kettenbruchentwicklung hyperelliptischer und ähnlicher Integrale," *American Journal of Mathematics* **16** (1894):1–91.

Veblen, Oswald, "George David Birkhoff (1884–1944)," *Yearbook of the American Philosophical Society* (1946):279–285.

———, "Henry Burchard Fine—In Memorium," *Bulletin of the American Mathematical Society* **35** (1929):726–730.

———, *Invariants of Quadratic Differential Forms*, Cambridge Tracts in Mathematics and Mathematical Physics, vol. 24, Cambridge: University Press, 1927.

———, *Projektive Relativitätstheorie*, Ergebnisse der Mathematik und ihrer Grenzgebiete, vol. 2, no. 1, Berlin: Springer-Verlag, 1933.

———, "A System of Axioms for Geometry," *Transactions of the American Mathematical Society* **5** (1904):343–384.

Veblen, Oswald and Whitehead, J. H. C., *The Foundations of Differential Geometry*, Cambridge Tracts in Mathematics and Mathematical Physics, vol. 29, Cambridge: University Press, 1932.

Veblen, Oswald and Young, John Wesley, *Projective Geometry*, vol. 1, Boston: Ginn and Co., 1910.

Veysey, Laurence R., *The Emergence of the American University*, Chicago: University of Chicago Press, 1965.

Volkert, Klaus T., *Die Krise der Anschauung*, Studien zur Wissenschafts-, Sozial-, und Bildungsgeschichte der Mathematik, vol. 3, Göttingen: Vandenhoeck & Ruprecht, 1986.

Walsh, Joseph L., "William Fogg Osgood," in *A Century of Mathematics in America—Part* II, ed. Peter Duren *et al.*, Providence: American Mathematical Society, 1989, pp. 79–85.

Webber, Samuel, *Mathematics, Compiled from the Best Authors and Intended to Be a Text-book of the Course of Private Lectures on These Sciences in the University at Cambridge*, Boston: Harvard College, 1801.

Weber, Heinrich, *Lehrbuch der Algebra*, 2nd ed., 3 vols., Braunschweig: F. Vieweg & Sohn, 1899-1912.

Wedderburn, Joseph H. M., "A Theorem on Finite Algebras," *Transactions of the American Mathematical Society* **6** (1905):349–352.

———, "On Hypercomplex Numbers," *Proceedings of the London Mathematical Society, 2nd ser.*, **6** (1907):77–118.

Weierstrass, Karl, *Mathematische Werke*, 7 vols., Berlin: Mayer & Müller, 1894–1927, reprint ed., Hildesheim: Georg Ohms Verlagsbuchhandlung and New York: Johnson Reprint Corporation, 1967.

Weil, André, *Number Theory: An Approach though History from Hammurapi to Legendre*, Boston: Birkhäuser Verlag, 1984.

Weimann, Jeanne Madeline, *The Fair Women: The Story of The Woman's Building, World's Columbian Exposition, Chicago* 1893, Chicago: Academy, 1981.

Weltausstellung in Chicago, 1893. *Reichskommissar für die amtlicher Bericht über die Weltausstellung in Chicago,* 1893, Berlin: Reichsdruckerei, 1894.

Weyl, Hermann, *Classical Groups: Their Invariants and Representations*, Princeton: University Press, 1939.

———, *Die Idee der Riemannschen Fläche*, Leipzig: B. G. Teubner, 1913.

———, *Die Idee der Riemannschen Fläche*, 3rd ed., Stuttgart: B. G. Teubner Verlagsgesellschaft, 1955.

Wheeler, Lynde Phelps, *Josiah Willard Gibbs: The History of a Great Mind*, New Haven: Yale University Press, 1952.

White, Henry Seely, "Abel'sche Integrale auf singularitätenfreien, einfach überdeckten, vollständigen Schnittcurven eines beliebig ausgedehnten Raumes," *Nova Acta der kaiserlichen Leopoldinisch-Carolinischen deutschen Academie der Naturforscher* **57** (1891):41–128.

———, "Autobiographical Memoir of Henry Seely White (1861–1943)," *Biographical Memoirs of the National Academy of Sciences* **25** (1944):16–33.

______, "A Brief Account of the Congress on Mathematics held at Chicago in August, 1893," in *Mathematical Papers Read at the International Mathematical Congress Held in Connection with the World's Columbian Exposition Chicago 1893*, ed. E. H. Moore *et al.*, New York: Macmillan & Co., 1896, pp. 350–366.

______, "Ueber zwei covariante Formen aus der Theorie der abel'schen Integrale auf vollständigen singularitätenfreien Schnittcurven zweier Flächen," *Mathematische Annalen* **36** (1890):597–601.

White, Henry S.; Woods, Frederick S.; Van Vleck, Edward B., *The Boston Colloquium*, AMS Colloquium Publications, vol. 1, New York: Macmillan and Co., 1905.

Whitman, Betsey S., "Mary Frances Winston Newson," *Mathematics Teacher* **76** (1983):576–577.

Whittaker, E. T. and Watson, G. N., *A Course of Modern Analysis: An Introduction to the General Theory of Infinite Process and of Analytic Functions; With an Account of the Principal Transcendental Functions*, 4th ed., Cambridge: University Press, 1927, reprint ed., Cambridge: University Press, 1963.

Wiener, Norbert, *Generalized Harmonic Analysis and Tauberian Theorems*, Cambridge, MA: MIT Press, 1964.

______, *I Am a Mathematician*, Garden City, NY: Doubleday, 1956.

______, *Selected Papers of Norbert Wiener*, Cambridge, MA: MIT Press, 1964.

Wilczynski, Ernest J., "The First Meeting of the San Francisco Section of the American Mathematical Society," *Bulletin of the American Mathematical Society* **8** (1902):429–437.

Wilder, Raymond L., "The Mathematical Work of R. L. Moore: Its Background, Nature, and Influence," *Archive for History of Exact Sciences* **26** (1982):73–97.

______, "Robert Lee Moore, 1882–1974," *Bulletin of the American Mathematical Society* **82** (1976):417–427.

Williams, L. Pearce, *Michael Faraday: A Biography*, New York: Basic Books, 1965.

Wilson, Edwin Bidwell, "The Contributions of Gibbs to Vector Analysis and Multiple Algebra," in *Commentary on the Scientific Writings of J. Willard Gibbs ...*, ed. Arthur Haas, 2 vols., New Haven: Yale University Press, 1936, pp. 127–160.

______, *Vector Analysis: A Text-book for the Use of Students of Mathematics and Physics Founded upon the Lectures of J. Willard Gibbs ...*, New York: Charles Scribner's Sons, 1902.

Wilson, Louis N., *Clark University Directory of Alumni, Faculty and Students*, Worcester: Clark University Press, 1915.

Wilson, Woodrow, "The Preceptorial System at Princeton," *Educational Review* **39** (1910):385–390.

Winston, Mary F., "Eine Bemerkung zur Theorie der hypergeometrischen Function," *Mathematische Annalen* **46** (1895):159–160.

______, *Ueber den Hermiteschen Fall der Laméschen Differentialgleichung*, Göttingen: W. Fr. Kästner, 1897.

Woods, Frederick S. and Bailey, Frederick H., *Course in Mathematics for Students of Engineering and Applied Science*, 2 vols., Boston: Ginn and Co., 1907–1909.

Wussing, Hans, *The Origin of the Abstract Group Concept*, trans. Abe Shenitzer, Cambridge, MA: MIT Press, 1984.

Yaglom, I. M., *Felix Klein and Sophus Lie: Evolution of the Idea of Symmetry in the Nineteenth Century*, trans. Sergei Sossinsky, Boston: Birkhäuser Verlag, 1988.

Young, Jacob W. A., "On the Determination of Groups Whose Order Is a Power of a Prime," *American Journal of Mathematics* **15** (1893):124–178.

Ziegler, Renatus, *Die Geschichte der geometrischen Mechanik im* 19. *Jahrhundert*, Stuttgart: Franz Steiner Verlag, 1985.

Subject Index